# COMPARATIVE PERCEPTION

## Volume I
## BASIC MECHANISMS

# Wiley Series in Neuroscience

*Editor: Josef P. Rauschecker*

# COMPARATIVE PERCEPTION

## Volume I
## BASIC MECHANISMS

Edited by

## Mark A. Berkley
*Florida State University*

## William C. Stebbins
*University of Michigan*

**WILEY**

A WILEY-INTERSCIENCE PUBLICATION

**JOHN WILEY & SONS**
**NEW YORK    CHICHESTER    BRISBANE    TORONTO    SINGAPORE**

*Library of Congress Cataloging in Publication Data:*

Comparative perception/edited by Mark A. Berkley, William C.
    Stebbins.
        p. cm. — (Wiley series in neuroscience)
    "A Wiley-Interscience publication."
    Includes bibliographies and indexes.
    Contents: V. 1. Basic mechanisms—v. 2. Complex signals.
    ISBN 0-471-63167-1 (v. 1).
    ISBN 0-471-52428-X (2 vol. set)
    1. Perception—Physiological aspects. 2. Psychophysics.
    I. Berkley, Mark A., 1936– II. Stebbins, William C., 1929–
    III. Series.
    QP441.C66   1989
    591.1'82—dc19                                          89-30627
                                                              CIP

Printed in the United States of America
10 9 8 7 6 5 4 3 2 1

*To K. B., L., & T.*

# CONTRIBUTORS

Whitlow W. L. Au
Naval Ocean Systems Center,
    Hawaii

Robert L. Balster
Virginia Commonwealth
    University

Joseph Bastian
University of Oklahoma

Michael D. Beecher
University of Washington

Mark A. Berkley
Florida State University

Donald S. Blough
Brown University

Patricia M. Blough
Brown University

Ronald G. Boothe
Yerkes Regional Primate Research
    Center, Emory University

Richard F. Braaten
Johns Hopkins University

Charles H. Brown
University of South Alabama

Susan D. Brown
University of Maryland

Sheryl Coombs
Loyola University

Brian Y. Cooper
University of Florida

Peter DeWeerd
Katholieke Universiteit Leuven,
    Belgium

Velma Dobson
University of Pittsburgh

Robert V. Dooling
University of Maryland

Apostolos P. Georgopoulos
Johns Hopkins University

Henry E. Heffner
University of Toledo

Rickye S. Heffner
University of Toledo

Stewart H. Hulse
Johns Hopkins University

John Janssen
Loyola University

Martha F. Jay
Northwestern University

Robert H. LaMotte
Yale University

Peter Marler
Rockefeller University

R. Bruce Masterton
Florida State University

Bradford J. May
Johns Hopkins University

A. J. Mistlin
University of St. Andrews,
Scotland

David B. Moody
University of Michigan

Douglas A. Nelson
Rockefeller University

Kazuo Okanoya
University of Maryland

David S. Olton
Johns Hopkins University

Guy A. Orban
Katholieke Universiteit Leuven,
Belgium

Suzanne C. Page
Johns Hopkins University

Thomas J. Park
University of Maryland

Tatiana Pasternak
University of Rochester

D. I. Perrett
University of St. Andrews,
Scotland

Duane M. Rumbaugh
Georgia State University

E. Sue Savage-Rumbaugh
Georgia State University, Yerkes
Regional Primate Research
Center, Emory University

Burton M. Slotnick
American University

David W. Smith
University of Toronto

David L. Sparks
University of
Alabama-Birmingham

William C. Stebbins
University of Michigan

Philip K. Stoddard
University of Washington

Sandra Trehub
University of Toronto

Eric Vandenbussche
Katholieke Universiteit Leuven,
Belgium

Charles J. Vierck, Jr.
University of Florida

Frances Wilkinson
McGill University, Montreal

# PREFACE

This volume is concerned with the sensory capacities of animals, a research area that has come to be called animal psychophysics. The book's aim is not only to provide interested readers with comparative sensory data but also to acquaint them with the sophisticated and ingenious methods that have been brought to bear in studying the sensory capacities of nonverbal animals, including human infants. The chapters represent a sampling of different sensory capacities of many different animals obtained with state of the art quantitative behavioral methods. The coverage is not meant to be comprehensive, as it might be in a handbook, and is limited to vertebrates. Some chapters were included to demonstrate the progress that has been made in techniques and the sophistication of the questions being asked rather than for the data they obtained. Unlike its mentor volume, *Animal Psychophysics,* which focused on methodology, this book is concerned with broader issues. In many cases, the work described addresses complicated aspects of sensory stimuli rather than first-order sensory questions such as absolute thresholds. In a few cases, however, the sensory data are secondary in that the questions being addressed concern the use of sensory stimuli rather than the detection of the stimuli themselves. The collection of works represented in this book is intended as a sampling of the field and some omissions will be evident. For example, sensory data obtained with classical conditioning methods have been omitted, as well as data obtained from invertebrates. These areas are not represented, not because they were unworthy, but simply because we had to limit the scope of what would be covered to maintain a theme and still stay within a length limit. Even with these constraints, we were forced to separate the chapters into two volumes.

What is animal psychophysics? The word psychophysics was coined by Fechner in 1860 in his work *Elemente der Psychophysik* in which he set forth his ideas on how it might be possible to quantify mental events. Fechner believed that mental events are manifestations of physical events and thus should be measurable. Hence, the term psychophysics. However, since nonhumans cannot describe their sensations, and thus indicate a state of self-awareness, it is reasonable to ask whether it is appropriate to use the prefix *psycho-* in describing the process by which such data are obtained from animals. Perhaps the best answer to this question can be given within the context of evolutionary biology. Thus, if one accepts the Darwinian view of the evolution of the nervous system, it follows that consciousness and awareness must also follow such a route. Thus one can consider animal sensation and self-awareness on a continuum on which the human level of awareness is but one point. Placed within such a context, the question of whether there is such a thing as animal psychophysics becomes moot.

What is special about animal psychophysics? Does it really differ from field studies, discrimination learning, and so forth? Interest in the capacities of animals has a long history (see chapter by Stebbins), much of which is anecdotal. Observations of animals in their natural habitat or within a research colony have yielded valuable information about what animals respond to and has guided more formal laboratory work. However, such observations do not permit describing the limits and modes of sensory capacity. To do that, rigorous, controlled measurements must be made not only of the animal's behavior but of the stimuli being employed. It is the rigorous control of the stimuli that makes psychophysics different from field studies or simple discrimination learning tests.

How is animal psychophysics different from human psychophysics? In a typical human psychophysical study, the exigencies of the experiment are described to the subject and the subject is instructed as to which aspects of the stimulus to attend to, what details to ignore, how to respond, and so on. Obviously this is not possible with nonverbal subjects. Consequently, a lengthy process of communication using essentially a two-word vocabulary (yes–no) is employed to guide the animal to the procedure desired. But how does one know that the animal is attending to the stimulus or dimension desired? Here one relies on a truism derived from the operant conditioning literature: If systematically changing the stimulus, or a dimension of it, systematically changes (controls) the animal's behavior, then that stimulus is being attended to. So here is one major difference between animal and human psychophysics. Another difference concerns the selection of a response that is appropriate for the animal. With nonhuman or preverbal human subjects, the choice is dictated by the subject's motor repertoire and the practical considerations of monitoring the response. The use of different responses, however, is of little practical consequence in that while the actual motor responses may be different in studies of different species, they are assumed to be equivalent when employed in a psychophysical context.

Perhaps more important than the issues raised above is the problem of the criteria used to define a threshold. In human psychophysics, common threshold definitions were developed partially based on statistical considerations, but they are wholly arbitrary and are not used by all workers in the field. In animal studies, many researchers employ the criteria used in human studies, for example, a 75% detection level for threshold determination. It is common, however, for researchers working with animals to employ criteria they have developed based on the nature of their data and strict statistical considerations. As long as one is consistent, the choice is not too important except when cross-comparisons are made. It is easily appreciated that comparing threshold values from different studies or different species in which different threshold criteria were used is meaningless.

Why are animal sensory capacities studied? There is no one reason that researchers study animal psychophysics. As will be seen in the chapters that follow, there are many rationales. They extend from gathering basic knowledge to evaluating one's pet theory. Some of the rationales represented in this volume are (a) pure knowledge, (b) cross-species comparisons, and (c) model testing. Of these, model testing, especially in the neurosciences, has produced an explosion in animal psychophysical studies. These studies search for isomorphisms between psychophysics and physiology, test brain structure–function relations, and examine the effects of various manipulations such as chemical or environmental intervention and genetic screening on sensory capacity.

Where has the study of the senses of animals gotten us? It certainly has contributed significantly to a much better understanding of ourselves and the world around us. Insofar as such knowledge will ultimately lead to the betterment of the human condition, the pursuit of such knowledge is well worth the hard work it requires. We are sure you will agree after reading the contributions to this volume.

The editors gratefully acknowledge the help of Jane Schoonmaker Rodgers in all of the many steps necessary to prepare the manuscript for publication.

Mark A. Berkley

*Tallahassee, Florida*
*September, 1989*

# CONTENTS

## LOCALIZATION

## DEVELOPMENT

# COMPARATIVE PERCEPTION

## Volume I
## BASIC MECHANISMS

# 1

# PERCEPTION IN ANIMAL BEHAVIOR

*William C. Stebbins*

*Department of Psychology and the Kresge Hearing Research Institute,
University of Michigan, Ann Arbor, Michigan*

## I. PERCEPTION DEFINED

An animal's behavior is largely determined by the information it extracts from its environment by means of its various senses. Perception is used here in one of its original meanings from the Latin *percipere*—to catch, get hold of, gather in; but also in the more generic and metaphorical sense of awareness or cognizance of the environment. For a thoughtful discussion of the definitional problem in human perception, see Uttal (1981, pp. 9–14). His conclusion, that perception is not a term easily and quickly defined and that a definition is best adapted to one's objectives, is appropriate and a strategy that I will adopt here in writing about animal perception. In fact, perception may best be defined operationally by what investigators do in field and laboratory rather than limited to some abstract philosophical statement. In the study of animal perception there has been and continues to be a significant concern with reception—with the early stages in the processing of sensory information directly from the environment, and with the detection and discrimination of that information. Basic measures of sensory acuity for simple stimuli such as pure tones or monochromatic light have characterized the first 75 years of perceptual research on animals other than humans. In animals, standard psychophysical methods are used in conjunction with operant conditioning

Supported in part by a Program Project Grant from the National Institute of Neurological and Communicative Disorders and Stroke.

procedures in the determination of environment–behavior relations in the laboratory. While the conditioning procedures supply the critical communicative link between animal subject and human experimenter, and ensure the necessary degree of stimulus control over the animal's behavior, the psychophysical methods provide the objective means of quantifying the stimulus–response relations. It is these relations that reveal the fundamental properties of a sensory system in terms of the minimum detectable and minimum discriminable levels of stimulation to which an animal can respond (see Stebbins, 1970a). It is important to have a clear idea of the nature of these relations and their parameters as we begin increasingly to consider the effects of stimuli, clearly above the threshold region, that are complex, varying in more than one dimension, and often of biological significance, together with the nonstimulus variables that influence an animal's perception of its environment. The latter may be developmental, historical, training-related, contextual, or biological.

On the one hand, an animal perceives when it discriminates the differences between stimuli at the limits of its ability to resolve those differences. At one end of a perceptual continuum the phenomenon has been referred to as a predominantly (but never exclusively) stimulus-driven or bottom-up process (for a discussion of the bottom-up, top-down distinction, see Gardner, 1985, p. 97; Gregory, 1987, p. 601; Rock, 1984, p. 129). Here the predominant, but not exclusive, focus is apt to be on peripheral rather than more central physiological mechanisms. On the other hand, an animal also perceives when it responds similarly to differences in stimulation (as in perceptual constancy)—differences that it can discriminate if the conditions warrant or demand it. For example, a predator alarm call given by a conspecific will vary considerably in acoustic structure depending on, among other things, the vocal characteristics of the sender. The receiver's response may be similar across many individual variants of the call, yet, if forced to do so, the animal can discriminate among these variants, as in individual recognition. The distinction is important and reflects the difference between what an animal can do (based on the resolving power of its sensory system) and what it does do under natural or specifically arranged laboratory conditions (see also Herrnstein, 1985, p. 141; Nelson & Marler, Vol. II, Ch. 14.). The latter, in contrast to the former, lies toward the other end of a perceptual continuum. More top-down processing relative to bottom-up is indicated, and nonstimulus variables play a greater role in influencing the behavior. In addition, a greater share of the control is evident in the higher levels of the central nervous system, including the cerebral cortex. The "can do"–"does do" distinction also has methodological implications, and the questions asked are very different in the two instances.

Herrnstein, in discussing animal cognition, makes a similar distinction between "performance at the limit" and "behavior in natural settings" which is "rarely driven to any sort of limit" (1985, p. 142). Perception is employed here to include both kinds of phenomena. Both proceed from the same theoretical and systematic base. The experimental framework and design that have proven so useful in the early and continuing studies of sensory thresholds and discrimination in animals (see Gilbert & Sutherland, 1969; Stebbins, 1970a; Stebbins, Brown, & Petersen, 1984) are also useful in those experiments that attempt to evaluate perception under conditions where complex stimuli varying along many dimensions and nonstimulus variables are significant determinants of the behavioral response. An important caveat is that the same degree of experimental rigor that has characterized the earliest sensory threshold experiments with animals since those of Yerkes and Watson in 1911 be maintained as we examine those many and complex events that influence the way that animals experience the world about them.

When the perceptual response can be shown to be strongly affected by an animal's previous developmental and phylogenetic history and, perhaps relatively less so, by the stimulus to which its attention is directed, top-down or cognitive processing is often invoked and in fact, at this level of analysis, cognition and perception are likely to be inextricably intertwined (Shepard & Podgorny, 1978). But in its 100-year history there have been arguably at least two reasonably distinct approaches to and subject matter areas within comparative psychology. In perceptual experiments, the physical stimulus remains a dominant focus and is most often an independent variable in the experiment; its direct relation to behavior continues to be of major interest. For example, in studies of pattern recognition, in the perceptual classification of animal calls, and in the attempt to relate the perceptual response to physiological mechanisms, the emphasis continues to be on stimuli, although recently in more varied and elaborate contexts. Experience and history may assume far greater importance, chronological stages in development may be a major determinant of the behavior, or a "hard-wired" system may direct the perceptual response. In studies of animal cognition, on the other hand, communication (signal output) rather than perception (input), decision making, learning and problem solving, comparative intelligence (since the nineteenth century, the major driving force in comparative psychology), and memory are among those issues that students of animal cognition find challenging (see Hulse, Fowler, & Honig, 1978; Roitblat, Bever, & Terrace, 1984; Wasserman, 1981; Weiskrantz, 1985). While the concern with the physical stimulus is seldom lost in studies of animal cognition, it must share the spotlight with other variables and is of interest primarily because it forms a basis for the action and the effects of these other variables. These investigators (see above) by their research have defined the field of animal cognition; it is one objective of this chapter and book to do similarly for animal perception.

## II. HISTORICAL PERSPECTIVE

### A. Changing Focus

The earliest laboratory studies in the twentieth century were involved with technique, measurement, and sensitivity in parallel with human psychophysics and later changed dramatically as the technology of instrumentation and the experimental methods changed. Animal psychophysics, in chronicling and cataloging sensory thresholds, was never forced to deal with more than a S–R account of perception. Stimuli were invariably precisely defined and behavior was tightly controlled. Complexity in the stimulus was eschewed and other disconcerting or distracting nonstimulus variables were unwelcome. This was an important and necessary stage in the development of a rigorous science of animal perception. The procedures for these simpler experiments are now well developed and further experiments in this line of inquiry are important and will continue. In addition, many investigators are now turning their attention to those more complex relations where the stimuli are no longer simple and no longer stand alone, and the behavior is often less tightly regulated by the conditions of the experiment.

### B. Early Influences

The history of animal perception is, in large part, the history of the comparative method applied to the study of animal behavior, in general, and to the experimental analysis of that behavior in the laboratory, in particular. The comparison implies that the behavior of any one animal will be better understood when compared with the behavior of others, particularly in the context of evolutionary theory. But the comparative method was described long before the appearance of evolutionary theory. Its introduction, together with observational methods and the inductive approach to animal biology, has been credited to Aristotle (Warden, Jenkins, & Warner, 1935) who wrote extensively in his *De Anima* of his own observations of animal behavior and of the lore from earlier times. It was not until the early nineteenth century that naturalistic observations of animal behavior became more elaborate and far more common, but reliance on anecdotal evidence continued until nearly the end of that century. The stage was set for change when, in midcentury, Darwin published his *Origin of Species*, and for the first time a truly comparative psychology was possible. Humans were placed on a continuum with other animals, and the difference between them could be considered quantitative rather than qualitative. The classical distinction in kind between rational man and instinctive brute was rapidly disappearing as a viable scientific credo, and the comparative method applied to animal behavior began to gather momentum. Unfortunately, early efforts were completely lacking in scientific rigor, and in the short term produced a sort of escalating "I can top this" parade of remarkable, idiosyncratic, and, for the most part, unsubstantiated feats of intellectual prowess by domestic and captive animals (Romanes, 1883).

Romanes (1882, 1883) pursued the animal mind relentlessly. In demanding what he called the use of the objective method in the study of animal intelligence, he suggested that an animal's mind could be inferred from its observable behavior. In drawing the line at insects, however, he helped to establish the cultural tradition of a natural scale topped by humans that continues to plague us. It has led to a serious and prevalent misunderstanding of the theory of evolution, and more than half a century later helped to fuel an identity crisis of no small proportions in comparative psychology (see Hodos & Campbell, 1969). Unfortunately, although some of his own observations and those of his colleagues were carefully documented, they were hardly objective and served to prolong the era of "just so" stories which provided only the most tenuous support for the Darwinian notion of continuity of species.

Morgan (1894), in promulgating his famous canon, was justifiably critical of anecdotal evidence and of unwarranted inference of mental abilities based on that evidence. His famous statement of parsimony, "In no case may we interpret an action as the outcome of the exercise of a higher psychical faculty if it can be interpreted as the outcome of the exercise of one which stands lower in the psychological scale," probably helped to trigger the considerable reaction (perhaps even overreaction) to subjectivism and mentalism that followed in both biology and psychology. It also perpetuated the strongly anthropocentric notion of a natural scale. The reaction to the anecdotal period was slow in coming but effective and represented the beginnings of the scientific study of animal behavior at the end of the nineteenth century. The beginnings of the scientific revolution in comparative psychology could be found in the work of Morgan on birds (1894) and of Lubbock (1888) principally on insects, and later in the experiments of Loeb (1905) and Jennings (1906) on invertebrates.

The outlook of some of the biologists like Loeb was almost atheoretical. Concepts such as orientation, tropisms, and forced movements were taken from the botanists and applied to the study of invertebrate behavior. It was a highly operational system, if somewhat mechanical in an almost Cartesian sense. These tropistic reactions were instinctive if not reflexive, and learning or experience was generally ruled out, to say nothing of cognition or perception. The diffuse and ungrounded speculation of the earlier anecdotal period gave way to a highly objective, rigorous, and somewhat inflexible interpretation of behavior, at least as applied to the invertebrates. Warden aptly described the contrast as follows: "It was, indeed, a far cry from the notion of Romanes that the insect flies into the candle flame out of an innate curiosity to the contention of Loeb that it is forced to do so in a very literal sense when presented with the appropriate external stimulus"

(Warden et al., 1935, p. 24). Jennings' approach to behavior was less radical and restrictive than Loeb's. For Jennings, concepts like perception, attention, or choice were attributes that could be inferred from the behavior of "lower animals" on the basis of analogy to similar behavior in humans. In this way mental events were not different in kind but only in degree within the animal kingdom, and Darwin's notion of continuity of species could be preserved. But even Jennings, who considered the possibility of consciousness in the amoeba, made it very clear that, for science, the question of its existence in other animals or other humans is indeterminate. With the work of the physiologists, much of the basis for twentieth century behaviorism was now in place, and it fell to the psychologists to bridge the gap between the coelenterates and the primates. In so doing, they modified the mechanical and physiological model for behavior that had been established by Loeb and others.

## C. The Comparative Psychological Laboratory

Laboratory research on vertebrates, particularly on mammals, did not get underway with any force or frequency until the early twentieth century with the experimental programs developed by Watson (1903), Yerkes (1907), and Thorndike (1911). It is the former two that are of particular interest here for their important early research on animal perception. In 1907 Yerkes, then at Yale, was commissioned by the American Psychological Association to design and construct a standardized apparatus for testing color vision in animals. Together with Watson, he did just that; their monograph published in 1911 is a paragon of thoroughness and explicit detail (Yerkes & Watson, 1911). For the next twenty years with only slight modification it was used in the study of brightness, size, shape, and color discrimination in all manner of experimental animals ranging in size from mice to cattle. It is shown here in Fig. 1 in its original form as depicted in a two-page foldout from the 1911 monograph. It was about four years in the making, and for over forty years, if it was not the only model, it was at least typical of the technology that prevailed in the study of animal perception in the laboratory. Two stimuli differing in intensity, size, shape, or wavelength were presented simultaneously, and the animal was then always reinforced for approaching that member of the pair that was larger, smaller, brighter, or that differed in whatever other characteristic had been selected by the experimenter. The correct choice was always followed by food; the incorrect choice, under most adaptations of the procedure, was followed by electric shock.

Although there was a rich variety of structures and floor plans for testing an animal's perceptual skills, including simple T mazes and jumping stands, the nature of the stimulus presentation and particularly of the behavioral experiment itself was established for many years to come. The required presence of the experimenter for the manual operation of

**FIGURE 1.** Standardized apparatus developed by Yerkes and Watson for measuring visual discrimination in animals. After Yerkes and Watson (1911).

every detail of the experiment and the sequence of individually separated trials were hallmarks of the process. It was, in certain respects, an objective and rigorous but cumbersome and time-consuming methodology, flawed particularly by the experimenter's presence, which could never be standardized and which raised the specter of a Clever Hans effect (that is, the conscious or unconscious cuing of the subject by the experimenter). On the other hand, it is always easy to fault our scientific predecessors, and it should be remembered that the research of the early investigators like Yerkes and Watson represented a quantum leap forward over what had gone before and set the stage for what was to follow.

A certain creative inventiveness characterized the early experimenters in animal perception. What they lacked in behavioral technique and sophisticated instrumentation they made up in simple but often effective and intriguing devices or protocols that were designed simply to get the job done without sacrificing rigor. Sheperd (1910), for example, used a harmonica to measure a monkey's differential sensitivity to acoustic frequency. When a high note was sounded the animal ascended a platform to obtain food, but was not so reinforced after a lower note. Shepard notes (1910, p. 26) that care was taken to sound the notes at the same intensity and that the "usual precautions" were followed to avoid cuing the subject by appearance or action on the part of the experimenter. His device for measurement of sound-intensity discrimination consisted simply of a board attached by a leather hinge to a flat box and two sticks of differing lengths. The hinged board could be raised and dropped on the box; the intensity of the resulting sound was precisely related to the force of its impact on the box which was, in turn, a function of its height above it. The sticks then ensured that the height and thus the intensity of the sound was the same on repeated trials; two sticks meant two sound levels. As another example, Englemann's outdoor laboratory for evaluating the acuity of horizontal sound localization in dogs is pictured in Fig. 2 (Englemann, 1928; also cited in Warden, Jenkins, & Warner, 1936, p. 298). The subjects were surrounded by a circle of small screens concealing electric buzzers; they were reinforced with food for approaching the screen behind which

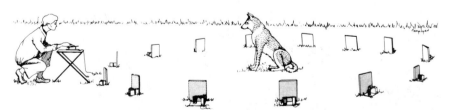

**FIGURE 2.** Outdoor laboratory for measuring horizontal sound localization in dogs by Englemann. Screens placed in circle conceal individual electric buzzers as acoustic stimuli. After Englemann (1928).

the buzzer had just sounded. Head or pinna movement as an aid in localization or sound reflectance were minimized by the brevity of the presented sound. A scaffolding and steps (not shown) were constructed for the dog to climb to indicate its choice in a subsequent experiment on vertical localization. The often-expressed concerns regarding experimenter cuing were heeded in so far as possible.

There were other means employed for studying animal perception in the early twentieth century. Preference tests, for example, were used for a period of time but were later rejected on the grounds that an animal's preference was not a reliable indicator of its perceptual abilities. The Pavlovian conditioned-reflex method has never been very popular for perceptual testing because of its lack of reliability and its limitation to fairly simple discriminations between stimuli. It continues to appear from time to time in various forms as a time-saving and simpler substitute for other more technically complicated and sophisticated procedures. Unfortunately, the simplicity is more apparent than real, and many of the specific applications of this procedure have been abandoned because of the instability of the behavioral baseline and often because of the lack of agreement with findings obtained with other procedures (see Stebbins, 1970b). A related methodology has developed around the use of stimuli that evoke unconditioned reactions in certain animals such as the pinna reflex to sound in the guinea pig. In general, with few exceptions (for example, see Ison, 1984), results from these efforts have not been reliable or in good agreement with results based on other procedures; they are idiosyncratic and therefore limited to use with certain subject species. Adaptation to the stimulus is frequently a problem. When these caveats have been carefully considered, natural responses can be used effectively and can reveal those features of complex signals that evoke the behavioral response (see Perrett & Mistlin, this book, Vol. II, Ch. 6).

In another important way the Yerkes and Watson monograph, among others, established a different kind of precedent. In this new and developing methodology for the study of animal perception it became important, in describing one's procedure, to dot every i and cross every t, for there were few to whom one could refer in the matter of apparatus construction or experimental design, and standardization and the potential for replication became essential, perhaps in good measure a reaction to the earlier, often undocumented, anecdotal format. Wendt's published experiments on the auditory acuity of several species of monkey illustrated again this penchant for detail (1934); it was a protocol the length of which would horrify present-day journal editors. Unlike the method section in current scientific publications, Wendt's description of his procedure is so complete as to permit its exact replication. Scale drawings of different views of the experimental apparatus are presented, together with sample protocols for trial sequences with copious notation, and tables and graphs of individual data. Every detail of the equipment, the training, and the procedures is meticulously described. His experiments are a tour de force—and all the more impressive because his findings were obtained under such arduous and primitive conditions.

## D. The Design of Experiments

From these early experiments there evolved a constellation of generally accepted and significant features common to their design. The purpose was to achieve the necessary degree of control over the relevant variables in the experiment, and therefore the adoption of these features was judged essential to the validity and reliability of the experimental findings. For organizational convenience, five classes of such features may be considered: those concerned with (1) behavioral training and testing, (2) psychophysical methods, (3) stimulus measurement and equipment calibration procedures, (4) data treatment, and (5) care and treatment of the subject species. Psychophysical methods, treatment of data, and the details of stimulus measurement were modeled after those used in the early human psychophysical experiments. Because of the somewhat primitive nature of the behavioral training and testing and the difficulty in carrying out such tests (see above), modifications of these psychophysical procedures were frequent and often without much justification. Accurate measuring instruments, at least in the field of hearing, were either both rare and expensive or unavailable. Consequently, "normal" human subjects were often employed to calibrate the apparatus. Thus, in hearing threshold experiments, for example, the obtained animal thresholds were relative to the threshold function for human hearing. Perhaps the most important advances in experimental design were made in the area of behavioral testing. Consistent motivation (deprivation level), the use of a ready signal and a preparatory response that served to place the animal (and especially the receptor organ) in the same fixed position with regard to the source of stimulation on repeated trials, the punishment of incorrect anticipatory responses, and the use of catch trials were features that permitted experimenters like Wendt to obtain findings that could be replicated many years later in spite of the limited facilities with which he had to operate (see Stebbins, 1971).

Significant changes in the strategies for behavioral training and testing and in the instrumentation for control of the experiment and for the measurement and calibration of the equipment would follow the remarkable advances in both behavioral and engineering science that were soon to come (see below). It would no longer be necessary to use the normal human ear as a standard reference for hearing of other animals. With these methodological improvements, the psychophysical procedures and the subsequent treatment of the data that had been developed and refined so satisfactorily for human subjects could usually be transferred intact for use with other animals, thus rendering cross-species comparisons far more feasible. Finally, these changes had significant consequences for the nature, complexity, and time course of laboratory experiments in animal perception, and for their application to a wide variety of subject species. As a result there was an increasing awareness of and attention to the care of these experimental animals that could be ignored only at the risk of the experiment.

## E. Emergence of the New Technologies

From the 1930s until the late 1950s the number of research papers on animal perception slowed to a trickle. As I have suggested, the experiments were time-consuming and laborious, and there was never a guarantee of success. It is likely that they were avoided by many who found a shorter route to scientific productivity in human psychophysics where experimenter and subject were often interchangeable, or even in the increasingly popular animal learning and conditioning experiments without the attendant preoccupation with the stimuli and the lengthy procedures necessary in training the animals to select certain features of those stimuli. Substantial progress and new discoveries in the field of electronics that led to the oscilloscope and later to the transistor and eventually to the small laboratory computer, and profound developments in the experimental analysis of behavior (Skinner, 1938) would have important and far-reaching effects on the nature of the next generation of laboratory experiments on animal perception. The same kind of switching circuitry that had revolutionized communication by telephone made it possible, for the first time, to automate behavioral experiments with animals and to dispense with that final subjective element in the conduct of the experiment—the human operator. At the same time Skinner's experimental treatise on animal behavior brought forth key concepts such as the free operant, differential reinforcement or shaping, and operant discrimination, and thus permitted an experimenter to observe and record in the laboratory the continuous process of an animal's interaction with its experimental environment. Both of these developments in the technology of instrumentation and in behavioral methodology had significant implications for the complexity of the questions that could now be asked and for the accuracy and the reliability of the answers that could be obtained. Both also, for the first time, freed the experimenter from the drudgery of manual labor in the conduct of the experiment.

## III. ANIMAL PSYCHOPHYSICS

A new period in animal perception, in which the recently acquired technologies in instrumentation and in behavior analysis came together, began in the late 1950s with a series of elegant experiments by Blough

(1955; 1958) on the visual sensitivity of the pigeon. All of the conditions in the original experiments were controlled and the data collected automatically by electromagnetic switching circuits with relays, timers, and counters, and by motor-controlled optical wedges and remote-controlled shutters. The behavior of the subjects was established and maintained by operant conditioning procedures with positive reinforcement. The dark-adaptation function in the pigeon (shown here in Fig. 3) was one of the first examples of a record of the moment-to-moment changes in the perceptual behavior of a nonhuman subject that reflected the underlying physiological changes in a sensory system. In this example, the photochemical process in the dark-adapting avian retina could be followed by observing the changes in visual threshold displayed in the bird's key-pecking responses to a light stimulus. The psychophysical procedure was taken from one that had been developed earlier for hearing testing in humans (Békésy, 1947).

The procedure set up a form of feedback loop between subject and stimulus whereby the bird's responding controlled the intensity of the light, and the light in turn controlled the bird's responding. In fixating the light stimulus, the bird faced two response keys; pecking on one increased the intensity of the light while pecking on the other attenuated it. In its original training the animal learned to peck on one key while the light was on; light off was the cue for switching to the second key where responding was reinforced with food. Subsequently, pecks on the first key decreased the light intensity in fixed steps, but on occasion the light was extinguished completely, and it was only on those occasions that the bird was reinforced with grain for switching to the second key. Responding on the second key was thus intermittently reinforced, but also served to increase the intensity of the light, thereby directing the bird to switch to the first key when

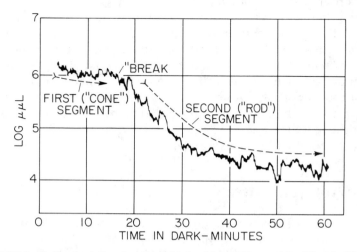

**FIGURE 3.** Dark adaptation function obtained for a pigeon. After Blough (1955).

the light became visible. Blough argued with impeccable logic that the pigeon is unable to discriminate "light off" from light below threshold and therefore would switch to the second key whenever the light fell below its threshold. The behavior was well maintained by the intermittent reinforcement schedule. The response keys were connected to a recording attenuator such that responses on the first key stepped the recording pen in one direction on the paper while at the same time decreasing the intensity of the light; responses on the second key drove the pen in the opposite direction while increasing the intensity of the light. The resulting function provided a continuous tracing in time of the animal's visual threshold (see Fig. 3).

Blough's experiments heralded a significantly new and powerful approach to the study of animal perception. The advantages of automation and of removing the experimenter from hands-on control of the experiment have been mentioned. The increases in complexity and speed that could be achieved by electronic circuitry created new possibilities for experiments that could not have been done previously. Later, these possibilities would be substantially increased by the introduction of the small laboratory computer. But perhaps most important of all, this conjunction of technologies permitted a degree of reliability and a level of precision not previously found in the experimental study of animal perception, and suggested an exciting array of options for the future. With the hardware and much of the basic software (the behavioral conditioning techniques) in place, the experimenter could focus on a variety of important questions in animal perception that could now be reasonably addressed for the first time.

In the 1960s there was a renewed effort in an expanding field that was then referred to as "animal psychophysics," culminating in a book by that name (Stebbins, 1970a). It was an edited book with contributions by many investigators describing their own research programs in animal perception, and had a strong methodological orientation. Psychophysics, almost by tradition, has attempted to establish a framework or set of procedures for examining sensory experience by determining the resolving power of a sensory system in behavioral terms. One objective of the book lay in illustrating how these procedures, originally developed for use with humans and heavily dependent on language, could be extended to other animals. The approach in the book is bottom-up, for throughout, its contributors for the most part were examining basic measures of sensory acuity—of stimulus detection and discrimination functions, how these might be determined, and how they might reflect changes in physiological processes. At that time, there was little known of a precise nature about the basic sensory characteristics of other animals. Much of what was available was still based on folklore or on a few unsubstantiated experiments over a limited range of stimulation. It was important, in addition to developing a viable set of procedures, to establish a solid foundation of empirical findings that were both objective and reliable (capable of being replicated), and valid (measuring what they purported to measure). Stimuli were simple (typically pure tones or monochromatic lights) and were varied along single continua. The data provided the beginnings of a biological catalog of species-typical values of minimum detectable and minimum discriminable energy levels for a variety of forms of sensory stimulation (see Fay, 1988). As bottom-up implies, there was a strong emphasis on stimulus–response functions and little interest in more complex processes of the top-down variety. The focus was decidedly practical and not much of a theoretical nature was offered. Such considerations would have been premature at that time; they would follow in the next decade.

The opportunities for precise and rigorous laboratory research in this field, that had been suggested by experiments such as Blough's, were successfully exploited and documented in *Animal Psychophysics*. Although very effective, Blough's procedure had been elaborate and complicated. There was a move toward simplification of the experimental design while retaining the important advantages and necessary controls, so that the procedures might be easily adapted for use with other animals and for answering other kinds of sensory and perceptual questions. The methods and technology employed by the investigators were carefully and thoroughly described, together with the potential pitfalls, and the necessary caveats were issued. The findings provided ample evidence that the general procedural and technological approach to the study of animal perception that characterized Blough's early experiments could be applied to a wide variety of questions regarding the basic nature of sensory perception in animals other than man.

Dalland's (1970) work on high-frequency hearing in the microchiropteran bat, described in *Animal Psychophysics*, provides a particularly good example of a method tailored to the demands of a highly specialized animal. Dalland carefully balanced the requirements of the animal in his selection of a suitable behavioral response and reinforcer with his method of presenting and measuring the ultrasonic signal which was always difficult at such short wavelengths. Although there was compelling circumstantial (Griffin, 1958) and indirect physiological (Wever & Vernon, 1961) evidence that bats could hear and respond to ultrasound, the confirming behavioral data and the characteristics of the auditory sensitivity function were missing. In supplying these data, Dalland described the evolution of an effective behavioral technique for measuring hearing in a most problematic species. An issue of some import raised by his experiments and those of others is whether there might be simpler, quicker, and yet equally valid ways of measuring sensory acuity in animals (including humans) by electrophysiological or other procedures. The answer depends, to a considerable extent, on the objectives of the experiment. In Dalland's experiment the answer was "no" because the investigator was measuring hearing, which, being a perceptual and not a physiological response, can be evaluated only by behavioral means. If the physiological data are shown to be isomorphic with behavioral findings, then one might argue for dispensing with the more difficult and time-consuming behavioral experiments. However, it is seldom known a priori whether the two sets of data will converge. The documented instances of such agreement are rare even today. Dalland's juxtaposition of his behavioral audiogram with the cochlear microphonic sensitivity function determined in the same animal (see Dalland, 1970, p. 24) make it very clear that the two sets of data are not measuring the same phenomenon.

Before 1970, what little research had been done in animal perception had been carried out in hearing and vision. Smith (1970) described a successful program of research and a procedure applicable to other sensory systems as well, and reported some of the earliest observations on olfactory and thermal as well as auditory and visual stimuli in animals as unusual, challenging, and diverse as the bushbaby and the tree shrew (see also Masterton, Heffner, & Ravizza, 1969), and the common pigeon. In the procedure the stimulus to be detected precedes brief, low-level, but unavoidable electric shock. The pairing of stimulus with shock is superimposed upon a baseline of food- or water-reinforced responding. After several pairings, responding is suppressed to the stimulus preceding shock, and suppression then serves as the reporting or indicator response to the stimulus to be detected. The procedure is a simple and effective one (see also Ray, 1970) and in its current modified form is still widely used (see Heffner & Heffner, this book, Vol. II, Ch. 9).

## IV. FROM PSYCHOPHYSICS TO PERCEPTION

While in 1970 a concern with simpler measures of detectability and discriminability predominated, there were the beginnings of an interest in measuring other forms of sensory experience such as visual illusions (Malott & Malott, 1970; Scott & Milligan, 1970), sensory magnitude estimation, for example, brightness and loudness (Moody, 1970; Miller, Kimm, Clopton, & Fetz, 1970), and information processing (Reynolds, 1970). The difficulties inherent in this new and substantially different class of experiments are related to their deeply rooted basis in human psychophysics and an absolute dependence on linguistic compatibility between subject and experimenter (see Stebbins et al., 1984). The language barrier had been less of an obstacle in simpler threshold experiments where "stimulus off" or physical equality of stimuli are conditions specifiable prior to the experiment. The appropriate use of reinforcement in instructing the nonverbal subject requires that the relation between the stimulus-reporting response and reinforcement be clearly established by the experimental protocol. Thus, for example, in Blough's experiment on the visual sensitivity of the pigeon, reporting responses when the light was off were reinforced; those that occurred when the light was on were not. In a study of difference thresholds for line length, for example, reinforcement might be contingent upon responding to either physically equal or physically unequal line lengths. But a variety of sensory experiences, particularly those involving observer judgment, such that the scale or continuum is psychological rather than physical, fail to fit this paradigm: for example, the extent of an illusion, sensory magnitude estimates, or stimulus matching on a given attribute. In a similar context, Blough (1984, p. 277) has suggested a significant distinction between two kinds of experiments based on stimulus definition. In experiments on sensory thresholds or concept formation, for example, the experimenter labels some stimuli as "correct" (i.e., light off or line lengths equal), but in others, which Blough refers to as perceptual, the subject defines the stimulus. It then remains for the experimenter to determine what elements, features, complexes, or aggregates of the stimulation introduced by the experimenter are controlling the subject's response.

In the typical human experiment on loudness, the human subject is asked to match two tones so that they appear equal in "loudness." There is no way of directly confirming the subject's judgment without the experimenter making a prejudgment of what constitutes equal loudness for the subject. This becomes critical in an experiment with nonhumans since it is now impossible to instruct the subject directly to make such a judgment; to do so would very likely determine the outcome of the experiment before it had been conducted—that is, to establish an experimenter-devised loudness scale for the subject by the manner in which the reinforcement contingencies were specified. Such putative psychological attributes of a stimulus as loudness are tightly tied to language. To determine such functions in animals requires the use of indirect procedures that will circumvent the reinforcement dilemma, finesse the language problem, and thus overcome the indeterminacy inherent in the direct method. One such procedure for determining loudness or brightness functions in animals is described by Moody (1970; see also Pfingst, Hienz, Kimm & Miller, 1975; Stebbins, 1966) and uses a subject's reaction time to varying intensities of stimulation at differing acoustic frequencies or wavelengths of light as a measure of sensory magnitude estimation. Equal reaction times at different frequencies or wavelengths are assumed to represent measures of equal sensory effect and thus equal loudness or brightness. The use of reaction time has proven effective in resolving the indeterminacy problem and providing reliable sensory magnitude estimates; it is not without certain drawbacks, such as the increased variance in reaction time at low levels of stimulation.

The reaction time procedure is characterized by its simplicity and generality and has been applied in a variety of experimental contexts: in the measurement of sensory thresholds and discriminative acuity, in sound localization, and in the discrimination of species communication calls (see Moody et al., this book, Vol. II, Ch. 10, Smith, et al., this book,

Vol. I, Ch. 3). Because it is illustrative of procedures now in common use, it will be described briefly here. In response to a signal (usually not in the sensory modality being tested), the animal makes a prolonged and stationary observing response (for example, continued manual contact with a small metal cylinder) that often serves the additional purpose of locating the receptor organ in a fixed position with respect to the stimulus source. If the response is maintained, it is followed after a few seconds by the stimulus to be detected. The time interval from the initiation of the response to the onset of this stimulus is varied from one trial to the next. The stimulus is on for about two seconds; the consequence for responding (the reporting response) during that time, that is, by breaking contact with the cylinder, is food or liquid reinforcement. Failure to respond sets the program for the next trial. Responding prematurely or late usually results in a short time-out from the experiment prior to the next trial. Trials are spaced at intervals of a few seconds and begin with the onset of the preparatory signal for the observing response. Thresholds or discriminative acuity are usually determined by one of two psychophysical procedures: the tracking or staircase method similar to the one that Blough used so successfully with the pigeon, or the method of constant stimuli in which the stimuli are presented randomly with regard to intensity or whatever dimension of the stimulus is being evaluated. Threshold is usually that value of stimulation to which 50% of the reporting responses occur (see Stebbins, 1970b, for a more complete description). The reaction time is measured from the second stimulus to the reporting response.

The examples taken from *Animal Psychophysics* reveal the diversity of active research programs in animal perception in 1970. These were to provide much of the basis for the research to come and that which is described in this book. The increasing use of computers in the 1970s and 1980s would permit greater complexity and almost unlimited variety with regard to the experimental questions that could be asked. The processing speed and capability of the new generation of laboratory computers provided a substantial advantage over the earlier electromagnetic and even digital modular switching circuitry. The relatively easy and rapid acquisition of graphic displays, mathematical models, and the synthesis of complex natural signals through newly devised software are only three examples of the many advantages that computer use bestowed on the laboratory investigator. Interestingly enough, the behavioral methodology had stabilized; the changes that did occur in the conditioning procedures were most often modifications or variations on a basic theme earlier developed with the objective of increased precision and reliability. Often, however, these changes were arranged in the hope of decreasing subject training time, which remained one of the greatest bugaboos in laboratory research on animal perception. The important changes that were occurring in the experimental approach to animal perception were directed toward application or problem solving in other disciplines such as neurophysiology, or ethology and evolutionary biology, or toward a better understanding of perceptual experience per se above and beyond the threshold region and with signals that were structurally more complex and typically of more natural origin than pure tones or monochromatic lights.

For example, behavioral baselines that measured precise sensory or perceptual effects in nonhumans had much to offer sensory physiologists. Comparisons of human sensory behavioral data with physiological recordings from other animals lack even face validity. Direct comparisons in the same species or even in the same subject were now possible with a degree of precision and reliability in the behavioral data that matched that obtained in psychophysical experiments with human subjects and in the physiological recordings taken from anesthetized physiological preparations.

In a related effort in our own research, we have worked closely with morphologists in examining auditory perception in animals whose hearing and peripheral auditory system were selectively impaired by controlled administration of an antibiotic known to be toxic to the inner ear (Stebbins, Hawkins, Johnsson, & Moody, 1979). A select group of antibiotics such as kanamycin and neomycin act to destroy the receptor cells (inner and outer hair cells) and nerve fibers in the basal part of the cochlea. With continued administration of these drugs, the pattern of cellular loss spreads in an inexorable but orderly manner toward the opposite or apical end of the cochlea. It is characteristic of the mammalian inner ear that the acoustic stimulus is coded in an acoustic frequency-to-place transform so that the more basal hair cells respond most strongly to high frequencies while the more apically located hair cells are most sensitive to low frequencies. An animal (a macaque monkey, for example) treated with kanamycin and tested daily for pure tone thresholds over its entire frequency range of hearing reveals an initial high-frequency hearing loss that will progress to the low frequencies with continued treatment with the drug. The hearing impairment, which is permanent, can be severe (greater than 90 dB) and is highly correlated with the degree of loss of inner and outer hair cells. It is the morphological organization of the cochlea in the form of a tonotopically arranged spiral that is reflected in the drug-treated animal's behavior—the further the spread of inner and outer hair cell loss in an apical direction, the further the progression of the animal's hearing loss to the lower frequencies. By contrast, in guinea pigs the cochlear outer hair cells are far more sensitive to kanamycin than the inner hair cells. A kanamycin-treated guinea pig displays a more moderate hearing impairment at high frequencies (50–60 dB) correlated with a complete loss of outer hair cells but retention of the inner hair cells in the cochlear base (Prosen, Petersen, Moody, & Stebbins, 1978). The relation is displayed in Fig. 4. In this example the two morphologically and physiologically distinct types of receptor cells are importantly distinguished in the animal's behavioral threshold response to pure tones. The antibiotics provide a way of making

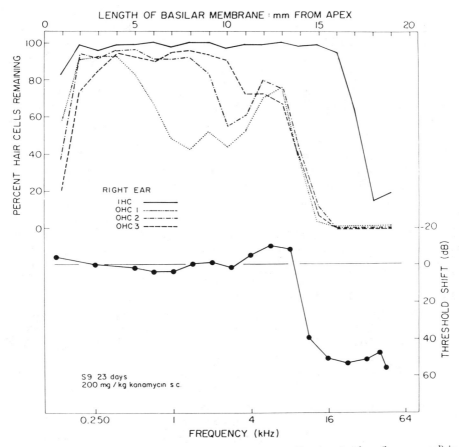

**FIGURE 4.** Receptor cell loss (upper panel) and corresponding hearing loss (lower panel) in guinea pig after kanamycin treatment. Cytocochleogram (upper panel) indicates location of missing inner (solid line) and three rows of outer (dashed lines) hair cells along the length of the basilar membrane. Audiogram shows extent of high-frequency hearing loss above 8 kHz. After Prosen, Petersen, Moody, and Stebbins (1978).

a discrete lesion in the sensory receptor tissue while the perceptual test offers the opportunity to determine the relation between sensory structure and perceptual function. In more recent findings (see Smith et al., this book, Vol. I, Ch. 3) this obviously complex relation is further defined by the perceptual changes that occur in acoustic frequency selectivity in animals coincident with outer hair cell loss, while the striking resemblance between electrophysiological recordings from individual hair cells or auditory nerve fibers and behavioral measures of frequency selectivity argue for the origin of this particular perceptual function in the auditory periphery.

A major advantage of the comparative approach to perception is that it enables the examination of the complex relation between an animal's perception and its ecology, and in turn engages the study of the evolution of the perception and processing of information from the environment. But the success of such a strategy presumes a substantial data base that includes an extensive sampling of a broad range of species to determine variation in the perceptual dimension under study. One such dimension that has been the object of scrutiny in this regard is high-frequency hearing in mammals, its relation to the external morphological feature of head size or interaural distance, and its implications for the importance of sound localization acuity (Masterton et al, 1969; Heffner & Heffner, 1985; Heffner & Masterton, this book, Vol. II, Ch. 9). On the basis of a data base that includes audiograms that have been determined in more than forty species, Masterton and his colleagues have argued that in the course of evolution the selective pressure for accurate sound localization in mammals has been especially intense. Such pressure, they argue further, has led to the successful adaptation of high-frequency hearing particularly in small mammals with small heads and short interaural distances.

Sound localization, which depends on binaural hearing, relies heavily on at least two binaural cues—the difference in the time of arrival of a sound and the disparity in the frequency–intensity spectrum at the two ears. Both cues are reduced in animals with small heads and ears that are close together. The use of the time-difference cue for relatively brief interaural time disparities is limited by the resolution of the nervous system, whereas the viability of the intensity–frequency cue depends on the sound shadow cast by the head and pinna. Small heads effectively shadow shorter wavelengths, so that high-frequency hearing permits the use of the intensity-difference cue for accurate sound localization. It is partly on this basis that the argument rests. The data support the argument, for there is a high and inverse correlation ($r = 0.85$; $P < 0.001$) between interaural distance and high-frequency hearing limit in a sample based upon 42 audiograms representing 40 mammalian genera (see Heffner & Heffner, 1985).

Among the most complex of naturally occurring stimuli is the human speech signal. The manner of its processing and perception by humans is still incompletely understood and has generated controversial theories, some of which require special, perhaps even species-specific processing mechanisms that distinguish humans from other animals in kind, not merely in degree. Central to the motor theory of speech perception, for example, is the assumption that the ability to categorize speech sounds such as phonemes depends upon the existence of motor control and articulatory mechanisms for the production of those same sounds (Liberman, Cooper, Shankweiler, & Studdert-Kennedy, 1967). It follows that animals that are unable to produce these sounds should be unable to perceive speech in a categorical manner. The work of Kuhl and colleagues (Kuhl, 1986; Kuhl & Miller, 1975; see also Moody et al., this book, Vol. II, Ch. 10) has been directed at evaluating this assumption by determining whether other animals perceive speech sounds categorically as we do or in some other fashion. Phonemic signals such as \da\ and \ta\, for example, vary in the time of voicing onset and can be placed at opposite ends of a continuum of computer-synthesized signals spaced at equal intervals of voice onset time. Human subjects bisect or categorize the continuum by identifying signals as only either \da\ or \ta\, and in addition have great difficulty in discriminating between signals within each of the two categories, but not in discriminating signals across category boundaries.

Kuhl's extensive research program, first with chinchillas and later with nonhuman primates, makes it abundantly clear that these animals categorize our speech in the same way that we do, thus obviating the need for an appeal to a special set of uniquely human processing mechanisms for speech. Before postulating a more general set of mechanisms based on the properties of the mammalian ear and auditory system, we must consider the more recent findings of Dooling, Park, Brown, Okanoya, and Soli (1987) and Kluender, Diehl, and Killeen (1987) that birds can learn phonetic categories. Although Kluender et al. report that based on extensive acoustic analysis, the signals that make up a given category, that is, \d\, do not all share one common acoustic feature. Comparative research of this nature may cast light upon the general properties of an auditory system that has evolved, at least in part, for the processing and perception of communication sounds. It also should begin to tell us something about the origins of human language and its basis in the communication sounds of other animals. It was thus a natural extension of this research, using the speech perception model, to the study of the perception by other animals of their own communication sounds (see Beecher & Stoddard, this book, Vol. II, Ch. 12; Dooling et al., this book, Vol. II, Ch. 11; Moody et al., this book, Vol. II, Ch. 10).

The interest in the perception of complex, biologically meaningful signals, such as speech, communication sounds, or musical patterns, while it represents a major focus of current interest in animal perception, is by no means limited to the auditory system. Communication signals are commonly transmitted in the visual mode. Bodily movement, particularly of the arm, hand, or digit, may provide directional information, express a greeting or threat, or denote positive or negative affect. Perhaps even more complex, extensive, and subtle or finely graded information can be conveyed by means of the orientation of the head, the direction of the gaze, or the expression of the face (see Perrett & Mistlin, this book, Vol. II, Ch. 6). While some of these forms of communication have been observed in countless species, others, such as facial expression, appear to be limited to Old World primates (including humans) who possess the necessary musculature for varying facial expression. The phenomenon has been described and the context identified in natural settings. Laboratory experiments have then pinpointed more precisely those dimensions of the stimulus that evoke the behavioral response. When confronted with an accurate model of the head in the laboratory, stumptail macaques more frequently exhibited teeth chattering, which is an appeasement gesture, than when the model was sytematically jumbled or distorted (Mistlin, cited in Perrett & Mistlin, this book, Vol. II, Ch. 6). An important dimension of this and other similar experiments is the effective use of a natural response by the animal as an index of perceptual discrimination. Perrett and Smith (cited in Perrett & Mistlin, this book, Vol. II, Ch. 6) have reported that these monkeys give appeasement gestures, including both teeth chattering and lip smacking, to pictures of faces in which the eyes are directed straight ahead both when the face is oriented frontally and at an angle of 45° but not when the eyes are averted laterally. Additional research by Perrett and others has provided information on the ontogeny of this form of visual perception and on its neural substrate. There are interesting and instructive parallels between the perception of visual communication signals and those transmitted by the auditory channel as in speech or animal calls.

## CONCLUSIONS

1. In addition to well-established procedures for measuring sensory thresholds in other animals, we now have a varied and powerful methodology for asking complex perceptual questions of them. The complexity is in the signal, in the behavior that is being measured, and in the many and diverse influences other than stimuli that determine that behavior (see particularly the other chapters in this book). In application, however, the procedures continue to be time-intensive but not overly labor-intensive; the workhorse is the small laboratory computer.

2. There are, however, no alternative quick and easy means of evaluating an animal's perceptual experience. Methods based upon conditioned or unconditioned reflexes or physiological recordings taken from anesthetized, comatose subjects are rarely equivalent.

3. The comparative approach remains a dominant theme in animal perception. Sensory and perceptual systems are better understood when perceptual behavior is clearly delineated in different and widely varying species. Furthermore, comparisons of such behavior contribute importantly and increasingly to our knowledge of the evolution of the perception of complex, naturally occurring signals such as those used in communication and human language.

4. The experimental findings in studies of animal perception are as precise and reliable as those obtained in psychophysical or perceptual experiments with human subjects; therefore, a solidly based research literature on animal perception is slowly building.

5. The research in animal perception is highly interactive and multidis-

ciplinary in the sense that it is often carried out in close collaboration with other psychological and biological disciplines, particularly neuroscience, with a view toward developing a better understanding of basic psychological, phylogenetic, and physiological mechanisms involved in an animal's processing and perception of the stimuli from its environment.

6. In the bottom-up domain—that is, in the study of sensory thresholds—consistently strong relations have been demonstrated between various measures of sensory acuity and physiological and morphological mechanisms.

7. With respect to top-down processing, we find that animals perceive in ways similar and dissimilar to our own, and in ways once thought to be unique to our species (for example, in categorical perception of human speech and in lateralization of communication sounds to the brain's left hemisphere. See Moody et al., this book, Vol. II, Ch. 10; Heffner & Heffner, this book, Vol. II, Ch. 9).

8. In spite of the strong interdisciplinary character of much of the current research in animal perception, the field has its own identity. Its objectives are distinct from those of human psychophysics and cognition and from animal cognition. A strong concern with the nature of the stimulus continues to prevail.

9. While theory and research in human cognition have had a substantial effect on research in animal cognition, they have had little effect on research in animal perception, in part because of the broad interdisciplinary focus in animal perception, and in part because of its continuing concern with the rigorous operational definition of those stimulus features that drive the perceptual response. As yet there has been little need in the field to invoke or to contend with such concepts as representation or consciousness.

10. For these many reasons there is an important story to tell about how an animal extracts information from its environment and how it makes use of that information in directing its behavior. And, of course, there is a continuing preoccupation on the part of the human investigator with the means of extracting that story from the animal perceiver.

## ACKNOWLEDGMENTS

I am grateful to Mark Berkley, John Jonides, and Joseph Hawkins for their comments on an earlier version of this chapter, to Bill Uttal for a healthy exchange of ideas by the electronic mail system, and to the Cambridge Center for Behavioral Studies in Cambridge, Massachusetts, and its director, Robert Epstein, for a fellowship that greatly facilitated writing final copy.

## REFERENCES

Békésy, G. von. (1947). A new audiometer. *Acta Oto-Laryngologica*, **35**, 411–422.

Blough, D. S. (1955). Method for tracing dark adaptation in the pigeon. *Science*, **121**, 411–422.

Blough, D. S. (1958). A method for obtaining psychophysical thresholds from the pigeon. *Journal of the Experimental Analysis of Behavior*, **1**, 31–43.

Blough, D. S. (1984). Form recognition in pigeons. In H. L. Roitblatt, T. G. Bever, & H. S. Terrace (Eds.), *Animal cognition* (pp. 277–289). Hillsdale, NJ: Erlbaum.

Dalland, J. I. (1970). The measurement of ultrasonic hearing. In W. C. Stebbins (Ed.), *Animal psychophysics: The design and conduct of sensory experiments* (pp. 21–40). New York: Appleton-Century-Crofts.

Dooling, R. J., Park, T. J., Brown, S. D., Okanoya, K., & Soli, S. D. (1987). Perceptual organization of acoustic stimuli by budgerigars (Melopsittacus undulatus). II. Vocal signals. *Journal of Comparative Psychology*, **101**,(4), 367–381.

Englemann, W. (1928). Untersuchungen über die Schallocalisation bei Tieren. *Zeitschrift für Psychologie und Physiologie der Sinnesorgane*, **105**, 317–370.

Fay, R. R. (1988). *Hearing in vertebrates: A psychophysics data book*. Winnetka, IL: Hill-Fay Press.

Gardner, H. (1985). *The mind's new science, a history of the cognitive revolution*. New York: Basic Books.

Gilbert, R. N., & Sutherland, N. S. (1969). *Animal discrimination learning*. New York: Academic Press.

Gregory, R. L. (Ed.). (1987). *The Oxford companion to the mind*. London & New York: Oxford University Press.

Griffin, D. R. (1958). *Listening in the dark*. New Haven, CT: Yale University Press.

Heffner, R. S., & Heffner, H. E. (1985). Hearing in the least weasel. *Journal of Mammalogy*, **66**, 745–755.

Herrnstein, R. J. (1985). Riddles of natural categorization. In L. Weiskrantz (Ed.), *Animal intelligence* (pp. 129–144). London & New York: Oxford University Press (Clarendon).

Hodos, W., & Campbell, C. B. G. (1969). Scala Naturae: Why there is no theory in comparative psychology. *Psychological Review*, **76**, 337–350.

Hulse, S. H., Fowler, H., & Honig, W. K. (1978). *Cognitive processes in animal behavior*. Hillsdale, NJ: Erlbaum.

Ison, J. R. (1984). Reflex modification as an objective test for sensory processing following toxicant exposure. *Neurobehavioral Toxicology and Teratology*, **6**, 437–445.

Jennings, H. S. (1906). *Behavior of the lower organisms*. New York: Columbia University Press.

Kluender, K. R., Diehl, R. L., & Killeen, P. R. (1987). Japanese quail can learn phonetic categories. *Science*, **237**, 1195–1197.

Kuhl, P. K. (1986). Theoretical contributions of tests on animals to the special-mechanisms debate in speech. *Experimental Biology*, **45**, 233–265.

Kuhl, P. K., & Miller, J. D. (1975). Speech perception by the chinchilla: Voiced-voiceless distinction in alveolar plosive consonants. *Science, 190,* 69–72.

Liberman, A. M., Cooper, F. S., Shankweiler, D. P., & Studdert-Kennedy, M. (1967). Perception of the speech code. *Psychological Review, 74,* 431–461.

Loeb, J. (1905). *Studies in general physiology.* Chicago, IL: University of Chicago.

Lubbock, J. (1888). *On the senses, instincts, and intelligence of animals, with special reference to insects.* London: Kegan Paul, Trench & Co.

Malott, R. W., & Malott, M. K. (1970). Perception and stimulus generalization. In W. C. Stebbins (Ed.), *Animal psychophysics: The design and conduct of sensory experiments* (pp. 99–124). New York: Appleton-Century-Crofts.

Masterton, B., Heffner, H., & Ravizza, R. (1969). The evolution of human hearing. *Journal of the Acoustical Society of America, 45,* 966–985.

Miller, J. M., Kimm, J., Clopton, B., & Fetz, E. (1970). Sensory neurophysiology and reaction time performance in nonhuman primates. In W. C. Stebbins (Ed.), *Animal psychophysics: The design and conduct of sensory experiments* (pp. 303–327). New York: Appleton-Century-Crofts.

Moody, D. B. (1970). Reaction time as an index of sensory function. In W. C. Stebbins (Ed.), *Animal psychophysics: The design and conduct of sensory experiments* (pps. 99–124). New York: Appleton-Century-Crofts.

Morgan, C. L. (1894). *Introduction to comparative psychology.* New York: Scribners.

Pfingst, B. E., Hienz, R., Kimm, J., & Miller, J. M. (1975). Reaction-time procedure for measurement of hearing. 1. Supra-threshold functions. *Journal of the Acoustical Society of America, 57,* 421–430.

Prosen, C. A., Petersen, M. R., Moody, D. B., & Stebbins, W. C. (1978). Auditory thresholds and kanamycin-induced hearing loss in the guinea pig assessed by a positive reinforcement procedure. *Journal of the Acoustical Society of America, 63,* 559–566.

Ray, B. A. (1970). Psychophysical testing of neurologic mutant mice. In W. C. Stebbins (Ed.), *Animal psychophysics: The design and conduct of sensory experiments* (pp. 99–124). New York: Appleton-Century-Crofts.

Reynolds, R. W. (1970). The use of reaction time in monkeys for the study of information processing. In W. C. Stebbins (Ed.), *Animal psychophysics: The design and conduct of sensory experiments* (pp. 329–339). New York: Appleton-Century-Crofts.

Rock, I. (1984). *Perception.* New York: Scientific American Books.

Roitblat, H. L., Bever, T. J., & Terrace, H. S. (1984). *Animal cognition.* Hillsdale, NJ: Erlbaum.

Romanes, G. J. (1882). *Animal intelligence.* London: Kegan Paul, Trench & Co.

Romanes, G. J. (1883). *Mental evolution in animals.* London: Kegan Paul, Trench & Co.

Scott, T. R. & Milligan, W. L. (1970). The psychophysical study of visual motion aftereffect rate in monkeys. In W. C. Stebbins (Ed.), *Animal psychophysics: The design and conduct of sensory experiments* (pp. 341–361). New York: Appleton-Century-Crofts.

Shepard, R. N., & Podgorny, P. (1978). Cognitive processes that resemble perceptual processes. In W. K. Estes (Ed.), *Handbook of learning and cognitive processes: Vol. 5. Human information processing* (pp. 189–237). Hillsdale, NJ: Erlbaum.

Shepherd, W. T. (1910). Some mental processes of the rhesus monkey. *Psychological Review Monographs, 12* (No. 52).

Skinner, B. F. (1938). *The behavior of organisms.* New York: Appleton-Century-Crofts.

Smith, J. (1970). Conditioned suppression as a psychophysical technique. In W. C. Stebbins (Ed.), *Animal psychophysics: The design and conduct of sensory experiments* (pp. 125–159). New York: Appleton-Century-Crofts.

Stebbins, W. C. (1966). Auditory reaction time and the derivation of equal loudness contours for the monkey. *Journal of the Experimental Analysis of Behavior, 9,* 135–142.

Stebbins, W. C. (Ed.). (1970a). *Animal psychophysics: The design and conduct of sensory experiments.* New York: Appleton-Century-Crofts.

Stebbins, W. C. (1970b). Principles of animal psychophysics. In W. C. Stebbins (Ed.), *Animal psychophysics: The design and conduct of sensory experiments* (pp. 1–19). New York: Appleton-Century-Crofts.

Stebbins, W. C. (1971). Hearing. In A. M. Schrier & F. Stollnitz (Eds.), *Behavior of nonhuman primates* (Vol. 3, pp. 159–192). New York: Academic Press.

Stebbins, W. C., Brown, C. H., & Petersen, M. R. (1984). Sensory processes in animals. In I. D. Smith, J. M. Brookhart, & V. B. Mountcastle (Eds.), *Handbook of physiology: Sensory functions* (Vol. 1, pp. 123–148). Washington, DC: American Physiological Society.

Stebbins, W. C., Hawkins, J. E., Jr., Johnsson, L.-G., & Moody, D. B. (1979). Hearing thresholds with outer and inner hair cell loss. *American Journal of Otolaryngology, 1,* 15–27.

Thorndike, E. L. (1911). *Animal intelligence.* New York: Macmillan.

Uttal, W. R. (1981). *A taxonomy of visual processes.* Hillsdale, NJ: Erlbaum.

Warden, C. J., Jenkins, T. N., & Warner, L. H. (1935). *Comparative psychology: Vol. 1. Principles and methods.* New York: Ronald Press.

Warden, C. J., Jenkins, T. N., & Warner, L. H. (1936). *Comparative psychology: Vol. 3. Vertebrates.* New York: Ronald Press.

Wasserman, E. A. (1981). Comparative psychology returns: A review of Hulse, Fowler, and Honig's cognitive processes in animal behavior. *Journal of the Experimental Analysis of Behavior, 35,* 243–257.

Watson, J. B. (1903). *Animal education.* Chicago, IL: University of Chicago.

Weiskrantz, L. (Ed.). (1985). *Animal intelligence.* London & New York: Oxford University Press.

Wendt, G. R. (1934). Auditory acuity of monkeys. *Comparative Psychology Monographs, 10* (No. 4).

Wever, E. G., & Vernon, J. A. (1961). Hearing in the bat, Myotis lucifugus, as shown by the cochlear potentials. *Journal of Auditory Research, 1,* 158–175.

Yerkes, R. M. (1907). *The dancing mouse.* New York: Macmillan.

Yerkes, R. M., and Watson, J. B. (1911). Methods of studying color vision in animals. *Behavior Monographs, 1* (No. 2).

# DISCRIMINATION

# 2

# EPICRITIC SENSATIONS OF PRIMATES

*Charles J. Vierck, Jr., and Brian Y. Cooper*

*Department of Neuroscience and Center for Neurobiological Sciences, University of Florida College of Medicine, Gainesville, Florida*

The ascending spinal pathways are organized similarly among many primate species, but the primate pattern differs considerably from that of carnivores and rodents. Hence, Macaque monkeys have been the subjects for psychophysical investigations of somatosensation, where generalization of the results to humans has been a major goal. Companion investigations of normal human subjects have been required, to decide upon the specific tasks that would be presented to the monkeys. This chapter will describe these studies of the discriminative capacities of humans and monkeys that have revealed the sensitivities of primates for different tests of spatiotactile acuity. In this context, the effects of interrupting a major somatosensory pathway (the dorsal spinal columns) will be presented.

Our investigations began with tests of spatiotactile capacities of monkeys and with demonstrations that the classical tests for discriminative (epicritic) somatosensations were not affected by interruption of the dorsal columns. The early results questioned either the importance of this pathway for spatial resolution or the validity of the classical tests for assessing spatial coding of somatic sensations. Later investigations supported the latter possibility and reaffirmed the assumption that the dorsal columns subserve somatosensations that require sophisticated neural codes. This amounts to a redefinition of epicritic sensations, and it forces reinterpretations concerning which anatomical and physiological characteristics of the dorsal columns support unique coding operations of functional significance. Also, the survival of a variety of spatiotactile capacities after interruption of the dorsal columns requires that we recognize the importance of other spinal pathways for support of localized tactile sensations.

The three cord sectors containing ascending pathways to the thalamus and cerebral cortex (see Fig. 1) are the ipsilateral dorsal column (DC), the ipsilateral dorsolateral column (DLC), and the contralateral lateral column (containing the spinothalamic tract, STT). There is a wealth of anatomical and physiological descriptions of these pathways (see Willis & Coggeshall, 1978), but considerably less information exists on the functional relevance of the different characteristics of the neuronal populations that comprise each pathway. The relative lack of behavioral investigations of somatosensory capacities that are supported by the spinal pathways has prompted the research that is reviewed here.

Early descriptions of the spinal somesthetic pathways set up a clear dichotomy between the ipsilateral dorsal column and the contralateral spinothalamic pathway. Based upon a presumption that each spinal pathway subserved entirely separable functions, the spinothalamic tract was considered to be the exclusive channel for pain and temperature sensations, and the dorsal column was characterized as the only channel

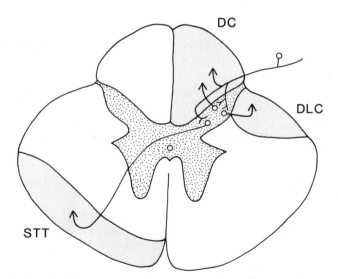

**FIGURE 1.** Diagram of a cross section of the spinal cord, showing the origins and locations of the major ascending somatosensory pathways for light tactile sensations. Large myelinated afferents to the spinal cord form a medial division of the dorsal roots and send long processes that ascend the ipsilateral dorsal column to synapse in the medulla as the DC–lemniscal pathway. In addition, collaterals of these afferents enter the dorsal horn to synapse on secondary cells which project axons rostrally via the ipsilateral dorsal column (DC), the ipsilateral dorsolateral column (DLC), and the contralateral ventrolateral column, as the spinothalamic tract (STT). Each of these pathways contributes input to the ventrobasal thalamus, which projects to the primary somatosensory cortex.

capable of supporting discriminative or epicritic somesthesis. Epicritic sensations were described by Head (1918) in terms of spatial acuity for tactile or proprioceptive sensations that do not have aversive qualities. These well-localized sensations are dependent upon activation of low-threshold receptors that are supplied by large myelinated afferents which project directly into the dorsal columns (Fig. 1). The dorsal columns have been shown to have a high degree of spatial resolution for tactile and proprioceptive stimulation (Mountcastle, 1961), and interruption of one dorsal column has been assumed to be responsible for ipsilateral impairments of epicritic sensibility that can result from lateralized spinal injuries in humans (see Nathan, Smith, & Cook, 1986).

Although the anatomical, physiological, and clinical evidence has presented a compelling case for a critical contribution of the dorsal columns to epicritic sensations, there have been weak points in the argument. The spinal lesions in humans are usually large, involving more than the dorsal columns; they rarely have been defined by postmortem histology; and the sensory testing of these patients has been simplistic, crude, and generally devoid of control measurements. The anatomical and physiological investigations cannot serve as direct tests for sensory capacity and could demonstrate an exclusive mediation of epicritic sensations by the dorsal column pathway if *only* the dorsal columns received a spatially organized input from large myelinated sensory neurons. However, the dorsal horn of the spinal cord contains a large number of neurons that (1) are spatially ordered in a somatotopic map (Brown & Culberson, 1981), (2) receive collateral input from large afferents that enter the dorsal columns, and (3) project axons rostrally within the ipsilateral dorsal *and* dorsolateral columns and the contralateral spinothalamic tract. Therefore, the potential exists for some spatial discriminations to survive interruption of the dorsal columns, and this possibility can be adequately tested only by psychophysical testing of trained subjects that receive lesions that are confirmed histologically to be restricted to the dorsal columns.

## SPATIOTACTILE DISCRIMINATION

Ordinarily, discriminative somesthesis has been investigated in terms of spatial resolution on the skin, and neurobiologists have repeatedly mapped somatoresponsive areas of the spinal cord, brain stem, thalamus and cerebral cortex for evidence of spatial (somatotopic) maps of the skin surface. In physiological experiments, spatial resolution has been defined by recording the responses of single cells to cutaneous stimulation and noting (a) the sizes of receptive fields, (b) the progression of receptive field locations across a pathway, nucleus, or cortical area, and (c) the number of neural units that represent different body regions. These mapping experiments appear to account for spatiotactile capacities, because body regions that exhibit the best psychophysical resolution for localization or discrimination of points (the fingertips and the tongue; Weinstein, 1968) are represented by a relatively high number of neural units with small receptive fields for punctate stimulation (Werner & Whitsel, 1971).

The ascending spinal pathway with the highest degree of topographic order and spatial resolution is the dorsal column. The dorsal columns project to the dorsal column nuclei, and thalamus (via the medial lemniscus), and then the primary somatosensory cortex (SI). This *DC–lemniscal system* has long been considered necessary for spatiotactile resolution (Mountcastle, 1961). That is, psychophysical tests of cutaneous localization have been thought to reveal the most sophisticated spatiotactile capacities and to be diagnostic for integrity of the dorsal columns. These tests have utilized punctate stimuli that have also been the dominant tools for neurophysiological descriptions of tactile receptive fields and of spatiotactile resolution.

Focal stimulation of the skin offers advantages for modeling mechanisms of spatial coding by the somatosensory system. Variations in the velocity, frequency, duration, depth, and location of indentations can be used to differentially activate the major categories of cutaneous sensory neurons (Burgess & Perl, 1973). These parametric manipulations sculpture profiles of central neural activity in time and determine the spatial distribution of neural responses within a topographic map. Because it is likely that tactile localization is given by the position of elicited activity a topographic map (Erickson, 1968; Mountcastle, Davies, & Berman, 1957; Gardner & Spencer, 1972; Békésy, 1967), localization accuracy should depend upon activation of many neurons that are clustered within a small portion of the map. These conditions have been assumed to be met as follows. Slight indentation of the skin (i.e., from 0.1 to 2 mm) by a single probe at a moderate velocity (e.g., 1 cm/sec) is an adequate stimulus for all categories of touch receptors, and therefore a large number of neurons are activated. When such a stimulus is applied to an area with small receptive fields and high innervation density (e.g., the tip of one digit), then localization accuracy should be high.

### Absolute Localization

As a test of whether the dorsal columns are required for accurate localization, *Macaca nemestrina* monkeys have been trained to indicate the location of punctate stimulation of various points on the skin (Vierck, Favorov, & Whitsel, 1988). The general approach for this and the other tests of somatosensory discrimination was to obtain stabilized preoperative thresholds for some attribute of stimulation delivered to either hindlimb. The discriminative responses were made with either hand so that the responding limb would not be affected by a spinal lesion at a thoracic level. Also, the spinal lesion was unilateral, so that ipsilateral versus contralateral effects of the lesion could be compared. Interruption of the dorsal column on one side produces strictly ipsilateral effects on somatosensation, and therefore thresholds for contralateral stimulation provide postoperative control values that are important for evaluating recovery from an ipsilateral deficit.

For the localization task, the monkeys were trained to initiate a trial by pressing a lever with either hand and holding it down until they received tactile stimulation at a single spot on the glabrous skin of either foot. After onset of the cutaneous stimulus, the monkeys could release the lever and press a button on a panel (early release of the lever terminated the trial without reinforcement). Six available buttons were positioned on drawings of monkeys' feet at points corresponding to skin locations that were stimulated on different trials (Fig. 2). The monkeys were reinforced only for pressing the button at the position that was stimulated on that trial (a correct localization). The locus of an incorrect (unreinforced) press of any

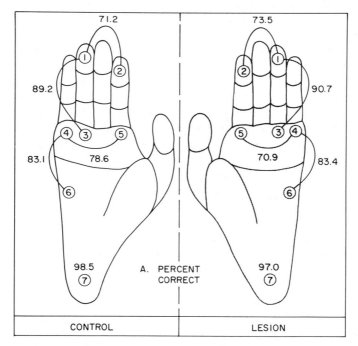

**FIGURE 2.** Localization accuracy on the feet of monkeys. The data from two monkeys are combined to show the percentage of correct responses when either of two points was stimulated on one foot. The two points that were stimulated within different blocks of trails are connected by lines. For example, when a monkey was stimulated at either point 1 or point 2, the animals responded correctly on 71.2% of the trials with stimulation of the control side (contralateral to a DC lesion) and 73.5% correct responses were emitted with stimulation ipsilateral to a DC lesion.

other button was noted, and the monkeys were required to initiate another trial to make a reinforced response.

A priori, we had some doubts as to whether the animals could learn to relate the points stimulated to buttons on a figurine, but with careful shaping of the correct response sequence, the animals became quite proficient at the task. The principle of the shaping procedure was to progress through a series of small additions to the response requirements and to gradually demand reliance upon somatosensory cues. The animals were trained in separate sessions to (a) press a bar and hold it down until after the onset of tactile stimulation, or (b) press a single button only when it was illuminated. During this early training, the button was recessed in the response panel, so that the animals learned to press the small button with a single finger. Then, both the bar and the button were available in each session, and the button was illuminated only when the animal correctly pressed and then released the lever. When an animal became proficient at chaining responses to the lever and a single button, the response panel was changed to one containing buttons corresponding to three points of stimulation: at the tip of one toe on either foot and at the heel of one foot. At this point, the button corresponding to the stimulated point was illuminated when the bar was correctly released. Then, over a series of sessions, button illumination was delayed after bar release. Finally, illumination of the buttons was discontinued, forcing the animals to select the correct button on the basis of the location of tactile stimulation. After this task was mastered, the points of stimulation were moved closer, using different response panels. Each panel contained six buttons, and any of three corresponding points were stimulated within blocks of 100 trials. Within a block of trails, either two or three of the stimulated points were on one foot.

Before considering the postoperative performance of these animals with stimulation ipsilateral to a dorsal column lesion, it is useful to consider their normal capacity for localization and to relate this to the spatial resolution of the somatotopic cortical map that is likely to subserve localization. The monkeys consistently responded briskly (within 500 msec from stimulus onset to button pressing), and they were not disrupted by changes in the locations of stimulators within blocks of 100 trials. Thus, they clearly were responding to the perceived location of each stimulus and not to some other cue (e.g., some characteristic of a particular stimulator that could be associated with a button within a block of trials). Although we were impressed with the monkeys' proficiency on the task, it was surprising that they consistently committed errors when stimulated on the tip of either of two digits (toes 2 or 4). For example, on blocks of trials with stimulation at the tip of either toe 2 or toe 4 on one foot, more than 25% of the responses were errors to the button located on the unstimulated toe (Fig. 2).

The performance of the monkeys on the localization task was worse than predicted from physiological mapping experiments that have revealed a preponderance of small receptive fields within the topographic map in the primary somatosensory cortex (e.g., Nelson, Sur, Felleman, & Kaas, 1980). Given the likelihood that absolute localization involves detection of the location of an ensemble of active neurons, somatotopic arrangements of neurons with receptive fields confined to a portion of one digit should support nearly perfect localization to different digits. This argument is especially compelling for maps that segregate the representations of the digits, as is the case for the postcentral gyrus of primates. How then, can the psychophysical data from monkeys be reconciled with the results of neurophysiological mapping studies?

A variety of considerations suggest that the cortical somatotopic maps represent absolute location less precisely than has been generally depicted, and absolute localization appears not to be a high-resolution function of the somatosensory system. Considering first the behavioral evidence for imprecise specification of which digit tip is stimulated, it is instructive to review investigations of tactile localization capacities of humans. With testing of fingers 3 and 4 (where errors can be made to a finger on either side), Marshall (1956) reported 10.2% errors to other digits; Halnan and Wright (1960) observed 7.3% digital mislocalizations; and Elithorn, Piercy, and Crosskey (1953) obtained 29.9% of these errors. Less attention has been directed to the feet, but Halnan and Wright (1960) obtained 47.1% mislocalizations to unstimulated digits with stimulation of toes 3 or 4. Thus, although there is considerable variability in accuracy, depending upon the site stimulated, and probably on characteristics of the stimulus (intensity, size, duration, and velocity), touch of one digit is not invariably localized correctly to that finger or toe. This result is of particular interest in comparison with other spatiotactile tests that reliably generate thresholds of less than 0.1 mm on a fingertip (*see below*).

Tests of absolute localization are conducted, of course, under conditions that prevent the subjects from observing the stimulations, and this is a situation that rarely occurs naturally. Ordinarily, cutaneous localization is aided considerably by the availability of visual cues and by awareness of our motoric activities. Tactile stimulation of the digits commonly occurs in the context of purposive movement and is consistent with expectations that are based on motor programs. When contact occurs at unexpected locations and/or times, motor reprogramming quickly provides confirmatory information, and visual orientation presents highly accurate localization and characterization of the stimulus. Often, relatively crude tactile localization provides sufficient information for adjustive motor programming and visual orientation. For example, for adapative withdrawal of the hand from an unexpected or potentially dangerous stimulus, the motor response would ordinarily involve the entire hand and arm, rather than a single finger, and imprecise localization to any fingertip or to the medial, lateral, anterior, or posterior surface of the hand would suffice.

Even if highly accurate tactile localization is not of critical adaptive significance, it seems likely that precise location would be specified as a matter of course by somatotopic maps with a high degree of spatial precision, even if that organization has evolved and developed to serve other functions. However, we may have been misled concerning the topographic details of somatosensory cortical maps by recordings from single units in anesthetized animals. Often, recordings have been restricted to layer IV (the thalamocortical input layer), where receptive fields are small relative to deeper and more superficial layers (Chapin, 1986) which contain neurons projecting to other cortical and subcortical regions (Friedman, Murray, O'Neill, & Mishkin, 1986). Furthermore, receptive fields can be reduced in size considerably by anesthetics (McKenna, Whitsel, & Dreyer, 1982), indicating that receptive fields from anesthetized preparations represent a subcomponent of the area of effective excitation in a behaviorally competent animal. For awake primates, many of the receptive fields in the somatosensory cortex cover the tips of more than one digit (Favorov & Whitsel, 1988b; Iwamura, Tanaka, Sakamoto, & Hikosaka, 1985), providing a basis for confusion concerning which finger or toe is stimulated. For example, a sample of receptive fields from layers 2–6 of area 1 of the somatosensory cortex of awake monkeys has revealed that more than 50% of the units activated by stimulation at the tip of one digit also responded to another digit (Vierck, Favorov, & Whitsel, 1988).

If absolute localization were dependent upon detecting the position of a single focus of activity within a topographic map, disruptions of the map should deteriorate the capacity for localization. According to investigations of human patients, absolute localization is severely impaired following lesions of the postcentral gyrus (the SI cortex; Corkin, Miller, & Rasmussen, 1970; Denny-Brown, Meyer, & Horenstein, 1952). These results confirm expectations that the high-resolution SI map subserves localization or provides a critical source of input to another region (or regions) that subserves localization. However, interruption of the dorsal spinal columns did not impair absolute localization by the monkeys (Fig. 2), and this lesion severely deafferents the SI cortex.

Following a dorsal column lesion, single unit recordings in SI of awake monkeys reveal a near absence of tactile driving of neurons in a "core" region of the distal limb representations (area 3b and anterior portions of area 1), and anatomical investigations indicate that the remainder of SI is partially deafferented by dorsal column interruption (Dreyer, Schneider, Metz, & Whitsel, 1974). Therefore, the absence of a deficit in absolute localization following dorsal column lesions suggests the following conclusions: (a) Accurate localization is not dependent upon functional integrity of the entirety of the SI map; (b) tactile localization does not rely upon input from the spinal pathway with the highest degree of spatial resolution; and (c) localization is subserved relatively crudely over a large extent of somatosensory cortex. These conclusions are consistent with recordings from SI of awake monkeys (e.g., Favorov & Whitsel, 1988a; Iwamura et al., 1985) and with metabolic mapping of cortical 2-deoxyglucose (2-DG) distribution after punctate stimulation of the skin (Juliano & Whitsel, 1985). These important findings from awake animals reveal a surprising extent of cortical activity that is elicited by a single punctate stimulus, suggesting that the primary somatosensory cortex is not organized primarily to serve functions such as localization of punctate stimuli. That is, if absolute localization were of crucial adaptive significance and were maximized by the organization of SI, then it seems likely that all neurons with overlapping receptive fields would be highly segregated, rather than widely dispersed.

If the dorsal columns do not provide a degree of spatial resolution that is required for localization on the distal extremities, what is the functional contribution of this pathway with small receptive fields and an orderly projection to the cerebral cortex? Does the DC–lemniscal system contribute not to epicritic sensations but rather to other functions, such as guidance of motor activities? This possibility has been proposed (Semmes, 1969; Wall, 1970), and the DC–lemniscal system is required for fine motor control of the distal extremities (Vierck, Cooper, & Leonard, 1987). However, it has been our working hypothesis that integrity of the DC–lemniscal system is necessary for the most sophisticated aspects of spatial discrimination. If this is correct, then absolute localization is subserved by relatively trivial neural codes and does not put maximal demands on somatotopic resolution.

### Two-Point Discrimination

In the attempt to determine which somatosensory capacities require an intact DC–lemniscal projection system, it is hoped that certain attributes of the dorsal columns can be identified which dictate whether a deficit will be observed on a psychophysical task. For example, discrimination of the separation between several points of stimulation could depend upon DC–lemniscal features that are not critical for absolute localization. A unique physiological characteristic of the DC–lemniscal system that has been proposed to contribute to spatial discriminations is precise afferent inhibition, which was described initially as having a center (excitatory)–surround (inhibitory) organization (Mountcastle & Powell, 1959). Thus, application of two points to the skin is expected to generate strong inhibitory influences on neurons with receptive fields located between the loci stimulated, sharpening the spatial contrast between two peaks of neural activity elicited by the stimuli.

A task that requires distinction of two points from one point has been used as a standard clinical test for integrity of the DC–lemniscal system (Nathan et al., 1986). Therefore, to evaluate the effect of verifiable dorsal column lesions, *Macaca arctoides* monkeys were trained to generate two-point threshold estimates by responding differentially to contact of the glabrous skin of either foot by two nylon monofilaments that were adjacent (one point) or separated by varing distances (M. Levitt, C. Vierck, & R. Schwartzman, unpublished observations). Correct responses to one point consisted of pressing a manipulandum on the left, and reinforcement was delivered for pressing a manipulandum on the right during stimulation by two points. Thresholds were determined by a method of limits that employed descending series to maintain a high level of responding and permit tracking of thresholds that might be elevated by the spinal lesions. That is, two-point and one-point trials were presented in pseudorandom sequences, and the separation between points was decreased in 2-mm steps from a high value (e.g., 20 mm) until an error occurred on a two-point trial. After a two-point error (defining one threshold estimate) the separation was increased to begin another descending series. Following extensive overtraining on the task, thresholds stabilized at 4 to 8 mm for a group of 8 monkeys. Thus, two points can be resolved from one point at separations much less than the 20 mm or more that was required for threshold discrimination on the test of absolute localization. This result suggests that the somatosensory system may be organized preferentially for discriminations of relations between multiple contours.

Following complete interruption of the ipsilateral dorsal column, two-point thresholds of the monkeys were not significantly elevated. Investigations of human patients with dorsal column lesions also have shown that two-point discrimination can remain essentially normal (Cook & Browder, 1965; Wall & Noordenbos, 1977). Thus, the requirement that human or nonhuman primates discriminate two contacts from a single point does not reveal the presence of a dorsal column lesion. These results appear to contradict the notion that the small excitatory and inhibitory receptive fields of the DC–lemniscal projection contribute to spatial resolution of nearby contacts, but perhaps two-point thresholds do not provide an adequate test of this hypothesis. Before generalizing the results of the two-point test to indicate that all discriminations of relations between punctate stimuli are not impaired by dorsal column lesions (e.g., their separation, relative position or pattern), it seemed advisable to compare two-point thresholds with other spatial discriminations in which different parameters inherent to the two-point test were selectively varied. This is most easily accomplished with human subjects.

When human subjects are tested for two-point sensitivity, they generally are instructed to report that two contacts have occurred only when two distinct and separate tactile sensations are present, which makes this a test of gap perception. However, as the gap is widened by moving two points apart, the spatial extent of a unitary sensation might increase (before the points are sensed as separate). Stated in neural terms, the extent of a topographic map that is activated by two points will increase with distance between the stimuli (see Fig. 3A), providing the subjects with a cue of

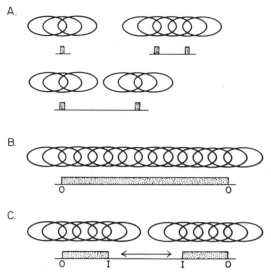

**FIGURE 3.** A diagrammatic representation of overlapping receptive fields (shown as ellipses) that would be activated by several varieties of stimulation that have been tested psychophysically. (*A*) A single point activates a population of cells with RFs supplying the spot stimulated. Two points close together would activated a larger population of cells, and two points that are perceived as separate would presumably activate cells with RFs that do not overlap. (*B* and *C*) Stimuli are depicted that permit an unambiguous determination of the threshold for detection of a single gap. The outside dimensions (O) of the stimuli are identical, and the size of a gap formed by inside edges (I) is varied.

length that can be utilized and might influence performance on the test. Thus, if the two-point threshold is intended to represent the capacity for gap detection, an unconfounded test of this capacity is needed. This was accomplished by evaluating the capacities of subjects to discriminate sizes of gaps within lines of constant overall length (comparing the stimulus in Fig. 3C with that in Fig. 3B). These stimuli were applied to the volar surface of either forearm for several reasons: (a) The forearm skin is free of calluses and creases and is highly pliable, permitting the skin to conform reliably to the contours of the stimuli, and (b) receptive fields on the forearms of primates have been reported to be elongated in the proximodistal dimension (Mountcastle & Powell, 1959), providing a test of the correspondence of thresholds to receptive field dimensions.

Two-point thresholds along the proximodistal axis of the forearms of humans have been reported to be 40 mm, on the average (Weinstein, 1968), when the subjects have received little or no pretraining with stimulation of the forearm. For our subjects, who were well trained and tested repeatedly, two-point thresholds ranged from 25 to 35 mm (Vierck & Jones, 1969). Similarly, when these and other subjects were tested for detection of a gap (e.g., in a 152-mm line placed proximodistally on the forearm), the length of unstimulated skin at threshold ranged from 26 to 38 mm (Jones, Vierck, & Graham, 1973). Based upon this correspondence of two-point thresholds and pure gap detection thresholds, the classical test of two-point thresholds can assess the capacity for detection of a gap. However, if gap detection is to survive as a standard test for spatial acuity, two-point sensitivity should be replaced by stimulation methods that do not covary the distance between inside and outside contours.

To detect the presence of a gap, it may be necessary to space two points or two inside edges beyond the borders of a certain number of receptive fields, leaving a population of unstimulated neurons between two populations of active neurons (Fig. 3A). If this is a necessary condition for gap detection, then gap thresholds should be closely related to the average receptive field size for a given skin region. In general, two-point thresholds do appear to correspond to receptive field sizes (Békésy, 1967), and this is a useful and interesting relation. However, it does not seem likely that spatiotactile resolution is restricted entirely by receptive field size. Other factors, such as innervation density, must permit spatial resolution of contours at separations less than the minimal (or average) receptive field size for a given skin region (Erickson, 1968). To evaluate this possibility, human subjects were tested for spatial resolution on a task that might well be sensitive to the pattern of overlap of receptive fields supplying the skin.

### Length or Size Discrimination

The number of activated neurons in a somatotopic map should vary directly with the extent of a tactile stimulus, and the units recruited by each increment in size or length will depend upon the number of overlapping receptive fields with partially shifted borders. This is not to say that receptive field size is irrelevant for these discriminations, or that innervation density is unimportant for gap detection, or that these factors can be separated absolutely. However, discriminations of length should be determined primarily by the distance between the borders of receptive fields supplying a skin region, and this distance is certainly less than the average receptive field length along any given orientation of the stimulus. This expectation was confirmed by thresholds for discriminating differences in overall length of lines impressed proximodistally on the forearms of the subjects that were tested for gap detection. Thresholds for length discrimination ranged from 3.2 to 9.5 mm. These thresholds are considerably less than the two-point thresholds (and estimates of the dimensions of receptive fields) for that region.

If increments in length are identified on the basis of the number of neurons activated by the stimulus, then thresholds for length discrimination should vary in proportion to both receptive field size and innervation density for different skin regions. Recordings from the somatosensory cortex indicate that innervation density remains constant for skin regions that are supplied by receptive fields of different sizes (Sur, Merzenich, & Kaas, 1980). That is, the number of neurons that respond to a punctate stimulus remains constant; the number of neurons with overlapping receptive fields is consistent; and remarkably similar areas of cortex are responsive to punctate stimulation of different skin sites, even if the receptive field sizes for the sites differ greatly. One implication of this principle is that thresholds for measures that are sensitive to innervation density will be determined by receptive field size. In other words, the distance between receptive field borders of nearby neurons will remain a constant fraction of receptive field size, because the number of overlapping fields remains constant. Thus, the relative sizes of gap and length thresholds should remain constant for different test sites or orientations. Alternatively, if the number of neurons with overlapping receptive fields changed for different skin regions (i.e., if innervation density did not remain constant), then thresholds for tests that are sensitive to receptive field size versus overlap would not remain in the same proportion for these test sites.

There are few experiments that permit a direct comparison of the relations between gap and size thresholds, but the limbs provide interesting regions to test, because the receptive fields are consistently asymmetrical, with the long axis oriented proximodistally (e.g., Mountcastle & Powell, 1959; Favorov & Whitsel, 1988b). Weber (1835) reported that two-point thresholds on the forearm were 3.2 cm when the points were alligned proximodistally, and 2.0-cm thresholds were obtained in the mediolateral orientation. Jones and Vierck (1973) obtained length thresholds of 7.8 versus 5.6 mm when stimuli were presented in the proximodistal versus mediolateral orientation. Thus, both tests are influenced to approximately the same degree by receptive field asymmetry. Similarly, gap thresholds vary from 2 mm on the fingertip (Phillips & Johnson, 1981) to 40 mm on the forearm (Weinstein, 1968), and length thresholds vary from 0.4 mm on the fingertips (Loomis, 1981) to 7.8 mm on the forearm (Jones & Vierck, 1973). These comparisons should be made in the same subjects and can be sensitive to factors other than receptive field geometry (e.g., the dermatomal composition of afferents activated; Fuchs & Brown, 1984), but available data suggest that gap and length thresholds vary with receptive field size and overlap in a manner that is consistent with a constant innervation density.

Based upon the supposition that length (or size) thresholds are sensitive primarily to innervation density, it seemed likely that a test of size discrimination would be more sensitive than gap detection to interruption of the dorsal columns. Although the DC–lemniscal system contains a high percentage of neurons with small receptive fields, other pathways to the thalamus from the dorsal horn of the spinal cord also are composed of neurons with small receptive fields. Thus, a dorsal column lesion will substantially reduce the *density* of cortical innervation by afferents with small receptive fields, but neurons with small receptive fields remain in the somatosensory cortex after dorsal column section (Dreyer et al., 1974). This leads to the prediction that size discrimination would be elevated by a reduction of afferent innervation of the cortical somatosensory maps. Furthermore, size discrimination is a more sensitive measure of spatiotactile resolution than gap detection, and this may reflect an especially effective sharpening of outside edge contours by spatially precise inhibitory influences (e.g., see Ratliff, 1965). Because DC–lemniscal neurons appear to have especially small inhibitory receptive fields (Janig, Schoultz, & Spencer, 1977), animals with dorsal column lesions might well reveal a spatiotactile deficit on a discrimination that is sharpened by inhibiton.

To test these hypotheses, monkeys were trained to discriminate the sizes of disks that contacted the glabrous skin of either foot (Vierck, 1973). Disks were used in place of lines so that orientation of the stimulus would not be a factor. The diameter of the standard disk was 6.3 mm, and the animals responded to the left side when stimulated by the standard. Responses to the right side were reinforced when the stimulus was a larger disk (the comparison stimulus). For this task, the method of tracking thresholds was different from the within-sessions method of limits that was utilized for estimating two-point thresholds. For the within-sessions tracking method, the progression of two-point separations was determined by whether or not an error was committed on the previous two-point trial. This presented difficulties in evaluating the possibility of a response bias. For example, a bias in favor of responses to the manipulandum on the left (correct for one-point trials) would shorten the descending series of two-point separations, artificially elevating thresholds. It is difficult to detect such a bias that would create errors for a high percentage of easily discriminable values of the comparison (two-point) stimulus. This problem can be circumvented by establishing stimulus parameters on the basis of a between-sessions progression. That is, for the size discrimination task, the size of the comparison disk was constant throughout each session and was decreased or increased for the next session, depending upon the percentage of correct responses (greater or lesser than 75%). In this way, any bias of an animal to respond preferentially to the right or left manipulandum can be evaluated by comparing the percentage of errors to the standard and comparison stimuli. Between sessions adjustment of the comparison value provides no advantage to the animal for a response bias, and the appealing features of threshold tracking are retained.

For investigations in which thresholds are expected to be elevated by an experimental treatment, it is important to maintain the animals' performance on the task. Threshold tracking is ideal for this purpose, because the stimulus parameters are adjusted to a level that is discriminable. For example, setting 75% correct responses as the criterion for increasing or decreasing the size of the comparison stimulus generates performance levels well above chance (50%). The criterion for tracking can be set above the classical definition of discrimination threshold, if the animals appear to be frustrated by tracking at 75%. Another advantage of threshold tracking is that the time-course of functional recovery can be described by following shifts in threshold as they occur (see Vierck, 1982).

Using the between-sessions tracking method on the size discrimination task, stabilized preoperative difference thresholds of 5–6 mm were obtained; that is, 75% performance or better was obtained with disks of 6.3 versus 12.7 mm. In contrast, for 70 to 92 days following unilateral dorsal column section, the monkeys performed at chance when the comparison stimulus was 31.7 mm in diameter. This result appeared to confirm expectations that (a) intact dorsal columns are required for normal performance on a sensitive test of spatiotactile discrimination, and (b) deficits from dorsal column interruption are apparent when the psychophysical task is sensitive to innervation density and/or when performance is dependent upon spatially precise inhibitory influences.

To be confident that such conclusions are correct, it is necessary to train the animals for an extended postoperative period, to demonstrate that thresholds for size discrimination are irrevocably increased following a dorsal column lesion. It is possible that sufficient cues are available for spatial discrimination after surgery but that the animals must relearn the discrimination on the basis of altered sensory information. Accordingly, the monkeys were retrained for months on the task. With extensive testing, the ipsilateral thresholds gradually lowered and eventually stabilized at normal levels. Therefore, the size-discrimination task revealed an aspect of spatial coding that is normally supplied by the dorsal columns but is not necessarily dependent upon integrity of the DC–lemniscal system. The recovery of size thresholds after dorsal column section can be accounted for on the basis of (a) a transient reduction of innervation density that is restored with time by neural reorganization (e.g., by reactive sprouting from intact axons to vacated synaptic sites) or (b) relearning the task, using cues that were not attended to preoperatively and are not supplied uniquely by the dorsal columns. To evaluate the latter possibility, we conducted psychophysical and neurophysiological investigations of sensory cues that might be available for size discrimination.

When a disk is pressed onto the skin, the most obvious sensation for the subjects is that of an annulus that corresponds in size to the circumference of the disk. There is a vague sensation of the surface of a disk when it is large enough to be discriminated from an annulus of the same outside diameter, but still the edge sensation is most salient. This introspective experience generates the prediction that size discrimination is based upon the distance between edge contours and not upon the surface area of skin contacted by the disk. This prediction can be evaluated to a certain extent by directly comparing growth functions for subjective sensations produced by increasing line lengths or disk sizes. If the sensations of line length and disk size are both determined on a linear basis (i.e., the distance between edge contours), then disk sizes (expressed in diameters) should be equally discriminable as line lengths (that are orientated transversely on the arm), and power functions based on line length or disk diameter should have identical exponents. However, these exponents should differ if area is a relevant sensory cue. This is because the area of a disk increases as a squared function of the radius; whereas for lines, area is a linear function of length. When human subjects estimated the sizes of lines and disks, disk sizes were discriminated more easily than lines on the basis of edge separation, and disks produced power function exponents that indicated a contribution by sensations of area (Jones & Vierck, 1973).

The psychophysical comparisons of sensations for length and size provide indirect evidence to support the following explanation for the effects of dorsal column section on size discrimination. Interruption of the dorsal columns could have reduced the clarity of edge sensations by eliminating spatially precise inhibitory influences. Large elevations of threshold would result from such a disruption of a salient cue that was relied upon preoperatively to judge size. Subsequent recovery could have resulted from relearning the discrimination on the basis of sensations related more directly to the area of skin contacted by the disks. In this case, recovery of performance on the size discrimination task would not mean that the full spectrum of somatosensory functions had been reacquired by some form of neural reorganization. A permanent deficit of edge resolution might have been produced, but if so, it was not revealed in the long run by the psychophysical task of size discrimination.

If edge detection is dependent upon spatially precise inhibitory interactions within the DC–lemniscal system, then edge sensitivity should first appear centrally (in the dorsal column nuclei or at thalamic or cortical sites of projection of the DC–lemniscal pathway). As a first step in directly evaluating this possibility, peripheral neurons were evaluated for edge sensitivity. For this purpose, recordings were obtained from single peripheral afferents innervating the skin of cats. Afferents with slowly adapting cutaneous receptors were isolated for study for several reasons: (a) Their receptive field borders could be defined clearly. (b) Also, the slowly adapting afferents contribute primarily to pathways emanating from the dorsal horn (Whitsel, Petrucelli, & Sapiro, 1969). Thus, if these afferents are preferentially sensitive to edges, then this is not a characteristic that can be attributed exclusively to the DC–lemniscal system.

Recordings from slowly adapting peripheral afferents clearly revealed preferential activation by edge contours (Vierck, 1979). This was tested by defining the receptive fields with punctate stimulation and then utilizing a feedback-controlled stimulator to precisely regulate indentations applied to the receptive fields by each of three stimulus configurations: (a) a conical point that was centered on the receptive field of a slowly adapting unit, (b) the edge of a disk, or (c) the surface of a disk. In the last condition, the circumference of the disk was outside the receptive field. The units discharged vigorously to the point or the edge, and both the onset transient and the steady-state responses of the slowly adapting units were monotonically related to indentations from 100 to 800 $\mu$m. Without exception, these units were activated less effectively by the surface of the disk, and the growth functions for indentation by the surface were nearly

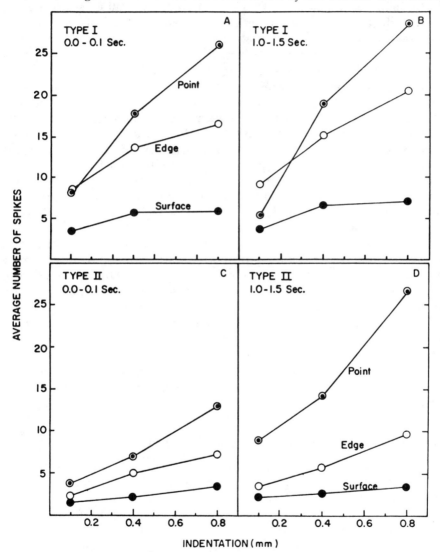

**FIGURE 4.** Activity elicited in peripheral slowly adapting tactile afferents (SA I or SA II) by application of a point, an edge or a surface to the receptive field. Panels A and C depict the number of action potentials elicited during the first 100 msec of stimulation at indentations ranging from 200 to 800 $\mu$m. Panels B and D present the number of spikes elicited during maintained indentations (from 1 to 1.5 sec after stimulus onset). For both types of SA afferent, indentation of the entire RF by a surface produced minimal activity, relative to stimulation by a point or edge. From Vierck (1979).

flat. At every depth, the discharge produced by the point or edge was greater than that produced by the surface, and this difference increased considerably as a function of indentation (Fig. 4). A preferential sensitivity to edges of all classes of peripheral touch receptors on hairy and glabrous skin has now been demonstrated (Johansson, Landstrom, & Lundstrom, 1982; Phillips & Johnson, 1981). Therefore, edge detection is provided by the transduction properties of peripheral cutaneous afferents, and it is not necessary to invoke central inhibitory mechanisms to account for enhancement of edge over surface sensations.

Finding edge detection at the peripheral level of the somatosensory system suggests that the dramatic recovery of size discrimination after dorsal column interruption is the result of reorganization within the DC–lemniscal system, and these processes of neural reactivity have been discussed elsewhere (Berkley & Vierck, 1987; Vierck, 1982, 1984). The present review will persevere with a description of attempts to define sensory capacities that are supplied uniquely by the dorsal columns. So far, permanent deficits from dorsal column interruption have not been observed for psychophysical tests for detection of a single gap or for discriminations of location or size. The most sensitive of these spatial tasks (size discrimination) was only temporarily disrupted by dorsal column interruption and appears to be based upon the *relative number* of neurons activated by tactile stimuli. Possibly a spatial task that would demand more of topographic resolution would require specification of the *relative positions* of neurons activated by different patterns, shapes, or orientations. Discriminations of relative position should require neural coding operations that also differ fundamentally from those required for absolute localization. Localization is likely to depend upon identification of the location of activity within a map, regardless of the spatial relations among the neurons that are activated. The DC–lemniscal system might support fine-grained discriminations of the relative positions of tactile stimuli, regardless of the absolute location of active neurons within the map.

## Discriminations of Orientation or Relative Position

To develop tests of acuity for spatial relations, human subjects were required to specify the orientation of different stimuli that were applied either proximodistally or mediolaterally on the volar surface of one forearm. When the stimuli were lines, their lengths at threshold for orientation discrimination (the *critical lengths*) were 16.8 mm. The importance for this discrimination of the end positions of the lines against the relative positions of all receptive fields stimulated by the entire line was then tested by aligning two points proximodistally or mediolaterally and determining critical lengths for orientation. The average threshold for two-point orientation was 13.1 mm, indicating that continuous lines were not advantageous for specifying orientation. Finally, when two points were applied sequentially in one orientation or the other, the critical lengths were reduced to 8.7 mm. Thus, with sequential application of a punctate stimulus to two locations, reasonably low thresholds could be obtained for a truly spatial discrimination of relative position. Such a test might be dependent upon the dorsal columns, because the somatotopy of the DC–lemniscal system and the somatosensory cortex represents the relative location of nearby skin sites with particular faithfulness (Werner & Whitsel, 1968).

Based upon the results of the human studies, an investigation of the effects of dorsal column lesions on a discrimination of relative location was initiated. Monkeys were presented with sequential stimulation of different points on either lateral calf, and they were required to discriminate whether the contacts occurred in proximodistal or distoproximal sequences (Vierck, Cohen, & Cooper, 1983). This *spatial sequence* task differs from the classical test of point localization (a misnomer), which requires only that the subjects determine whether different spots were stimulated. The test for the monkeys required that they sense two separate contacts *and* that they discriminate the relative position of the points stimulated.

Preoperative thresholds for spatial sequence recognition were 10.4 mm, and early postoperative thresholds averaged 37.4 mm. However, with retraining over 6 postoperative months, normal discrimination recovered. With this disappointing result, the conclusion was reached that the DC–lemniscal system is not critical for spatiotactile discriminations that depend upon the following representations within a somatotopic map: (a) the location of activity elicited by a point stimulus, (b) the number of neurons activated by a single stimulus, (c) a gap between populations of neurons activated by two points, or (d) the relative positions of neurons activated by sequential stimuli.

The normal thresholds for these discriminations that are not permanently deteriorated by dorsal column lesions appear to be determined by the receptive field characteristics of cortical cells, as defined by punctate stimulation. The thresholds vary appropriately with the sizes, configurations, overlap, and distribution of cortical receptive fields for stimulation of different body regions by a single stimulus. Surprisingly, the DC–lemniscal system appears not to elaborate upon these receptive fields for orthogonal indentation to provide unique receptive field organizations that cannot be derived from the output of the spatially ordered dorsal horn. This conclusion from psychophysical studies is reinforced by single-unit recordings from the somatosensory cortex of monkeys with dorsal column lesions; many cortical neurons have apparently normal receptive fields for a single punctate stimulus (Dreyer et al., 1974).

The finding that the dorsal columns do not provide information that is especially suited for derivation of simple spatial relations should not be taken to mean that the DC–lemniscal system cannot support spatial discriminations. This review is concentrating on the effects of dorsal

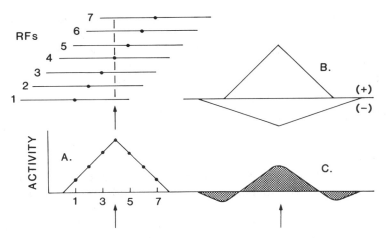

**FIGURE 5.** A model of responsivity to a punctate stimulus by a population of neurons distributed across a topographic map of the body surface. If 7 (columns of) neurons are lined up along one dimension of the map, and their RFs are partially shifted along one dimension of the skin (upper left diagram), then a punctate stimulus will produce a profile of activity that is maximal at the location of neurons with RFs centered at the stimulus. The activity will tail off for neurons that are stimulated within relatively more peripheral portions of their RFs (A). (B) If coincident inhibitory RFs are assumed, and the activity of each neuron is dictated by an approximate algebraic summation of excitation and inhibition, then a punctate stimulus will elicit a slightly smaller peak of activity that is surrounded by a reduction of activity below spontaneous levels (C).

influences, the available cue for discriminating different numbers of stimulus points (within the same overall length) is the magnitude of activity across the population of neurons activated (compare Fig. 6A with B and C with D). However, inclusion of inhibitory receptive fields (Fig. 6E–H) reveals a sculpturing of the population profile that is dependent upon the number and spacing of the points. Stimulation by 2 or 3 points within a distance less than the average receptive field diameter (Fig. 6E and F) produces peaks of activity that vary slightly in amplitude and width but differ little from the profile generated by a single point (Fig. 5C). In contrast, the same spacings between longer series of points (Fig. 6G and H) provide peaks and troughs of activity that occur at spatial frequencies that correspond to the distances between points.

The sculpturing of population activity into distinct peaks and troughs by a linear array of points appears to depend upon stimulation over an extent that exceeds the diameter of the receptive fields supplying the stimulated region (Fig. 6E–H). Also, the critical length for pattern discrimination approximates the threshold for resolution of two points from one, and two-point thresholds are related to the sizes of receptive fields supplying different skin regions. Thus, the two-point threshold does not define the minimal gap size that can be perceived; it appears instead to correspond to a minimal distance that is required for interactions that subserve tactile

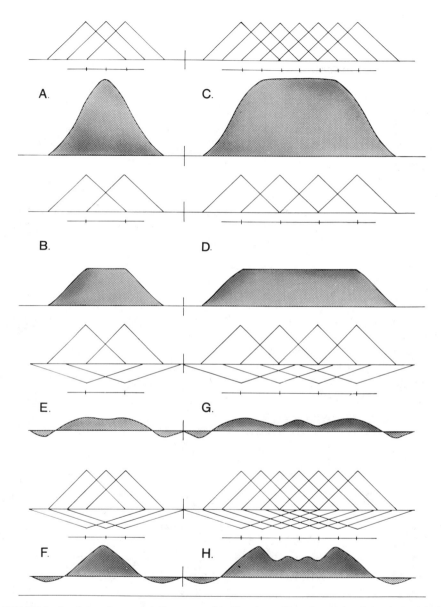

**FIGURE 6.** Applying the model diagrammed in Fig. 4, profiles of neural activity that might be elicited by patterns of points are shown for different spacings and for excitation alone (*A–D*) or excitation and inhibition (*E–H*). The shaded profiles are generated by albegraic summation of excitatory and inhibitory influences (shown at the top of each panel) of spatial patterns that are depicted as ticks on the intermediate lines. In each case, the stimuli are spaced at intervals less than the average RF diameter of the neurons represented. These diagrams result in spatial profiles that are maximally sculptured in the presence of inhibition and with stimuli of critical lengths that extend beyond the diameter of a single RF (panels *G* and *H*).

pattern perception. The possibility that inhibition is an important component of the interactive process that requires spatial replication is indicated in Fig. 6.

Once again, human psychophysical experiments have given us a lead for investigating the functions of the DC–lemniscal system, and we are presently testing monkeys on discriminations of different patterns of replicated gaps. As usual, we expect deficits from dorsal column interruption, based upon the notion that the precise inhibitory receptive fields within the DC–lemniscal system make unique contributions to spatiotactile coding. These experiments will provide further information on the importance of establishing a spatial pattern over an appropriate critical length. In the absence of results on discriminations of patterns by lesioned animals, we present the results of another series of studies that was based upon a related logic: that high spatial acuity is revealed when stimuli progress *across* a critical length of skin (Johansson & LaMotte, 1983; Lamb, 1983).

### Direction Sensitivity

For skills that involve discriminations of spatiotemporal progressions of stimuli across the skin surface, the combination of high spatial *and* temporal resolution in the DC–lemniscal system (Blum, Bromberg, & Whitehorn, 1975; Bystrzycka, Nail, & Rowe, 1977; Ferrington, Rowe, & Tarvin, 1987) might provide unique cues. The first problem in testing this possibility was to define a discrimination that could be learned by the monkeys and is sensitive to a high-order sensory code that underlies spatially acute sensations of movement on the skin. For this purpose, we chose a discrimination of direction, which is simplistic but fundamental to more complex tests of graphesthesia (i.e., perception of forms "drawn" on the skin).

To evaluate the importance of the dorsal columns for tactile direction sensitivity, monkeys were trained to discriminate either the presence of motion across the skin or the direction of movement (Vierck, 1974). Light tactile stimulation was provided by a brush, and detection of movement was tested as follows: The soft bristles contacted the skin of the shaved lateral calf or the glabrous skin of the foot, and after 1 sec the brush was withdrawn, with or without a flick that moved the bristles across the skin. Responses to a manipulandum on the right were rewarded for orthogonal withdrawal, and responses to the left were rewarded following withdrawal that included movement across the skin. Interruption of the ipsilateral dorsal column did not impair performance on this task, and it was concluded that the lesioned animals could detect the simple presence of a brief and spatially limited movement across the skin.

When monkeys were trained to discriminate the direction of movement on the lateral calf, a different result was obtained. For this test the brush was applied orthogonally, as before, but withdrawals consisted of proximally or distally directed flicks. The animals were required to respond differentially to the direction of movement on withdrawal. In a variant of the test, one monkey received proximal or distal rotation of a cylindrical bottle brush; in this case, the region of skin stimulated was identical for movements in either direction. For either manner of stimulation, dorsal column interruption eliminated direction sensitivity, and recovery was not observed over testing periods in excess of a year. This result has been confirmed in human patients with traumatic interruption of the dorsal columns (Wall & Noordenbos, 1977). Similarly, deficits of graphesthesia occur following spinal lesions that include the dorsal columns (Nathan et al., 1986).

The test for direction sensitivity reveals a fundamental importance of the dorsal columns for appreciation of spatiotemporal progressions of stimuli across the skin. Direction sensitivity may in turn be dependent upon activity in subsets of cortical neurons which exhibit preferential sensitivities for movement in certain directions at speeds that match psychophysical functions for discrimination of direction (Essick & Whitsel, 1985b). The direction sensitivity of individual somatosensory neurons appears de novo at the cortical level (Whitsel, Roppolo, & Werner, 1972) and is not a property of neurons at other levels of any somatosensory pathway. Thus, the DC–lemniscal system appears to provide input to the somatosensory cortex that is required to determine the directionally selective responses of a subset of cortical cells. Possibly this represents a general principle—that somatosensory functions that depend upon the dorsal columns are derived from specialized patterns of convergence onto single cells (i.e., feature detecting neurons), as distinct from sensations that are derived from patterns of activity that progress across a population of neurons. This has not been tested directly, and it would require a very large sample to conclude that single cells no longer coded stimulus direction after a dorsal column lesion.

Another approach to understanding why the DC–lemniscal system is required for direction sensitivity is to determine the cues that are most effective in deriving direction. When a stimulus is moved across the skin surface, the parameters that may contribute to direction sensitivity are (a) the location and orientation of the traverse, (b) the extent of excursion, (c) the velocity of movement, (d) the spatiotemporal pattern of skin contact(s) (e.g., a continuous, single contact versus multiple, sequential contacts), and (e) skin stretch. In an investigation of these factors, a variety of stimuli were placed upon or moved across the volar surface of the forearms of human subjects (Gould, Vierck, & Luck, 1979), and thresholds for critical length were obtained. That is, we determined the minimal length of a stimulus or of a traverse that would support 75% correct identifications of orientation or direction.

To gain an appreciation of the importance of moving the stimulus across the skin, subjects were required to identify whether the stimulus (or its traverse) was oriented mediolaterally or proximodistally on the forearm. Three types of stimulation were delivered in separate tests, and the thresholds for critical length were (a) 16.8 mm for lines that were applied orthogonally but not stroked across the skin, (b) 10.5 mm for the length of an arc that was rolled over the skin, and (c) 4.3 mm for the traverse of a weighted (5 g) probe that was stroked across the skin. Each of these thresholds was significantly different, demonstrating that sequential stimulation by the arc (providing cues of direction and orientation) provided better acuity than simultaneous contact, and stroking the skin provided a much greater advantage. When the skin was stroked with the weighted probe, there was a frictional component (and skin stretch) not present for the arc, and the directional cue was purely sequential rather than cumulative. That is, the probe contacted and left each point on the skin in sequence, but the arc retained contact with each skin point stimulated.

In an attempt to determine whether the spatial acuity imparted by stroking across the skin is provided entirely by sequential and focal contact or depends also on friction, the discrimination was changed from orientation to direction (medial vs. lateral progressions on the volar forearm). Again, three types of stimulation were utilized in separate tests of the same subjects, and the thresholds for critical length were (a) 11.3 mm for sweeping across the skin with an air jet (700 mg), (b) 4.1 mm for stroking with the 5-g probe (corresponding to orientation thresholds with the same stimulus), and (c) 0.6 mm for stretch by a probe that was glued to the skin. Each of these comparisons tells us that friction and/or stretch is an important factor in the high degree of spatial acuity that is provided by stimuli that move across the skin surface.

This result puzzled us in relation to the enduring deficit of direction sensitivity that was produced by dorsal column section. It seemed likely that the combination of spatial and temporal resolution within the DC–lemniscal system would be important for coding directions of movement, but stretch sensitivity has been attributed to only one type of slowly adapting peripheral afferents (SA II) that do not contribute substantially to the long dorsal column projection from the lumbosacral cord to the medulla (Whitsel et al., 1969). This view of stretch sensitivity has arisen from recording experiments that have applied a restrictive criterion that the peripheral receptor respond to stretch from outside the receptive field for punctate stimulation. However, tests of neural coding for stretch and direction sensitivity should include direct tangential distortion of receptive fields at different velocities. When this has been done, it has been shown that QA afferents do respond to lateral deformation of glabrous skin (LaMotte & Srinivasan, 1987).

In the psychophysical test of stretch sensitivity, with the probe glued to the skin, the lateral or medial stretch was delivered by a servo-controlled stimulator at velocities of 10 mm/sec (producing the 0.6-mm threshold for critical length) or 1 mm/sec (which required a substantial excursion of 2.5 mm for direction to be discriminated). Since slowly adapting receptors should have been activated effectively at low or high rates of excursion, this result suggests that QA afferents are brought in at the velocities that are optimal for directional discrimination. Support for this possibility comes from a recent demonstration that QAs account for detection of a small dot that is stroked across the skin at rates of 10 mm/sec or higher (Lamotte & Whitehouse, 1986). Also, psychophysical discrimination of the direction of brush strokes is maximal at rates of movement from 50 to 300 mm/sec (Essick & Whitsel, 1985a). Thus, direction sensitivity is optimal at rates of movement that preferentially activate afferents that represent the dominant component of the DC–lemniscal system.

The velocity dependence of direction sensitivity reinforces the conclusion that feature detection by single neurons is important for directional coding. That is, if direction sensitivity were best for low rates of movement, it would favor the idea that a progression of activity across somatotopically organized neurons generates sensations of movement direction. Furthermore, the parameter of velocity has interesting implications for the role of inhibition in determining the directional sensitivity of single neurons. Inhibitory influences appear to be crucial for the encoding of direction by somatosensory neurons at the cortical level (Warren, Hamalainen, & Gardner, 1986), and DC–lemniscal inhibition depends upon receptors with preferential sensitivities for certain velocities. For example, inhibitory influences within the DC–lemniscal system have been shown to be activated only by afferents with a sensitivity for high velocities of stimulation (Bystrzycka et al., 1977). An additional property of DC–lemniscal inhibition is that it outlasts excitation that is generated by a brief stimulus (Janig et al., 1977). Therefore, as a stimulus moves across the skin, excitation of a DC–lemniscal cell might be confined to an early period of contact with its receptive field, because inhibition would build up. Depending upon the rate of stimulus progression in relation to the time course of inhibition, such a mechanism could produce a variety of receptive field characteristics that differ from those observed with isolated punctate stimuli.

For the visual system, inhibitory influences also have been demonstrated to be important for the directional specificity of single neurons (Mikami, Newsome, & Wurtz, 1986; Sillito, 1975), and the spatial selectivity of the inhibition can be as important as the topography of excitatory inputs (Blakemore & Tobin, 1972). Evidence in the visual system indicates that intracortical inhibition can be entirely responsible for feature detection by single cells, or it can act to reinforce a selectivity that has been built in at another cortical or subcortical region (Innocenti & Fiore, 1974). The latter (reinforcement) process appears necessary to explain the importance of the DC–lemniscal system for cutaneous directional sensitivity.

## TEMPORAL DISCRIMINATION

The time course of inhibition in the DC–lemniscal system has implications for tactile capacities other than direction sensitivity. For example, with repetitive stimulation at a single locus, the second and subsequent contacts in the sequence would be affected by inhibition that does not act on the first stimulus in the train. Therefore, the temporal characteristics of activity by neurons receiving DC–lemniscal input would be expected to depend upon the frequency of stimulation (as related to the time course of excitatory and inhibitory influences). This and other considerations led us to explore capacities for temporal coding by animals with dorsal column lesions.

As a test of purely temporal discrimination, monkeys were trained to respond following presentation of the higher frequency of two trains of tactile stimulation (Vierck, Cohen, & Cooper, 1985). Each stimulus train lasted 1 sec and consisted of 11-msec-duration half-sinusoids (pulses) that were separated by different intervals, defining frequencies from 10 to 35 pulses per second (pps). That is, the intervals between pulses varied, but the characteristics of the sinusoids did not. This procedure differs from the more common technique of varying the period (and therefore the frequency) of continuous sinusoids, which changes the velocity and depth of indentation, altering the proportion of distinct tactile receptors that are activated (LaMotte & Mountcastle, 1975). In the present study, the velocity and depth of the pulses remained constant and activated each type of tactile receptor equally at all frequencies. Such a stimulus produces equal sensation magnitudes across a wide range of frequencies (Rothenberg, Verrillo, Zahorian, Brachman, & Bolanowski, 1977).

A signal detection paradigm was utilized, and the stimulus–response contingencies were as follows:

(a) The standard stimulus (10 pps for 1 sec) was presented as the first train of each trial, followed by a 3-sec interstimulus interval. Lever presses during this first interval constituted anticipatory responses and were not reinforced.

(b) On half the trials, a comparison stimulus was presented as the second train (e.g., 20 pps for 1 sec). When this occurred, a lever press in the second interval was reinforced as a hit, and failure to respond was scored as a miss.

(c) On half the trials, the standard stimulus was presented second, and the animal was required to withhold responding (a correct rejection) until after a third stimulus (the comparison) was presented. Unreinforced responses to the standard as the second stimulus were scored as false alarms.

This was our initial experience with a signal detection procedure, and the paradigm created some difficulties for training the animals. Monkeys are impatient by nature, and the task demanded that they withhold responding through 2 or 3 stimulation intervals. This proved to be very difficult for most of the animals, and a great deal of time was expended training them to withhold responding until after the comparison stimulus was presented. In retrospect, a two-alternative forced-choice paradigm would have been simpler and equally as valid. In the forced-choice paradigms (such as those used in the spatial tests), response biases can be eliminated, and spinal lesions are not expected to have selective effects on biases that require the analytic power of the signal detection procedure.

After stabilization of performance on the signal-detection task, many sessions were conducted with four standard-comparison pairs that generated discrimination which bracketed 85% correct responding. For this paradigm, between-sessions tracking of the thresholds was utilized until performance stabilized. Then the threshold-bracketing comparison values were presented in a sequence that generated equal numbers of each comparison value. Thresholds were estimated from $d'$ values that were computed for the standard-comparison pairs. With stimulation of the glabrous skin of either foot, thresholds for frequency discrimination of normal monkeys averaged 2.8 pps. In distinct contrast, following interruption of the ipsilateral dorsal column, these monkeys could not discriminate 35 pps from 10 pps (thresholds in excess of 25 pps). These severe deficits of frequency discrimination did not recover (over more than one year of postoperative testing).

For the frequency-discrimination task, interpulse intervals of 100 msec or less were chosen, to subject the 2d to the 11th pulse of each train to inhibition from the preceding pulse (Laskin & Spencer, 1979). Thus, the discharge of DC–lemniscal neurons to these pulses should be limited (i.e., brief). This would facilitate temporal discrimination by reducing background noise. To explore this and other possible explanations for the deficit in temporal resolution after dorsal column section, it would be useful to test discrimination between a variety of frequencies and with different sinusoids that would selectively stimulate different categories of tactile receptors. However, in the course of attempting to retrain the DC-lesioned animals on the frequency discrimination, an additional deficit was revealed, suggesting a more fundamental role of the dorsal columns in temporal resolution.

In an attempt to bring the lesioned animals back on task, the comparison stimulus was not only increased in frequency to 35 pps, it also was increased in duration to several seconds. This provided the animals with cues of frequency and duration, and it was our intention to bring the comparison duration gradually back to 1 sec after performance improved. Much to our surprise, the monkeys remained at chance levels of performance over months of testing, indicating that they could not discriminate frequency *or* duration. We were uncomfortable with this result, because the animals had not been trained with variations in duration before the lesion. Therefore, a study of duration discrimination is in progress. Three animals have been trained to discriminate 3 pulses at 10 pps (200-msec-duration standard) from longer trains of 10 pps, and normal threshold performance fell between comparison durations of 500- to 600-msec duration. Following dorsal column lesions, a severe deficit is apparent. Postoperatively, these monkeys cannot discriminate the 200-msec train from one of 3.7 sec (i.e., 3 pulses versus 38 pulses). This result suggests several possibilities. In the absence of the dorsal columns either (a) each pulse elicits activity that persists for seconds, obscuring the presence or absence of further stimulation, or (b) tactile stimulation is ineffective in maintaining neural activity with repetitive stimulation; that is, after the first (or first few) stimuli at certain frequencies, the tactile sensation may disappear.

Consistent with the possibility that the critical functions of the dorsal columns depend upon specialized inhibitory influences, it seems likely that the deficits of frequency and duration discrimination result from excessive neural activity that outlasts the eliciting stimuli. However, in a recent review of the literature on human spinal cord injuries that include the dorsal columns, Nathan et al. (1986) report evidence that tactile sensations of these patients diminish or disappear with continued stimulation. This observation provides compelling support for the interpretation that the dorsal columns are required to maintain neural activity with repetitive stimulation of a single locus. These conflicting hypotheses (that dorsal column lesions promote either prolonged afterdischarge or rapid habituation) are presently being evaluated in our attempt to determine the functions of the pathway that is considered to provide the epicritic somatic sensations.

## SUMMARY

The purpose of the research covered in this review has been to directly relate sensory capacities to the neurobiological characteristics of a somatosensory pathway, and therefore the psychophysical findings are discussed in mechanistic terms. More important, we feel strongly that mutual consideration of neurobiological and behavioral findings is beneficial to either type of experimental program. For example, from our point of view, physiological investigations of the DC–lemniscal system have made excessive use of the punctate stimulus, based on an untested assumption that the specialized coding operations of this system could be revealed by simple, relatively static indentations. However, investigations of monkeys with dorsal column lesions did not reveal deficits on the classical psychophysical tests of spatial sensitivity for punctate stimuli. Lesioned animals were able to identify the locus of stimulation, detect the presence of a gap in two stimuli, and discriminate the size or relative position of simple stimuli. An interpretation of these findings is that the DC–lemniscal system is not required for spatiotactile resolution which depends primarily upon innervation density and receptive field sizes within a topographic map. For several of the tasks (localization and two-point discrimination) the partially deafferented somatotopic map supported normal discrimination early after the lesion, but for the more sensitive tests (discriminations of size and relative position) retraining and/or neural reorganization were required.

The results from lesioned monkeys suggested that the essential characteristics of epicritic sensations were not known. As a first step toward evaluating this possibility—that the classical tests of discriminative somethesis were inadequate to reveal epicritic capacities that do indeed depend upon the dorsal columns—different spatiotactile tests were required. The working assumption was that tests that revealed the highest degree of spatial resolution would be the most sensitive to interruption of the DC–lemniscal pathway. However, the stimulus features that promote spatial resolution had not been well described, and it would not be feasible to determine these with monkeys as subjects, because of the training time required for each test. Thus, human psychophysical investigations were initiated. These experiments permitted efficient screening of a number of tests and provided an appreciation of stimulation methods that generated consistent discrimination on the basis of a limited number of salient cues. They were an essential accompaniment of the studies with monkeys.

The psychophysical experiments with human and nonhuman primates have identified stimulus features that are important for spatiotactile resolution (e.g., spatial replication and spatiotemporal progression). Modeling of neural responses to these parametric manipulations is consistent with a special contribution by the DC–lemniscal system to pattern recognition and direction sensitivity. In addition, the paucity of purely spatial deficits from dorsal column interruption has prompted an examination of temporal resolution, which appears to be the essence of epicritic sensibilities. Some of the factors that have been considered in these analyses are as follows.

*A. Critical Length.* When all portions of a stimulus contact the skin simultaneously, a high degree of spatial acuity depends upon a minimal length or size. This was seen clearly for discriminations of punctate patterns, suggesting that spatial replication is an important factor. Thus, the classical tests for epicritic sensations were of insufficient spatial extent to benefit from integrative processing which is crucial for the capacities these tests were designed to reveal. For example, the two-point test appears to reflect the critical length for gap detection but not the capacity of the system for gap detection. The generality of this finding has been reinforced by a recent demonstration that estimates of the length of a tactile sensation on different skin regions are constant for extents below the two-point threshold, but sensory length increases linearly with stimulus extents beyond the two-point threshold (Sherrick & Cholewiak, 1986). Thus, a variety of spatiotactile discriminations depend upon a critical length in excess of some approximation of the average receptive field size for a given skin region. Similarly, spatial replication has been shown to be an important factor for spatiotemporal discriminations of visual stimuli (McKee & Welch, 1985).

*B. Spatially Precise Inhibition.* Because edge detection is provided at the periphery by a selective sensitivity of tactile receptors to focal indentation, inhibitory influences are not crucial for sensing the location, limits, or size of simple tactile stimuli. However, the spatial configuration of excitatory and inhibitory receptive fields for DC–lemniscal neurons appears to facilitate resolution of tactile patterns with a critical length in excess of the average receptive field sizes (see Figs. 5 and 6). Thus, discriminations of tactile patterns have a potential for demonstrating a purely spatial capacity that is dependent upon integrity of the DC–lemniscal system.

*C. Sequential Stimulation.* A variety of factors contribute to high acuity for spatial progressions of stimulation across the skin, and these have been only partially defined. Simple resolution of the relative locations of punctate stimuli is facilitated by sequential application (e.g., point localization thresholds are lower than two-point thresholds). This effect occurs over long delays, and it does not reveal special functions of the DC–lemniscal system (e.g., spatial sequence recognition survives section of the dorsal columns). The retention of spatial sequence recognition after dorsal column section is consistent with demonstrations that low frequencies of sequential stimulation support a "long-range" process of spatial resolution that depends upon population coding and is distinct from a "short-range" process of feature detection by certain neurons in response to rapid movement (e.g., Newsome, Mikami, & Wurtz, 1986).

*D. Velocity, Frequency, and Duration.* Previous descriptions of epicritic sensations and of the DC–lemniscal system have focused on strictly spatial characteristics, but each sensory test that has been shown to be sensitive to dorsal column interruption is influenced strongly by temporal factors: (a) Directional sensitivity is affected by the velocity of stimulation, even when the only available cue is skin stretch; and (b) frequency and duration discriminations are purely temporal tests that were impaired for a stimulus that activated receptors sensitive to the full physiological range of velocities.

*E. Temporal Coding.* The primary afferents to DC–lemniscal projection neurons are quickly adapting and are sensitive to moderate to high velocities of tactile stimulation. Similarly, inhibitory influences on DC–lemniscal neurons are elicited by high velocities and influence subsequent responses to stimuli that occur in rapid succession. These features, combined with high spatial resolution, are clearly relevant to the stimulus characteristics that are required for direction sensitivity. More generally, these inputs may be crucial for feature detection by central somatosensitive neurons that have a variety of specialized requirements for spatially organized inputs that are precisely timed in onset and duration. The deficits in *both* frequency and duration discrimination are more difficult to comprehend in mechanistic terms. It seems likely that some temporal modulation (of frequency and duration) must remain for responsive neurons after dorsal column section. If this is the case, then DC–lemniscal input must provide important time-marking or time-setting functions.

*F. Adaptive Significance.* In retrospect, a redefinition of epicritic sensations is consistent with the typical strategies for somatosensory sampling by mammals with highly mobile distal extremities. The manipulative capabilities of primates are especially demanding in terms of dynamic sensitivities to stimuli that are sampled briefly, repetitively, and in the context of purposive movements. Therefore, it is not surprising that timing is a cardinal feature of epicritic sensibilities. Also, because localization and gross identification of environmental objects are usually provided or aided by visual cues, tactile acuity is relied upon for more fine-grained discriminations of subtle variations in the surface features of objects that are grasped, palpated, or stroked. Pattern recognition is one of many sensory tasks that are presented by these methods of tactile exploration that typically engage a large proportion of the skin of the distal extremity. Therefore, specialized sensory capacities that require certain critical lengths and velocities of stimulation are in keeping with the most sophisticated somatosensory capacities of mammals.

## ACKNOWLEDGMENTS

Research of the authors that is presented in this chapter has been supported by Grants NS 07261 and NS17474 from the National Institute of Neurological and Communicative Disorders and Stroke.

## REFERENCES

Békésy, G. von. (1967). *Sensory inhibition.* Princeton, NJ: Princeton University Press.

Berkley, K. J., & Vierck, C. J., Jr. (1987). Transient changes in retrograde labeling of diencephalic-projecting neurons in the monkey dorsal column nuclei following removal of their dorsal column input. In L. Pubols & B. Sessle (Eds.), *Effects of injury on somatosensory systems: Neurology and neurobiology* (Vol. 30, pp. 429–436). New York: Liss.

Blakemore, C., & Tobin, E. A. (1972). Lateral inhibition between orientation detectors in the cat's visual cortex. *Experimental Brain Research*, **15**, 439–440.

Blum, P., Bromberg, M. B., & Whitehorn, D. (1975). Population analysis of single units in the cuneate nucleus of the cat. *Experimental Neurology*, **48**, 57–78.

Brown, P. B., & Culberson, J. L. (1981). Somatotopic organization of hindlimb cutaneous dorsal root projections to cat dorsal horn. *Journal of Neurophysiology*, **45**, 137–143.

Burgess, P. R., & Perl, E. R. (1973). Cutaneous mechanorecptors and nociceptors. In A. Iggo (Ed.), *Handbook of sensory physiology* (Vol. 2, pp. 29–78). Berlin: Springer.

Bystrzycka, E., Nail, B. S., & Rowe, M. (1977). Inhibition of cuneate neurones: Its afferent source and influence on dynamically sensitive "tactile" neurons. *Journal of Physiology (London)*, **268**, 251–270.

Chapin, J. K. (1986). Laminar differences in sizes, shapes and response profiles of cutaneous receptive fields in the rat SI cortex. *Experimental Brain Research*, **62**, 549–559.

Cook, A. W., & Browder, E. J. (1965). Function of posterior columns in man. *Archives of Neurology and Psychiatry*, **12**, 72–79.

Corkin, S., Milner, B., & Rasmussen, T. (1970). Somatosensory thresholds. Contrasting effects of postcentral-gyrus and posterior parietal-lobe excisions. *Archives of Neurology (Chicago)*, **23**, 41–58.

Craig, J. C. (1983). Some factors affecting tactile pattern recognition. *International Journal of Neuroscience*, **19**, 47–58.

Denny-Brown, D., Meyer, J. S., & Horenstein, S. (1952). The significance of perceptual rivalry resulting from parietal lesions *Brain*, **75**, 433–471.

Dreyer, D. A., Schneider, R. F., Metz, C. B., & Whitsel, B. L. (1974). Differential contribution of spinal pathways to body representation in postcentral gyrus of *Macaca mullata*. *Journal of Neurophysiology*, **37**, 119–145.

Elithorn, A., Piercy, M. G., & Crosskey M. A. (1953). Tactile localization. *Quarterly Journal of Experimental Psychology*, **5**, 171–182.

Erickson, R. P. (1968). Stimulus coding in topographic and non-topographic modalities: On the significance of the activity of individual sensory neurons. *Physiological Reviews*, **75**, 447–465.

Essick, G., & Whitsel, B. L. (1985a). Assessment of the capacity of human subjects and SI neurons to distinguish opposing directions of stimulus motion across the skin. *Brain Research Reviews*, **10**, 187–212.

Essick, G., & Whitsel, B. L. (1985b). Factors influencing cutaneous directional sensitivity: A correlative psychophysical and neurophysiological investigation. *Brain Research Reviews*, **10**, 213–230.

Favorov, O., & Whitsel, B. L. (1988a). Spatial organization of the peripheral input to area 1 cell columns. I. The detection of "Segregates." *Brain Research Reviews*. **13**, 25–42.

Favorov, O., & Whitsel, B. L. (1988b). Spatial organization of the peripheral input to area 1 cell columns. II. The forelimb representation achieved by a mosaic of segretates. *Brain Research Reviews.* **13**, 43–56.

Ferrington, D. G., Rowe, M. J., & Tarvin, R. P. C. (1987). Actions of single sensory fibres on cat dorsal column nuclei neurones: Vibratory signalling in a one-to-one linkage. *Journal of Neurophysiology,* **386,** 293–309.

Friedman, D. P., Murray, E. A., O'Neill, J. B., & Mishkin, M. (1986). Cortical connections of the somatosensory fields of the lateral sulcus of Macaques: Evidence for a corticolimbic pathway for touch. *Journal of Comparative Neurology,* **252,** 323–347.

Fuchs, J. J., & Brown, P. B. (1984). Two-point discriminability: Relation to properties of the somatosensory system. *Somatosensory Research,* **2,** 163–169.

Gardner, E. P., & Spencer, W. A. (1972). Sensory funneling. II. Cortical neuronal representation of patterned cutaneous stimuli. *Journal of Neurophysiology,* **35,** 954–977.

Gordon, G. (1978). *Active touch: The mechanisms of recognition of objects by manipulation. An interdisciplinary approach.* Oxford: Pergamon.

Gould, W. R., Vierck, C. J., Jr., & Luck, M. M. (1979). Cues supporting recognition of the orientation or direction of movement of tactile stimuli. In D. R. Kenshalo (Ed.), *Sensory functions of the skin in humans* (pp. 63–78). New York: Plenum.

Halnan, C. R. E., & Wright, G. H. (1960). Tactile localization. *Brain,* **83,** 677–700.

Head, H. (1918). Sensation and the cerebral cortex. *Brain,* **41,** 57–253.

Innocenti, G. B., & Fiore, L. (1974). Post-synaptic inhibitory components of the responses to moving stimuli in area 17. *Brain Research,* **80,** 122–126.

Iwamura, Y., Tanaka, M., Sakamoto, M., & Hikosaka, O. (1985). Diversity in receptive field properties of vertical neuronal arrays in the crown of the postcentral gyrus of the conscious monkey. *Experimental Brain Research,* **58,** 400–411.

Janig, W., Schoultz, T., & Spencer, W. A. (1977). Temporal and spatial parameters of excitation and afferent inhibition in cuneothalamic relay neurons. *Journal of Neurophysiology,* **40,** 822–835.

Johansson, R. S. & LaMotte, R. H. (1983). Tactile detection thresholds for a single asperity on an otherwise smooth surface. *Somatosensory Research,* **1,** 21–32.

Johansson, R. S., Landstrom, U., & Lundstrom, R. (1982). Sensitivity to edges of mechanoreceptive afferent units innervating the glabrous skin of the human hand. *Brain Research,* **144,** 27–32.

Jones, M. B., & Vierck, C. J., Jr. (1973). Length discrimination on the skin. *American Journal of Psychology,* **86,** 49–60.

Jones, M. B., Vierck, C. J., Jr., & Graham, R. B. (1973). Line-gap discrimination on the skin. *Perceptual and Motor Skills,* **36,** 563–570.

Juliano, S., & Whitsel, B. L. (1985). Metabolic labeling associated with index finger stimulation in monkey SI: Between animal variability. *Brain Research,* **342,** 242–251.

Lamb, G. D. (1983). Tactile discrimination of textured surfaces: Psychophysical performance measurements in humans. *Journal of Physiology (London),* **338,** 551–565.

LaMotte, R. H., & Mountcastle, V. B. (1975). Capacities of humans and monkeys to discriminate between vibratory stimuli of different frequency and amplitude: A correlation between neural events and psychophysical measurements. *Journal of Neurophysiology*, **38**, 539–559.

LaMotte, R. H., & Srinivasan, M. A. (1987). Tactile discrimination of shape: Responses of rapidly adapting mechanoreceptive afferents to a step stroked across the monkey fingerpad. *Journal of Neuroscience*, **7**, 1672–1681.

LaMotte, R. H., & Whitehouse, J. (1986). Tactile detection of a dot on a smooth surface: Peripheral neural events. *Journal of Neurophysiology*, **56**, 1109–1128.

Laskin, S. E., & Spencer, W. A. (1979). Cutaneous masking. II. Geometry of excitatory and inhibitory receptive fields of single units in somatosensory cortex of the cat. *Journal of Neurophysiology*, **42**, 1061–1082.

Lederman, S. J. (1983). Tactual roughness perception: spatial and temporal determinants. *Canadian Journal of Psychology*, **37**, 498–511.

Loomis, J. M. (1981). Tactile pattern perception. *Perception*, **10**, 5–27.

Marshall, J. (1956). Studies in sensation: Observations on the localization of the sensations of touch and prick. *Journal of Neurology, Neurosurgery and Psychiatry*, **19**, 84–87.

McKee, S. P., & Welch, L. (1985). Sequential recruitment in the discrimination of velocity. *Journal of the Optical Society of America A*, **2**, 243–251.

McKenna, T. M., Whitsel, B. L., & Dreyer, D. A. (1982). Anterior parietal cortical topographic organization in macaque monkey: A reevaluation. *Journal of Neurophysiology*, **48**, 289–317.

Mikami, A., Newsome, W. T., & Wurtz, R. H. (1986). Motion selectivity in Macaque visual cortex. I. Mechanisms of direction and speed selectivity in extrastriate area MT. *Journal of Neurophysiology*, **55**, 1308–1327.

Mountcastle, V. B. (1961). Some functional properties of the somatic afferent system. In W. A. Rosenblith (Ed.), *Sensory communications* (pp. 403–436). Cambridge, MA: MIT Press.

Mountcastle, V. B., Davies, P. W., & Berman, A. L. (1957). Response properties of neurons of Cat's somatic sensory cortex to peripheral stimuli. *Journal of Neurophysiology*, **20**, 374–407.

Mountcastle, V. B., & Powell, T. P. S. (1959). Neural mechanism subserving cutaneous sensibility, with special reference to the role of afferent inhibition in sensory perception and discrimination. *Bulletin of the Johns Hopkins Hospital*, **105**, 201–232.

Nathan, P. W., Smith, M. C., & Cook, A. W. (1986). Sensory effects in man of lesions of the posterior columns and of some other afferent pathways. *Brain*, **109**, 1003–1041.

Nelson, R. J., Sur, M., Felleman, D. J., & Kaas, J. H. (1980). Representation of the body surface in postcentral parietal cortex of *Macaca fascicularis*. *Journal of Comparative Neurology*, **192**, 611–643.

Newsome, W. T., Mikami, A., & Wurtz, R. H. (1986). Motion selectivity in Macaque visual cortex. III. Psychophysics and physiology of apparent motion. *Journal of Neurophysiology*, **55**, 1340–1351.

Phillips, J. R., & Johnson, K. O. (1981). Tactile spatial resolution. II. Neural representation of bars, edges and gratings in monkey primary afferents. *Journal of Neurophysiology*, **46**, 1192–1203.

Ratliff, F. (1965). *Mach bands: Quantitative studies on neural networks in the retina.* San Francisco, CA: Holden-Day.

Rothenberg, M., Verrillo, R. T., Zahorian, S. A., Brachman, M. L. & Bolanowski, S. J., Jr. (1977). Vibrotactile frequency for encoding a speech parameter. *Journal of the Acoustical Society of America*, **62**, 1003–1012.

Semmes, J. (1969). Protopathic and epicritic sensation, a reappraisal. In A. L. Benton (Ed.), *Contributions to clinical neuropsychology* (pp. 142–171). Chicago, IL: Aldine.

Sherrick, C. E., & Cholewiak, R. W. (1986). Princeton Cutaneous Research Project Report No. 48.

Sillito, A. M. (1975). The contribution of inhibitory mechanisms to the receptive field properties of neurones in the striate cortex of the cat. *Journal of Physiology (London)*, **250**, 305–329.

Sur, M., Merzenich, M. M., & Kaas, J. H. (1980). Magnification, receptive-field area, and "hypercolumn" size in areas 3b and 1 of somatosensory cortex in owl monkeys. *Journal of Neurophysiology*, **44**, 295–311.

Vierck, C. J., Jr. (1973). Alterations of spatio-tactile discrimination after lesions of primate spinal cord. *Brain Research*, **58**, 69–79.

Vierck, C. J., Jr. (1974). Tactile movement detection and discrimination following dorsal column lesions in monkeys. *Experimental Brain Research*, **20**, 331–346.

Vierck, C. J., Jr. (1979). Comparisons of punctate, edge and surface stimulation of slowly adapting somatosensory afferents of cats. *Brain Research*, **175**, 155–159.

Vierck, C. J., Jr. (1982). Plasticity of somatic sensations and motor capabilities following lesions of the dorsal spinal columns in monkeys. In A. R. Morrison & P. L. Strick (Eds.), *Changing concepts of the nervous system* (pp. 151–169). New York: Academic Press.

Vierck, C. J., Jr. (1984). The spinal lemniscal pathways. In R. Davidoff (Ed.), *Handbook of the spinal cord* (pp. 673–750). New York: Dekker.

Vierck, C. J., Jr., Cohen, R. H., & Cooper, B. Y. (1983). Effects of spinal tractotomy on spatial sequence recognition in Macaques. *Journal of Neuroscience*, **3**, 280–290.

Vierck, C. J., Jr., Cohen, R. H., & Cooper, B. Y. (1985). Effects of spinal lesions on temporal resolution of cutaneous sensations. *Somatosensory Research*, **3**, 45–56.

Vierck, C. J., Jr., Cooper, B. Y., & Leonard, C. M. (1987). Motor capacities and deficiencies after interruption of the dorsal spinal columns in primates. In L. Pubols & B. Sessle (Eds.), *Effects of injury on somatosensory systems* (pp. 419–428). New York: Liss.

Vierck, C. J., Jr., Favorov, O., & Whitsel, E. L. (1988). Neural mechanisms of absolute tactile localization in monkeys. *Somatosensory Research*. **6**, 41–62.

Vierck, C. J., Jr., & Jones, M. B. (1969). Size discrimination on the skin. *Science*, **158**, 488–489.

Wall, P. D. (1970). The sensory and motor role of impulses traveling in the dorsal columns toward cerebral cortex. *Brain*, **93**, 505–524.

Wall, P. D., & Noordenbos, W. (1977). Sensory functions which remain in man after complete transection of dorsal columns. *Brain*, **100**, 641–653.

Warren, S., Hamalainen, H. A., & Gardner, E. P. (1986). Coding of the spatial period of gratings rolled across the receptive fields of somatosensory cortical neurons of awake monkeys. *Journal of Neurophysiology*, **56**, 623–639.

Weber, E. H. (1835). Uber den Tastinn. *Archives für Anatomie, Physiologie und Wissenshaftliche Medicin*, 152.

Weinstein, S. (1968). Intensive and extensive aspects of tactile sensitivity as a function of body part, sex and laterality. In D. Kenshalo (Ed.), *The skin senses* (pp. 195–222). Springfield, IL: Thomas.

Werner, G., & Whitsel, B. L. (1968). Topology of the body representation in somatosensory area 1 of primates. *Journal of Neurophysiology*, **311**, 856–869.

Werner, G., & Whitsel, B. L. (1971). The functional organization of the somatosensory cortex. In A. Iggo (Ed.), *Handbook of sensory physiology*. New York: Springer.

Whitsel, B. L., Petrucelli, L. M., & Sapiro, G. (1969). Modality representation in the lumbar and cervical fasciculus gracilis of squirrel monkey. *Brain Research*, **15**, 67–78.

Whitsel, B. L., Roppolo, J. R., & Werner G. (1972). Cortical information processing of stimulus motion on primate skin. *Journal of Neurophysiology*, **35**, 691–717.

Willis, W. D., & Coggeshall, R. E. (1978). *Sensory mechanisms of the spinal cord*. New York: Plenum.

# 3

# AUDITORY FREQUENCY SELECTIVITY

*David W. Smith,\* David B. Moody, and William C. Stebbins*

Kresge Hearing Research Institute and the Department of Psychology,
The University of Michigan, Ann Arbor, Michigan

The auditory environment of animals is filled with acoustic information, the physical characteristics of which change rapidly from moment to moment. Few of the successful adaptations of living mammals are as distinctive as those of the system that provides the basis for the processing of these auditory stimuli. Acoustic communication signals represent some of the most important and complex of these stimuli, having numerous frequency components which vary independently of one another over time. Relative changes in these frequency components provide the primary basis for identification, discrimination, and evaluation of these complex signals. Similarly, one important cue for the binaural localization of sound in space is the relative filtering of the higher frequency components of these signals to the distal ear by the head's "sound shadow." These two auditory behaviors—the perception of conspecific vocalizations and the localization of sound in space—are critical to the survival of the animal, and represent significant evolutionary adaptations. Given the importance of these behaviors to the organism, it is not surprising that frequency selectivity, or the process by which the ear resolves the presence of, or relative changes in, the individual frequency components of an acoustic complex, is one of the critical functions of the auditory system.

Frequency selectivity of the auditory system as a whole can be addressed by use of psychophysical masking paradigms. The most commonly used of these procedures are those that measure either the critical

\*Present address: Department of Otolaryngology, University of Toronto, Toronto, Ontario.

bandwidth or psychophysical tuning curves. The critical bandwidth corresponds to that region of the basilar membrane over which stimulation is integrated within an auditory filter (Weber, 1977). The width of this band can be determined empirically by measuring the masked threshold of a pure tone centered in a noise band as the bandwidth of the noise is increased while the energy level per cycle is held constant. Each doubling of bandwidth produces an increase in overall sound pressure of 3 dB, and, when the noise band is less than one critical bandwidth, a 3-dB elevation of threshold. A continued doubling of bandwidth produces a monotonically increasing threshold function up to a certain value, the critical bandwidth, beyond which further increases have no effect. The interpretation of these results is that only energy that falls within the filter, or critical bandwidth, is effective in masking the tone. This procedure theoretically provides an estimate of the width of the internal auditory filter. Unfortunately, the critical band assumes a symmetric filter and is therefore limited in its ability to act as an effective descriptor of the actual filter, which is dependent upon the selective characteristics of the basilar membrane and hair cells, and which has been shown to be highly asymmetric. Since this asymmetric filter is a product of different processes acting in concert, a measure of an auditory selectivity function which assumes filter symmetry is necessarily inaccurate as a descriptor of filter action (cf. Evans, 1974; Evans & Wilson, 1973; Smith, Moody, Stebbins, & Norat, 1987).

This objection can be met satisfactorily by the use of the psychophysical tuning curve (PTC), which is a measure of the ability of the system to separate the patterns of basilar membrane excitation produced simultaneously by two or more different stimuli, that is, to detect the presence of one component of an acoustic complex, to the relative attenuation or exclusion of the other components (Zwicker, 1974). PTCs are generally measured by asking a subject to detect the presence of a pure-tone probe signal in the presence of masking tones that are varied in frequency and level. In a highly selective system, the subject would be capable of detecting the pattern of excitation produced by the test signal even when that of the maskers is spatially quite similar. Detection of the test signal would not be possible when the pattern of excitation produced by the maskers interferes or "covers up" that produced by the probe tone. At low signal levels, when the probe tone produces a very narrow region of excitation, the PTC is considered the psychoacoustic analog of the physiological or single-neuron frequency tuning curve (FTC). In this situation, the PTC is assumed to describe the filter characteristics of the limited region on the basilar membrane stimulated by the probe tone, just as an FTC describes the filter characteristics of the very narrow region of the cochlear partition innervated by that neuron (for example, the FTC for a given auditory neuron can be determined by measuring the combinations of frequency and intensity which produce a specified increment in response). The left panel of Fig. 1 shows three psychophysical tuning curves obtained

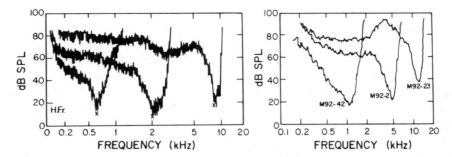

**FIGURE 1.** PTCs measured in humans at frequencies of 630 Hz, 2 and 8 kHz (from Zwicker, 1974) (left panel). FTCs from 3 normal cat auditory nerve fibers (right panel). From Zwicker (1974) after Kiang and Moxon (1974).

from humans in such a paradigm. For comparison, the right panel shows FTCs measured from three single auditory nerve fibers in cats at about the same frequencies. A comparison of psychophysical and physiological tuning curves reveals the strikingly similar appearances of tuning curves measured physiologically and behaviorally: Both have sharply tuned tips, gradually sloping low-frequency tails, and extremely sharp high-frequency cut-off slopes.

Numerous studies have shown that animals are readily trained for testing in the various psychophysical tuning curve paradigms. Figure 2 is a comparison of behavioral tuning curves in humans (Small, 1959) and in monkeys (Serafin, Moody, & Stebbins, 1982); their similarity is readily apparent. Psychophysical tuning curves measured in monkeys (Serafin et al., 1982; Smith, Moody, & Stebbins, 1987), cats (Nienhuys and Clark, 1979), chinchillas (Clark & Bohne, 1986; Dallos, Ryan, Harris, McGee, &

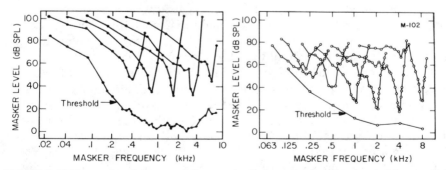

**FIGURE 2.** PTCs measured at several frequencies in humans over the threshold curve (left panel) (from Small, 1959). PTCs measured at several frequencies in the macaque (right panel). From Serafin, Moody, and Stebbins (1982).

Özdamer, 1977; Ryan, Dallos, & McGee, 1979), goldfish (Fay, Ahroon, & Orawski, 1978), and parakeets (Saunders, Rintelman, & Bock, 1979), are all qualitatively and quantitatively similar to those measured in humans. Moreover, the similarity of PTCs from human and monkey to eighth nerve FTCs measured in cats is remarkable (see right panel of Fig. 1).

Because of the similarity of the PTC to its physiological counterpart, it is useful as an analytic tool for describing auditory system function in the behaving organism. For example, a number of studies in humans have shown that psychoacoustic tuning curves are both elevated and broadened with increases in threshold resulting from pathology of cochlear origin (Carney & Nelson, 1983; Hoekstra & Ritsma, 1977; Jesteadt, 1980; Wightman, McGee, & Kramer, 1977; Zwicker, 1983; Zwicker & Schorn, 1978). These results are similar to those seen in FTCs from damaged ears in experimental animals (cf. Evans, 1974, 1975; Evans & Harrison, 1976; Evans & Wilson, 1975; Liberman & Dodds, 1984; Wilson, 1977). In general, the two sets of data are in good agreement, showing a relation between threshold elevation and increases in tuning curve tip elevation and breadth. With threshold elevations of 20–30 dB, a discernible, but elevated, tip remains. With losses greater than 30 dB, the curves lose tip tuning and are significantly broadened. The consensus among the studies is that the pathological mechanisms responsible for the increases in threshold in humans also account for deterioration of selectivity, although these mechanisms can only be surmised by correlation with analogous physiological observations in experimental animals.

These data are salient in several respects. First, the similarities between normal physiological and behavioral measurements suggest that the selectivity mechanisms are common to both, and are well established at the initial stages of peripheral activation (Small, 1959; Zwicker, 1974). Second, the similarity of the changes in response at both the initial processing levels and of the entire system to pathology indicates interruption of those same shared processes. For obvious reasons, those mechanisms cannot be accurately determined in humans since it is not possible to identify the responsible pathology. This limitation suggests the value of studying these same psychoacoustic processes in animal models where pathology can be controlled, accurately specified, and correlated with changes in hearing function measured behaviorally in the same individual.

A review of the literature reveals few studies of changes in frequency selectivity in animals in response to controlled pathology. Changes in psychophysical tuning curves following noise exposure have been measured in monkeys (Moody, Rarey, Norat, Stebbins, & Davis, 1984) and in chinchillas (Clark & Bohne, 1986). Moody and colleagues demonstrated that decreases in frequency selectivity following both temporary and permanent threshold shift were a flattening of the low-frequency tail, a truncation of the PTC tip region, and a decrease in $Q_{10}$ ($Q_{10}$ is a measure of tuning defined as the characteristic frequency, or CF, divided by the width of a tuning curve 10 dB above CF; therefore, a decrease in $Q_{10}$ indicates a loss of selectivity). These changes were well correlated with the threshold shift at the probe-tone frequency. Clark and Bohne (1986) showed little change in PTCs from chinchillas traumatized with low-frequency noise. Unfortunately, the well-documented variability in and relative nonspecificity of the effects of noise on the cochlea introduce difficulties in interpreting related functional changes in frequency selectivity. These difficulties may be overcome to a great extent by taking advantage of the selective ototoxicity of the aminoglycoside antibiotics. Different aminoglycosides produce varying patterns of sensory hair cell loss within the cochlea, in such a way that the extent of the damage varies systematically with species and with treatment duration. Thus, to a certain extent it is possible to select a particular animal, type of antibiotic, and dosage to produce consistent cochlear receptor lesions of the desired type.

Dallos et al. (1977) measured both FTCs and PTCs in chinchillas with selective outer hair cell (OHC) loss in the cochlea. The animals were given kanamycin, which in the chinchilla produces a receptor cell lesion primarily restricted to OHCs. As can be seen in Fig. 3 from Dallos et al., post-drug analyses of hearing from regions of the basilar membrane with selective damage to OHCs showed broadly tuned FTCs with selective loss of tip tuning. These physiological tuning curve data are consistent with systematic changes in frequency selectivity following selective loss of, or damage to, outer hair cells noted in other studies (cf. Liberman & Dodds, 1984). However, the associated PTCs in the Dallos et al. study were elevated by approximately 40 dB, yet retained sharp tuning. Dallos and colleagues have suggested that one possible explanation for the differences seen in the chinchilla between the behavioral and physiological data is that tuning undergoes additional sharpening beyond the level of the eighth nerve. This suggestion seems at odds with the congruence of the normal pre-drug psychophysical and physiological tuning curves, and Dallos et al. did not offer a possible mechanism for such a sharpening process.

A similar finding was reported by Nienhuys and Clark (1979), who compared critical bands in cats before and after administration of kanamycin. Critical bands measured at frequencies corresponding to regions of selective outer hair cell loss were normal. Broadening of critical bands was realized only when OHC loss was accompanied by an inner hair cell (IHC) loss of greater than 40%. Nienhuys and Clark's data agree well with those of Dallos et al. (1977) and they, too, conclude that normal frequency selectivity is mediated by IHCs alone.

The behavioral data of Dallos et al. (1977) and Nienhuys and Clark (1979) are not in agreement with the physiological tuning curves in animals showing losses in frequency selectivity with loss of normal OHC function (cf. Evans, 1975; Evans & Wilson, 1973; Harrison & Evans, 1977; Liberman & Dodds, 1984). Moreover, these data do not agree with the human psychophysical data which show deterioration in selectivity associated

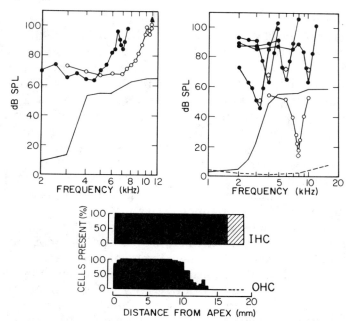

**FIGURE 3.** Upper left panel: FTCs recorded from two single auditory nerve fibers in the chinchilla from a region of selective OHC loss (○, ●) following kanamycin poisoning. The post-drug threshold curve from one chinchilla (—) is also presented. Upper right panel: Averaged pre-drug threshold curve for 9 chinchillas (– – – line) and post-drug threshold curve for one animal (—) with normal PTCs (●) measured at frequencies corresponding to regions of selective OHC loss. An average normal PTC is shown for comparison (○). Lower panel: The percentage of remaining hair cells from the subject shown in the upper right panel is plotted as a function of distance along the basilar membrane. All data from Dallos et al. (1977). The threshold shifts, and alterations in FTCs, can be seen to be tonotopically related to regions of OHC loss.

with hearing losses of cochlear origin (cf. Jesteadt, 1980; Wightman et al., 1977; Zwicker, 1983). Therefore, identification and definition of the role of various structures in determining the frequency-selective abilities of the ear require further systematic study in animal models where changes in psychoacoustic frequency selectivity can be correlated with controlled lesions of various structures in the same ear.

## AUDITORY FREQUENCY SELECTIVITY—AN ANIMAL BEHAVIORAL MODEL

The following sections will describe experiments that were performed to determine the effects of alterations of outer hair cell function on a behavioral measure of auditory sensitivity and frequency selectivity. The subjects in these studies were four male juvenile patas monkeys (*Erythrocebus patas*). Administration of dihydrostreptomycin (DHSM) in this species induces a progressive OHC loss beginning at the cochlear base. The resulting high-frequency hearing loss usually begins between the seventh and the ninth week of drug administration. These losses continue to progress after termination of drug administration, sometimes for up to 6 months, and will eventually include middle and, in the extreme, apical OHCs. The result is a chochlea denuded of outer hair cells with apparently normal inner hair cells over much of its length (Stebbins, Moody, Hawkins, Johnsson, & Norat, 1987).

## Behavioral Training Procedures

The behavioral procedures used in these studies were chosen for several reasons. Most importantly, to assess changes in auditory function, the psychophysical index must be sufficiently sensitive to characterize any changes in receptor function. The psychophysical tuning curve is a precise indicator of normal cochlear function and as such can tell us much about changes in both sensitivity and frequency selectivity with losses of outer hair cell function. Moreover, exploitation of the similarity of psychophysical tuning curves to their physiological counterpart FTCs will allow for comprehensive comparisons of changes seen behaviorally with those from analogous electrophysiological measures detailed in the abundant physiological literature. Another important feature of these procedures is that the basic behavioral tasks involved in measuring PTCs share many common features with those used in simpler psychoacoustic testing procedures in our laboratory so that animals are readily switched from one task to another. This flexibility allows for characterization of the effects of the particular pathology on different aspects of hearing.

During test sessions, the monkeys were seated in specially designed primate-restraint chairs (Moody, Stebbins, & Miller, 1970). A feeder chute and a small metal cylinder were positioned in front of the subject. The cylinder contained a cue light that served to signal experimental conditions to the animal and the behavioral response consisted of making contact with the cylinder. The subject's head was restrained and earphones were fitted over the external ears, centered as closely as possible over the ear canal. Figure 4 shows a subject seated in the test chamber with earphones and feeder trough mounted as described. The subject is shown with his hand on the response cylinder.

The beginning of a trial sequence was signaled to the subject by the flashing cue light. The subject was then required to make an observing response by reaching out and making contact with the metal cylinder. When contact was made, the cue light remained on without flashing for the duration of the trial sequence. If the animal maintained contact for a variable hold period of 1–9 sec, a 2.5-sec-test trial was presented. Releases of the cylinder during test-signal presentations were followed by delivery of a banana-flavored food pellet through the feeder chute. Following a

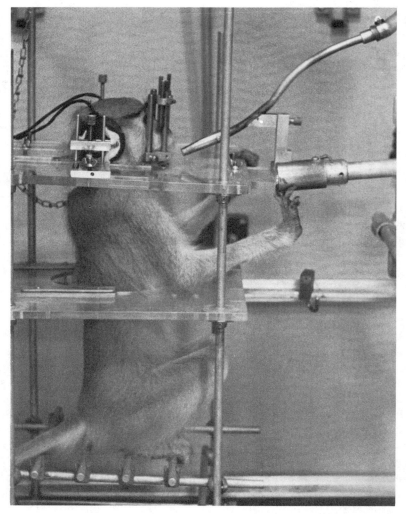

**FIGURE 4.** *E. patas* shown seated in restraint chair inside sound-insulated booth during test session. The animal's head is restrained and is fitted with headphones. The feeder trough can be seen directly in front of the animal's head. The subject's hand is touching the response tube.

3-sec intertrial interval, in which the cue light remained off, the light would begin flashing to signal the start of another test sequence. A missed response (holding through the test trail) resulted in another variable hold and test trial presentation.

Release of the cylinder at any time within the test sequence, except during the test trial, resulted in a 5-sec time-out period during which the subject was required to wait before again touching the cylinder. A response during the time-out period resulted in the initiation of another time-out period. The time out was mildly aversive to the animals and served to decrease inappropriate responding.

## Pure-Tone Threshold Procedures

Absolute thresholds to pure tones were measured using both steady and pulsed-tone stimuli at frequencies from 63 Hz to 40 kHz in $\frac{1}{2}$-octave steps. Steady-tone thresholds were measured to allow for comparison with subjects from other studies and to determine the functional status of the experimental ear prior to drug treatments. Thresholds to pulsed tones were measured daily in either quiet or noise to allow setting of the PTC probe-tone level relative to threshold.

Steady-tone thresholds were measured for 2.5-sec tones, with rise–fall times of 20 msec. Pulsed tones were 25 msec in duration with rise-fall times of 10 msec. An interpulse interval of 235 msec separated each subsequent pulse, and a train of 10 pulses, a total duration of 2.6 sec, defined a test signal interval.

A tracking or "staircase" procedure was used to vary the level of the test tone from trial to trial. The initial level of the signal was typically set about 40–50 dB above threshold. Correct detection of the test tone resulted in a 10-dB decrease in signal level for the subsequent trial. This decrease in intensity continued until the subject failed to report the presence of the test tone. At this time, the level of the test tone was increased by 10 dB for the next trial presentation. This change from a detection to a miss, or vice versa, was called a "transition." Seven transitions were measured at each frequency and threshold was defined as the average sound pressure level at the last 5 transitions. Following determination of threshold at a given frequency, the procedure was repeated at another frequency. A minimum of 9 frequencies were measured daily. Depending upon the status of the hearing loss during and following drug administration, the frequencies tested in each session were selected to include those frequencies where threshold changes were occurring.

## Psychophysical Tuning Curve Procedure

Forward masking psychophysical tuning curves were obtained with the probe-tone set at a constant level above the pulsed-tone threshold measured immediately prior to the tuning curve determination. Probe-tone levels were maintained at a constant sensation level as absolute thresholds changed to allow for meaningful comparisons across conditions and across subjects with changes in threshold.

The stimulus configuration used to measure PTCs is the same as that used in the pulsed-tone threshold procedure, except for the presence of masking tones. In this procedure, initial contact with the response tube produced a train of pure-tone masking pulses. The pulses were 130 msec in duration at the half-amplitude points, with rise–fall times of 10 msec. Following a variable number of masker presentations (i.e., 1- to 9-sec variable hold), 25-msec probe-tone pulses (i.e., the same pulses used in the pulsed-tone absolute threshold procedure) followed each of 10 consecutive maskers. As with the threshold procedures, the total test-trial duration was 2.6 sec.

In the tuning curve procedure, the level of the masker (rather than the probe tones) was varied using a tracking method. The initial level of the masker was set at a value considerably below levels required to mask the test tone pulses. Correct detection of the test signals by the subject produced a 10-dB increase in the level of the masker for the subsequent trial. When the subject failed to detect the presence of the probe tones on a test trial, the masker level was decreased by 10 dB. A minimum of 9, and up to 18, masker frequencies were tested for each PTC.

Tuning curve shape has been shown to be sensitive to the absolute sound pressure level at which they are measured (Nelson & Freyman, 1984; Smith, Moody, & Stebbins, 1987). In pathological ears where increases in thresholds are evident, it becomes necessary to differentiate between changes resulting from increases in test levels necessary to compensate for hearing losses, and those changes brought on by pathology, per se. Because loss of OHC function is associated with a 40- to 60-dB loss of sensitivity, it was necessary to determine the effects of test-tone level on psychoacoustic tuning curves, and whether any changes observed might be controlled by the use of masking noise. To this end, comparisons were made of PTCs measured at 2, 4, and 8 kHz at 10, 30, and 60 dB SL in quiet and in two levels of continuous background noise. For the noise PTC condition, pulsed-tone thresholds were measured exactly as they were for PTCs in quiet except for the addition of continuous broadband masking noise to the test paradigm. The noise served to control the growth of the pattern of excitation on the basilar membrane with increases in absolute signal level (Green, Shelton, Picardi, & Hafter, 1981; Nelson, 1980). Psychophysical tuning curves were measured in the noise backgrounds by simply setting the PTC probe-tone level 10 dB above masked threshold.

### Drug Administration and Histological Preparation

Following collection of baseline data, each subject received daily intramuscular injections of dihydrostreptomycin sulfate at the human clinical dose of 20 mg/kg/day. Injections were discontinued when a threshold shift of 20 dB was evident at 16 kHz.

Following completion of post-drug data collection, the animals were sacrificed and temporal bones taken for histological evaluation. The procedures used in preparing the ears for examination were carried out using standardized histological techniques (cf. Hawkins & Johnsson,

1976). Cytocochleograms were constructed in which the percentages of remaining inner and outer hair cells were plotted as a function of their position on the basilar membrane.

## RESULTS AND DISCUSSION

Changes in PTC shape will be described in qualitative rather than quantitative terms for two reasons: First, mathematical measures of filter shape such as $Q_{10}$ or bandwidth indices yield a symmetric description of filter function. However, changes in PTCs in response to pathology, like changes in FTCs, are highly asymmetric, therefore these measures are inappropriate. Second, psychophysical tuning curves lose their bandpass characteristics rapidly with loss of outer hair cells, which precludes use of the typical quantitative measures of filter shape.

### Effects of Test-Tone Level on PTC Shape

While the actual shapes of the tuning curves varied across subjects, the systematic changes in PTC shape associated with increases in signal level were very similar in three of four subjects in both quite and noise. Each animal showed a consistent loss of selectivity with increases in stimulus level. The tuning curves from the fourth animal showed no consistent changes across conditions. Since the effects of level were of primary concern in these studies, the data from all subjects have been averaged and only mean data are presented.

Average psychophysical tuning curves are shown in the upper panel of Fig. 5. At the lower stimulus levels of 10 and 30 dB SPL, the plots show the characteristic increase in sharpness associated with increase in frequency from 2 to 8 kHz. Also evident are decreases in selectivity associated with increases in signal level from 10 to 60 dB SL. PTC depth, as measured from the lowest masker frequency in the low-frequency tail to the tip frequency, decreases, and the width of the tip section increases. In general, a change from 10 to 30 dB SL resulted in little or no change in PTC shape, while a change from 10 to 60 dB SL resulted in decreased selectivity.

These relative changes become more obvious when PTCs at 10, 30, and 60 dB SL are superimposed. The lower panel of Fig. 5 contains the same data replotted on a normalized ordinate on which the zero point is the masker level at the frequency of the test tone. Again, the progressive loss of PTC depth and systematic broadening of the curves with increases in measurement level are observed. While the extent of the broadening varied across frequency, the tendency to broaden and decrease in depth was consistent across all frequencies with increases in absolute test signal level.

At both 4 and 8 kHz, the relative shifts in the slopes of the high- and low-frequency tail were asymmetric, with the low-frequency tail showing

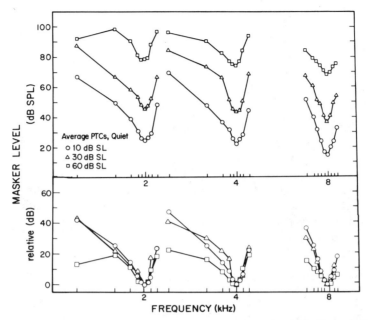

**FIGURE 5.** Average PTCs measured in quiet at probe-tone levels of 10, 30, and 60 dB SL (upper panel). Same PTCs as in upper panel superimposed on a normalized ordinate (lower panel). From Smith, Moody, and Stebbins (1987).

the greatest shift in slope. This finding is consistent with the suggestion of Glasberg and Moore (1982) and Nelson and Freyman (1984) that the symmetry of the filter may change with level.

Similar level-dependent nonlinear detuning has been previously observed in psychophysical tuning curves from humans (cf. Green et al., 1981; Nelson and Freyman, 1984). For example, Fig. 6, from Nelson and Freyman (1984), presents tuning curves measured in two human subjects at 1 kHz at several probe-tone levels ranging from 15 to 35 dB above threshold. The data clearly indicate that as the absolute level of the signals was increased, tuning curves exhibited a decrease in tip-to-tail depth and a broadening of the PTC tip. While the tails exhibit only a 20- to 25-dB shift, the tip region was shown to shift by 60 dB or more. These data are consistent with the notion that as the level of the signals is increased, both the probe and masker produce broader excitation patterns on the basilar membrane and, therefore, broadened tuning curves (Nelson, 1980; Nelson & Freyman, 1984; Zwicker, 1974).

## Effects of Level on PTC Shape in Noise Backgrounds

Thus far, we have presented data that demonstrate that tuning curves exhibit level-dependent changes. Green et al. (1981) and Moore, Glasberg,

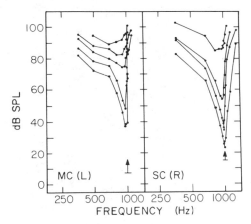

**FIGURE 6.** 1-kHz PTCs measured in two human subjects at probe tone levels of 15, 20, 25, 30, and 35 dB SPL for subject MC and 20, 25, 30, and 35 dB SPL for subject SC. From Nelson and Freyman (1984).

and Roberts (1984) showed with human subjects that if masking noise is added to the test procedure to control the broadening of the excitation pattern with increase in SPLs, tuning curve shape remains relatively constant with increases in test-tone level up to 34 dB SL. These results are interpreted as indicating that the difference in the excitation pattern produced by the test tone and that produced by the masker (the "residual excitation pattern") remained constant across levels, and therefore the shape of the resulting PTC remained relatively invariant with increases in absolute test signals levels.

We set out to confirm those findings in monkeys by measuring tuning curves at two levels of broadband masking noise. The tuning curves were measured at a constant probe tone level of 10 dB above masked threshold. Averaged data from these studies are shown in Fig. 7. As seen with those measured in quiet, PTCs obtained in the presence of masking noise show the characteristic increase in sharpness associated with the increase in frequency, and both a loss of depth and a broadening of the tip region with increases in test-tone level.

When the PTCs obtained with masking are superimposed on a normalized ordinate, as seen in the lower panel of Fig. 7, the relative loss of selectivity associated with the increase in level is evidenced by both a broadening of the tip at 2 and 8 kHz, and a decrease in tip-to-tail depth at all frequencies. At both 2 and 8 kHz, the slope of the high–frequency tail was either unchanged or steepened with increases in test signal level. These shifts in symmetry are qualitatively similar to those seen in PTCs measured in quiet, but differ from those of Green et al. (1981) and Moore et al. (1984). Our data demonstrate that at high test-tone levels, even with the use of broadband masking, PTCs show a loss of selectivity. If we assume

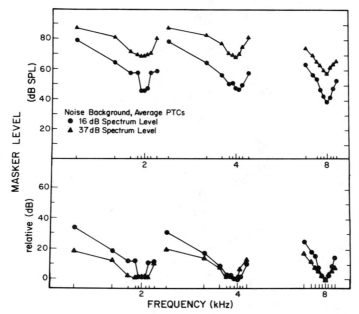

**FIGURE 7.** Average PTCs measured in two levels of continuous background masking noise (upper panel). Same PTCs as in upper panel superimposed on a normalized ordinate (lower panel). From Smith, Moody, and Stebbins (1987).

that the masking was effective in producing uniform residual excitation patterns at different levels, then the loss of selectivity cannot be explained by broadened excitation patterns. An alternative explanation involves nonlinear transduction processes at the receptor level.

It has been suggested (Khanna, 1984; Kim, 1984) that the nonlinear tuning of the basilar membrane may arise in the mechanics of the outer hair cell system. When measured at higher signal levels, physiological tuning curves (recorded as isoresponse curves) taken from individual receptor cells show a loss of selectivity similar to that found in the present psychophysical data (Selleck & Russell, 1978), suggesting that the response of the hair cell is nonlinear at levels of greater than 30 dB above threshold. Khanna (1984) attributes this effect to a nonlinear mechanical response in the hair cell at these higher sound pressure levels. This suggests that had Green et al. (1981) and Moore et al. (1984) tested at higher SPLs, as was done in the present studies, they might have seen these same nonlinearities. This nonlinearity is also evident in the input/output functions of auditory nerve fibers (cf. Geisler, Rhode, & Kennedy, 1974; McGee, Walsh, & Javel, in press). Since input/output functions differ from frequency to frequency in the vicinity of CF for a given fiber—being relatively steeper at frequencies in the low-frequency tail—increases in measurement level produce a nonlinear growth of response and, hence, a broadening of the tuning function. If the behavior of the system at higher stimulation levels (i.e., broadening of the filter function) is based upon a nonlinear mechanical response property of the receptor rather than the excitation pattern on the basilar membrane, then attempts to control the effective excitation pattern by the addition of background maskers should not produce PTCs that are invariant with level. The broadening associated with changes in signal level observed in the data presented here are consistent with an interpretation based on mechanical nonlinearities in the transduction process.

## Effects of Outer Hair Cell Loss on Selectivity

Of the four patas monkeys studied, two showed very slow progressive hearing losses with similar patterns of receptor lesions (M-155 and M-157); the third (M-156) showed a very rapid loss of sensitivity, although the extent of cochlear damage was similar to that for M-155 and M-157; and the fourth (M-154) required substantially longer drug treatment to produce initial loss of sensitivity, and subsequently experienced a significantly greater loss of both outer and inner hair cells. However, all four animals showed identical systematic changes in tuning function. To illustrate the main findings of the study, we will discuss in detail the data from only one of these subjects. To serve as comparisons with the post-drug results, the normal tuning curves measured at probe tone levels of 30 and 60 dB SL for each animal are also presented. These test-tone levels represent the highest absolute SPLs that are likely to be encountered with a complete OHC loss. Comparisons of pathological PTCs with abnormally elevated thresholds and normal PTCs taken at roughly equivalent SPLs will allow differentiation of those changes due to alteration of tuning mechanisms from those associated with increases in test signal levels necessary to compensate for hearing losses.

The upper panel of Fig. 8 presents the cytocochleogram for subject M-155, plotted as the percentage of the hair cells remaining as a function of distance along the basilar membrane from the base of the cochlea. Subject M-155 had a slow, progressively developing hearing loss. It was possible to monitor daily changes in threshold at frequencies of 8, 4, and 2 kHz. The lower panel presents the corresponding audiogram, showing steady-tone threshold shift as a function of frequency. The abscissae have been aligned to illustrate the frequency-to-place transformation characteristic of the tonotopic organization of the cochlea. The alignment is based upon our previous experience with the relation between pure-tone threshold shifts and the location of the hair cell loss (Stebbins, Hawkins, Johnsson, & Moody, 1979).

The hair cell count for M-155 shows an absence of IHCs in the basal 15% of the cochlea with a rapid transition to near-complete retention of IHCs throughout the apical 80%. The outer hair cells are absent in the basal-most

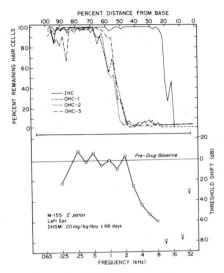

**FIGURE 8.** The cytocochleogram or final hair cell count (upper panel) for M-155, plotted as percentage of remaining hair cells as a function of distance along the basilar membrane from the base. Corresponding final audiogram or threshold contour (lower panel) for M-155 is given as steady tone threshold shift as a function of frequency. From Smith, Moody, Stebbins, and Norat (1987).

50% of the cochlea with a transition to near-normal populations along the remainder of the ear. In this ear, there is a region with normal-appearing IHCs and no OHCs over approximately one-third of the basilar membrane. This region is responsible for transduction of frequencies between 4 and 16 kHz. Light microscopic evaluation of remaining IHC condition showed they were normal in all respects: The stereocilia were of normal number and configuration, the cell body was of normal shape and size, there was no evidence of intracellular vacuolization, and the cell nucleus was of normal location and appearance. The frequencies of the accompanying threshold shifts correspond closely to the regions of receptor damage. A loss of 22 to 60 dB was evident at frequencies between 2.8 and 8 kHz, corresponding to the regions of selective IHC retention. The threshold shifts were so large as to preclude measurement at frequencies above 8 kHz. Thresholds were normal at 2 kHz and below, except for a 20–dB shift at 125 Hz.

Figure 9 presents a direct comparison of baseline and post-drug pulsed-stimulus thresholds and tuning curves for M-155. At 8 kHz, corresponding to a region of complete OHC loss and normal IHC populations, the threshold shift was 65 dB and the resulting tuning curve resembled a low-pass filter. At 4 kHz, the PTC is elevated approximately 40 dB, yet still exhibits some tuning since the low-frequency tail overlaps the 2-kHz region where post-drug sensitivity was normal. The high-frequency tails of

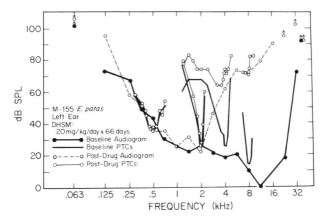

**FIGURE 9.** Direct comparison of baseline and post-drug pulsed-tone threshold curves and PTCs for M-155. From Smith, Moody, Stebbins, and Norat (1987).

both 4- and 8-kHz tuning curves parallel the pulsed-stimulus threshold curve, elevated by approximately 10 dB. The post-drug PTCs at 500 Hz and 2 kHz, where thresholds to both steady and pulsed stimuli were unaffected, overlie the baseline functions.

The progressive deterioration in tuning associated with outer hair cell damage can be observed by comparison of individual tuning curves. The left panel of Fig. 10 shows the systematic changes in tuning evident for M-155 during degeneration of OHCs at 8 kHz as measured on various days post-drug. The right panel shows normal PTCs for M-155 measured at probe-tone levels of 10, 30, and 60 dB SL, for comparison. The initial changes in tuning curve shape were consistently seen as two distinct phenomena. The first was observed as an elevation and shift of the tip toward lower frequencies. Concomitant with threshold elevations of up to 50 dB was either no change in, or a hypersensitivity of, the low-frequency tail. At days 14, 15, and 17, the low-frequency tail at 6.8 kHz was either equal to or below baseline values, while there was a greater-than-40 dB loss in tip sensitivity within those 3 days. By days 22–23, the function was elevated nearly 60 dB and resembled a low-pass filter.

Changes caused by increases in SPL are seen as a decrease in tip-to-tail ratio and a slight broadening of the filter function. The changes seen with loss of outer hair cells are a complete detuning and transition from band-pass to low-pass function. These two types of changes are qualitatively different and are easily distinguished, suggesting that the loss of tuning in pathological tuning curves is not due to an increase in test tone levels.

Figure 11 presents the systematic changes in 4 kHz PTCs from M-155. An average of the data from days 36–39 showed a 30-dB elevation and a shift toward lower frequencies at the CF, while hypersensitivity was

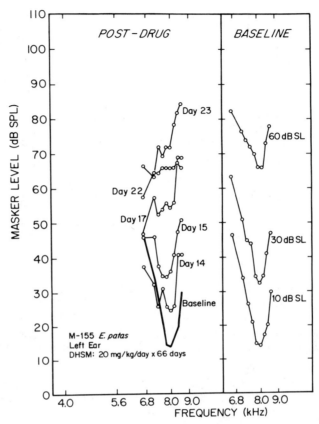

**FIGURE 10.** Systematic progression of changes in 8 kHz PTC for M-155 measured on various days post-drug (left panel). Normal PTCs measured at high SPLs are presented for comparison at 30 and 60 dB (right panel). From Smith, Moody, Stebbins, and Norat (1987).

present in the low-frequency tail. At day 40, the tip is elevated by over 40 dB, but little effect is seen at low frequencies. By day 41, with a 50-dB decrease in sensitivity at CF, thresholds at the highest masker frequencies of 4.4 kHz and above were not measurable, and an elevation at low frequencies was present. Although levels of greater than 110 dB SPL were not measured, extrapolation from the 4.2-kHz masker at 84 dB SPL to these higher levels at 4.4 kHz yields a low-pass filter function.

Again, comparisons of normal PTCs at 10, 30, and 60 dB SL (right panel), and post-drug PTCs at roughly equivalent SPLs fail to explain these changes on the basis of increases in test-signal levels. Even at the highest test level of 60 dB SL, the normal PTCs exhibit sharp band-pass tuning. This characteristic disappears quite rapidly with increases in threshold for pathological PTCs.

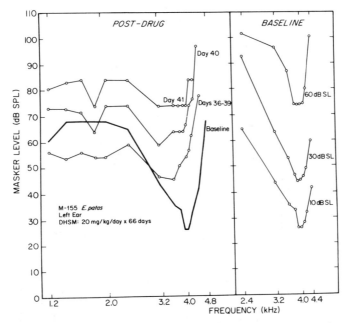

**FIGURE 11.** Systematic changes in 4 kHz PTC for M-155 measured on various days post-drug (left panel). Normal PTCs for same subject at 30 and 60 dB SL are presented for comparison (right panel). From Smith, Moody, Stebbins, and Norat (1987).

## The Detuning of PTCs Associated with Alterations in OHC Function

Initial discussion will focus on comparisons of changes in psychophysical tuning curve shape in response to outer hair cell loss with changes in filter function due to increased measurement levels. These comparisons will allow those changes brought on by pathology to be differentiated from those associated with higher measurement levels per se. The systematic changes in PTC response during destruction of the OHC subsystem can be seen as two distinct phenomena. These correspond to behavior of the tip and low-frequency tail regions of the PTC response. Each will be described independently in the following two sections.

For purposes of simplifying discussion of the role of various processes upon psychophysical frequency selectivity, the general phenomena evident in the data from the four subjects have been incorporated into the idealized tuning curves presented in Fig. 12. Changes associated with outer hair cell loss are depicted in the left panel. For purposes of comparison, similar idealized data are presented in the right panel for changes in PTC shape associated with changes in absolute measurement levels. Subsequent discussions will refer to this figure.

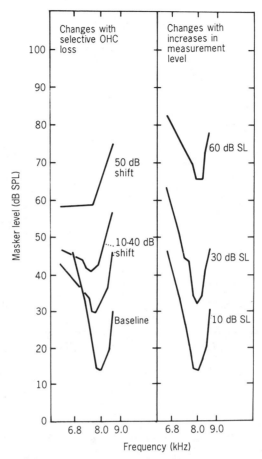

**FIGURE 12.** Idealized PTCs for changes associated with selective loss of OHCs (left panel). Idealized PTCs for changes associated with increases in measurement level (right panel). From Smith, Moody, Stebbins, and Norat (1987).

**Comparison of Changes in Psychophysical Tuning Curves Due to Increased SPLs with Those Associated with OHC Damage.** As was shown earlier in this chapter, measures of frequency selectivity have been shown to be sensitive to changes in the absolute signal level at which measurements are taken and indicate that tuning is a level-dependent, nonlinear process. As can be seen in the right panel of Fig. 12, the shape of the PTC broadens and decreases in tip-to-tail depth (as measured in dB from the level of the lowest frequency in the low frequency tail to the level of the tip) as the absolute signal level is increased. However, unlike changes brought on by pathology, even at the highest levels of 60 dB SL, significant selectivity was observed, and normal tuning curves retain band-pass function. The changes we have measured in PTCs are qualitatively similar to those observed in single inner hair cell tuning curves when the signal amplitude criterion is increased (Russell & Selleck, 1978).

When comparisons are made between normal psychophysical tuning curves taken at high signal levels (right panel) and pathological PTCs taken at roughly equivalent SPLs (left panel), it becomes clear that the changes in tuning curves following outer hair cell loss cannot be accounted for on the basis of increased measurement levels. Following systemic drug treatment, tuning curves undergo a rapid increase in tip sensitivity and nearly complete detuning in response to OHC loss. In the present study, this transition from low-threshold sharply tuned tuning curve to high–threshold low-pass filter can only be explained as a change in the tuning mechanism brought on by loss of outer hair cell function.

**Changes in PTC Tip Response Associated with Loss of Outer Hair Cells.** In these studies, increases in threshold associated with DHSM treatment were invariably followed by a rapid and nearly complete loss of psychophysical frequency selectivity. The rate of increase in threshold varied across frequency and length of time post-drug, but could be as rapid as 10–20 dB/day, yet the pattern of change in tuning was systematic and entirely dependent upon threshold, and therefore is presumably a function of orderly alterations in the processing mechanism.

As seen in the idealized PTCs presented in the left panel of Fig. 12, with increases of up to 40 dB, there was a selective elevation and broadening of the tip region in the PTC response. The tip was also shifted slightly toward lower frequencies. The slope of the high-frequency tail remained relatively constant, yet was elevated in a more or less linear fashion with increases in threshold at CF. Following threshold shifts of greater than 50 dB, the highly selective tip was completely absent and the filter took on low-pass characteristics; that is, it showed a sharp high-frequency cutoff and an approximately flat low-frequency pass band. As can be seen from comparisons of the cytocochleograms and corresponding final threshold curves (Fig. 8), threshold shifts to both steady and pulsed tones were on the order of 40–60 dB at frequencies corresponding to regions completely lacking outer hair cells. This suggests that the final transition of the tuning curve to a low-pass filter function concomitant with threshold shifts of greater than 40–50 dB is probably a result of complete removal of OHC influence. Because of the orderly covariation of threshold and tuning, it is likely that these two processes are dependent upon the same mechanisms, namely the outer hair cells.

The results of these studies are in both qualitative and quantitative agreement with the data from numerous physiological studies indicating the dependence of the sensitivity and sharply tuned response of the physiological tuning curve on normal outer hair cell function (cf. Evans, 1974, 1975; Evans & Harrison, 1976; Evans & Wilson, 1975; Liberman &

Dodds, 1984; Wilson, 1977). This correspondence is most striking in comparisons of the left panel of Fig. 12 with the schematized physiological tuning curve data of Liberman and Dodds (1984) in Fig. 13. The correspondence between normal psychophysical tuning curves (10 dB SL PTCs in right panel of Fig. 12) and physiological tuning curves (top panel Fig. 13) has been noted previously. However, the degree to which the pathological tuning curves resemble each other is of particular interest here. These data are taken from recordings of single auditory neurons made in noise-traumatized cats. Following the recording of FTCs, each fiber was labeled and then traced to its cochlear origin. By noting the condition of the hair cells in that localized region of the organ of Corti, it was possible to determine the relation between hair cell condition and physiological tuning curve response characteristics. These data match very well the psychophysical tuning curves measured in our animals during and following loss of outer hair cells, as shown in the middle and lower panels of Fig. 13. The low-pass broad tuning evident in the behavioral data and in single eighth nerve tuning curves (cf. Evans, 1975; Liberman & Dodds, 1984) following complete loss of OHCs is also observed in direct measurements of basilar membrane motion (Békésy, 1960; Johnstone & Boyle, 1967; Rhode, 1971, 1973, 1980), and thus likely reflect the passive mechanical interactions of the basilar membrane and surrounding fluids. Because of the remarkable similarity of physiological and psychophysical tuning curves in both normal and pathological ears, it is suggested that those processes that determine the tuning of the auditory system as a whole are primarily peripheral phenomena, and very likely can be defined as functions of certain cellular and subcellular mechanisms.

These data are also in agreement with those from human psychophysical analyses of changes in auditory frequency selectivity with cochlear pathology (cf. Pick, Evans, & Wilson, 1977; Wightman, 1982; Zwicker, 1983; Zwicker & Schorn, 1978). The human data have consistently shown a relationship between increases in threshold and loss of frequency selectivity similar to that shown with moderate losses of sensitivity in the present study. However, in this study, the mechanisms for increases in threshold and loss of specificity are controlled and identified as outer hair cells. Assuming equivalent pathology, it is difficult to reconcile these data with those of Dallos et al. (1977) and Nienhuys and Clark (1979) which show no decrease in selectivity in areas of the basilar membrane with complete loss of OHCs. Their data are also inconsistent with the human literature.

The psychophysical data presented here corroborate suggestions of Evans (1974, 1975), Evans and Wilson (1973, 1975), Khanna (1984), and Kim (1984) that the observed physiological filter function is a composite of two interacting processes, the first being a low-pass filter process dependent upon the passive mechanical properties of the basilar membrane and dynamics of the surrounding fluids. This mechanical system is relatively immune to trauma. The second process, the so-called "second filter," is a

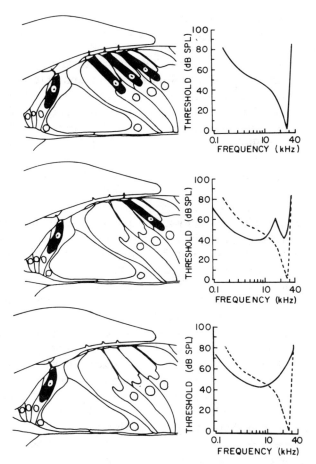

**FIGURE 13.** Schematic drawing of the organ of Corti with various lesion conditions (darkened hair cells represent those present) (left side): normal (upper panel); partial loss of OHCs (middle panel); complete loss of OHCs (lower panel). The corresponding schematic FTC for each condition is given on the right. Redrawn from Liberman and Dodds (1984).

sharply tuned response near the high-frequency cutoff of the first filter. This active process reflects the function of outer hair cells which is physiologically labile.

The frequency-resolving characteristics of the normal ear are determined by the mechanical influence of the sharply tuned outer hair cells on the low-pass, broadly tuned basilar membrane motion. The frequency response of the inner hair cells, and subsequently of the auditory afferents and system as a whole, directly reflects the influence of these interactions. When the ear is subjected to temporary or permanent insult, the sharply tuned segment of the response contributed by the outer hair cells is especially vulnerable and rapidly disappears. What remains following removal of the outer hair cells, then, is the low-pass response of the basilar membrane alone. This residual tuning function is the same reported in basilar membrane tuning curves from dead ears in living preparations or from human cadavers (Békésy, 1960; Johnstone & Boyle, 1967; Rhode, 1971, 1973), inner hair cell responses during anoxia and hypothermia (Brown, Nuttall, Masta, & Lawrence, 1983; Brown, Smith, & Nuttall, 1983), eighth-nerve responses from areas of selective outer hair cell loss or damage (Liberman & Dodds, 1984), and now from the behavior of the system as a whole following outer hair cell degeneration. This residual low-pass response is determined by the passive mechanical properties of the basilar membrane and hydrodynamics of the surrounding fluids (Khanna, 1984; Kim, 1984).

**PTC Low-Frequency Tail Response to Selective Lesions of OHCs.** The data also indicate that moderate increases in threshold at the characteristic frequency produce a hypersensitivity in the response of the low-frequency tail of the psychoacoustic tuning curve. This increase in sensitivity, which always accompanies a hyposensitivity of the tip response, is similar to that recently observed by Liberman and Dodds (1984) (see middle panel, Fig. 13) in single-unit physiological tuning curves arising from identified inner hair cells innervating areas of selective outer hair cell damage or loss. Zwislocki (1984) suggests that a disruption in the normal shearing motion between the tectorial and basilar membranes, caused by depletion of OHCs, explains this finding. Since these same processes form the basis for psychophysical frequency selectivity, he suggests that this same hypersensitivity should be obtainable in psychophysical tuning curves. These data, showing an increase in sensitivity at the low-frequency tail accompanying increases in threshold of 10–40 dB, indicating initial selective insult to OHCs, probably represent this same phenomenon.

## The Use of Animal Models to Study Sensory Function and Perception

It has been over 300 years since DuVerney (1683) first suggested that the initial processing of acoustic information by the auditory system was carried out in the frequency domain. Research of the past 20 years into this process has provided a more complete understanding of its physiological underpinnings. However, until quite recently, there have been significant discrepancies between the physiological and behavioral findings. The qualitative differences in physiological and behavioral measurements have led to the suggestion that processing must occur at higher levels within the auditory system. However, the development and refinement of animal behavioral models has provided new insights into the processing of frequency information by the awake, intact, behaviorally trained animal. The use of animals allows precise monitoring of functional changes associated with controlled lesions of various anatomical structures which obviously are not feasible in humans. Study of these processes has traditionally been carried out by physiologists in isolated preparations. However, these processes do not exist independently, and their study must at some point be related to their service to the system as a whole—to the behavior of the intact animal.

Changes in psychophysical tuning curves in response to controlled lesions of the outer hair cells of the organ of Corti are, for all intents and purposes, indistinguishable from those seen in measurements of basilar membrane motion or in single-unit recordings of inner hair cells or auditory neurons. These findings suggest that the analysis of acoustic frequency is performed at the lowest centers of the auditory system and relayed with great veridicality to the behavior of the organism.

## CONCLUSIONS

Based on our analyses of physiological processes using behavior, several points are worthy of particular note:

1. The similarity between physiological measures in the auditory periphery and behavioral measures of frequency selectivity in normal ears and in response to controlled trauma suggests a common peripheral origin for frequency selectivity at the cellular level.

2. A good qualitative and quantitative agreement exists between changes in physiological and psychophysical indices of frequency selectivity in normal ears and in response to cochlear trauma resulting in interruption or loss of normal outer hair cell function. The behavioral data are sufficiently sensitive to characterize accurately those processes that have been shown to be mediated at the cellular level.

3. The similarity of tuning curves measured in humans and animals suggests that this behavioral paradigm offers a powerful model for the study of animal as well as human sensory function.

## ACKNOWLEDGMENTS

The research described in this chapter was supported by NINCDS Grant NS05785 and a Horace H. Rackham Dissertation Grant to David W. Smith. The authors thank Cynthia A. Prosen for her comments on this chapter, and Richard A. Altschuler and the Otopathology Laboratory for the histological analyses upon which these studies are based. We also thank Charlotte Leiser and Jane Rodgers for their assistance in the preparation of this manuscript. This chapter was prepared while David W. Smith was a NINCDS Post-Doctoral Fellow at Boys Town National Institute, Omaha, NE.

## REFERENCES

Békésy, G. von (1960). *Experiments in hearing.* New York: McGraw-Hill.

Brown, M. C., Nuttall, A. L., Masta, R. I., & Lawrence, M. (1983). Cochlear inner hair cells: Effects of transient asphyxia on intracellular potentials. *Hearing Research, 9,* 131–144.

Brown, M. C., Smith, D. I., & Nuttall, A. L. (1983). The temperature dependency of neural and hair cell responses evoked by high frequencies. *Journal of the Acoustical Society of America, 73,* 1662–1670.

Carney, A. E., & Nelson, D. A. (1983). Analysis of psychophysical tuning curves in normal and pathological ears. *Journal of the Acoustical Society of America, 73,* 268–278.

Clark, W. W., & Bohne, B. A. (1986). Cochlear damage: Audiometric correlates? In M. J. Collins, T. J. Glattke, & L. A. Harker (Eds.), *Sensorineural hearing loss: Mechanisms, diagnosis, and treatment* (pp. 59–82). Iowa City: University of Iowa Press.

Dallos, P., Ryan, A., Harris, D., McGee, T., & Özdamer, Ö. (1977). Cochlear frequency selectivity in the presence of hair cell damage. In E. F. Evans & J. P. Wilson (Eds.), *Psychophysics and physiology of hearing* (pp. 249–258). New York: Academic Press.

DuVerney, J. F. (1683). Traite de l'organe de l'ovie, Paris. As cited in E. G. Wever, *Theory of Hearing* (pp. 12–14). Wiley, New York, 1949; also cited in G. von Békésy and W. A. Rosenblith (1948), The early history of hearing-observations and theories. *Journal of the Acoustical Society of America, 20,* 727–748.

Evans, E. F. (1974). Auditory frequency selectivity and the cochlear nerve. In E. Zwicker & E. Terhardt (Eds.), *Facts and models in hearing* (pp. 118–129). New York: Springer-Verlag.

Evans, E. F. (1975). The sharpening of cochlear frequency selectivity in the normal and abnormal cochlea. *Audiology, 14,* 197–201.

Evans, E. F., & Harrison, R. V. (1976). Correlations between cochlear outer hair cell damage and deterioration of cochlear nerve tuning properties in the guinea pig. *Journal of Physiology* (London), *256,* 43–44.

Evans, E. F., & Wilson, J. P. (1973). The frequency selectivity of the cochlea. In A. R. Møller (Ed.), *Basic mechanisms of hearing* (pp. 519–551). New York: Academic Press.

Evans, E. F., & Wilson, J. P. (1975). Cochlear tuning properties: Concurrent basilar membrane and single nerve fiber measurements. *Science, 190,* 1219–1221.

Fay, R. R., Ahroon, W. A., & Orawski, A. A. (1978). Auditory masking patterns in the goldfish (Carassius auratus): Psychophysical tuning curves. *Journal of Experimental Biology, 74,* 83–100.

Geisler, C. D., Rhode, W. S., & Kennedy, D. T. (1974). Responses to tonal stimuli of single auditory nerve fibers and their relationship to basilar membrane motion in the squirrel monkey. *Journal of Neurophysiology, 37,* 1156–1172.

Glasberg, B. R., & Moore, B. C. J. (1982). Auditory filter shapes in forward masking as a function of level. *Journal of the Acoustical Society of America, 71,* 946–949.

Green, D. M., Shelton, B. R., Picardi, M. C., & Hafter, E. R. (1981). Psychophysical tuning curves independent of signal level. *Journal of the Acoustical Society of America*, **69**, 1758–1762.

Harrison, R. V., & Evans, E. F. (1977). Cochlear nerve tuning and OHC loss in the guinea pig. In E. F. Evans & J. P. Wilson (Eds.), *Psychophysics and physiology of hearing* (pp. 25–29). New York: Academic Press.

Hawkins, J. E., Jr., & Johnsson, L.-G. (1976). Microdissection and surface preparations of the inner ear. In C. A. Smith & J. A. Vernon (Eds.), *Handbook of auditory and vestibular research methods* (pp. 5–52) Springfield, IL: Thomas.

Hoekstra, A., & Ritsma, R. J. (1977). Perceptive hearing loss and frequency selectivity. In E. F. Evans & J. P. Wilson (Eds.), *Psychophysics and physiology of hearing* (pp. 263–271). New York: Academic Press.

Jesteadt, W. (1980). Frequency analysis in normal and hearing-impaired listeners. *Annals of Otology, Rhinology, & Laryngology*, **89** (Suppl. 74), 88–95.

Johnstone, B. M., & Boyle, A. F. J. (1967). Basilar membrane vibration examined with the Mössbauer technique. *Science*, **158**, 389–390.

Khanna, S. M. (1984). Inner ear function based on the mechanical tuning of the hair cells. In C. I. Berlin (Ed.), *Hearing science: Recent advances* (pp. 213–239). San Diego, CA: College Hill Press.

Kiang, N. Y. S., & Moxon, E. C. (1974). Tails of tuning curves of auditory-nerve fibers. *Journal of the Acoustical Society of America*, **55**, 620–630.

Kim, D. O. (1984). Functional roles of the inner and outer hair cell subsystem in the cochlea and brainstem. In C. I. Berlin (Ed.), *Hearing science: Recent advances* (pp. 241–262), San Diego, CA: College Hill Press.

Liberman, M. C., & Dodds, L. W. (1984). Single-neuron labeling and chronic cochlear pathology. III. Stereocilia damage and alteration of threshold tuning curves. *Hearing Research*, **16**, 55–74.

McGee, J., Walsh, E. J., & Javel, E. (in press). Discharge rate and synchronization in cat auditory nerve fibers. *Journal of the Acoustical Society of America*.

Moody, D. B., Rarey, K. E., Norat, M. A., Stebbins, W. C., & Davis, J. A. (1984). *Psychophysical tuning curves following permanent noise-induced threshold shift in monkeys*. Abstracts of the Association for Research in Otolaryngology, St. Petersburg Beach.

Moody, D. B., Stebbins, W. C., & Miller, J. M. (1970). A primate restraint and handling system for auditory research. *Behavioral Research Methods and Instrumentations*, **2**, 180–182.

Moore, B. C. J., Glasberg, B. R., & Roberts, B. (1984). Refining the measurements of psychophysical tuning curves. *Journal of the Acoustical Society of America*, **76**, 1057–1066.

Nelson, D. A. (1980). Comment on: The use of psychophysical tuning curves to measure frequency selectivity. In G. van den Brink & F. A. Bielsen (Eds.), *Psychophysical, physiological, and behavioral studies in hearing* (pp. 116–117). Delft: Delft University Press.

Nelson, D. A., & Freyman, R. L. (1984). Broadened forward-masked tuning curves from intense masking tones: Delay-time and probe-level manipulations. *Journal of the Acoustical Society of America*, **75**, 1570–1577.

Nienhuys, T. G. W., & Clark, G. M. (1979). Critical bands following the selective destruction of cochlear inner and outer hair cells. *Acta Oto-Laryngologica,* **88,** 350–358.

Pick, G. F., Evans, E. F., & Wilson, J. P. (1977). Frequency resolution in patients with hearing loss of cochlear origin. In E. F. Evans & J. P. Wilson (Eds.), *Psychophysics and physiology of hearing* (pp. 273–282) New York: Academic Press.

Rhode, W. S. (1971). Observations of the vibration of the basilar membrane in squirrel monkeys using the Mössbauer technique. *Journal of the Acoustical Society of America,* **49,** 1218–1231.

Rhode, W. S. (1973). An investigation of post-mortem cochlear mechanics using the Mössbauer effect. In A. R. Møller (Ed.), *Basic mechanisms in hearing* (pp. 49–63). New York: Academic Press.

Rhode, W. S. (1977). Some observations of two-tone interactions measured using the Mössbauer effect. In E. F. Evans & J. P. Wilson (Eds.), *Psychophysics and physiology of hearing* (pp. 27–38). New York: Academic Press.

Rhode, W. S. (1980). Cochlear partition vibration—recent views. *Journal of the Acoustical Society of America,* **67,** 1696–1703.

Russell, I. J., & Selleck, P. M. (1978). Intracellular studies of hair cells in the mammalian cochlea. *Journal of Physiology (London),* **284,** 261–290.

Ryan, A., Dallos, P., & McGee, T. (1979). Psychophysical tuning curves and auditory thresholds after hair cell damage in the chinchilla. *Journal of the Acoustical Society of America,* **66,** 370–378.

Saunders, J. C., Rintelman, W. F., & Bock, G. R. (1979). Frequency selectivity in bird and man: A comparison among critical ratios, critical bands, and psychophysical tuning curves. *Hearing Research,* **1,** 303–323.

Selleck, P. M., & Russell, I. J. (1978). Intracellular studies of cochlear hair cells: Filling the gap between basilar membrane mechanics and neural excitation. In R. F. Naunton & C. Fernandez (Eds.), *Evoked electrical activity in the auditory nervous system* (pp. 113–139). New York: Academic Press.

Serafin, J. V., Moody, D. B., & Stebbins, W. C. (1982). Frequency selectivity of the monkey's auditory system: Psychophysical tuning curves. *Journal of the Acoustical Society of America,* **71,** 1513–1518.

Small, A. M. (1959). Pure-tone masking. *Journal of the Acoustical Society of America,* **31,** 1619–1625.

Smith, D. W., Moody, D. B., & Stebbins, W. C. (1987). The effects of changes in absolute measurements level on psychophysical tuning curves in quiet and noise in patas monkeys. *Journal of the Acoustical Society of America,* **82,** 63–68.

Smith, D. W., Moody, D. B., Stebbins, W. C., & Norat, M. A. (1987). Effects of selective outer hair cell loss on the frequency selectivity of the patas monkey auditory system. *Hearing Research,* **25,** 125–138.

Stebbins, W. C., Hawkins, J. E., Jr., Johnsson, L.-G., & Moody, D. B. (1979). Hearing thresholds with outer and inner hair cell loss. *American Journal of Otolaryngology,* **1,** 15–27.

Stebbins, W. C., Moody, D. B., Hawkins, J. E., Jr., Johnsson, L.-G., & Norat, M. A. (1987). The species-specific nature of the ototoxicity of dihydrostreptomycin in the patas monkey. *Neurotoxicology,* **8,** 33–44.

Weber, D. L. (1977). The growth of masking and the auditory filter. *Journal of the Acoustical Society of America,* **62,** 424–429.

Wightman, F. L. (1982). Psychoacoustic correlates of hearing loss. In R. P. Hamernik, D. Henderson, & R. Salvi (Eds.), *New perspectives on noise-induced hearing loss* (pp. 375–393). New York: Raven Press.

Wightman, F. L., McGee, T., & Kramer, M. (1977). Factors influencing frequency selectivity in normal and hearing-impaired listeners. In E. F. Evans & J. P. Wilson (Eds.), *Psychophysics and physiology of hearing* (pp. 295–306). New York: Academic Press.

Wilson, J. P. (1977). Towards a model for cochlear frequency analysis. In E. F. Evans & J. P. Wilson (Eds.), *Psychophysics and physiology of hearing* (pp. 115–124). New York: Academic Press.

Zwicker, E. (1974). On a psychophysical equivalent of tuning curves. In E. Zwicker & E. Terhardt (Eds.), *Facts and models in hearing* (pp. 132–141). New York: Springer-Verlag.

Zwicker, E. (1983). On peripheral processing in human hearing. In R. Klinke & R. Hartmann (Eds.), *Hearing—Physiological bases and psychophysics* (pp. 104–109). New York: Springer-Verlag.

Zwicker, E., & Schorn, K. (1978). Psychoacoustic tuning curves in audiology. *Audiology, 17,* 120–140.

Zwislocki, J. J. (1984). How OHC lesions can lead to neural cochlear hypersensitivity. *Acta Oto-Laryngologica, 97,* 529–534.

# 4

# VISUAL AFTEREFFECTS IN THE CAT

*Mark A. Berkley*

*Department of Psychology, Florida State University, Tallahassee, Florida*

This chapter describes a series of studies in which adaptation of the visual system to contoured targets was studied in the cat. What is contour-dependent adaptation? Why study it in the cat? The answer to the first question is that several types of adaptation are very important in vision. Light adaptation, for example, allows the visual system to operate over an extraordinary range of light intensities, from the bright sunlight on a snow-covered mountainside to the dimness of a moonless night, a range spanning some 12 log units of intensity. This extraordinary dynamic range is achieved by a feedback-controlled gain control system which operates primarily in the retina (Barlow, 1972). What is not as well appreciated is the fact that under such varying conditions of intensity, contrast (defined as the difference between the intensities of adjacent dark and light portions of an image divided by the average intensity of the image) also varies greatly. To function under these conditions, the visual system adjusts its sensitivity to contrast, so called contrast gain (Ohzawa, Sclar, & Freeman, 1982, 1985). The change in contrast gain allows the system to detect contrasts of less than 1% essentially over the entire brightness range. Unlike brightness adaptation, which occurs over tens of minutes, contrast adaptation occurs over a period of a few seconds and persists only a minute or two upon returning to the original contrast condition (e.g., Bodinger, 1978; Hammer, 1949). You can appreciate the phenomenon of contrast gain control yourself by looking at the high-contrast target marked *A* in Fig. 1 for about 30 sec, and then looking at the target marked *C*. Note that your ability to see target *C* is greatly reduced (and its apparent contrast is reduced) but quickly returns. (See Fig. 1 legend for more about the demonstration,

**FIGURE 1.** Demonstration of the effects of adaptation to contrast. To observe the effect, scan the vertical bar target (grating) labeled *A* (top left) by looking around the circle in the target for about 30 sec, then quickly look at the central test target labeled *C*. With good adaptation, the faint vertical bars in the test target (*C*) should at first be invisible but slowly return over 5–10 sec. Repeating the demonstration by first scanning the horizontally oriented grating labeled *B* (top right) before viewing *C* should have no effect on the visibility of the test target *C*. Similarly, adapting to either of the two lower grating targets will have little effect on the subsequent visibility of the test target *C*. The effects observed demonstrate the specificity of the adaptation effect for the size and orientation of contours. From Maffei (1978).

which is adapted from a paper by Blakemore & Campbell, 1969.) Early studies of contrast adaptation demonstrated that the effect is contour specific, as you can see for yourself in the demonstration shown in Fig. 1. For example, after viewing a target consisting of vertically-oriented high-contrast light and dark bars (grating), the amount of contrast necessary to detect vertically oriented low-contrast bars is greatly increased while no such change in sensitivity is seen after adapting to the high-contrast, horizontally oriented bars (Blakemore & Campbell, 1969; Sekuler, Rubin, & Cushman, 1968) (also see Fig. 1). This finding suggested that the visual system's adaptation to contrast level could be used to determine if the visual system is made up of a number of tuned channels, and if so, to obtain estimates of the tuning characteristics of the underlying neural processors. For example, to determine if the system has filters or "channels" selective for contour orientation, the adapting target orientation could be varied and the effects on sensitivity to a standard oriented target measured. (Alternatively, the adaptation target orientation can be kept constant and the test target varied.) If variations in orientation (or any other selected parameter) produce variations in the subsequently measured threshold of the test target, the underlying process is considered to be selective (tuned) for that parameter. If, on the other hand, there are no changes in threshold or the changes do not vary with changes in the adapting parameter, the underlying process is considered not to be tuned for that parameter. Many studies using this technique have demonstrated that adaptation (1) occurs rapidly, (2) is relatively specific to the contours present in the adaptation stimulus, and (3) is specific to the region of the visual field stimulated (Blakemore & Campbell, 1969; Blakemore & Nachmias, 1971; Campbell & Kulikowski, 1966).

Contrast adaptation is thus important both as a phenomenon and as a technique of study. This being true, why choose the cat as a psychophysical subject in which to study it? Surely there are more cooperative animals to work with. The major reason, of course, is the existence of the extensive body of knowledge about the anatomy and physiology of the cat visual system. This vast compendium makes the cat an especially good model system for testing the relation of anatomy and physiology to function (Berkley, 1976). With regard to contrast adaptation, for example, neurophysiological data derived from single neurons in the cat show properties similar to those observed in the psychophysical data (Albrecht, Farrar, & Hamilton, 1984; Maffei, Fiorentini, & Bisti, 1973; Ohzawa et al., 1982, 1985; Vautin & Berkley, 1977). Thus, it has been shown that the response of cortical neurons to contours (contrast borders) are orientation specific and change with the level of contrast in the contour, becoming less sensitive after being stimulated with high-contrast contours and more sensitive after being stimulated with low-contrast contours (Ohzawa et al., 1985). Similar adaptation effects have been observed using an interocular paradigm in which a cortical neuron is adapted via stimulation of one eye and sensitivity changes observed when tested via the other eye (Ohzawa et al., 1985). Neurons earlier in the visual pathway do not exhibit these characteristics (Movshon & Lennie, 1979; Ohzawa et al., 1985), consistent with human psychophysical data, which suggests that contrast gain control is a process mediated by visual cortex.

Whether the cortical neurons described above are the substrate of the contrast gain change phenomenon in the cat or in humans depends heavily on how well the characteristics of the neurons match the psychophysical data. Since the only strategem available for establishing a neural substrate of behavior is the demonstration of correlations between neural and psychophysical data, that is, establishing isomorphisms between levels of observation, certain types of experiment become particularly important. For example, animal psychophysical experiments that closely approximate human psychophysical studies permit direct comparison with human data and with electrophysiological data obtained from the same animal species. These types of experiment provide firmer ground for neurobehavioral comparisons, but, more importantly, they permit extrapolations from animal neurophysiological data to presumed underlying neural mechanisms in the human brain.

Is the cat a good model system for deriving the neural substrate of contrast adaptation in humans? The answer to this question rests on the discovery of (1) how similar contrast adaptation as measured behaviorally in the cat is to adaptation measured in cat cortical neurons, and (2) how similar the adaptation process in the cat is to that observed in humans.

The major link in the chain of evidence tying contrast adaptation to cortical neurons has been the comparison of human psychophysical measures of adaptation with the response characteristics of single neurons as measured in animal models, especially the cat. The weak link in the chain of evidence has been the lack of psychophysical data on the contrast adaptation process in the species in which the neurophysiological data was gathered, namely the cat. The work described in this chapter attempts to provide the necessary cat psychophysical data.

## STATE OF KNOWLEDGE IN THE FIELD

Before getting to into the details of the issues that frame the studies to be described in this chapter, it is necessary to define a few terms that are idiosyncratic to this area of research. (If the reader is already familiar with the general area, this section can be skipped). Almost all visual targets used in this research consist of linear, alternating light and dark bars. Such a target is called a grating with a single light and dark bar pair defining one bar cycle. The number of cycles subtending 1° of visual angle at the eye is called the spatial frequency of the grating, for example, 5 cycles/deg or 0.75 c/deg, higher numbers meaning finer gratings and lower numbers coarser gratings. The difference in intensity between the light and dark bars (divided by the sum of the intensity of the light bar and the darker bar) is called the contrast of the grating and can vary from 0.0 to 1.0. A grating with zero contrast appears as a uniform field whose intensity is equal to the average intensity (i.e., the mean intensity of the light and dark bar) of the grating. In most experiments, the average luminance of the grating is held constant and only the contrast, spatial frequency, or orientation is varied. This has the advantage of keeping the intensity adaptation level constant. Typically, the reciprocal of the minimal contrast necessary (threshold) to detect the presence of a grating is plotted as a function of the spatial frequency of the grating and is called a contrast-sensitivity function (CSF) (Campbell & Robson, 1968). It can be thought of as similar to an audiogram and, because of this similarity, is sometimes called a visuogram. As is the case for the audiogram, the function is U-shaped and believed to represent the envelope of the sensitivities of a number of relatively narrow-tuned spatial frequency-selective filters, one of the hypotheses addressed by the studies to be described. Of particular importance here is the fact that the analysis described above has also been applied to single cells in the visual system of the cat (Albrecht & Hamilton, 1982, Campbell, Cleland, Cooper, & Enroth-Cugell, 1968; Campbell, Cooper, & Enroth-Cugell, 1969; Maffei & Fiorentini, 1973; Movshon, Thompson, & Tolhurst, 1978) so that orientation and spatial frequency tuning curves are available for single cells.

## EVIDENCE FOR CONTOUR-TUNED CHANNELS IN HUMAN VISION

### Human Psychophysical Studies

A variety of human psychophysical studies have been directed at revealing the contour-selective properties of the visual system and its putative underlying neural substrate. In its simplest formulation, the rationale for these studies assumes that there are independent feature "channels" in the visual system which can be activated by a stimulus containing the feature (Braddick, Campbell, & Atkinson, 1978; Maffei, 1978; Sekuler & Ganz, 1963). To determine the tuning characteristics of the channel being sampled by the test stimulus, some dimension of a conditioning stimulus is manipulated, for example, spatial frequency, orientation, contrast, or movement, and the subsequently increased (or reduced) thresholds to a test stimulus are noted. Plots of threshold changes to the test stimulus as a function of the dimension manipulated are presumed to depict the tuning characteristics of the channel sampled (Blakemore & Campbell, 1969; Blakemore & Nachmias, 1971). Other experiments have used similar paradigms to demonstrate the existence of various other stimulus analyzers in the visual system, such as movement (Pantle & Sekuler, 1969; Sekuler et al., 1968) and spatial frequency tuning (Blakemore & Campbell, 1969; Blakemore, Nachmias, & Sutton, 1970; Sachs, Nachmias, & Robson, 1971). (For review, see Braddick et al., 1978; Maffei, 1978.)

In a parametric study of orientation sensitivity, Campbell and Kulikowski (1966) used a simultaneous masking method not only to demonstrate the existence of orientation-selective mechanisms but to estimate the orientation bandpass of these channels in the human visual system. In this experiment, the contrast necessary to detect a stationary test grating was measured as a function of the angular difference between the test grating and a simultaneously presented masking grating of constant contrast. These authors found that as the angular difference between the test and the masking grating increased, the threshold contrast necessary to detect the test grating decreased. By plotting threshold contrast against the angular difference between the test and masking grating, these authors found declining angular selectivity function which had a half-width of approximately 12°

Coincident with the studies of orientation sensitivity, studies of spatial frequency selectivity were also done (Blakemore & Campbell, 1969; Sachs et al., 1971; Stromeyer & Julesz, 1972). Studies of spatial frequency tuning using the aftereffects paradigm (Blakemore & Campbell, 1969) have reported a bandpass of about 1.5 octaves with selectivity being somewhat narrower after adaptation to higher frequencies. At spatial frequencies below about 3 c/deg, the maximum adaptation effect did not occur at the same spatial frequency as the adaptation target. The tuning curves they obtained with low-frequency adaptation targets peaked at higher frequencies. Using a lightly different paradigm Stromeyer and Julesz (1972) have reported slightly different values but generally in the same range (e.g., 0.5–2 octaves bandpass). These disparities point up the importance of using comparable methods for obtaining comparable data.

One important feature of the aftereffect paradigms concerns the temporal course for both induction and recovery of the change in sensitivity. A few studies have examined this aspect of adaptation systematically (e.g., Blakemore & Campbell, 1969; Bodinger, 1978; Hammer, 1949; Keck, Pallela, & Pantle, 1976; Rose & Lowe, 1982; Sekuler, 1974). While there are some differences in the paradigms and parameters employed, there is general agreement that adaptation (when a moderate-to-high-contrast induction target is used) grows rapidly during the first few seconds of viewing, saturating at about 1 min. Recovery from the adaptation proceeds rapidly as well, being more or less complete in 1–2 min. (Blakemore & Campbell, 1969; Bodinger, 1978; Keck & Pentz, 1977) although recovery may take as long as 30 min after prolonged adaptation (10 min) to a high-contrast target (Blakemore et al., 1970; Bodinger, 1978; Rose & Lowe, 1982).

On the other hand, Daughman (1983) has found that when a test procedure keeps response time to a minimum (e.g., 2 brief test probes presented during adaptation and recovery) adaptation and recovery is very rapid (time constants 5–7 sec). While adaptation time was independent of the spatial frequency of the stimulus, recovery time constants were related to spatial frequency of the adaptation target with the higher the spatial frequency, the longer the recovery time constant (J. G. Daughman, unpublished dissertation).

In summary, psychophysical studies employing adaptation (or masking) paradigms have demonstrated spatial frequency and orientation-selective channels and their tuning. In addition, the adaptation and recovery characteristics of those channels have been described. How do these properties compare with physiological estimates of the same features measured in visual system neurons?

## Human Physiological Studies

Neural evidence for the existence of orientation- and size-selective elements in the human visual system is, for the most part, indirect. For example, Campbell and Maffei (1970) and Blakemore and Campbell (1969) have shown that the presence of a grating in one orientation will reduce the amplitude of an evoked potential only to gratings of similar orientation. It is assumed that the first grating activates and adapts a pool of neurons sensitive to orientation, leaving a pool of less sensitive neurons to respond to the test grating thus producing a smaller response when stimulated with the test grating. These and similar data are taken as a demonstration of the existence of orientation-selective neurons in the human visual system. Single-cell spatial frequency tuning curves have been obtained in primates (e.g., DeValois, Albrecht, & Thorell, 1982) and have tuning curves similar to those observed from human psychophysical studies. A few cells have been recorded in area 17 of humans while the patient was undergoing a neurosurgical procedure (Marg, Adams, & Rutkin, 1968) and, although they were not studied parametrically, they appear to have properties similar to those described by Hubel and Wiesel in the cat (1959, 1962) and monkey (Hubel & Wiesel, 1968, 1977). Based on this evidence, most investigators feel that the human visual system is similar to the visual system found in the cat and monkey in that it possesses cells uniquely tuned to the orientation and spatial frequency of contours.

## NEURAL STUDIES OF CONTOUR-SELECTIVE ADAPTATIONS IN ANIMALS

Electrophysiological measures of adaptation have been made in rabbit retinal ganglion cells (Barlow & Hill, 1963), cat cortical neurons (Maffei et al., 1973; Mansfield & Simmons, 1979; Movshon & Lennie, 1979; Vautin & Berkley, 1977; von der Heydt, Hanny, & Adorjani, 1978) cat evoked potentials (Bonds, 1984), tectal neurons of the pigeon (Woods & Frost, 1977), lobular plate neurons of the sheep blow fly (Srinivasan & Dvorak, 1979), as well as in human evoked potentials (Blakemore & Campbell, 1969; Clarke, 1974; Mecacci & Spinelli, 1976; Ochs & Aininoff, 1980; Tyler & Kaitz, 1977). These studies also show that neurons in the visual system become less reponsive after periods of continuous stimulation. In studies of adaptation of cortical neurons, precortical adaptation, that is, adaptation of neurons in the retina and lateral geniculate nucleus which provide the input to the cortex, has been ruled out by several control procedures. For example, adaptation of a cortical neuron induced via stimulation of one eye is still in effect when the neuron is tested via stimulation of the other eye (Maffei et al., 1973). Since the cortex is the first site of binocular interaction, interocular transfer of adaptation is taken as evidence that the cortex was the site of adaptation. In other studies, the adaptation effects were shown to be specific for stimulus contour orientation (Maffei et al., 1973; Vautin & Berkley, 1977), and size (Movshon & Lennie, 1979), both characteristics also known to be cortical in origin. Perhaps most important was the observation that responses recorded from precortical neurons showed no adaptation effects under the same conditions in which such effects were observed in cortical neurons (Movshon & Lennie, 1979). These results, taken together with human psychophysical studies showing orientation specificity (Blakemore & Campbell, 1969; Campbell & Kulikowski, 1966) and interocular transfer of adaptation (Bjorklund & Magnussen, 1981; Blake & Fox, 1972; Day, 1958; Gilinsky & Doherty, 1969; Mitchell & Ware, 1974) strongly support the view that the cortex is the locus of contour-selective channels.

Human psychophysical studies using the aftereffects paradigm, as well as others, suggest a model of visual processing in which various aspects of the visual scene are processed by more or less independent channels tuned to specific stimulus features. How do the tuning characteristics of cortical neurons compare to the tuning of the feature-selective channels suggested by the psychophysical studies? Maffei and Fiorentini (1973) measured the contrast response and spatial selectivity of single cells at several levels of the cat visual system. In agreement with previous studies, they found relatively broad spatial tuning for retinal ganglion cells and lateral geniculate cells. However, they found that one class of cortical cells called simple cells exhibited a narrow spatial frequency bandpass with broader tuning for other cell classes. The spatial tuning characteristics of simple cells have been studied by several investigators (Albrecht & Hamilton, 1982; Andrews & Pollen, 1979; Bisti, Maffei, Clement, & Meccaci, 1977; Holub & Morton-Gibson, 1981; Maffei and Fiorentini, 1973; Movshon et al., 1978). In general, a spatial frequency bandpass of about 1 octave is reported. It is noteworthy that the tuning characteristics of cortical neurons (Albrecht & Hamilton, 1982; Andrews & Pollen, 1979; Bisti et al., 1877; Holub & Morton-Gibson, 1981; Maffei and Fiorentini, 1973; Movshon et al., 1978) bear a striking resemblance to the spatial frequency channels derived in human psychophysical studies (e.g., Blakemore and Campbell, 1969; Sachs et al., 1971). (While the validity of the channels model does not depend on the discovery of its possible neural substrate, the resemblance described above adds considerable support to this model of visual processing.)

The time course of adaptation and recovery from prolonged stimulation of cortical neurons in cats has also been measured (Albrecht et al., 1984;

Ohzawa et al., 1985; Vautin & Berkley, 1977). The time-constants obtained ranged from 6 to 12 sec and are similar to those observed in human psychophysical studies of aftereffects reported by Blakemore and Campbell (1969) and Bodinger (1978). A recent evoked-potential study in cats reported a value of 13 sec (Bonds, 1984), although it was only a suggested value, and a comprehensive study by Albrecht et al. (1984) reports values ranging between 2 and 80 sec for adaptation, and 2 sec to several minutes for recovery. Overall, the single-cell studies demonstrate considerable variability in adaptation and recovery characteristics between neurons.

Do animal models of vision show similar tuning as measured in psychophysical tests? A serious difficulty in behavioral studies of complex sensory events in animals is the limited level of communication between subject and experimenter. Thus, is is not possible to verbally instruct the subject nor for the subject to verbally report its sensations. Thus, class B studies (Brindley, 1970), in which reports of stimulus appearance are made, are essentially impossible to do. A notable exception is a study of the movement aftereffect in monkeys (e.g., Scott & Powell, 1963). More recently, a behavioral study of color-contingent contour aftereffects (Mc-Cullough effect) has been reported in monkeys (Maguire, Meyer, & Baizer, 1980) in which indirect measures of the appearance of the test target were taken (class B rather than class A type study).

## CAT ADAPTATION STUDIES

While single-cell data from cats are consistent with "channel" models and with human psychophysical studies designed to test such models, the question of the generality of the "channel" model to nonprimates and the relation of neuronal sensitivity profiles to channel selectivity is still open. A behavioral study of cats that employed a masking paradigm has suggested that cats, like humans, possess channels tuned to spatial frequency (Blake & Martens, 1981) That study, however, employed a paradigm that yielded data not directly comparable to available electrophysiological data. To compare channel properties, it is important to obtain tuning curves from cats using a paradigm similar to that used with humans to demonstrate the presence of channels and to compare such tuning curves with the neurophysiological data. Since the aftereffects paradigm is the one most widely used in human studies, it was adapted for use with the cat.

When the current studies were undertaken, an aftereffects paradigm had never been used with cats and there was some doubt as to the possibility of carrying out such studies. Thus, the experiments began with an evaluation of several training procedures and stimulus parameters. The initial attempts in my laboratory employed a paradigm modified from a reaction time procedure devised for use with monkeys (Stebbins, 1966). It required the cat to keep a key depressed with its nose in order to present the adaptation target and to release the key when the test stimulus appeared. This method did not work well. It was difficult to teach the cats the response, and the false positive rate rose to unacceptably high levels. A conditioned suppression scheme (Loop & Berkley, 1972) was tried but also proved unsuitable because too few trials could be run each day. After numerous pilot studies were run trying out various techniques that would ensure the cat's viewing the adaptation target for the required time, the simple two-choice paradigm described below was devised. The procedure required that the cat break an infrared photobeam with its head to present itself with an adaptation target, then maintain its viewing position for a mimimum time, after which a bipartite test field was presented. The cat then was required make a response indicating the location of a test target which varied in left–right position (e.g., 2 AFC procedure in which trials are presented only after completion of an adaptation period). Details of the procedure are described in the next section.

### Experimental Methods

The same basic behavioral test apparatus and procedure was used in all of the studies described below. The testing situation employed a two-alternative, forced-choice (2 AFC) testing paradigm in which pairs of test stimuli, generated on a special microprocessor-controlled oscilloscope display device, were presented to the cats as a left–right pair. All stimuli were viewed by the cats by placing their heads in a small Plexiglas head chamber (see Fig. 2). The stimuli could be viewed by looking through two transparent keys which make up the outer wall of the head chamber. Responses consisted of pressing the right or left panel with the nose. A food terminal, located just below the response keys, delivered 1-cm$^3$ dollops of beef baby food as a reward. Two infrared photocells in the head chamber monitored the animal's head position. The photocells were used to determine (1) when the cat's head was in the head chamber, and (2) when the cat's head was elevated above a specified level permitting it to view the stimuli (e.g., not licking the food terminal). Head position as determined by the photocells controlled the presentation of the adaptation target (e.g., "on" if the head was in the correct position; "off" if the head was out of position). The adaptation target consisted of a single grating, whose orientation, size (spatial frequency), or contrast was varied between blocks of test trials.

A test trial consisted of the presentation of an adapting stimulus when the upper photobeam was broken and the lower photobeam was unbroken. A trial was aborted if either photobeam condition was violated before the adaptation time was completed or a premature response was made. If the required time interval for viewing the adaptation target was successfully completed, it was followed by presentation of the test stimu-

TOP VIEW

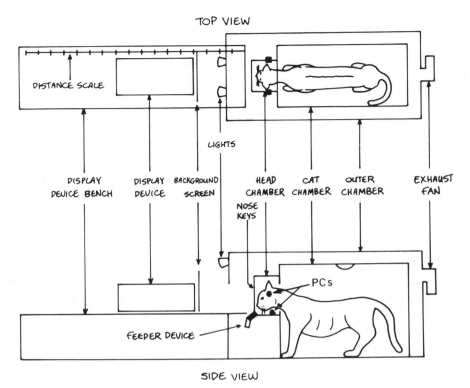

**FIGURE 2.** Diagram of cat testing apparatus, top view and side view. Note the location of the two sets of photocells (labeled PC) used to detect the position of the cat's head. Modified from Berkley, 1979.

lus pair. The cat indicated its choice of the left or right stimulus by depressing the left or right nose-key panel. Correct choices produced delivery of the food reward at the food terminal. Incorrect responses produced a time-out. A summary of the program of events constituting a test trial are shown in Fig. 3.

Because the presentation of the adaptation target as well as a test trial was dependent upon the cat's behavior, intertrial intervals were variable and therefore the state of adaptation would depend on the level of recovery from a previous trial. Thus, only the minimal level of adaptation could be specified during a test trial, rather than the actual state of adaptation. This situation is depicted in Fig. 4 which shows a hypothetical test trial and the presumed changes in adaptation state that were assumed to occur during a trial sequence. Thus, while the cat is looking at the adaptation target, contrast adaptation is increasing, as shown by the rising solid line. When the test target appears, recovery from the adaptation begins because the test target is usually of lower contrast. After a response

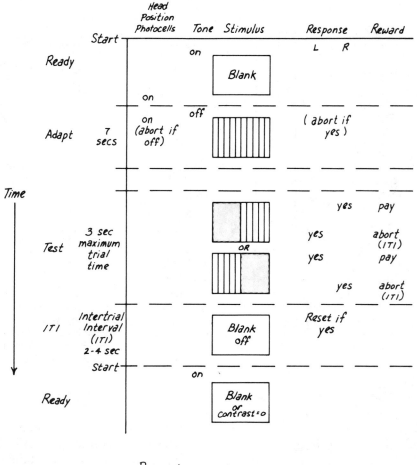

**FIGURE 3.** Diagrammatic representation of the test paradigm used in the selective adaptation studies. The flow of the program can be followed by starting at the top and following the time line at the left of the figure. The various stimulus and response conditions are listed across the top. Possible outcomes at each condition are listed at the appropriate places on the figure.

is made, no target is displayed and recovery from adaptation continues. The duration of the recovery period and thus the residual level of adaptation is variable in that the cat can initiate a trial either at the minimal intertrial interval or sometime later. A similar series of fluctuations in adaptation level will also occur if the cat aborts a trial by not maintaining the required head position. In this case, the adaptation target is immedi-

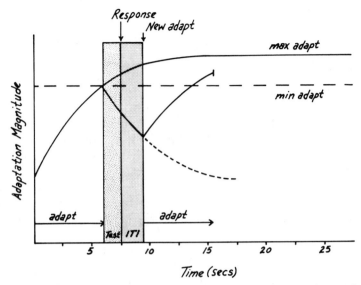

**FIGURE 4.** Hypothetical time course of the adaptation level changes that might occur at various points during a test trial. The solid line depicts the changing levels of adaptation. The horizontal dashed line denotes the "minimal" adaptation level if all the viewing conditions are met by the subject. The lightly stippled bar indicates the period during which the test stimuli are presented, while the densely stippled bar indicates an intertrial interval during which no stimuli are present. The condition depicted is for the case where the adaptation interval was not interrupted by a premature response or breaching of the head position requirement.

ately turned off, which would initiate recovery. However, as soon as the required head position is achieved again, the adaptation target would be redisplayed and the adaptation level increased. While such variability is not desirable, it was permitted to keep the test paradigm simple and allow the cat to work at its own pace. Some estimates of the actual state of adaptation were made during the study and will be described later. Using the paradigm described above, the following data were recorded: (1) the percentage of correct choices on each key made in a session consisting of 100 trials; (2) the number of aborted trials; (3) the number of correct/incorrect responses on each response key; (4) the response latencies; and (5) the total session time. Several 100-trial blocks were run each day with several hours intervening between each session. Within any one block of trials, the test stimulus parameters were kept constant but were varied from block to block. During the pre-threshold training, random interleaving of different suprathreshold values of the test stimulus were used to accustom the animal to seeing different targets and thus prevent the adoption of a maladaptive response strategy. Maladaptive response strategies often reappear during threshold testing and are not easily changed at that stage. In particular, if too few easily discriminable trials are given, that is, too many test values near threshold, the animals adopt a guessing strategy which is virtually impossible to change (see Berkley & Sprague, 1979, for further discussion).

Several other special features were also placed in the test program to control or eliminate common pesky behavior difficulties such as position habits, anticipatory responses, and inappropriate viewing strategies. For example, to eliminate the formation of position habits, that is, as repeated responding to one key regardless of the position of the correct stimulus, a limit of 8 consecutive responses to the same side were permitted at which time the correct stimulus was displayed on the side opposite the position habit (Berkley, 1970). Two consecutive responses to this side were required to return the program to its quasi-random stimulus presentation mode. (Responses made in position habit mode were not used in computation of thresholds.)

To prevent too rapid responding during initial training (and thus insufficient viewing of the test targets), the prethreshold training program required a minimum response latency of 1.0 sec before a response could be made after the test stimuli appeared.

The preliminary training described above typically required 3–4 months of daily training but yielded stable data as well as being effective in controlling undesirable behaviors (see Berkley, 1970; Berkley, Warmath, & Tunkl, 1978; Bloom & Berkley, 1977; Tunkl & Berkley, 1977).

**Threshold Estimation Procedures.** When stable performance was achieved on the preliminary training tasks, the threshold estimation testing was started. The first threshold data set obtained simply determined the amount of contrast necessary to detect a grating after adapting to a 0.0-contrast grating, that is, a uniform adaptation field (see Fig. 5). To compute contrast thresholds, psychometric plots were constructed (mean percentage of correct choices plotted against stimulus contrast) and a threshold estimate taken as the contrast value at the intercept of the psychometric function at 57% correct choices as shown by the dashed line in Fig. 5. Variances and confidence intervals were also computed. The 57% correct value represents 2 standard deviations from chance performance. (The mean and variance of chance performance levels was estimated using an insoluble problem with at least 10 cats. A more detailed description of the rationale and methods can be found in Bloom and Berkley [1977] and Berkley and Sprague [1979]. Typically, in the plots of sets of such threshold data, the inverse of the contrast (sensitivity) necessary to detect a particular grating size is plotted against the size of the test grating (spatial frequency). The resulting plot represents a detection function and is known as a contrast-sensitivity function. An example of such a function for the cat is shown in Fig. 6. Such contrast-sensitivity

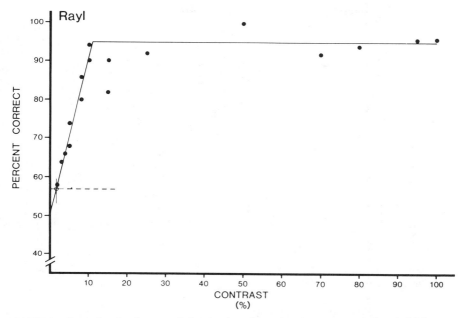

**FIGURE 5.** Example of a data set and procedure for estimating a contrast threshold for one spatial frequency. Percentage of correct detections is plotted against the contrast of the test grating. Threshold contrast is taken as the intercept of the best-fit line at chance (e.g., two standard deviations from chance shown as dashed line at 57%).

functions were determined for each cat so that it would be possible to select appropriate contrast values in the adaptation phase of the studies.

After completing the initial contrast-sensitivity measurements, the contrast-adaptation measures were taken. The training and testing procedures employed during this phase of the experiments were essentially the same as those used in the prethreshold training period described above except that the adaptation stimulus now consisted of a high-contrast grating target rather than the uniform (0.0-contrast) field. Several suprathreshold stimulus values (3–5) as well as near-threshold values were used within a day's block of sessions to avoid too many sessions in which the animal received few rewards (and thus was more likely to adopt a guessing strategy). The testing procedure continued until at least 10 daily sessions (of 3–5 100-trial blocks) had been run at each stimulus value (7–10 values). Thus, a complete threshold function for one stimulus dimension (e.g., spatial frequency or orientation) required from 3 to 4 months, each point representing 3–10 sessions of 100 trials. (From initial training to completion of 1 tuning curve usually required 7–10 months.)

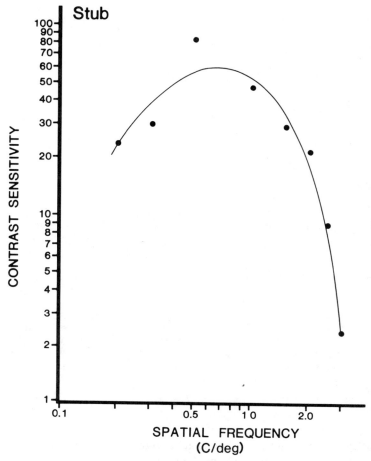

**FIGURE 6.** Plot of the contrast necessary to detect gratings of various spatial frequencies after adapting to a 0.0-contrast grating (uniform field) whose average luminance was equal to the average luminance of the test grating (i.e., contrast sensitivity function for one cat (Rayl) ).

**Reliability and Control Procedures.** Because a relatively long time was necessary to accumulate threshold data in animal studies (see above), long-term, experience-dependent changes in performance could influence the threshold estimates. As a means of determining whether there had been a change in level of performance over the testing period, threshold estimates obtained early in testing were compared with later ones. This was possible because all stimulus values that enter into threshold estimation were used throughout the threshold testing period. Testing continued if there was more than a 10% difference.

As a check against inadvertent cueing (e.g., auditory cues, possible stimulus artifacts, possible program cues (lef–right runs, etc.), two procedures were routinely used. In the first, an occasional test session was run in which the stimulus pairs were made identical and the performance level of the cats determined. Under such circumstances, the cats' performance should not deviate from chance levels if the cue they were using was the appropriate stimulus dimension. The second procedure for testing inadvertent cuing was the same except that rewards were delivered for responses to what was the "incorrect"stimulus in the previous (control test above) test. This test permitted the evaluation of asymmetries in key "preference."

## Adaptation Effects in Cats

Since no data were available defining the optimal parameters for producing adaptation, for example, adaptation time and contrast of adaptation target, that would be appropriate for the cat, these data were gathered next. Thus, the percentage of correct choices of a standard, high-contrast (0.9) grating test target was measured (1) as a function of the contrast of an adapting grating of the same orientation and spatial frequency as the test grating with viewed at a constant viewing (adaptation) time (7 sec) (Fig. 7); (2) as a function of the orientation of an adapting grating of the same spatial

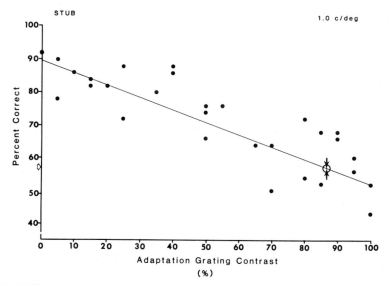

**FIGURE 7.** The percentage of correct detections of the left–right position of a vertically oriented 1.0-c/deg, 0.90-contrast test grating after viewing a 1.0-c/deg adapting grating of various contrasts for at least 7 sec for 1 cat (Stub). The test grating appeared as half the field, in random left-right positions, in a bipartite field in which the other half of the field was a 0.0-contrast grating of equal average luminance (see Fig. 3). The point labeled with arrows (i.e., 86%) indicates the adapting target contrast producing chance detections of the 0.90–contrast test grating.

frequency and contrast as the test grating (Fig. 8); and (3) as a function of the adaptation time required to view a grating of the same spatial frequency, orientation, and contrast as the test grating (Fig. 9).

As can be seen from the lengthy description given above, a considerable amount of preliminary work was required before the measurements of interest could be made. While the pilot studies strongly suggested that adaptation effects would be observed in cats, it was not until the actual testing was done that the issue was decided. Several surprises emerged. First, it was found that detection of even a high-contrast (0.9) grating was severely reduced after viewing a high-contrast adaptation target of the same spatial frequency and similar orientation. These observations established that contrast adaptation occurred in cats and increased the probability that it would be possible to establish that visual aftereffects can be observed in the cat (see Figs. 7 and 8). While these initial observations were what the study was looking for, the magnitude of the effect after 7 sec of adaptation was, however, much larger than expected (see Fig. 9). It often approached a 2-log unit change in threshold. Even moderate-contrast adapting targets produced very large reductions in probability of detection of the test grating (Figs. 7 and 8). Compared to human adaptation studies, the threshold elevation effect was very large. The difference, however, was likely due to the higher contrasts of the adaptation targets that were used with the cats as well as the shorter response times permitted. Human studies rarely employ an adapting contrast 1 log unit or more above

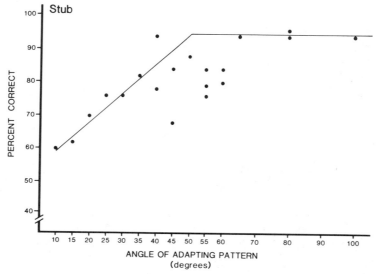

**FIGURE 8.** The percentage of correct detections of 1.0-c/deg, vertically oriented grating plotted as a function of the orientation of a 1.0-c/deg, 0.90-contrast adapting grating.

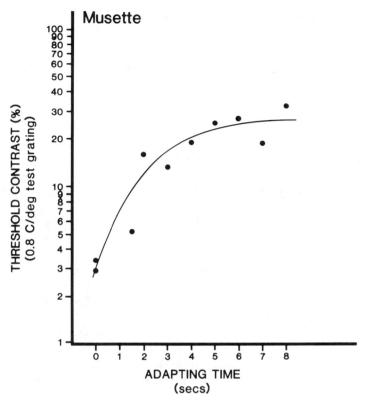

**FIGURE 9.** The threshold contrast necessary to detect a 0.8-c/deg grating as a function of the required duration of prior viewing of an adapting target consisting of a 0.8-c/deg, 0.90-contrast grating of the same orientation. Lowest point represents a 200-msec adaptation time. Curve is best-fit line fitted to an exponential function.

threshold, whereas in the cat studies much higher contrasts were employed, and in human studies some recovery may occur while the subjects make their response (e.g., adjusting a potentiometer). A second surprise in the cat aftereffects data was the rapidity with which the aftereffects developed (Fig. 9). Thus, as shown in Fig. 9, after just 2 sec of adaptation to a 0.9 contrast grating, very large changes in detection of the test grating were seen. The data showed that after only 5–7 sec of adaptation, the major threshold elevation was over. Compared with the values reported for humans, the adaptation rate in the cat is very fast. However, close reading of the studies of the rate of adaptation and recovery in humans (Blakemore & Campbell, 1969; Bodinger, 1978; Rose & Lowe, 1982) reveals that the time constants may have been effected by the methods used to measure threshold. Thus, in Blakemore and Campbell (1969) and Rose and Lowe (1982) a method of adjustment was employed in which several seconds are needed by the subjects to set the contrast control and during which time the adaptation stimulus is off. This procedure tends to confound recovery rate and adaptation state. In the cat studies reported here, a single response had to be made within 2 sec of termination of the adapting field or the trial was terminated and a new adaptation period required. Measurement of the cats reaction time showed that their response latencies were between 0.5 and 1.5 sec, with a mean value of about 700 msec. Thus, it is possible that the more rapid time constants observed in the cat studies are more representative of the instantaneous adaptation state because there was little time for recovery to take place during the threshold test trial. Additional support for this view comes from unpublished observations in humans by Daughman (1983, and personal communication) in which a 2 AFC paradigm similar to the cat paradigm was used. That is, in Daughman's studies, on each trial, after the adaptation period, the subject simply had to report whether a test grating of variable contrast was visible or not. Daughman found adaptation time constants in the same range I have observed in the cats, 2–6 sec. The issue of adaptation and recovery times in cats is an important one because their accurate measurement permits direct comparisons with the parametric neuronal data now available for this aspect of stimulus processing in the cat (Albrecht et al., 1984; Vautin & Berkley, 1977).

Because trial presentation rate and total adaptation time are determined by the cat's behavior, the actual adaptation level cannot be precisely specified except for its minimal value. This is true because it was possible for the cats to present themselves with a new trial before they recovered from the effects of the previous trial, or because they sometimes aborted a trial during the adaptation period by moving their heads out of position or responding prematurely and then immediately initiating a new trial. In these circumstances, the effects of the previous adaptation period had not subsided and, since the recovery time was variable because it depended upon the cat's behavior, adaptation state varied.

For these reasons, it was not possible to directly measure the state of adaptation continuously during a test session. To obtain an estimate of the momentary adaptation state, an electronic simulation of adaptation and recovery was carried out in which the cat's behavior during a test session controlled the simulator. The simulator consisted of an electronic integrator–differentiator constructed with a time constant equal to the average value reported for "adaptation" and recovery effects in cortical neurons (e.g., 9 sec) (Albrecht et al., 1984; Movshon & Lennie, 1979; Vautin & Berkley, 1977). The input to this device was connected to the behavioral test apparatus and its output to a linear pen recorder. When the cat initiated a trial, presenting itself with the adaptation target, the integrator was started and continued as long as the cat fulfilled the requirements for displaying the adaptation target (see Methods). The simulator began its discharge upon a trial abort or successful trial completion, for example, whenever the adaptation target was turned off. A sample of a typical record produced by this device is shown in Fig. 10. The bottom segment shows the record produced by a full charging and discharging of the device emulating a cycle of complete adaptation and complete recovery. Other segments of the record show the level of charge–discharge of the device produced by a cat as it turns the adaptation target on and off during several real trials. As can be seen, during most of a test run the cats reach about 50% of the maximum adaptation possible (assuming the time-constant selection and the maximum required adaptation time was appropriate). While such records were instructive, they do not give an accurate estimate of the adaptation level of the cat because the adaptation time constant that was subsequently determined was shorter than that used in the integrator device (e.g., Fig. 9 suggests that adaptation effects asymptote after about 6 sec). Thus, the cats were probably more adapted than these records indicate.

In the final set of experiments representing the actual goal of the studies, the threshold contrast for detecting a test target of one spatial frequency and orientation was determined after adaptation to stimuli of (1) various spatial frequencies, and (2) various orientations. An example of the spatial frequency tuning data is shown in Fig. 11. In this figure, the data for cat Rayl are plotted for when the test target was 2 c/deg. Note that the threshold for detecting the 2-cycle test grating was maximally elevated about 1 log unit when the spatial frequency of the adapting target was the same as the test grating, that is, 2 c/deg. Note also that the threshold elevation effect declined as the adapting frequency differed from the test grating. Unadapted threshold (i.e., threshold after adapatation to a zero-contrast target of matched brightness) is indicated on the figure by the circle with crosshair. Similar data were obtained for several other cats at other spatial frequencies. An interesting feature of almost all the adaptation functions obtained so far is that at some adapting frequencies, the threshold for detecting the test grating was actually enhanced, for example 0.5 c/deg in Fig. 11. That is, after adaptation to some spatial frequencies, less contrast was necessary to detect the test target than was required after adaptation to a zero-contrast target.

Another important feature of the adaptation data is their bandpass characteristics which provide an estimate of the selectivity of the adapted channels. Spatial frequency bandwidths can be estimated from the plotted data by drawing a horizontal line at half the contrast of the maximum threshold elevation, determining the two intercepts of this line with the data function, and dividing by 2 (half-width at half-height). Such measurements yielded values ranging from 0.5 to 1.5 octave. At the present time, there does not appear to be a relation between bandwidth and spatial frequency, but too few data are currently available to be certain that such a relation does not exist. Considered together, the data demonstrate that (1) grating adaptation in cats is spatial frequency-selective (spatial frequency-tuned); (2) the width of the spatial frequency tuning curve is about 0.5–1.5

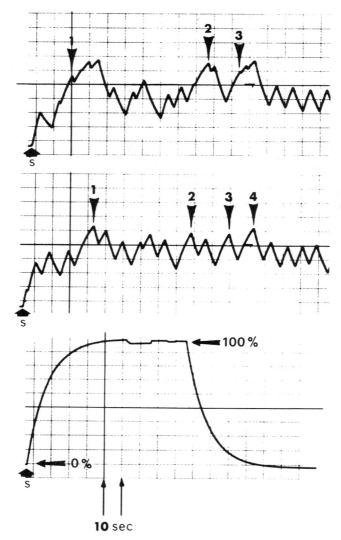

**FIGURE 10.** Electronic simulation to estimate the level of adaptation achieved by a cat during a test session. Time is represented on the horizontal scale and simulated level of adaptation on the vertical scale. *Bottom trace:* Behavior of the electronic integrator–differentiator simulator with a time constant of 9 sec allowed to totally charge and discharge. Point marked 0% indicates totally discharged (unadapted) state of simulator; 100% level indicates maximum charge of simulator (maximum adaptation). The two upper traces are samples of the output of the simulator taken from different test sessions in which the simulator was charging when the adaptation target was being displayed (i.e., cat's head in correct position) and discharged when the adaptation target was not being displayed. At points indicated by arrow S, the test session started and cat presented itself with an adapting target. *Middle trace:* After several aborts, at 1 a trial was achieved (i.e., cat fulfilled the 7-sec adapation requirement), the test target was presented, the target position (L–R) was correctly detected, and a reward was delivered. Cat consumes reward and initiates but does not complete several adaptation

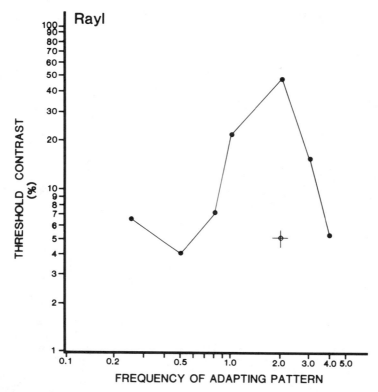

**FIGURE 11.** The threshold contrast necessary for detecting a 2.0-c/deg grating after completing 7 sec of adaptation to 0.90-contrast gratings of various spatial frequencies for cat Rayl. The point indicated with crosshairs represents the threshold contrast for detecting a 2.0-c/deg after 7 sec of adaptation to a 0.0-contrast grating (uniform field) of equal average luminance.

octaves (half-width at half-height); (3) the tuning curve is asymmetric being slightly elevated toward lower spatial frequencies; (4) there is often a sensitivity enhancement after adapting to a grating which differs in spatial frequency from the test grating frequency by a factor of 2–4.

Similarly, the effects of varying the orientation of the adapting grating on the subsequent contrast threshold for the test grating are shown in Fig. 12. The figure shows that adapting to a grating target of the same spatial frequency and orientation as the test target elevated the threshold for the test target approximately 25 times. The figure also shows that the

---

periods between 1 and 2. At 2, a successful trial was completed and reward collected, etc. *Upper trace:* Simulator record from another session in which after successful completion of the required adaptation trials (1,2,3) the cat selected the incorrect stimulus. Note that because of the slow recording speed (10 sec/division), the long-time constant of the integration (9 sec), and small scale, short breaks (e.g., <1 sec) in the adaptation time period cannot be seen on the record.

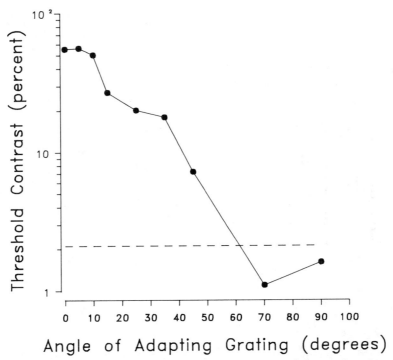

**FIGURE 12.** Plot of contrast necessary to detect a 1.0c/deg, vertically oriented grating after at least 7 sec of adaptation to a 1.0-c/deg, 0.9-contrast grating of various orientations. Dashed line represents the contrast threshold for detecting the test grating after adapting to 0.0-contrast field. Subject cat: Stub.

degree to which the threshold was elevated varied with the difference in orientation between the test and adaptation grating targets. Thus, contours whose orientation were similar to the test target produced large threshold elevations, with the magnitude of the threshold elevation effect declining as the difference in orientation between the adapting and test target increased. Such a finding demonstrates that the mechanism being adapted by prolonged viewing of a grating is contour orientation-selective (i.e., tuned for orientation). In addition to demonstrating orientation selectivity, a threshold enhancement effect was also observed as described for spatial frequency. Thus, after adaptation to a grating that differed by 60–90° from the test grating, the amount of contrast needed to detect the test grating is reduced, that is, sensitivity is enhanced. Finally, employing the procedure described above for estimating spatial frequency bandpass values, orientation bandpass values were calculated. These values ranged from 15° to 20° (half-width at half-height). Overall, the orientation tuning data show that (1) the threshold elevation after grating adaptation is orientation-selective (tuned for orientation); (2) the bandpass of the orientation tuning curve is about 15° (half-width at half-height); and (3) just as in the spatial frequency data, there is a threshold enhancement effect at orientations 60–90° from the test grating (see Fig. 12).

An important aspect of the data is that the channel model developed from human psychophysical data appears to apply equally well to cat and humans despite the differences in the neural organization of the two visual systems. A surprising outcome of these studies has been that the band-pass characteristics of the contour-selective channels as measured with an aftereffects paradigm are remarkablely similar to those observed in humans. Thus, even though the overall visual capacities of the cat are much poorer than those of humans in terms of acuity and ability to discriminate complex scenes, the basic analysis mechanisms appear to be very similar in both species. For example, in human studies, orientation tuning is reported to be about 10–15° (Blakemore & Campbell, 1969; Campbell & Kulikowski, 1966) and spatial frequency tuning bandwidth 1–1.5 octaves (Blakemore & Campbell, 1969; Stromeyer & Julesz, 1972), values very similar to those obtained in the cat. The most significant difference between the cat and humans appears to be in the speed and magnitude of adaptation, although, as pointed out earlier, the human speed estimates are probably confounded with response times.

How well does the cat data described support the view that the adaptation effects are cortical in origin? The available electrophysiological data are very clear: Contrast-adaptation effects are not seen in precortical neurons. However, the question of the origin of the contrast-adaptation effects is often raised in psychophysical aftereffects studies and is also appropriate to be considered here. Could the observed adaptation effects be peripheral (e.g., retinal afterimage) as opposed to central in origin? In human adaptation studies, peripheral adaptation is minimized by (1) having the subjects scan the adaptation target (see Fig. 1), (2) using drifting or temporally modulated targets, (3) employing relatively low-contrast adaptation targets, or (4) employing a dichoptic stimulus presentation method (i.e., presenting the adaptation stimulus to one eye and the test stimulus to the other eye). Since these conditions were not used in the cat studies, it could be argued that the effects observed are due to retinal adaptation. I do not believe, however, that the threshold elevation effects that were observed in the cats are peripheral in origin for the following reasons. First, continuous video monitoring of the cats during test sessions showed that while not specifically trained to do so, the cats looked from one side of the adaptation target to the other in anticipation of the presentation of the test target (and thus the opportunity to respond and collect a reward). Since it is highly unlikely that the left-right refixation positions were always in the same phase relation to the adaptation targets, the situation was similar to having human subjects scan the target. Second, the adaptation effects observed were spatial frequency- and orientation-dependent and showed asymmetries in their tuning curves that are not explicable on the basis of retinal adaptation effects. However, to completely resolve this issue empirically will require further study with flickering (counterphase) gratings, a project currently underway.

Given that the observed adaptation effects are cortical in origin, how do the behavioral data compare to neurophysiological studies of orientation- and spatial frequency-selective neurons in striate cortex? While a detailed discussion is beyond the scope of the present chapter, it can be said that the spatial frequency results are in excellent agreement with cortical neuron tuning curves (e.g., Albrecht & Hamilton, 1982; Movshon & Lennie, 1979). The observed orientation tuning curves are, however, broader than the tuning seen in orientation selective neurons (e.g., Rose & Blakemore, 1974; Watkins & Berkley, 1974). It is likely, however, that the paradigm employed to obtain the data from which to estimate orientation tuning is responsible since Kulikowski, Abadi, and King-Smith (1973) have reported that different test procedures, for example, subthreshold masking versus aftereffects, yield different estimates of tuning.

Several important goals were achieved with the present studies: (1) The feasibility of using cats in an aftereffects paradigm was demonstrated; (2) estimates of spatial frequency and orientation selectivity of contour-dependent aftereffects were obtained; and (3) the contour-selective channel model has been more directly supported. In summary, cats, like humans, have contrast-adaptation mechanisms that are spatial frequency- and orientation-selective. These observations coupled with other measures of feline vision (Berkley, 1976) suggest that while the cat's visual perceptual capacities may not be as rich as those of humans, the cat nevertheless sees the world essentially the same way we do.

## ACKNOWLEDGMENTS

The author thanks Susan Partington and Marybeth Skladany for data collection, Ross Henderson and Paul Hendrick for the design and construction of the stimulus production devices, and Lawrence Bloom and David Lowe for computer programming. The work described herein was supported by a grant from the National Science Foundation, BNS-8418832.

## REFERENCES

Albrecht, D. G.., Farrar, S. B., & Hamilton, D. B. (1984). Spatial contrast adaptation characteristics if neurons recorded in the cat's visual cortex. *Journal of Physiology (London)*, **347**, 713–739.

Albrecht, D. G., & Hamilton, D. B. (1982). Striate cortex of monkey and cat: Contrast response. *Journal of Neurophysiology*, **48**, 217–237.

Andrews, B. W., & Pollen, D. A. (1979). Relationship between spatial frequency selectivity and receptive field profile of simple cells. *Journal of Physiology (London),* **287,** 163–176.

Barlow, H. B. (1972). Single units and sensation: A neuron doctrine for perceptual psychology? *Perception,* 1, 371–394.

Barlow, H. B., & Hill, R. M. (1963). Evidence for a physiological explanation of the waterfall phenomenon and figural after-effects. *Nature (London),* **200,** 1345–1347.

Berkley, M. A. (1970). Visual discriminations in the cat. In W. Stebbins (Ed.), *Animal psychophysics* (pp. 231–248). New York: Appleton-Century-Crofts.

Berkley, M. A. (1976). Cat visual psychophysics: Neural correlates and comparisons with man In J. Sprague & A. M. Epstein (Eds.), *Progress in Psychobiology and Physiological Psychology* (pp. 63–120). New York: Academic Press.

Berkley, M. A. (1979). A system for behavioral evaluation of the visual capacities of the cat. *Behavioral Research Methods & Instrumentation,* **11,** 545–548.

Berkley, M. A., & Sprague, J. M. (1979). Striate cortex and visual acuity functions in the cat. *Journal of Comparative Neurology,* **187,** 679–702.

Berkley, M. A., Warmath, D. S., & Tunkl, J. E. (1978). Movement discrimination capacities in the cat. *Journal of Comparative and Physiological Psychology,* **92,** 463–473.

Bisti, S., Maffei, L., Clement, R., & Mecacci, L. (1977). Spatial frequency and orientation tuning curves of visual neurones in the cat: Effects of mean luminance. *Experimental Brain Research,* **27,** 335–345.

Bjorklund, R., & Magnussen, S. (1981). A study of interocular transfer of spatial adaptation. *Perception,* **10,** 511–518.

Blake, R., & Fox, R. (1972). Interocular transfer of adaptation to spatial frequency during retinal ischaemia. *Nature (London), New Biology,* **240,** 76–77.

Blake, R., & Martens, W. (1981). Critical bands in cat spatial vision. *Journal of Physiology (London),* **314,** 175–187.

Blakemore, C., & Campbell, F. W. (1969). On the existence of neurones in the human visual system selectively sensitive to the orientation and size of retinal images. *Journal of Physiology (London),* **203,** 237.

Blakemore, C., & Nachmias, J. (1971). The orientation specificity of two visual aftereffects. *Journal of Physiology (London),* **213,** 157–174.

Blakemore, C., Nachmias, J., & Sutton, P. (1970). The perceived spatial frequency shift: evidence for frequency-selective neurones in the human brain. *Journal of Physiology (London),* **210,** 727–750.

Bloom, M., & Berkley, M. A. (1977). Visual acuity and the near point of accommodation in the cat. *Vision Research,* **17,** 723–730.

Bodinger, D. M. (1978). The decay of grating adaptation. *Vision Research,* **18,** 89–91.

Bonds, A. B. (1984). Spatial adaptation of the cortical visual evoked potential of the cat. *Investigative Ophthalmology & Vision Science,* **25,** 640–646.

Braddick, O., Campbell, F., & Atkinson, J. (1978). Channels in vision: Basic aspects. In R. Held, H. Leibowitz, & H.-L. Teuber, (Eds.), *Handbook of sensory physiology* (Vol. 8, pp. 3–38). New York: Springer.

Brindley, G. S. (1970). *Physiology of the retina and visual pathway: Monograph of the Physiological Society* (2nd ed., pp. 132ff.). London: Edward Arnold.

Campbell, F. W., Cooper, G. F., & Enroth-Cugell, C. (1969). The spatial selectivity of the visual cells of the cat. *Journal of Physiology (London)*, **203**, 223–235.

Campbell, F. W., & Kulikowski, J. (1966). Orientation selectivity of the human visual system. *Journal of Physiology (London)*, **187**, 437–445.

Campbell, F. W., & Maffei, L. (1970). Electrophysiological evidence for the existence of orientation and size detectors in the human visual system. *Journal of Physiology (London)*, **207**, 635–652.

Campbell, F. W. & Robson, J. (1968). Application of Fourier analysis to the visibility of gratings. *Journal of Physiology (London)*, **197**, 551–556.

Campbell, F. W., Cleland, B., Cooper, G. F., & Enroth-Cugell, C. (1968). The angular selectivity of visual cortical cells to moving gratings. *Journal of Physiology (London)*, **198**, 237–250.

Clarke, P. (1974). Are visual evoked potentials to motion-reversal produced by direction-sensitive brain mechanisms. *Vision Research*, **14**, 1281–1284.

Daughman, J. G. (1983). *Dynamics of spatial channel adaptation*. Unpublished doctoral dissertation, Harvard University, Campbridge, MA.

Day, R. (1958). On interocular transfer and the central origin of visual after-effects. *American Journal of Psychology*, **71**, 784–789.

DeValois, R. L., Albrecht, D. G., & Thorell, L. G. (1982). Spatial frequency selectivity of cells in macaque visual cortex. *Vision Research*, **22**, 545–559.

Gilinsky, A., & Doherty, R. (1969). Interocular transfer of orientation effects. *Science*, **164**, 454–455.

Hammer, E. R. (1949). Temporal factors in figural after-effects. *American Journal of Psychology*, **62**, 337–354.

Holub, R. A., & Morton-Gibson, M. (1981). Responses of visual cortical neurons of the cat to moving sinusoidal gratings: Response-contrast functions and spatiotemporal interactions. *Journal of Neurophysiology*, **46**, 1244–1259.

Hubel, D., & Wiesel, T. (1959). Receptive fields of single neurones in the cat's striate cortex. *Journal of Physiology (London)* **148**, 574–591.

Hubel, D., & Wiesel, T. (1962). Receptive fields, binocular interaction and functional architecture in the cat's visual cortex. *Journal of Physiology (London)* **160**, 106–154.

Hubel, D., & Wiesel, T. N. (1968). Receptive fields and functional architecture of monkey striate cortex. *Journal of Physiology (London)* **195**, 215–243.

Hubel, D., & Wiesel, T. N. (1977). Functional architecture of macaque monkey visual cortex. *Proceedings of the Royal Society of London, Series B*, **198**, 1–59.

Keck, M. J., Palella, T. D., & Pantle, A. (1976). Motion aftereffect as a function of the contrast of sinusoidal gratings. *Vision Research*, **16**, 187–191.

Keck, M. J., & Pentz, B. (1977). Recovery from adaptation to moving gratings. *Perception*, **6**, 719–725.

Kulikowski, J., Abadi, R., & King-Smith, P. (1973). Orientation selectivity of grating and line detectors in human vision. *Vision Research*, **13**, 1479–1486.

Loop, M., & Berkley, M. A. (1972). Conditioned suppression as a psychophysical technique for the cat. *Behavioral Research Methods & Instrumentation*, **4**, 121–124.

Maffei, L. (1978). Spatial frequency channels: Neural mechanisms. In R. Held, H.

Leibowitz, & H.-L. Teuber (Eds.), *Handbook of sensory physiology* Vol. 8, pp. 39–66). New York: Springer.

Maffei, L., & Fiorentini, A. (1973). The visual cortex as a spatial frequency analyzer. *Vision Research,* **13,** 1255–1267.

Maffei, L., Fiorentini, A., & Bisti, S. (1973). Neural correlate of perceptual adaptation to gratings. *Science,* **182,** 1036–1038.

Maguire, W. M., Meyer, G. E., & Baizer, J. S. (1980). The McCollough effect in rhesus monkey. *Investigative Ophthalmology & Vision Science,* **19,** 321–324.

Mansfield, J., & Simmons, L. K. (1979). Intrinsic processing in the visual cortex of primates. *Neuroscience Abstracts,* **5,** 795.

Marg, E., Adams, J., & Rutkin, B. (1968). Receptive fields of cells in the human visual cortex. *Experientia,* **24,** 348–350.

Mecacci, L., & Spinelli, D. (1976). The effects of spatial frequency adaptation on human evoked potentials. *Vision Research,* **16,** 477–479.

Mitchell, D. E., & Ware, C. (1974). Interocular transfer of a visual after-effect in normal and stereoblind humans. *Journal of Physiology (London),* **236,** 707–721.

Movshon, J. A., & Lennie, P. (1979). Pattern-selective adaptation in visual cortical neurones. *Nature (London),* **278,** 850–852.

Movshon, J. A., Thompson, I. D., & Tolhurst, D. J. (1978). Spatial and temporal contrast sensitivity of neurones in areas 17 and 18 of the cat's visual cortex. *Journal of Physiology (London),* **283,** 101–120.

Ochs, A. L., & Aminoff, M. J. (1980). The effect of adaptation to the stimulating pattern on the latency and wave form of visual evoked potentials. *Electroencephalography and Clinical Neurophysiology,* **48,** 502–508.

Ohzawa, I., Sclar, G., & Freeman, R. D. (1982). Contrast gain control in the cat visual cortex. *Nature (London),* **298,** 266–268.

Ohzawa, I., Sclar, G., & Freeman, R. D. (1985). Contrast gain control in the cat's visual system. *Journal of Neurophysiology,* **54,** 651–667.

Pantle, A., & Sekuler, R. (1969). Contrast response of human visual mechanisms sensitive to orientation and direction of motion. *Vision Research,* **9,** 397–406.

Rose, D., & Blakemore, C. (1974). An analysis of orientation selectivity in the cat's visual cortex. *Experimental Brain Research,* **20,** 1–17.

Rose, D., & Lowe, I. (1982). Dynamics of adaptation to contrast. *Perception,* **11,** 505–528.

Sachs, M., Nachmias, J., & Robson, J. (1971). Spatial frequency channels in human vision. *Journal of the Optical Society of America,* **61,** 1176–1186.

Scott, T. R., & Powell, D. A. (1963). Measurement of a visual motion aftereffect in the rhesus monkey. *Science,* **140,** 57–59.

Sekuler, R. (1974). Spatial vision. *Annual Review of Psychology,* **25,** 195–232.

Sekuler, R., & Ganz, L. (1963). A new aftereffect of seen movement with a stabilized retinal image. *Science,* **139,** 419–420.

Sekuler, R., Rubin, E. L., & Cushman, W. H. (1968). Selectivities of human visual mechanisms for direction of movement and contour orientation. *Journal of the Optical Society of America,* **58,** 1146–1150.

Srinivasan, M. V., & Dvorak, D. R. (1979). The waterfall illusion in an insect visual system. *Vision Research, 19,* 1435–1437.

Stebbins, W. C. (1966). Auditory reaction time and the derivation of equal loudness contours for the monkey. *Journal of Experimental Analysis of Behavior, 9,* 135–142.

Stromeyer, C. F., III, & Julesz, B. (1972). Spatial-frequency masking in vision: Critical bands and spread of masking. *Journal of the Optical Society of America, 62,* 1221–1232.

Tunkl, J., & Berkley, M. A. (1977). The role of superior colliculus in vision: Visual form discriminations in cats with superior colliculus ablations. *Journal of Comparative Neurology, 176,* 575–588.

Tyler, C. W., & Kaitz, M. (1977). Movement adaptation in the visual evoked response. *Experimental Brain Research, 27,* 203–209.

Vautin, R. G., & Berkley, M. A. (1977). Responses of single cells in cat visual cortex to prolonged stimulus movement. *Journal of Neurophysiology, 40,* 1051–1065.

von der Heydt, R., Hanny, P., & Adorjani, C. (1978). Movement aftereffects in the visual cortex. *Archives Italiennes de Biologie, 116*(3–4), 248–254.

Watkins, D., & Berkley, M. (1974). The orientation selectivity of single neurons in cat striate cortex. *Experimental Brain Research, 19,* 433–446.

Woods, E. J., & Frost, B. J. (1977). Adaptation and habituation characteristics of tectal neurons in the pigeon. *Experimental Brain Research, 27,* 347–354.

# 5

# PERCEPTION OF DRUG EFFECTS

*Robert L. Balster*

*Department of Pharmacology and Toxicology, Medical College of Virginia, Virginia Commonwealth University, Richmond, Virginia*

Most readers will have had the opportunity to personally experience the effects of one or more psychoactive drugs. Those who have will undoubtedly agree that these effects can readily be perceived as states of intoxication. These perceptible drug effects are the subject of this chapter. Considerable laboratory research in psychopharmacology has been devoted to the study of perceived drug effects. Indeed, human research subjects can be explicitly trained to detect when they have received an active drug or placebo and correctly report their perception. Typical discrimination training procedures can be used. For example, in a study by Chait, Uhlenhuth, and Johanson (1984), volunteer subjects were trained to discriminate the effects of 10 mg of *d*-amphetamine from placebo. Subjects reported to the laboratory three days a week and were told that their task was to learn to discriminate between drug A and drug B. They were not told what the two drugs were and the capsules were identical in appearance. For some subjects drug A was *d*-amphetamine and drug B placebo; for others drug B was *d*-amphetamine and drug A placebo. For the first few sessions they were given both capsules, one each day, told it was drug A or drug B, and told to associate the effects with the name. During the subsequent six sessions, subjects were given drug A three times and drug B three times, but were not told which they had received. Six hours after each ingestion they phoned the laboratory and reported which drug they thought they had been given, A or B. If correct, they were told so and were credited $3. If incorrect, they were told so and only given $1. Thus the subjects were being trained to discriminate active drug from placebo. Many of the subjects learned the discrimination as evidenced by correct responses on at least 5 of the 6 training sessions.

Once subjects had been trained to discriminate 10 mg *d*-amphetamine from placebo using this procedure, they could be tested with other drug doses. On different days, they were given placebo, 2.5, 5, and 10 mg of *d*-amphetamine and were told to report their identification in the normal manner. The results are shown in Fig. 1. Placebo was identified as drug on only 10% of the occasions, whereas increasing doses of *d*-amphetamine resulted in a dose-dependent report of having received drug. With 10 mg, they were 100% correct.

This experiment exemplifies the use of explicit discrimination training as a means of allowing the laboratory investigation of perceived drug effects. One of the important aspects of drug discrimination training is that it can be carried out in nonhuman subjects. In fact, the training procedure just described was based closely upon a far more extensive research literature on the perceptual effects of drugs in animal subjects. The purpose of this chapter is to introduce the reader to the use of drug discrimination as a means of examining the perception of drug effects. A very substantial research literature has been devoted to this area, with over 600 publications through 1983 contained in a cumulative bibliography on the subject (Stolerman, Baldy, & Shine, 1982; Stolerman & Shine, 1985). For reasons that will become apparent, drug-discrimination research has been of

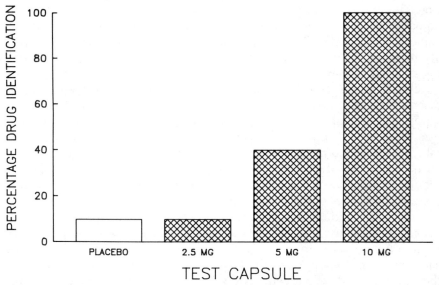

**FIGURE 1.** Results of tests with placebo capsules and various doses of *d*-amphetamine in four human subjects trained to discriminate 10 mg *d*-amphetamine from placebo. Subjects received each test capsule twice. Shown is the percentage of times subjects reported that they had received the capsule that had been associated with 10 mg *d*-amphetamine during training. Adapted from Chait, Uhlenhuth, and Johanson (1984). Used with permission.

considerable interest to pharmacologists and substance-abuse scientists. There has been less interest in these phenomena as basic perceptual processes; however, it is hoped that this introductory review may result in increased attention to this research literature by those interested in comparative perception.

## I. METHODS OF STUDYING DRUG DISCRIMINATION IN ANIMALS

Nonhuman subjects cannot make phone calls to the laboratory to indicate whether they had received drug or placebo administration; however, they are perfectly able to report the presence or absence of interoceptive stimuli by differential responding. The two approaches that have been most widely used are lever selection in a two-lever operant task and the T maze. The former will be used to illustrate the procedure. Illustrative data will be presented from our research with phencyclidine, although perception of the effects of nearly any psychoactive drug can be studied using this general approach. Phencyclidine has prominent central nervous system effects and has been widely abused ("angel dust"). We have used rats and monkeys for our studies, although pigeons and other animal species are widely used as well.

Subjects are trained to lever press under intermittent schedules of reinforcement for food presentation in two-lever operant chambers. On any given session, responding on only one lever is reinforced. To select the reinforced lever for each session, the subject must detect whether a drug or vehicle placebo injection was given prior to the session. Training is accomplished by pairing one of the levers with a drug injection and the other with a vehicle injection. On days when active drug is administered, responding on only the drug-associated lever is reinforced. On days when vehicle is administered, responding on only the vehicle-associated lever is reinforced. In all other respects, drug and vehicle training sessions are identical. Sessions generally last 30–60 min with correct responding reinforced under fixed-ratio schedules. It is common practice to provide multiple opportunities to reinforce correct lever selection in each training session, although this may not be necessary (Tomie, Loukas, Stafford, Peoples, & Wagner, 1985).

Typical acquisition data are illustrated in Fig. 2. For these data, 10 rats were trained to discriminate 3.0 mg/kg phencyclidine from saline vehicle. Training sessions were conducted daily (M–F) and every third session began with a 2-min test period for stimulus control during which responding on either lever was reinforced. As shown in the figure, the mean percentage of correct lever responses during these test periods gradually increased, reaching about 90% after either phencyclidine or vehicle injections by the tenth test. In this particular study the average number of training sessions for acquisition was 49.2. This amount of training is fairly

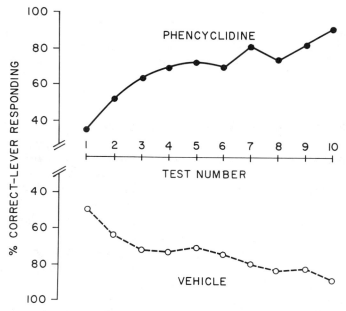

**FIGURE 2.** Acquisition of a discrimination between phencyclidine and vehicle. Ten rats were trained to press one lever after ip injections of 3 mg/kg phencyclidine and another lever after vehicle injections. Data are from tests for stimulus control conducted before every third training session. Tests with phencyclidine and vehicle injections alternated. Values are the mean percentage of responses on the lever associated with each injection during training.

typical for this type of research, although various procedural variations have been used to facilitate training (Harris, Wood, Lal, & Emmett-Oglesby, 1987; Overton & Hayes, 1984).

It should be apparent that training a drug discrimination is similar to multiple-schedule discrimination training procedures using more conventional stimuli such as lights or tones, where responding on only one lever is correct in each component. An important difference is that the components of the schedule alternate on a daily basis rather than within sessions to accommodate the protracted nature of the drug stimulus. Indeed, the temporal characteristics of the drug stimulus can be studied by conducting tests for stimulus control at various times after drug administration. Data showing the time course for the discriminative stimulus effects of phencyclidine are shown in Fig. 3. A 5-min test session conducted immediately after drug injection resulted in vehicle-lever selection. However, predominately phencyclidine-lever selection occurred during a 5-min test session conducted 10 min after injection. After 60 min some control over phencyclidine-lever responding is still apparent, with a return to predominately vehicle-lever selection 2 hr after phencyclidine administration. This presumably represents the time course for drug

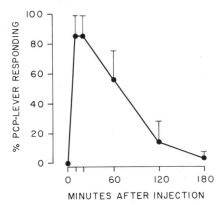

**FIGURE 3.**  Time course for the discriminative stimulus effects of phencyclidine. Five-minute tests for stimulus control were conducted at various times after ip injection of 2.0 mg/kg phencyclidine in 7 rats trained to discriminate this dose from vehicle given 15 min presession. Values are the mean (± SE) percentage of responses on the lever associated with phencyclidine (PCP) during training.

absorption distribution and subsequent elimination. The gradual onset and offset and long duration of drug stimuli represents an important difference from more traditional stimuli.

It should be obvious that experimental control of the important parameters of drug stimuli is difficult. The onset and offset is gradual, with the intensity necessarily changing over time. Stebbins, Smith, and Moody (1988) have argued that the relatively poor control that the experimenter has over the parameters of drug stimuli compared to exteroceptive stimuli such as lights and tones may make it very difficult to perform psychophysical research with drug stimuli; however, the differences are probably only one of degree. Technical means can probably be found to improve the relation between drug administration and stimulus control, such as using intravenous, inhalation, or intracerebral routes of administration. The problems of studying drug stimuli are similar to those of studying other interoceptive sensory systems such as proprioception or kinesthesis or for studying the chemical senses (Balster, 1988).

## II. GENERALIZATION

As in other areas of animal psychophysics, generalization testing plays a prominent role in drug-discrimination research. Two aspects of drug stimuli can be more or less systematically altered. Doses other than the dose the subjects were trained with can be tested. This is probably analogous to the exploration of stimulus intensity. In addition, drugs other than the training drug can be tested. This can be viewed as a change in the quality of the stimulus. It will be useful to examine the results of generalization experiments in some detail, but first some description of how generalization testing is carried out will be provided.

Lever selection during drug-discrimination training is under the control of two sets of stimuli. One is the effect of the drug or vehicle injection given before each session; the other is the information provided by meeting the reinforcement contingencies on the correct lever. Since in a typical drug-discrimination training session the same lever will remain correct for an entire training session, subjects could adopt a "win–stay, lose–shift" strategy to select the correct lever. Thus, overall lever-selection data may not be a clear measure of drug stimulus control when responding on only one lever is reinforced. Therefore, tests for drug stimulus control need to be carried out under conditions where reinforcement feedback cannot help determine lever selection. Many investigators use the first lever selected each session prior to the delivery of the first reinforcer as a measure of drug stimulus control. Others conduct periodic generalization test sessions. In these test sessions, there is no differential reinforcement for lever selection. Either responding on both levers is reinforced or tests are conducted during extinction, although the choice of which approach to use is not without consequences (Kaempf & Kallman, 1987).

To conduct a series of generalization tests in a group of trained subjects, test sessions are generally interspersed with continued training sessions. For most of our research, we conduct generalization test sessions on Tuesdays and Fridays with continued training on the remaining days of the week. Thus, if only one value of a test stimulus can be evaluated each test session, generalization test data can only slowly be obtained. This is another important difference between the types of procedures used to study drug stimuli and exteroceptive stimuli such as lights and tones, where multiple presentations of many different values of the stimulus can be achieved in a single test session. To reduce this problem somewhat, some investigators have developed generalization test procedures that allow for testing an ascending series of doses in a single test session (Bertalmio, Herling, Hampton, Winger, & Woods, 1982; Sannerud & Young, 1987). The problems of generalization testing with drug stimuli may be more similar to problems studying olfactory and gustatory stimuli.

## A. Quantitative Generalization

Manipulation of stimulus intensity can be carried out by varying the dose of the drug administered before generalization test sessions. When a large enough range of doses is tested, a generalization gradient results. A typical drug dose generalization gradient (i.e., a dose–effect curve) for phencyclidine is shown in Fig. 4. This is a typical manner of data presentation in drug discrimination research. The measure of stimulus control is the

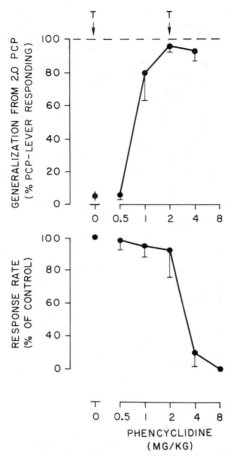

**FIGURE 4.** Generalization gradient for different ip doses of phencyclidine in rats trained to discriminate injections of 2 mg/kg phencyclidine from vehicle. Training stimuli are indicated at "T". The upper panel shows the mean percentage of responses on the lever that had been paired with 2 mg/kg phencyclidine during training as a function of test dose. The lower panel shows the overall rates of responding on both levers, expressed as the percentage of response rates after vehicle injections (0 mg/kg). Values are the means (± SE) of data from test sessions conducted with five rats.

percentage of responses on the lever paired with drug injection during training. Data were obtained from test sessions in which responding on either lever was reinforced with food presentation. At the dose of phencyclidine used as the training stimulus (2 mg/kg, ip), an average of over 95% of the responses occurred on the phencyclidine lever. When the vehicle was tested, about 5% of the responses occurred on the phencyclidine lever, that is, 95% of the responses occurred on the vehicle lever. By pharmacological tradition and because of some assumptions about the underlying quantitative basis of drug–receptor interactions (Goldstein, Aronow, &

Kalman, 1974), doses are usually plotted on a logarithmic scale. With a dose of 1 mg/kg phencyclidine, one-half the dose with which the subjects were trained, an average of about 80% of the responses occurred on the phencyclidine lever indicating that substantial generalization occurred to this lower intensity stimulus. Generalization gradients obtained with intermediate drug doses often take a sigmoid shape, although the shape of the curve at higher doses than the training dose is not well established. There are some conditions where a peak shift occurs, where doses higher than the training dose result in even higher levels of drug-lever selection. An example of this is shown in the gradient for phencyclidine included in Fig. 5. In other cases, doses higher than the training dose result in less drug-lever selection.

A number of variables in addition to drug stimulus control might enter into determining the shape of the generalization gradient, the chief of which is direct drug effects on the discriminative performance. The problem of direct drug effects on performance is somewhat unique to drug discrimination research and is exemplified in the data shown in the lower panel of Fig. 4. This panel shows the overall rates of responding on both levers during the generalization test sessions that generated the lever-selection data shown in the upper panel. What is shown is what would be expected; high doses of phencyclidine are seen to disrupt responding. Thus, two effects of drugs occur in generalization test sessions—their discriminative effects as revealed by lever-selection data and their direct effects on behavior as revealed in response rate data (or some other suitable measure, such as latency to respond). This is a problem that is not usually encountered in visual or auditory psychophysics, unless tone or light intensity was so great as to interfere with responding. A particularly troublesome aspect of separating discriminative and unconditioned direct effects of drugs is the possibility that one of the direct effects may be to interfere with discrimination behavior itself. Drug disruption of stimulus control has been well documented (Appel & Dykstra, 1977). These disruptive effects of drugs at high doses may be responsible for the biphasic nature of the generalization gradient obtained in some dose–effect studies. The animals' behavior may be sufficiently disrupted to interfere with lever selection, resulting in a more even distribution of responding at high doses of the drug.

An interesting method is available for increasing the separation of discriminative and direct response rate effects of drugs. Animals can be trained to detect successively lower doses of drug by gradually lowering the training dose over a number of sessions. For example, in a recent study Beardsley, Balster, and Salay (1987) initially trained rats to discriminate 3.0 mg/kg phencyclidine from vehicle. After the discrimination was learned and a drug dose generalization gradient obtained, the training dose was gradually lowered by nearly 10-fold to 0.375 mg/kg. When redetermined, the generalization gradient was shifted to the left about

10-fold. On the other hand, response rate effects of phencyclidine remained the same. Thus, at the end of training, stimulus control of lever selection occurred at doses of phencyclidine about 20 times lower than those that modified rates of responding. Indeed, fading procedures such as these can result in exquisite sensitivity to psychoactive drugs (Overton, 1979; F.J. White & Appel, 1982).

As will be discussed more fully later, the likely proximal event in drug stimulus control is the cellular action of the drug at some critical areas of the central nervous system. Thus, the parameter that is probably most important in determining drug stimulus intensity is the concentration of the active form of the drug at these sites of action. There is at best an imperfect relation between administered dose and concentration at sites of action. It should also be apparent that proximal stimulus intensity is in a constant state of variation, as the drug is absorbed, distributed, metabolized, and excreted. Many factors enter into the biodisposition of drugs to sites of action in the brain, not the least of which is route of administration. Some routes, for example, oral dosing, result in notoriously more variable drug levels at sites of action than routes such as intravenous administration. A complete knowledge of these dispositional factors is essential if a quantitative psychophysical study of drug stimuli is to be undertaken. Given the difficulty of directly manipulating drug concentrations at sites of action, it is not surprising that quantitative data on drug stimuli are difficult to obtain. On the other hand, it should be pointed out that investigators in drug discrimination have paid relatively little attention to quantitative psychophysical studies of drug stimuli. For example, methods for determining drug stimulus intensity thresholds have not been systematically evaluated.

## B. Qualitative Generalization

In addition to testing different doses of the training drug, different drugs can be tested for stimulus generalization from the training drug. Unlike the case for quantitative generalization, the underlying dimension in qualitative drug generalization is much less clear. Presumably, it is chemical structure. The problem is similar to studying the chemical senses. What are the relevant structural modifications of molecules that influence their smell and taste? Considerable research will be needed to understand the physical basis of drug stimuli. However, for some classes of drug stimuli, quite a bit is already known about the structure–activity relations for discriminative stimulus effects.

The results of qualitative generalization tests provide important insights into the basis of the perception of drug effects that result from discrimination training. One might have expected that animals trained to discriminate drug from nondrug might base the discrimination simply on the presence or absence of an intoxicating state, and that many different types of psychoactive drug stimuli would result in drug-lever responding. This is clearly not the case. In general, generalization testing with other drugs results in remarkable pharmacological specificity of drug stimulus control. Figure 5 provides an example of a failure to show cross-drug generalization. These data were obtained in monkeys trained to discriminate phencyclidine from saline vehicle as was described earlier, then tested with three doses of two other psychoactive drugs, morphine and mescaline. At no dose of either drug did the subjects respond on the drug lever to any appreciable extent, indicating a failure of these drugs to be generalized

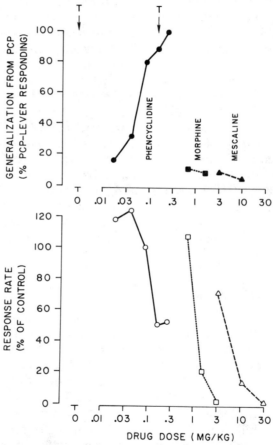

**FIGURE 5.** Results of generalization tests with various im doses of phencyclidine, morphine, and mescaline tested in squirrel monkeys trained to discriminate injections of phencyclidine from vehicle. Training stimuli are indicated at "T". The upper panel shows the mean percentage of responses on the lever that had been paired with 0.16 mg/kg phencyclidine during training as a function of test dose of each drug. The lower panel shows the overall rates of responding on both levers, expressed as the percentage of response rates after vehicle injections (0 mg/kg). Values are the means of data from test sessions conducted with four monkeys.

from the phencyclidine stimulus. It is clear that an adequate dose range of both drugs was tested since at high doses responding was markedly disrupted.

What is perhaps most interesting about the data depicted in Fig. 5 is that both these drugs were generalized from the vehicle stimulus, since responding occurred almost exclusively on the vehicle lever. One might have expected that mescaline and morphine would be perceived as similar to neither phencyclidine nor vehicle, a situation that might result in indiscriminate responding on both levers or some other variable result. The result shown here with morphine and mescaline are very typical of results when pharmacologically unrelated, yet clearly psychoactive, drugs are tested for generalization. Subjects often respond as if they had received vehicle. Mescaline and morphine are not the only psychoactive drugs that are not generalized from phencyclidine. Table 1 provides a partial list of a

**TABLE 1. Selected Drugs Not Generalized from Phencyclidine in Drug Discrimination Studies**[a]

| *Opioids* | *Depressants* |
|---|---|
| Morphine | Pentobarbital |
| Nalorphine | Methaqualone |
| Meperidine | Diazepam |
| Pentazocine | Althesin |
| Nalbuphine | |
| Ketocyclazocine | *Neuroleptics* |
| Ethylketocyclazocine | Chlorpromazine |
| Butorphanol | Thiothixine |
| Dextrometorphan | |
| | *Cholinergics* |
| *Hallucinogens* | Arecoline |
| LSD | Atropine |
| MDA | Benztropine |
| THC | Ditran |
| Quipazine | Mecamylamine |
| | Nicotine |
| *Dopaminergics* | Physostigmine |
| d-Amphetamine | Scopolamine |
| Methylphenidate | |
| Cocaine | *Others* |
| Amantadine | Propranolol |
| Apomorphine | Cyproheptadine |
| | Muscimol |
| | Yohimbine |
| | Veratradine |

[a]From Poling, White, and Appel (1979), Holtzman (1980), Browne, Welch, Kozlowski, and Duthu (1983), and Shannon (1983).

number of others showing the clear pharmacological specificity of the phencyclidine stimulus. In nearly every case, tests with these drugs resulted in responding almost exclusively on the vehicle lever.

The abundance of data showing predominately vehicle-lever responding when cross-drug generalization tests are conducted has led a number of investigators to suggest that what is being learned in a drug versus nondrug discrimination is really the presence versus the absence of a specific constellation of interoceptive stimuli produced by the training drug (Overton, Merkle, & Hayes, 1983; Swedberg & Järbe, 1986). An analogy with the perception of exteroceptive stimuli may help clarify why there is a lack of generalization among many drug stimuli. If animals were trained to discriminate a tone from no tone and then generalization tests with various colors of lights were conducted, the presentation of lights without the tone would result in no-tone responding since there is no basis for forming a generalization between sensory modalities. Generalization tests with drugs pharmacologically different from the drug training stimulus produce results similar to the results of testing lights in this example. Stated in a different way, animals trained to discriminate phencyclidine from vehicle when tested with morphine are reporting that they did not receive phencyclidine, not that they received vehicle. Thus, it could be concluded that there is no generalization gradient among the many classes of drug stimuli. This conceptualization of the nature of drug stimuli has many important implications for understanding the results of drug discrimination research.

Although most psychoactive drugs are not generalized from phencyclidine, some do have the ability to substitute completely for the phencyclidine stimulus. An example of data of this type is shown in Fig. 6. These data were obtained from squirrel monkeys trained to discriminate phencyclidine from saline vehicle (Brady & Balster, 1981). The test drugs were five structural analogues of phencyclidine, including ketamine (KET) which is used clincally as a dissociative anesthetic. Generalization gradients for phencyclidine and the analogs are shown. It can be seen that, although potency differences are apparent, all five analogues completely substitute for the phencyclidine stimulus and that the shapes and slopes of the generalization gradients are also similar. For example, after the administration of 2.0 mg/kg ketamine, responding occurred exclusively on the phencyclidine lever. Thus, these monkeys could be said to have perceived the effects of ketamine as being similar to those of phencyclidine. Perhaps this should not be surprising since the pharmacological profile of phencyclidine and ketamine are very similar (Chen, Ensor, & Bohner, 1969) and they produce similar subjective effects when given to humans (Ghoneim, Hinrichs, Mewaldt, & Petersen, 1985; Pollard, Uhr, & Stern, 1965). A large number of other arylcycloalkylamine analogs of phencyclidine have been evaluated for generalization from phencyclidine with the result that the structural requirements for producing phencyclidine-like discrimina-

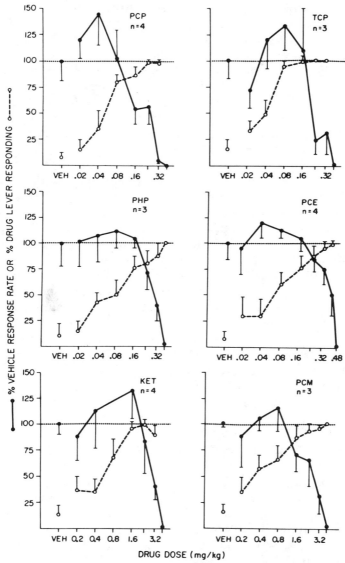

**FIGURE 6.** Results of generalization tests with various im doses of phencyclidine (PCP) and chemical analogues of phencyclidine in squirrel monkeys trained to discriminate injections of phencyclidine from vehicle. Dashed lines and open symbols show the mean percentage of responses on the lever that had been paired with 0.16 mg/kg phencyclidine during training. Solid lines and symbols show the overall rates of responding on both levers, expressed as a percentage of response rates after vehicle (VEH) injections. Values are the means (± SE) of data from one or two test sessions conducted with each dose in four monkeys. From Brady and Balster (1981) and used with permission. See orginal article for chemical names of PCP analogues indicated by the abbreviations TCP, PHP, PCE, KET, and PCM.

tive stimulus effects are quite well established (Brady & Balster, 1981; Cone, McQuinn, & Shannon, 1984; Kozlowski, Browne, & Vinick, 1986; Shannon, 1981; Solomon, Herling, Domino, & Woods, 1982).

Although this is an oversimplification of the results of hundreds of cross-drug generalization studies, it has usually been found that generalization occurs among drugs with similar central nervous system pharmacological effects. That is, drug classification by this method generally follows classification schemes found in central nervous system pharmacology textbooks (Barry, 1974). For example, cross-generalization usually occurs among amphetamine, cocaine, and other sympathomimetic stimulants (Huang & Ho, 1974; R. Young & Glennon, 1986), among barbiturates and other classical depressants, such as alcohol, volatile anesthetics, and the benzodiazepines (Barry & Krimmer, 1977; Colpaert, 1977; Overton, 1966; D. C. Rees, Knisely, Balster, Jordan, & Breen, 1987; York & Bush, 1982), and among classical morphine-like opiates (Colpaert, 1978; Herling & Woods, 1981; Holtzman, 1985). Indeed, the pharmacological specificity of opioid stimuli is particularly impressive. Using primarily the results of cross-drug generalization studies, subtypes of opioid drugs can be identified with qualitatively different discriminative stimulus effects, and these subtypes may act upon subtypes of opiate receptors identified biochemically (Herling & Woods, 1981; Holtzman, 1985). In fact, one of these subtypes, referred to as $\sigma$-agonist opioids, produce psychotomimetic effects in humans. These drugs share discriminative stimulus effects with phencyclidine (Brady, Balster, & May, 1982; Shannon, 1981) and have been found to have a common cellular site of action as phencyclidine (Balster & Willetts, 1988; Johnson, 1987).

## III. DISCRIMINATION BETWEEN DRUGS

Although considerable pharmacological specificity of drug stimuli can be demonstrated by cross-drug generalization studies, an even greater separation of drug stimulus effects can be achieved by explicitly training animals to discriminate between drugs. An example has been the research comparing the discriminative stimulus properties of depressant drugs. In animals trained to discriminate either ethanol, a barbiturate, or a benzodiazepine from vehicle, cross-generalization is often obtained among these subtypes of depressants (Barry & Krimmer, 1977; Colpaert, 1977; Overton, 1966). On the other hand, Krimmer and Barry (1979), using a typical two-lever drug-discrimination procedure, could train rats to discriminate between phenobarbital and chlordiazepoxide. Interestingly, when these animals were tested with vehicle injections, many did not respond at all. For this reason, investigators often train three-lever discriminations to include a vehicle-associated lever (e.g., J. M. White &

Holtzman, 1981). Under these conditions, drugs that are unlike either of the training drugs generally result in vehicle-lever selection. Overton (1982) has used an interesting variation on the drug versus drug training procedure to increase the specificity of drug stimuli. During training, responding on one lever was always reinforced after phenobarbital administration. Responding on the other lever was reinforced after the administration of any one of a number of other drugs or vehicle. For example, one group of rats were trained to discriminate between phenobarbital and either ethanol, chlordiazepoxide, ketamine, or vehicle. In other words, subjects were required to discriminate phenobarbital from "other" drugs. Although it took quite a lot of training for the subjects to acquire this discrimination, considerably greater specificity of the phenobarbital stimulus was observed in these animals compared with subjects trained only to discriminate phenobarbital from vehicle.

It should be apparent that there is a considerable qualitative variety of drug stimuli, and that procedures are available to differentiate them even further. If one accepts that these studies of the perception of drug effects by animals is a means of studying properties of drugs relevant to their production of subjective effects in humans, this should not be surprising. There is a wide variety of profiles of subjective effects produced by psychoactive drugs in humans and experienced drug users are readily able to discriminate between classes of drugs, even without the benefit of explicit discrimination training in the laboratory (Haertzen, 1966; Jasinski, 1977). The use of explicit discrimination training to explore subtle differences in the perception of drug effects has only begun to be exploited to distinguish the perception of closely related drug effects.

## IV. DISCRIMINATIVE AND SUBJECTIVE EFFECTS OF DRUGS

It might be useful at this point to consider the evidence that the perception of the discriminative stimulus effects of drugs in animals represents a model of subjective effects in humans. The most striking evidence for a relation between animal perception of drug effects and human subjective experience is the similarity in the subjective effects among drugs that show cross-generalization in animal experiments. For example, nearly all morphine-like opioids that show complete cross-generalization in animals (Colpaert, 1978; Herling & Woods, 1981; Holtzman, 1985) also have morphine-like subjective effects in humans when tested under laboratory conditions (Jasinski, 1977). Even when drugs produce multiple effects, the pattern of cross-generalization results usually follows what would be expected from comparisons of human subjective effects. For example, phencyclidine has many biochemical and behavioral effects similar to those of amphetamine-like drugs (Balster, 1987; Johnson, 1987; H. Y. Meltzer,

Sturgeon, Simonovic, & Fessler, 1981), yet the subjective effects of phency-clidine bear little resemblance to those of amphetamine (Pollard et al., 1965). In drug discrimination studies, amphetamine does not generalize from phencyclidine (McMillan, 1982; Shannon, 1981).

Further evidence for the relation between discriminative stimulus effects and subjective effects of drugs comes from studies of both phenomena in human subjects. As described at the outset of this chapter, humans can be explicitly trained to discriminate drugs from placebo. In such studies, subjective effects are often measured simultaneously. Two approaches can be used to compare discriminative and subjective effects (Schuster & Johanson, 1988). It can be determined whether drugs that are cross-generalized show common profiles of subjective effects. Second, the covariation of individual differences in discriminative and subjective effects can be analyzed to determine the degree of concordance. In a series of studies in human subjects trained to discriminate $d$-amphetamine from placebo, Chait, Uhlenhuth, and Johanson (1986a, 1986b) found that generalization occurred from $d$-amphetamine to phenmetrazine, phenyl-propanolamine, and mazindol, but not to fenfluramine or diazepam. At the same time that the subjects reported whether they had received $d$-amphetamine or placebo (i.e., drug A or drug B), they completed paper and pencil tests designed to profile subjective drug effects. In general, there were many similarities in the subjective effects of $d$-amphetamine and phenmetrazine as measured by these scales and a reasonable concord-ance in the effects of $d$-amphetamine and those of phenylpropanolamine and mazindol. As would be expected, the effects of $d$-amphetamine and diazepam were the opposite on a number of the scales, and the doses of fenfluramine used were essentially without subjective effects. Chait et al. (1986a, 1986b) also compared the subjective effects of $d$-amphetamine in those subjects who learned the $d$-amphetamine versus placebo discrimina-tion and in those who did not. In general, subjects who could learn the discrimination had greater effects on scales measuring subjective effects. Taken together, these results are an indication that the discriminative stimulus effects of $d$-amphetamine are based upon similar aspects of drug action to those that account for subjective effects.

## V. TRANSDUCTION OF DRUG STIMULI

I will turn now to an examination of the mechanisms for the perceptual effects of drugs. Where and how are drug stimuli transduced? This question has been the subject of considerable research in neurobehavioral pharmacology, and a lot is known about the cellular bases for the discriminative stimulus effects of many classes of drugs (Colpaert & Balster, 1988).

## A. Anatomical Location of Transduction Sites

Studies attempting to localize the site of action for drug stimuli have generally found that the stimuli result from central nervous system pharmacological actions, not local irritant or other effects associated with drug administration itself. Perhaps the best evidence for this is that the time course of drug stimuli nearly always follows the known absorption, distribution, and elimination of the drug in the central nervous system, as shown, for example, in Fig. 2. If local irritant effects were responsible, drug stimuli should be strongest immediately after administration and should not require a period of absorption. In addition, generalization often occurs when the same drug is given by various routes of administration, where similar local effects would not be expected to occur. For example, animals trained to discriminate intraperitoneal injections of tetrahydrocannabinol or phencyclidine from vehicle generalize to these drugs administered by smoke inhalations (Järbe, Johansson, & Henriksson, 1976; Wessinger, Martin, & Balster, 1985), and cross-generalization has been shown to occur between subcutaneously and intravenously administered nicotine (Schechter, 1973). Similarly, mice trained to discriminate intraperitoneal injections of toluene from vehicle, respond on the toluene lever after toluene inhalation (Rees, Knisely, Jordan, & Balster, 1987).

In addition to producing local irritant or other sensory effects that might serve as the basis of the discrimination of a drug from placebo administration, most psychoactive drugs also produce effects on the autonomic nervous system resulting in peripheral physiological changes which could also be discriminated from placebo. This possibility has been evaluated in a number of studies. One approach for addressing this question is to conduct studies with drugs, such as quaternary derivatives, that do not readily penetrate the blood–brain barriers. They would be expected to have all the peripheral effects of their centrally acting counterparts but lack direct effects on the brain. For example, in animals trained to discriminate atropine from vehicle, quaternary methyl atropine fails to substitute for atropine (Kubena & Barry, 1969; Overton, 1966). Similarly, a comparison of the effects of antagonists with differing access to the central nervous system can provide information on site of action of an agonist. For example, the nicotine stimulus can be antagonized by the centrally active nicotinic–cholinergic antagonist mecamylamine but not by hexamethonium, a peripherally active nicotinic–cholinergic ganglionic blocker with poor penetration to the brain (Schechter & Rosecrans, 1971; Stolerman, Pratt, Garcha, Giardini, & Kumar, 1983). Studies with the muscarinic–cholinergic agonist arecoline also provide evidence for a central site of action for its discriminative stimulus effects. For example, the arecoline stimulus is antagonized by the centrally active muscarinic antagonist atropine sulfate, but not by quaternary methyl atropine (L. T. Meltzer &

Rosecrans, 1981a). Similarly, the discriminative stimulus effects of morphine are readily antagonized by the competitive opiate antagonist naltrexone, but not by its quaternary derivative (Locke & Holtzman, 1985; Valentino, Herling, Woods, Medzihradsky, & Merz, 1981).

Another approach for ruling out peripheral sites of action for drug stimuli is to study the effects of drugs administered directly into the central nervous system through indwelling cannuli. When this has been done in animals trained to discriminate systemically administered drugs, generalization often occurs when much smaller quantities of drugs are administered intraventricularly. For example, in rats trained to discriminate subcutaneous injections of morphine from vehicle, complete generalization from morphine occurred when nearly 1000-fold lower doses of morphine were administered unilaterally into the lateral ventricle (Locke & Holtzman, 1985; Shannon & Holtzman, 1977). In addition, intraventricular administration of small doses of naltrexone, either in its tertiary or quaternary form, completely antagonized the discriminative stimulus effects of subcutaneous morphine (Locke & Holtzman, 1985). Intraventricular administration of phencyclidine is also generalized from an intraperitoneal phencyclidine stimulus, although in this case the potency difference is not as large as with morphine (Slifer & Balster, 1985).

The evidence provided here, as well as from a number of other similar studies with other drug classes, clearly suggests that drug-discriminative stimuli are generally mediated by actions on the central nervous system. On the other hand, the necessary studies to establish this are not routinely done with all examples of drug stimuli. For some drugs, or in some experimental situations, peripheral effects may play an important role. For example, we have trained animals to discriminate between behaviorally active concentrations of inhaled toluene and air exposure. In this case, it appears that the olfactory stimuli accompanying exposure to toluene vapor are an important component of the toluene stimulus, even though under other circumstances the same concentrations of inhaled toluene can produce disciminative stimulus effects that appear to be centrally mediated (Rees, Coggeshall, & Balster, 1985; Rees, Knisely, Jordan, & Balster, 1987).

Can the site of transduction of drug stimuli be localized more precisely within the central nervous system? One approach has been to inject drugs directly into nuclei within the brain, and test for generalization from, or antagonism of, systemically administered drug stimuli. In general, this approach has not been very successful in identifying specific brain regions that are uniquely responsible for drug stimuli, although it does appear that the periaquaductal gray region may be important in morphine's discriminative stimulus effects (Shannon & Holtzman, 1977) and the dorsal hippocampus and the mesencephalic reticular formation have been implicated in nicotine discrimination (L. T. Meltzer & Rosecrans, 1981b).

## B. Biochemical Characterization of Mechanisms of Transduction

Perhaps it should not be surprising that the transduction mechanisms for drug stimuli would not be anatomically localized within the brain. Since systemically administered drugs are widely distributed within the central nervous system, their stimulus effects would be expected to arise from the net result of actions throughout the brain. On the other hand, the selective actions of psychoactive drugs on pathways within the brain served by specific neurotransmitter systems have been found to be important in transducing their stimulus effects. Indeed, studies of the biochemical mechanisms for the stimulus effects of drugs is a major front in drug discrimination research, and was the subject of a recent international workshop (Colpaert & Balster, 1988).

A number of approaches can be used to investigate biochemical actions of drugs that may mediate their discriminative stimulus effects (Colpaert & Slangen, 1982; Rosecrans & Glennon, 1979). In general these approaches have much in common with other research in neuropharmacology directed toward studying mechanisms of drug action. Among the strategies that have been employed are (1) structure–activity correlations between stimulus potency and biochemical actions, (2) studies of the stereoselectivity of drug stimulus effects, (3) studies of the correspondence between the discriminative stimulus effects of drugs and their other behavioral actions whose neural basis is better understood, (4) the use of pharmacologically selective agonists and antagonists to determine if they produce or block drug stimuli, (5) studies of cross-tolerance among drug stimuli, (6) the use of selective neurochemical lesions to determine if they modify the acquisition or retention of drug stimulus control, and (7) the use of other pharmacological tools such as selective ion channel blockers to determine if they modify drug stimuli.

It is beyond the scope of this chapter to review the findings of studies that have used these approaches to investigate transduction mechanisms of drug stimuli. In general, specific mechanisms have not been identified that are responsible for only the discriminative stimulus effects of drugs. The same biochemical processes that have been implicated in these effects, also play a role in other behavioral and pharmacological effects of drugs. In fact, one of the important goals of this area of research is to try and separate desirable therapeutic actions of psychoactive drugs from their ability to produce intoxicating effects. For example, evidence so far suggests that the same receptors responsible for the analgesic effects of opiates also mediate their discriminative stimulus effects (Dykstra, Bertalmio, & Woods, 1988). On the other hand, it is certainly possible that the receptors responsible for these disparate effects of opiates may prove to have subtle differences, and they may likely be found in different neural pathways, leaving open the possibility that effective analgesics without the perceptual effects of morphine could be developed.

## VI. IMPLICATIONS AND CONCLUSIONS

### A. Stimulus Properties of Drugs

Comparative studies of the perception of psychoactive drug effects have become an important aspect of research in behavioral pharmacology. The enormous growth in this area has resulted from the use of explicit training in the discrimination of drug stimuli. The procedures used have their bases in experimental psychology and the experimental analysis of behavior, and are operationally analogous to methods used to study more traditional exteroceptive stimuli such as lights and sounds. In this conceptualization, drugs serve discriminative stimulus functions in the control of behavior. Drugs can serve other stimulus functions as well (Thompson & Pickens, 1971). Drugs can function as unconditioned stimuli in a respondent conditioning paradigm. For example, environmental stimuli repeatedly paired with drug administration can come to elicit drug effects, or in some cases, conditioned compensatory effects. A dramatic example of this is the research of Goldberg and Schuster (1967) in which morphine-dependent rhesus monkeys trained to lever press for food reinforcement were given intravenous injections of the opiate antagonist naloxone paired with a tone. In these dependent animals, naloxone administration immediately resulted in withdrawal signs and a suppression of food-maintained responding. After several pairings, the tone alone produced conditioned withdrawal effects, including emesis, excessive salivation, and suppression of operant behavior. Laboratory studies of conditioned drug effects are an important area of behavioral pharmacology, with relevance to such phenomena as placebo effects, drug abuse relapse, and tolerance (O'Brien, Testa, O'Brien, Brady, & Wells, 1977; Siegel, 1982).

Another important stimulus property of drugs is their ability to function as reinforcing stimuli (Henningfield, Lukas, & Bigelow, 1986; A. M. Young & Herling, 1986). These properties can be studied in drug self-administration experiments. A typical arrangement would utilize an animal with an intravenous catheter connected to an infusion pump. Lever-pressing responses would activate the pump and deliver response-contingent drug infusions. Most drugs of abuse readily serve as reinforcers under these conditions (Johanson & Balster, 1978), providing an opportunity to study drug-taking behavior under carefully controlled laboratory conditions.

Thus, research on the perception of drug stimuli using drug discrimination procedures is just one aspect of an approach to studying the behavioral effects of drugs that conceptualizes drugs as having stimulus functions in stimulus–response theory. These stimulus effects can be operationally defined in a manner analogous to more traditional exteroceptive stimuli. This conceptualization has allowed behavioral pharmacologists to utilize the discoveries and advanced technology of the behavioral sciences to study the extremely complex problems raised by the use and abuse of psychoactive drugs.

## B. Abuse Potential of Drugs

Many people self-administer psychoactive drugs to produce a desired state of intoxication. This state of intoxication is perceived as subjective drug effects. Since drug discrimination procedures in animals are thought to yield information relevant to the subjective effects of drugs in humans, the potential of drugs to be abused can be examined using this approach. New psychoactive drugs are continually being developed for various medical uses resulting in a need to evaluate the likelihood that they might be abused. Often, it is important to obtain an estimate of the abuse potential of drugs before undertaking the extensive safety evaluation necessary to test them in humans. In other cases, carefully conducted animal research can complement the results of human testing to provide data for regulatory purposes.

A recent example has been the development of the new antianxiety drug buspirone. Since buspirone had antianxiety effects similar to those of the benzodiazepines and other central nervous system depressants such as the barbiturates, there was a concern that it might also have the abuse potential of these classes of drugs. A number of studies were carried out to investigate the abuse potential of buspirone in both animals and humans, including studies of its discriminative stimulus effects. To help answer the question of whether high doses of buspirone would produce perceptual effects similar to those of benzodiazepines and barbiturates, it was tested for generalization in animals trained to detect these drug stimuli (Hendry, Balster, & Rosecrans, 1983). In rats trained to discriminate the benzodiazepine oxazepam from vehicle, buspirone did not substitute for the oxazepam stimulus. Similarly, in animals trained to discriminate pentobarbital from vehicle, tests with buspirone failed to produce generalization from pentobarbital. Because of these and other data indicating that buspirone was unlikely to be abused, the Food and Drug Administration recently approved it for use in the United States without drug abuse controls.

Another even more current example is provided by studies of MK-801, a drug that may have some neuroprotective effects in cases of stroke or head injury (Foster, Gill, Kemp, & Woodruff, 1987). Because MK-801 had been shown to have biochemical and pharmacological effects in common with phencyclidine (Wong et al., 1986), it was tested for generalization from phencyclidine in animals (Willetts & Balster, 1988). It was found to produce very potent and effective phencyclidine-like discriminative stimulus effects, raising the possibility that it might produce a phencyclidine-like intoxication in humans. These data from animal studies point to the need to carefully evaluate the ability of MK-801 to produce phencyclidine-like side effects or to result in phencyclidine-like abuse potential as it undergoes clinical trials.

## C. Experimental Basis for Studying Mood and Subjective Experience

Subjective drug effects in humans are often experienced as changes in mood, cognitive ability, and perception. Indeed, the patterns of changes in these experiences produced by different drugs form part of the basis of the ability of experienced users to distinguish one drug effect from another. A widely used instrument for measuring the subjective effects of drugs is the Profile of Mood States (POMS) (McNair, Lorr, & Droppleman, 1971). An experimental version of the POMS contains 72 adjectives describing momentary mood states. After being given a drug, subjects score how they feel at the moment in relation to these mood descriptors on a 5-point scale from "not at all" to "extremely." Scales have been developed empirically by factor analysis, and include such clusters as friendliness, elation, anxiety, and fatigue. Many drugs produce consistent changes in scores in one or more of these scales. For example, cocaine administration results in increased vigor, elation, confusion, friendliness, positive mood, and arousal (Fischman, Schuster, & Rajfer, 1983). Since drugs can reliably produce these mood changes, they are powerful tools to study the phenomenon of mood. For example, studies of the perceptual effects of cocaine provide an opportunity to investigate feelings of elation and friendliness. In animals, studies on the transduction of cocaine stimuli promise to provide important information on the neural substrates of euphoria. The training of increasingly more subtle discriminations among interoceptive states produced by drug stimuli should provide a remarkably powerful approach in this area. Studies of the perceptual effects of drugs such as LSD which dramatically alter the perception of other stimuli may also prove to be a useful strategy in perception research in general. That much of this work can be done in animals provides an opportunity for the exploration of many variables that would not be possible in human subjects. The remarkable concordance in discriminative drug effects in animals and subjective experience of drug intoxication in humans provides additional evidence, if more was needed, of the close relation of basic brain-behavior processes in humans and nonhuman species and the power of comparative research to study these phenomena.

## ACKNOWLEDGMENTS

Support for the preparation of this chapter was provided by National Institute on Drug Abuse Grants DA-01442, DA-03112, and DA-00490. The helpful comments of J. Willetts and J. S. Knisely on this paper are gratefully acknowledged.

## REFERENCES

Appel, J. B., & Dykstra, L. A. (1977). Drugs, discrimination, and signal detection theory. *Advances in Behavioral Pharmacology,* **1,** 139–166.

Balster, R. L. (1987). The behavioral pharmacology of phencyclidine. In H. Y. Meltzer (Ed.), *Psychopharmacology: The third generation of progress* (pp. 1573–1579). New York: Raven Press.

Balster, R. L. (1988). Drugs as chemical stimuli. In F. C. Colpaert & R. L. Balster (Eds.), *Transduction mechanisms of drug stimuli* (pp. 3–11). Berlin: Springer-Verlag.

Balster, R. L., & Willetts, J. (1988). Receptor mediation of the discriminative stimulus properties of phencyclidine and *sigma*-opioid agonists. In F. C. Colpaert & R. L. Balster (Eds.), *Transduction mechanisms of drug stimuli* (pp. 122–135). Berlin: Springer-Verlag.

Barry, H., III. (1974). Classification of drugs according to their discriminable effects in rats. *Federation Proceedings,* **33,** 1814–1824.

Barry, H., III, & Krimmer, E. C. (1977). Discriminable stimuli produced by alcohol and other CNS depressants. In H. Lal (Ed.), *Discriminative stimulus properties of drugs* (pp. 73–92). New York: Plenum.

Beardsley, P. M., Balster, R. L., & Salay, J. M. (1987). Separation of the response rate and discriminative stimulus effects of phencyclidine: Training dose as a factor in phencyclidine-saline discrimination. *Journal of Pharmacology and Experimental Therapeutics,* **241,** 159–165.

Bertalmio, A. J., Herling, S., Hampton, R. Y., Winger, G., & Woods, J. H. (1982). A procedure for rapid evaluation of the discriminative stimulus effects of drugs. *Journal of Pharmacological Methods,* **7,** 289–299.

Brady, K. T., & Balster, R. L. (1981). Discriminative stimulus properties of phencyclidine and five analogues in the squirrel monkey. *Pharmacology, Biochemistry and Behavior,* **14,** 213–218.

Brady, K. T., Balster, R. L., & May, E. L. (1982). Stereoisomers of N-allylnormetazocine: Phencyclidine-like behavioral effects in squirrel monkeys and rats. *Science,* **215,** 178–180.

Brady, K. T., Woolverton, W. L., & Balster, R. L. (1982). Discriminative stimulus and reinforcing properties of etoxadrol and dexoxadrol in monkeys. *Journal of Pharmacology and Experimental Therapeutics,* **220,** 56–62.

Browne, R. G., Welch, W. M., Kozlowski, M. R., & Duthu, G. (1983). Antagonism of PCP discrimination by adenosine analogs. In J. -M. Kamenka, E. F. Domino, & P. Geneste (Eds.), *Phencyclidine and related arylcyclohexylamines: Present and future applications* (pp. 639–666). Ann Arbor, MI: NPP Books.

Chait, L. D., Uhlenhuth, E. H., & Johanson, C. E. (1984). An experimental paradigm for studying the discriminative stimulus properties of drugs in humans. *Psychopharmacology,* **82,** 272–274.

Chait, L. D., Uhlenhuth, E. H., & Johanson, C. E. (1986a). The discriminative stimulus and subjective effects of phenylpropanolamine, mazindol and *d*-amphetamine in humans. *Pharmacology, Biochemistry and Behavior,* **24,** 1665–1672.

Chait, L. D., Uhlenhuth, E. H., & Johanson, C. E. (1986b). The discriminative stimulus and subjective effects of *d*-amphetamine, phenmetrazine and fenfluramine in humans. *Psychopharmacology,* **89,** 301–306.

Chen, G., Ensor, C. R., & Bohner, B. (1969). The pharmacology of 2-(ethylamino)-2(2-thienyl)-cyclohexanone-HCl (CI-634). *Journal of Pharmacology and Experimental Therapeutics,* **168,** 171–179.

Colpaert, F. C. (1977). Discriminative stimulus properties of benzodiazepines and barbiturates. In H. Lal (Ed.), *Discriminative stimulus properties of drugs* (pp. 93–106). New York: Plenum.

Colpaert, F. C. (1978). Discriminative stimulus properties of narcotic drugs. *Pharmacology, Biochemistry and Behavior,* **9,** 863–887.

Colpaert, F. C., & Balster, R. L. (Eds.). (1988). *Transduction mechanisms of drug stimuli.* Berlin: Springer-Verlag.

Colpaert, F. C., & Slangen, J. L. (Eds.). (1982). *Drug discrimination: Applications in CNS pharmacology.* Amsterdam: Elsevier Biomedical Press.

Cone, E. J., McQuinn, R. L., & Shannon, H. E. (1984). Structure-activity relationship studies of phencyclidine derivatives in rats. *Journal of Pharmacology and Experimental Therapeutics,* **228,** 147–153.

Dykstra, L. D., Bertalmio, A. J., & Woods, J. H. (1988). Discriminative and analgesic effects of mu and kappa opioids: *In vivo* $pA_2$ analysis. In F. C. Colpaert & R. L. Balster (Eds.), *Transduction mechanisms of drug stimuli* (pp. 107–121). Berlin: Springer-Verlag.

Fischman, M. W., Schuster, C. R., & Rajfer, S. (1983). A comparison of the subjective and cardiovascular effects of cocaine and procaine in humans. *Pharmacology, Biochemistry and Behavior,* **18,** 711–716.

Foster, A. C., Gill, R., Kemp, J. A., & Woodruff, G. N. (1987). Systemic administration of MK-801 prevents N-methyl-D-aspartate-induced neuronal degeneration in rat brain. *Neuroscience Letters,* **76,** 307–311.

Ghoneim, M. M., Hinrichs, J. V., Mewaldt, S. P., & Petersen, R. P. (1985). Ketamine: Behavioral effects at subanesthetic doses. *Journal of Clinical Psychopharmacology,* **5,** 70–77.

Goldberg, S. R., & Shuster, C. R. (1967). Conditioned suppression by a stimulus associated with nalorphine in morphine-dependent monkeys. *Journal of Experimental Analysis of Behavior,* **10,** 235–242.

Goldstein, A., Aronow, L., & Kalman, S. M. (1974). *Principles of pharmacology: The basis of pharmacology* (2nd ed., pp. 1–127). New York: Wiley.

Haertzen, C. A. (1966). Development of scales based upon patterns of drug effects, using the Addiction Research Center Inventory (ARCI). *Psychological Reports,* **18,** 163–194.

Harris, C. M., Wood, D. M., Lal, H., & Emmett-Oglesby, M. W. (1987). A method to shorten the training phase of drug discrimination. *Psychopharmacology,* **93,** 435–436.

Hendry, J. S., Balster, R. L., & Rosecrans, J. A. (1983). Discriminative stimulus properties of buspirone compared to central nervous system depressants in rats. *Pharmacology, Biochemistry and Behavior,* **19**, 97–101.

Henningfield, J. E., Lukas, S. E., & Bigelow, G. E. (1986). Human studies of drug as reinforcers. In S. R. Goldberg & I. P. Stolerman (Eds.), *Behavioral analysis of drug dependence* (pp. 69–122). Orlando, FL: Academic Press.

Herling, S., & Woods, J. H. (1981). Discriminative stimulus effects of narcotics: Evidence for receptor-mediated actions. *Life Sciences,* **28**, 1571–1584.

Holtzman, S. G. (1980). Phencyclidine-like discriminative effects of opioids in the rat. *Journal of Pharmacology and Experimental Therapeutics,* **214**, 614–619.

Holtzman, S. G. (1985). Discriminative stimulus properties of opioids that interact with mu, kappa and PCP/sigma receptors. In L. S. Seiden & R. L. Balster (Eds.), *Behavioral pharmacology: The current status* (pp. 131–147). New York: Liss.

Huang, J. -T., & Ho, B. T. (1974). Discriminative stimulus properties of *d*-amphetamine and related compounds in rats. *Pharmacology, Biochemistry and Behavior,* **2**, 669–673.

Järbe, T. U. C., Johansson, J. O., & Henricksson, B. G. (1976). Characteristics of tetrahydrocannabinol (THC)-produced discrimination in rats. *Psychopharmacology,* **48**, 181–187.

Jasinski, D. R. (1977). Assessment of abuse potentiality of morphine-like drugs (Methods used in man). *Handbook of Experimental Pharmacology,* **45**(1), 197–258.

Johanson, C. E., & Balster, R. L. (1978). A summary of the results of a drug self-administration study using substitution procedures in rhesus monkeys. *Bulletin on Narcotics,* **30**, 43–54.

Johnson, K. M. (1987). Neurochemistry and neurophysiology of phencyclidine. In H. Y. Meltzer (Ed.), *Psychopharmacology: The third generation of progress* (pp. 1581–1588). New York: Raven Press.

Kaempf, G. L., & Kallman, M. J. (1987). A comparison of testing procedures on the discriminative morphine stimulus. *Psychopharmacology,* **91**, 56–60.

Koslowski, M. R., Browne, R. G., & Vinick, F. J. (1986). Discriminative stimulus properties of phencyclidine (PCP)-related compounds: Correlations with $^3$H-PCP binding potency measured autoradiographically. *Pharmacology, Biochemistry and Behavior,* **25**, 1051–1058.

Krimmer, E. C., & Barry, H., III. (1979). Pentobarbital and chlordiazepoxide differentiated from each other and from nondrug. *Communications in Psychopharmacology,* **3**, 92–99.

Kubena, R. K., & Barry, H., III. (1969). Generalization by rats of alcohol and atropine stimulus characteristics to other drugs. *Psychopharmacologia,* **15**, 196–206.

Locke, K. W., & Holtzman, S. G. (1985). Characterization of the discriminative stimulus effects of centrally administered morphine in the rat. *Psychopharmacology,* **87**, 1–6.

McMillan, D. E. (1982). Generalization of the discriminative stimulus properties of phencyclidine to other drugs in the pigeon using color tracking under second order schedules. *Psychopharmacology,* **78**, 131–134.

McNair, D. M., Lorr, M., & Droppleman, L. F. (1971). *Manual for the profile of mood states.* San Diego, CA: Educational and Industrial Testing Service.

Meltzer, H. Y., Sturgeon, R. D., Simonovic, M., & Fessler, R. G. (1981). Phencyclidine as an indirect dopamine agonist. In E. F. Domino (Ed.), *PCP (Phencyclidine): Historical and current perspectives* (pp. 207–242). Ann Arbor, MI: NPP Books.

Meltzer, L. T., & Rosecrans, J. A. (1981a). Discriminative stimulus properties of arecoline: A new approach for studying central muscarinic receptors. *Psychopharmacology, 75,* 383–387.

Meltzer, L. T., & Rosecrans, J. A. (1981b). Investigations on the CNS sites of action of the discriminative stimulus effects of arecoline and nicotine. *Pharmacology, Biochemistry and Behavior, 15,* 21–26.

O'Brien, C. P., Testa, T., O'Brien, T. J., Brady, J. P., & Wells, B. (1977). Conditioning of narcotics abstinence symptoms in human subjects. *Science, 195,* 1000–1002.

Overton, D. A. (1966). State-dependent learning produced by depressant and atropine-like drugs. *Psychopharmacologia, 10,* 6–31.

Overton, D. A. (1979). Drug discrimination training with progressively lowered doses. *Science, 205,* 720–721.

Overton, D. A. (1982). Multiple drug training as a method for increasing the specificity of the drug discrimination procedure. *Journal of Pharmacology and Experimental Therapeutics, 221,* 166–172.

Overton, D. A., & Hayes, M. W. (1984). Optimal training parameters in the two-bar fixed-ratio drug discrimination task. *Pharmacology, Biochemistry and Behavior, 21,* 19–28.

Overton, D. A., Merkle, D. A., & Hayes, M. L. (1983). Are "no-drug" cues discriminated during drug-discrimination training? *Animal Learning and Behavior, 11,* 295–301.

Poling, A. D., White, F. J., & Appel, J. B. (1979). Discriminative stimulus properties of phencyclidine. *Neuropharmacology, 18,* 459–463.

Pollard, J. C., Uhr, L., & Stern, E. (1965). *Drugs and phantasy,* Boston, MA: Little, Brown.

Rees, D. C., Coggeshall, E., & Balster, R. L. (1985). Inhaled toluene produces pentobarbital-like discriminative stimulus effects in mice. *Life Sciences, 37,* 1319–1325.

Rees, D. C., Knisely, J. S., Balster, R. L., Jordan, S. L., & Breen, T. J. (1987). Pentobarbital-like discriminative stimulus properties of halothane, 1, 1, 1-trichloroethane, isoamyl nitrate and flurothyl in mice. *Journal of Pharmacology and Experimental Therapeutics, 241,* 507–515.

Rees, D. C., Knisely, J. S., Jordan, S., & Balster, R. L. (1987). Discriminative stimulus properties of toluene. *Toxicology and Applied Pharmacology, 88,* 97–104.

Rosecrans, J. A., & Glennon, R. A. (1979). Drug-induced cues in studying mechanisms of drug action. *Neuropharmacology, 18,* 981–989.

Sannerud, C. A., & Young, A. M. (1987). Environmental modification of tolerance to morphine discriminative stimulus properties in rats. *Psychopharmacology, 93,* 59–68.

Schechter, M. D. (1973). Transfer of state-dependent control of discriminative behavior between subcutaneously and intravenously administered nicotine and saline. *Psychopharmacologia, 32*, 327–335.

Schechter, M. D., & Rosecrans, J. A. (1971). C. N. S. effect of nicotine as the discriminative stimulus for the rat in a T-maze. *Life Sciences, 10*, 821–832.

Schuster, C. R., & Johanson, C. E. (1988). The relationship between the discriminative stimulus properties and subjective effects of drugs. In F. C. Colpaert & R. L. Balster (Eds.), *Transduction mechanisms of drug stimuli*. Berlin: Springer-Verlag.

Shannon, H. E. (1981). Evaluation of phencyclidine analogs on the basis of their discriminative stimulus properties in the rat. *Journal of Pharmacology and Experimental Therapeutics, 216*, 543–551.

Shannon, H. E. (1983). Discriminative stimulus effects of phencyclidine: Structure-activity relationships. In J. -M. Kamenka, E. F. Domino, & P. Geneste (Eds.), *Phencyclidine and related arylcyclohexylamines: Present and future applications* (pp. 311–335). Ann Arbor, MI: NPP Books.

Shannon, H. E., & Holtzman, S. G. (1977). Discriminative effects of morphine administered intracerebrally in the rat. *Life Sciences, 21*, 585–594.

Siegel, S. (1982). Opioid expectation modifies opioid effects. *Federation Proceedings, 41*, 2339–2343.

Slifer, B. L., & Balster, R. L. (1985). A comparison of the discriminative stimulus properties of phencyclidine, given intraperitoneally or intraventricularly in rats. *Neuropharmacology, 24*, 1175–1179.

Solomon, R. E., Herling, S., Domino, E. F., & Woods, J. H. (1982). Discriminative stimulus effects of N-substituted analogs of phencyclidine in rhesus monkeys. *Neuropharmacology, 21*, 1329–1336.

Stebbins, W. C., Smith, D. W., & Moody, D. B. (1988). Discrimination strategies in animal psychophysics and their role in understanding sensory receptor function. In F. C. Colpaert & R. L. Balster (Eds.), *Transduction mechanisms of drug stimuli* (pp. 199–214). Berlin: Springer-Verlag.

Stolerman, I. P., Baldy, R. E., & Shine, P. J. (1982). Drug discrimination procedure: A bibliography. In F. C. Colpaert & J. L. Slangen (Eds.), *Drug discrimination: Application in CNS pharmacology* (pp. 401–424). Amsterdam: Elsevier Biomedical Press.

Stolerman, I. P., Pratt, J. A., Garcha, H. S., Giardini, V., & Kumar, R. (1983). Nicotine cue in rats analyzed with drugs acting on cholinergic and 5-hydroxytryptamine mechanisms. *Neuropharmacology, 22*, 1029–1037.

Stolerman, I. P., & Shine, P. J. (1985). Trends in drug discrimination research analyzed with a cross-indexed bibliography, 1982–1983. *Psychopharmacology, 86*, 1–11.

Swedberg, M. D. B., & Järbe, T. U. C. (1986). Drug discrimination procedures: Differential characteristics of the drug A vs drug B and the drug A vs drug B vs no drug cases. *Psychopharmacology, 90*, 341–346.

Thompson, T., & Pickens, R. (Eds.). (1971). *Stimulus properties of drugs*. New York: Appleton-Century-Crofts.

Tomie, A., Loukas, E., Stafford, I., Peoples, L., & Wagner, G. C. (1985). Drug discrimination training with a single choice trial per session. *Psychopharmacology*, **86**, 217–222.

Valentino, R. J., Herling, S., Woods, J. H., Medzihradsky, F., & Merz, H. (1981). Quaternary naltrexone: Evidence for the central mediation of discriminative stimulus effects of narcotic agonists and antagonists. *Journal of Pharmacology and Experimental Therapeutics*, **217**, 652–659.

Wessinger, W. D., Martin, B. R., & Balster, R. L. (1985). Discriminative stimulus properties and brain distribution of phencyclidine in rats following administration by injection and smoke inhalation. *Pharmacology, Biochemistry, and Behavior*, **23**, 607–612.

White, F. J., & Appel, J. B. (1982). Training dose as a factor in LSD-saline discrimination. *Psychopharmacology*, **76**, 20–25.

White, J. M., & Holtzman, S. G. (1981). Three-choice drug discrimination in the rat: Morphine, cyclazocine and saline. *Journal of Pharmacology and Experimental Therapeutics*, **217**, 254–262.

Willetts, J., & Balster, R. L. (1988). Phencyclidine-like discriminative stimulus properties of MK-801 rats. *European Journal of Pharmacology*, **146**, 167–169.

Wong, E. H. F., Kemp, J. A., Priestly, T., Knoght, A. R., Woodruff, G. N., & Iversen, L. L. (1986). The anticonvulsant MK-801 is a potent N-methyl-D-aspartate antagonist. *Proceedings of the National Academy of Sciences of the U.S.A.*, **83**, 7104–7108.

York, J. L., & Bush, R. (1982). Studies on the discriminative stimulus properties of ethanol in squirrel monkeys. *Psychopharmacology*, **77**, 212–216.

Young, A. M., & Herling, S. (1986). Drugs as reinforcers: Studies in laboratory animals. In S. R. Goldberg & I. P. Stolerman (Eds.), *Behavioral analysis of drug dependence* (pp. 9–67). Orlando, FL: Academic Press.

Young, R., & Glennon, R. A. (1986). Discriminative stimulus properties of amphetamine and structurally-related phenalkylamines. *Medicinal Research Reviews*, **6**, 99–130.

# 6

# OLFACTORY PERCEPTION

**Burton M. Slotnick**

*The American University, Washington, D.C.*

Behavioral studies with animals have demonstrated that the sense of smell plays an important role in learning, social organization, reproductive behavior, control of neuroendocrine activity, feeding in the newborn, appetite, and perception of flavor. However, there are relatively few animal studies of basic olfactory processes and behavioral scientists have yet to exploit fully the various possibilities that have been revealed by anatomical and physiological investigations for encoding olfactory stimuli. Three areas in which behavioral studies have made significant contributions to our understanding of olfactory perception are reviewed in this chapter: olfactory learning, the role of olfactory projections to the forebrain in olfactory discrimination, and the mechanisms for neural coding of odors. There are important methodological problems in this area of research and some of these will be briefly considered. The topics covered in this chapter represent only a few of the many active areas of investigation in olfaction. They are still broad enough to make complete coverage of each impractical in a single chapter, and, thus, coverage is selective rather than exhaustive.

## I. CONTROL OF THE STIMULUS

Standards for the control of olfactory stimuli have not been established and the literature is replete with studies of olfactory behavior that, at best, would be difficult to replicate because the basic parameters of the stimulus (e.g., intensity, duration, point source, and quality) were not or could not be specified. The need to gain good control of the stimulus is as important in studies of olfaction as it is in research on the visual, auditory, and other sensory systems. In fact, one can argue that it is more important. In vision and audition, for example, we know enough about the nature of the effective stimulus, limits of receptor sensitivity, the changes in perception resulting from changes in stimulus complexity (e.g., color mixing, auditory masking) to make rational choices of stimuli and exert appropriate control over them. This is not the case with olfaction: The stimulus is only poorly understood, the limits of sensitivity have not been adequately assessed, and most of the factors determining perceptual experience are largely unexplored. For behavioral studies, the use of uncontrolled complex odors whose composition varies over time, strong odors that produce adaptation, odors that have irregular rates of onset or termination, or odors that are not adequately controlled in the subject environment can only decrease the signal value of the stimulus, degrade performance, and mask an animal's full capacity to utilize such stimuli. In short, some recognition of the problems and potential solutions for controlling odor stimuli is important in planning and evaluating studies of olfaction.

## A. Generating Odors

There is no general agreement among workers on a best method for generating pure odorants or odorant mixtures at known concentrations, and for delivering these to a subject. Solutions to such problems are varied and often determined by the ingenuity of individual investigators. Since the excellent, although now somewhat dated, reviews of Dravnieks (1975) and Moulton (1973) on the design and construction of odor generators many investigators have adopted, in one form or another, a positive-pressure, air dilution system to produce and control odors. Such systems are referred to as olfactometers. An olfactometer for behavioral studies can be viewed as three integral units: a vapor saturator, a vapor dilution system, and a system of valves for introducing the vapor stimulus to the subject (Fig. 1). As discussed below, there are difficulties with all methods that are used to generate and deliver olfactory stimuli. However, for behavioral research these shortcomings are less of a troublesome than the fact that too few investigators are exploiting the best methods that are available.

Controlled generation of a vapor generally involves saturating an air or nitrogen carrier stream with the vapor of an odorant and, subsequently, diluting the vapor with air to a desired concentration. This is accomplished most simply by passing a carrier gas over the extended surface of the odorant. As described by Dravnieks, air or nitrogen at flow rates of up to 100 mL/min over the surface of odorants contained in a 15-cm-long, 1.25-cm-wide tube will reach 95% or greater saturation for most odorant materials.

An olfactometer must also be capable of producing different concentra-

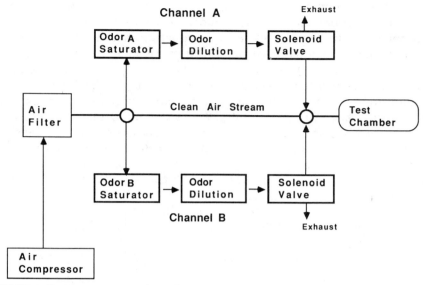

**FIGURE 1.** Basic components of an olfactometer for generating two stimulus vapors. Odors, continuously generated, are exhausted via the normally open port of the 3-way solenoid valves. Momentary valve operation adds an odor to the clean air stream. The circles represent manifolds. Additional details are shown in Figure 2.

tions of a stimulus. Obviously, a wide range of vapor dilutions are required for psychophysical studies. But even for simple two-odor discrimination tasks, dilutions of several orders of magnitude may be needed to minimize olfactory adaptation (Bennett, 1968) and stimulation of sensory receptors other than those of the first cranial (olfactory) nerve. Smell experience is generally considered to be mediated by olfactory receptors in the nasal epithelium, but other nasal receptors may be activated by high concentrations of vapors and these could provide cues for discriminative responding in a behavioral study (Doty et al., 1978; Silver & Moulton, 1982; Tucker, 1971).

The concentration of an odor may be varied by liquid dilution of the odorant source, gaseous dilutions of an odorant-saturated vapor, or a combination of these. Liquid dilution systems have the advantage of simplicity but have a host of disadvantages, as described in detail by Dravnieks. An important disadvantage is that unless the odorant and diluent form an ideal mixture, the concentration of the odorant in the vapor phase will be difficult to specify. Air dilution systems largely avoid these problems but are more difficult to construct and maintain because several dilution stages are generally needed and each requires flow controllers (needle valves) and monitors (flowmeters). Because of the limitations of commercially available flowmeters and the difficulty in handling large volumes of air, a single-stage dilution system is impractical for accurately producing the wide range of concentrations needed for psychophysical studies. It is feasible to produce dilutions as large as $10^{-3}$ at each dilution step. Absolute thresholds for many odors are in the range of $10^{-5}$ to $10^{-9}\%$ of vapor saturation and, thus, at least three dilution stages are generally required for psychophysical tests. In the first stage, vapor-saturated air is mixed with clean air. The odorant concentration can then be further reduced in additional dilution stages (Fig. 2). Because air from the final stage provides the stimulus to the animal, the airflows of this dilution are kept constant. Hence, only the dilutions upstream of this one are free to vary.

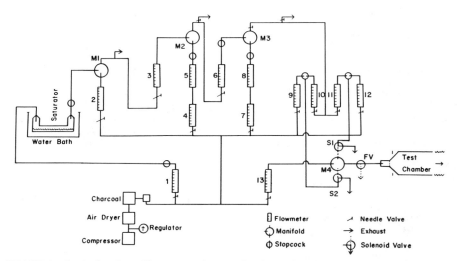

**FIGURE 2.** A single-odor olfactometer for psychophysical studies. Air is provided by an oil-free air compressor whose output is dehydrated with silica gel or a refrigerant dryer and filtered through fiber micropore filters and activated charcoal. Ductwork is made from sections of 8-mm (outside diameter) glass tubing connected by Teflon sleeves. Flowmeters 2, 4, 5, 7, 8, and 13 supply clean air to the dilution manifolds. Flowmeters ranges are 1–6 L/min (flowmeters 2, 4, 7, and 13); 0.1–1 L/min (flowmeters 4, 7, 9–12); 0.005–0.1 L/min (flowmeter 1). Flowmeters may be used in series (e.g., those supplying manifolds M2 and M3) or in parallel to extend their range. The vertical lines across the wind-tunnel test chamber indicate the position of a photobeam used to detect trial-initiating responses. The common ports of the odor control solenoids (S1 and S2) are connected to stimulus output flowmeters. The normally closed ports are connected to the clean air stream (regulated by flowmeter 13) and the normally open ports are connected to exhaust lines (arrows). Note that each stimulus-control solenoid is connected to both a clean air line and an odor line. For absolute threshold tests, either solenoid can be used to control the odor stimulus and, for intensity-difference threshold determinations, airflows in the $S^-$ (comparison stimulus) channel can easily and rapidly be adjusted to produce the desired fraction of the $S^+$ (standard stimulus) channel. The normally open port of the final valve (FV) is connected to manifold 4 (M4) and the normally closed port is connected to an exhaust line. The valve operates briefly at the beginning of each trial. This serves to divert air from the test chamber and signal the initiation of the trial, and, while the valve is operated, allows the odor to mix with the clean air stream.

## B. Olfactometers Used in Behavioral Studies

Several olfactometers currently used by the author are described here. The design and general principles of their operation are similar to those that have been used by other investigators (e.g., Bennett, 1968; Eichenbaum, Shedlack, & Eckmann, 1980; Moulton, Celebi, & Fink, 1970; Tucker, 1963) and comparisons among these illustrate some of the many solutions possible in constructing an automated positive pressure odor generator and dilution system for behavioral studies. Other types of olfactometers used in human psychophysical investigations might be adaptable for conditioning studies with animals (e.g., Benignus & Prah, 1980; Dravnieks, 1975; Stone, Pryor, & Steinmetz, 1969).

A multiple-stage air dilution system is used for psychophysical studies (Fig. 2). Charcoal-filtered air is divided into two channels. One supplies the odor saturator and the other provides clean air for dilution channels and the clean air stimulus channel. Odorized and clean airstreams are joined at manifolds which have a number of "push-ins" (glass surfaces protruding into the manifold) to increase air turbulence and enhance mixing of the two vapor streams. Other details of the system are given in the figure caption.

The system used for two-odor discrimination studies is shown in Fig. 3. For some investigations (e.g., studies of multiple-odor discrimination, odor mixtures, stimulus generalization) the system shown in Fig. 4 has been used to present many different odors within a single session. This olfactometer is relatively simple to construct and maintain but its simplicity

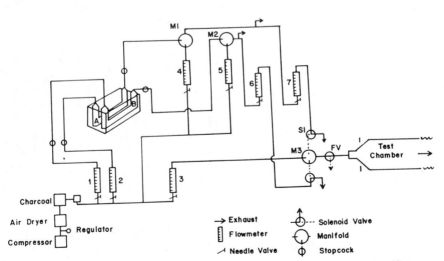

**FIGURE 3.** Two-odor multiple-dilution stage olfactometer. Odorants are contained in saturators A and B.

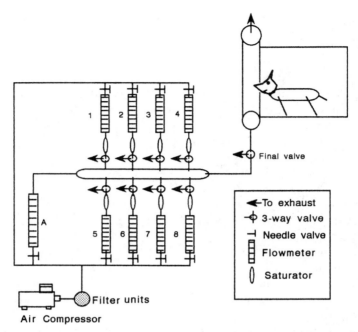

**FIGURE 4.** Eight-channel, single-dilution stage olfactometer. Each odor saturator is connected to a 3 L/min clean air stream (regulated by flowmeter A) by a 3-way solenoid. Arrows indicated exhaust paths. Airflows through the odor saturators can be varied from 0 to 100 mL/min. The vertical stimulus sampling tube is shown in this figure. A photocell across the sampling port serves to detect insertions of the animal's nose into the airstream.

is obtained at the cost of having only one air dilution stage. If necessary, lower concentrations could be produced by liquid dilution of the odorant material.

## C. Odor Delivery

In an automated computer-controlled system, electrically operated valves are most often used to switch a continuously generated vapor from an exhaust path into one connected to the animal chamber. Small Teflon-body solenoids with negligible dead space are available from a number of manufacturers. These valves operate in the millisecond range, are reasonably quiet, and can be cleaned with most solvents. As illustrated in Fig. 2, the common port of the solenoid is connected to a stimulus airstream, the normally open port to an exhaust line, and the normally closed port to a clean airstream. Operation of the solenoid introduces the odorant to the clean air (or carrier) stream and this constitutes the final dilution step of the stimulus. This method of controlling the stimulus streams has two potential difficulties. The first is that if the odorant and carrier stream do not mix completely before being introduced to the test chamber, the subject will experience an ascending concentration phase of the stimulus. This would be particularly undesirable in studies of intensity difference thresholds or odor mixtures. The second problem is that differences in the operation of valves may produce unintended auditory or pressure pulse cues. One solution to these problems is to use mechanically operated doors to close odor sampling ports and perform all switching of odors during the intertrial interval (Moulton & Marshall, 1976; Pietras & Moulton, 1974). A simpler solution is to use a common final valve (Fig. 2) to divert all air from the chamber for a brief period at the initiation of each trial (Slotnick & Schoonover, 1984). The odorant and carrier stream mix during this diversion of airflow and, when the valve is relaxed, the animal is presented with a sudden and sharp onset of the stimulus. The potential problems of sound and pressure pulse cues have proved negligible. In any case they are eliminated or adequately masked by the operation of the final valve.

Odors can be presented to the animal in a modified wind tunnel (Fig. 2) or introduced through a relatively narrow channel which has a sampling port (Fig. 4). In either case odor sampling can be enhanced by making the presentation of the stimulus contingent upon the animal inserting its snout into the stimulus airstream. A wind tunnel has the advantage of allowing almost complete control over the test environment. The subject is essentially immersed in the stimulus during odor presentation and the duration of odor exposure can be specified. However, because the subject is surrounded by the stimulus it is difficult to record attentive behavior or deliberate sampling. A narrow stimulus delivery tube fitted with an odor sampling port can be used to measure stimulus sampling. Sampling duration on each trial can be measured using a photocell or other detector. The disadvantages of this method are that the animal must be trained to sample from the port and, because it may not sample throughout the odor trial, the duration of odor exposure may be somewhat variable from trial to trial.

The method of choice depends upon the particular requirements of the study. For example, in a study of olfactory stimulus generalization by Braun and Marcus (1969), rats were reinforced on a variable-interval schedule for responding on an manipulandum during 1-min presentations of one odor but were not reinforced in the presence of a second odor. A wind-tunnel test chamber was used because it was essential to keep the duration of odor exposure constant on each trial. If the duration (or other parameters) of odor sampling is of primary interest then it is preferable to present odors in a narrow channel equipped with a sniffing port and detector.

A unique solution to the presentation of an odor has been described by Laing, Murray, King, and Cairncross (1974). A cannula connected by a flexible tube to an olfactometer was implanted into the nasal cavity of rats. This allowed odor stimuli to be presented directly to the olfactory epithelium. The cannula remained patent for periods up to six months and the animals could be trained to press a bar in response to odor delivery.

### D. Odor Purity

The olfactory detection threshold of macrosmatic animals (those with a well-developed sense of smell), such as rats and dogs, is generally lower than human threshold (Marshall & Moulton, 1977) or that of many instruments and, thus, a test subject may respond to odoriferous contaminants that are not perceived by the investigator but provide unintended cues for discriminative responding. For example, consider the problem of generating a single odor. Except for a few odorants that are available in highly purified form in tanks of compressed gas, most odorant sources used in experiments will be liquids. Only a small number of otherwise useful odorants is commercially available at high levels of purity and an odorant source with odoriferous contaminants will constitute an odor mixture. If the odorants in the mixture have appreciably different vapor pressures then a gradual change in the character of the odor may occur because differential evaporation will alter the composition of the original mixture.

Contaminants may also occur in the air used to generate and convey an odor and in the ductwork of the system. Ideally, air used in diluting and conveying an odorant should itself be free of any odors. Many contaminants can be removed by passing air through fine particle filters and activated charcoal but a more elaborate air purification method (such as a refrigerated trap) may be required to produce sufficiently pure air for critical work (e.g., for studies of odor deprivation; Laing, 1984). However, for behavioral studies absolutely odor free air is probably not necessary and probably not desirable because it would be novel stimulus for the subject. A filtering system can be considered adequate for most behavioral studies if it produces air that cannot be discriminated from reasonably clean background or room air.

Perhaps the most difficult contaminants to document and control stem from odor molecules that have been absorbed by surfaces of containers or ductwork and/or adsorbed onto those surfaces, only to be released at a later time. Absorption of odor molecules can be minimized by using smooth materials with low specific surface areas (e.g., glass). Even so, odor molecules will still be adsorbed onto the walls of ductwork; the amount of adsorption will vary as a function of the molecule, concentration of the odorant, nature and total surface area of the ductwork, rate of air flow, and temperature. Adsorption can affect the final concentration of an odor stimulus in two ways. First, if a very low concentration of a vapor is generated and the odor molecules are strongly adsorbed, then an appreciable amount of time may be required before adsorptive surfaces are saturated and the vapor concentration achieves a steady state at the exit port of the system. Second, desorption of odor molecules from duct walls may occur only slowly, especially with substances of low volatility, and could provide a detectable environmental stimulus for many minutes or even days after terminating or removing the stimulus source (e.g., Davis & Tapp, 1972). Also, desorption could significantly contribute to stimulus concentration when the same ductwork is used to convey a low concentration of the vapor after it has been exposed to a higher concentration. In practice, these problems can be minimized by using a system that continuously generates a fixed concentration of an odorant, and by selecting odors that have a reasonably high vapor pressure and that are not highly polar and thus less "sticky." In general, most of the pure odorants that have been used in behavioral research (esters of saturated fatty acids, low molecular weight alcohols and ketones) meet these criteria.

## II. STIMULUS CONTROL OF BEHAVIOR

Stimulus control of behavior refers to the extent to which an animal's response can be determined by an antecedent conditioned stimulus (Terrace, 1966). In the present context stimulus control is achieved using the procedures of operant conditioning. The extent to which an animal comes under stimulus control reflects, loosely speaking, how well it has been trained or what it has learned.

There are as many ways of achieving olfactory stimulus control of behavior as there are conditioning methods. A variety of instrumental positive reinforcement methods (free operant procedures or discrete trials with symmetrical or asymmetrical reinforcement), negative reinforcement methods (conditioned emotional responses), and classically conditioned odor aversion techniques have all been used successfully (Doty, 1975; Palmerino, Rusiniak, & Garcia, 1980; Panhuber, 1982; Pierson, 1974; Stevens, 1975). However, some methods are better than others. For example, the conditioned suppression method, although effective, is less efficient because trials must be widely spaced (Pierson, 1974). Classical conditioning using the aversion paradigm of Garcia appears to be distinctly less effective and reliable for producing olfactory stimulus control than are operant discrimination methods (Panhuber, 1982).

### A. Training Rats to Detect and Discriminate Odors

The discrete trials, positive reinforcement operant conditioning procedure used in our laboratory is described here because it was employed in many of the experiments reviewed below and because it produces reliable and efficient learning. Stimuli are designated as positive ($S^+$) or negative ($S^-$) and reinforcement is available only on $S^+$ trials. In our experience this asymmetrical reinforcement procedure is more efficient and produces better stimulus control than does a symmetrical reinforcement procedure in which two manipulanda are available and all correct responses are reinforced.

Adult male Wistar, Sprague–Dawley, or Long–Evans strain rats have been used. Training is begun after 12–14 days on a 8- to 10-mL/day water deprivation schedule. The entire training procedure is automated and under computer control. Standard operant "shaping" methods are used to create an association between the sound of the reinforcement solenoid and delivery of a 0.05-mL water reward, to reinforce breaking the photocell beam with the snout, contact with a manipulandum (metal bar or drinking tube; Field & Slotnick, 1987), and, finally, to reinforce responding during brief presentations of the $S^+$ odor. The behavior chain of breaking the photobeam with the snout (which initiates a trial) and responding on the manipulandum is acquired in one or two 30-min training sessions. The control program introduces reinforcement contingencies gradually but requires a high level of proficiency at each stage. In experiments using a stimulus sampling port, the final stage of training requires that the rat samples the stimulus (i.e., holds its snout in the photobeam) for at least 0.2 sec before a response is reinforced.

In the next training session 200 $S^+$-only trials are presented and the final trial parameters are gradually introduced. The sequence of events and response contingencies in effect at the end of the session are shown in Fig. 5.

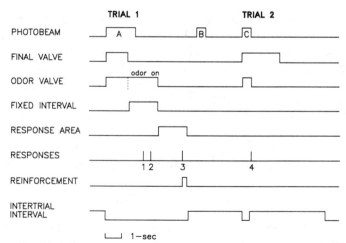

**FIGURE 5.** Sequence of events during a $S^+$ trial. When the animal breaks the photobeam after the end of the intertrial interval (trial 1, A) the odor and final valve are operated. During this time the odor and carrier streams mix and are diverted to an exhaust path. When the final valve is deenergized the odor is presented to the animal for 2 sec (region marked "odor on" in figure). Termination of the 2-sec odor-on interval initiates a response interval. Responses during the odor-on interval are ignored (e.g., trial 1, responses 1 and 2). The first response during the response interval of an $S^+$ trial is reinforced (trial 1, response 3). Responses made during the final valve interval abort the trial (trial 2, response 4). As described in the text, a minimum sample time required may also be imposed.

Typically, by the end of this S$^+$ only session, animals initiate each trial within a few seconds after the intertrial interval, vigorously sniff throughout the final valve period, sample the stimulus for approximately 0.5 sec, and then immediately respond on the manipulandum until reinforcement is delivered. The number of trial aborts (responding before the odor is introduced) and short sample trials (sampling the odor for less than 0.2 sec) rapidly decrease over the session to approximately 0–2 per block of 20 trials in the last 2–3 blocks of trials.

In the next session positive and negative trials are presented in a predetermined random order. The trial parameters and response termination contingencies are the same for both S$^+$ and S$^-$ trials. The two types of trials differ only in the stimulus presented and in the fact that only responses on S$^+$ trials are reinforced. Criterion performance is set at 90% correct responding in a block of 20 trials.

## III. BEHAVIORAL STUDIES

Although rats had been used in innumerable studies of learning, few of these studies employed odors as discriminative cues. Until recently, reports on olfactory learning did not indicate that rats were particularly adept at using olfactory cues or that they gained any special advantage when olfactory rather than, say, visual or auditory cues were available for discriminative responding. In fact, at least one report (Eayrs & Moulton, 1960) noted some difficulty in training rats to use odor cues. This sort of outcome was, perhaps, not unexpected given the then prevalent view that olfaction was a primitive sensory modality and important primarily in primitive or species-specific functions. In fact, behavioral findings such as the role of olfaction in neuroendocrine regulation (Bruce, 1960; review by Macrides, 1976) and the severe alterations in sexual and maternal behaviors in anosmic rodents (review by Murphy, 1976) were consonant with this view and, perhaps, helped foster the notion that olfaction played an important role in certain species-typical behaviors but a less important one in learning. Subsequently, conditioning studies have demonstrated that rodents not only readily learn to use olfactory cues but their performance on operant discrimination tasks may be superior when trained with odors than with visual, auditory, or taste stimuli (Slotnick & Brosvic, 1987). The results of many of these studies seem to have been anticipated by Engen who, in a 1973 review on the sense of smell stated, "It seems that the time has passed in psychology when animal psychologists study learning problems using visual cues in animals with poor vision."

The following account reviews studies of how rats use olfactory cues in simple and complex learning. Olfactory psychophysics, a topic not covered in this chapter, is reviewed by Passe and Walker (1985). Psychophysical studies employing the methods presented above are described by Slotnick and Ptak (1977), Slotnick and Schoonover (1984), and Marshall, Doty, Lucero, and Slotnick (1981).

## A. Rapid Acquisition of Stimulus Control: Errorless Learning

Under appropriate training conditions rats learn to detect and even discriminate between two odors within a single training session and often with few or no errors. After initial training to respond to a positive stimulus, rats generally do not respond to the negative stimulus when it is introduced immediately after a series of $S^+$-only trials (Nigrosh, Slotnick, & Nevin, 1975). This rapid acquisition of a detection or discrimination task may be specific for odors. Rats trained with similar procedures but with visual, auditory, or taste stimuli make many, even hundreds, of errors before reaching performance criterion (Nigrosh et al., 1975; Slotnick, 1984; Slotnick & Brosvic, 1987).

The rapid acquisition of odor detection or discrimination tasks is reminiscent of errorless learning, a phenomenon first studied systematically by Terrace (1963) in pigeons trained on a wavelength discrimination using a go/no-go procedure. Errorless learning occurred if the intensity and duration of $S^-$ was gradually increased during training. Other procedures, including abrupt introduction of the $S^+$ and $S^-$ stimuli resulted in the birds making many errors. In contrast, errorless performance of an olfactory discrimination by rats appears to be more robust; it occurs even when the negative stimulus is introduced abruptly and at full strength and it is obtained in almost all animals tested. There are, however, at least two conditions under which errorless acquisition is not obtained. First, if animals are trained initially on a nonolfactory modality and then abruptly transferred to an odor discrimination, acquisition is slower and typically 50–100 errors may be made in achieving criterion performance. Second, if the $S^-$ stimulus differs only in intensity from the $S^+$ stimulus or if it consists of a mixture containing the positive stimulus, errorless acquisition does not occur (B. M. Slotnick, unpublished observations). In the latter case the number of errors made is a function of the discriminability of the two stimuli.

## B. Selective Attention to Odors

The term attention is generally used as an intervening variable to describe or explain the well-known fact that animals do not respond equally to all stimuli within the environment and, even within a modality, will respond to certain dimensions or configurations of a stimulus and ignore others. Attention per se is difficult to measure directly but can be inferred from certain observing, orienting, or stimulus sampling behaviors. Some stimuli achieve control of behavior more rapidly or effectively than do others in spite of the fact that the stimuli do not differ greatly in discriminability

(Shettleworth, 1972). The rapid learning of odor detection and discrimination tasks suggest that rats may be particularly attentive to odor stimuli. Support for this comes from studies showing that the rat selectively attends to odor stimuli in situations where odors and lights or tones are available as discriminative cues. For example, consider the results of an experiment on selective attention reported by Nigrosh et al. Rats were first trained on a visual discrimination and then given 400 trials of training with compound olfactory and auditory stimuli. Ethyl acetate odor presented simultaneously with a 1-kHz tone served as the positive stimulus and amyl acetate odor presented with a 2.5-kHz tone served as the negative stimulus. During training, cue reduction trials were used to assess the extent to which elements of the compound had gained stimulus control. A cue reduction trial consisted of presenting only one element (the $S^+$ odor, the $S^-$ odor, the $S^+$ tone, or the $S^-$ tone) of the compound stimulus. Responses during test trials were not reinforced. In the next session animals were given alternating blocks of 40 trials on the tone stimuli alone or the odor stimuli alone until there was errorless or near errorless performance on both modalities. This was followed by a final session of compound stimulus training in which cue competition trials were inserted. A cue competition trial consisted of presenting the positive odor stimulus together with the negative tonal stimulus or the negative odor stimulus with the positive tonal stimulus. The results of the cue reduction test (Fig. 6) demonstrated that during compound stimulus training rats learned to respond to the positive odor stimulus but had not learned to respond to the positive tonal stimulus. Because, in the absence of odor cues, rats learn the tonal discrimination in 100–300 trials, these results suggest that during compound stimulus training the presence of odors interferes with the acquisition of the tonal discrimination. The cue competition test (Fig. 6) demonstrated that when cues of competing sign were presented (e.g., the positive odor plus the negative tone) much greater stimulus control was exerted by odors than by tones.

A second experiment by Nigrosh et al. yielded even stronger evidence for selective attention to odor stimuli. One group of four rats (group O–L) was initially trained on a positive odor stimulus and then, within the same session, the negative odor stimulus was introduced. When criterion performance was achieved redundant visual stimuli were presented. After 160 trials on the compound stimulus the odor stimuli were terminated and training continued with the visual stimuli alone. A second group of four rats (group L–O) were first trained on the visual discrimination, then given redundant odor cues and, finally, tested on the odors alone. The results demonstrated that rats in the O–L group acquired the olfactory discrimination with few or no errors but failed to acquire the visual discrimination during the compound odor–light trials. In contrast, the rats in group L–O made many errors in acquiring the visual discrimination but attended to and learned the odor discrimination during the compound light–odor

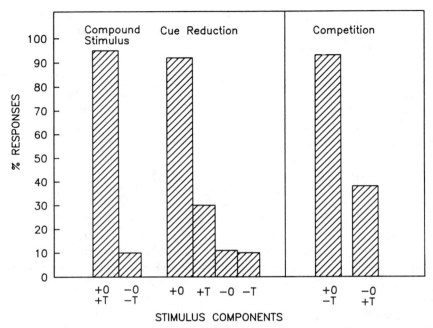

**FIGURE 6.** *Compound stimulus (left panel).* Mean percent responses of 3 rats trained to discriminate between a positive stimulus compound of tone plus odor (+O, +T) and a negative stimulus compound of a different tone and odor (−O, −T). *Cue reduction (left panel).* Mean percent responses in unreinforced test trials in which one of the four stimulus components (+O, +T, −O, −T) is presented. *Competition (right panel).* Mean percent responding to unreinforced test trials in which the positive odor and negative tone (+O, −T) or negative odor and positive tone (−O, +T) are presented. These test trials were given after all animals had been trained in separate sessions to discriminate tones and odors. See text for additional details. Figure based on data of Nigrosh, Slotnick, and Nevin (1975).

trials. The results are best illustrated in Fig. 7 which shows the complete test performance for one rat from each group. Note that the O–L rat (Fig. 7, top) had no decrement in performance when the negative odor stimulus was introduced. This illustrates the errorless acquisition of an olfactory discrimination described above. Performance was not disturbed when redundant visual cues were given but dropped to chance levels when the odor stimuli were terminated. In contrast, the L–O rat made many errors in achieving criterion performance in the visual discrimination task and its performance was disrupted when redundant odor cues were introduced. However, when the visual stimuli were terminated, its performance accuracy in the first block of 20 odor-only trials was 90%. Thus the animal had attended and learned to discriminate the redundant olfactory cues during compound light–odor training.

Two additional outcomes of this experiment are of interest. First, rats in the O–L group made as many errors in acquiring the visual discrimination

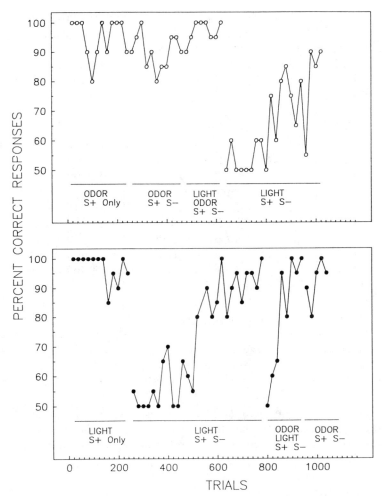

**FIGURE 7.** Complete training and test data for each of two rats. *Top panel:* The rat was initially trained to respond to 0.01% amyl acetate vapor (odor, S+, only) before introduction of a 0.1% ethyl acetate S− stimulus (odor, S+ S−). Next, redundant visual cues (S+, flashing light; S−, steady light) were added for 160 trials (light, odor, S+ S−). Finally, the odor stimuli were turned off and the animal was tested on the visual cues (light, S+ S−). *Bottom panel:* The same procedures were used except that initial training was on visual cues, odor was added as the redundant stimulus, and the animal was tested on the redundant odor cues. Figure based on data from Nigrosh, Slotnick, and Nevin (1975).

as did those in the L–O group. Thus, the visual discrimination performance of the O–L group did not benefit from the 160 trials of compound stimulus training in which lights were available as discriminative (albeit, redundant) cues.

Second, observation of the rats during the test session indicated that each rat in the L–O group was obviously distracted by odors when these were first introduced as redundant stimuli: In initial trials these rats sniffed vigorously at the novel odors and failed to respond to the $S^+$ visual stimulus. As a result, the performance accuracy of these rats dropped to chance on at least the first block of 20 trials of compound stimulus training. In contrast, none of the animals in the O–L group appeared to be disturbed by the onset of the redundant light stimuli during compound stimulus training.

These cue competition and stimulus blocking experiments indicate that rats selectively attend to components of a compound stimulus and that better stimulus control occurs for odors than for visual or auditory stimuli. The preferential use of olfactory cues has also been documented in studies of maze learning and other instrumental behavior in which rats came under the control of inadvertent odors from food reinforcement, the odors of visual stimuli, or odors left behind by other rats (e.g., Ludvigson & Sytsma, 1967; Phillips, 1968; Southall & Long, 1971). A particularly striking example of odor confounding in instrumental learning was reported by Thorne and O'Brien (1971). Rats were trained in a miniature Wisconsin General Test Apparatus to discriminate between white and black stimulus cards. The animals learned the discrimination readily but, surprisingly, blinding produced no deficits in retention. Blinded animals that were also olfactory bulbectomized had no retention and did not relearn the discrimination. Normal rats that were only bulbectomized had no retention but were able to relearn the discrimination, now, presumably, using visual cues. The inadvertent odor cue was eventually identified as the paint used to color the stimulus cards.

As described in the next section, rats not only gain a quantitative advantage when odors are used in an operant discrimination task, but their ability to acquire certain complex discriminations such as learning sets may depend upon the use of odor cues.

## C. Learning to Learn: A Measure of Intelligence or Perception?

The performance on any one discrimination task may be influenced by a large number of variables, such as motivational level, task difficulty, prior reinforcement history, conditioned and unconditioned inhibitory factors, and incentive value of the reinforcement, and may not accurately reflect an animal's ability to learn. To circumvent these problems learning capacity for a species or an individual can be measured not by the acquisition of a single task but the improvement in performance over a series of tests of the same general type. Assuming that the different problems in the series are approximately equal in difficulty then one of two outcomes might be expected: The animal will make approximately the same number of errors on each task or the number of errors made will decrease over the series of problems. The latter is possible if, over a number of problems, the subject acquires a rule that governs the solution to each problem. Animals that achieve errorless or near errorless performance over a series of problems of the same general class are said to have acquired a learning set; they have, so to speak, learned something about learning. Or, as Hodos states:

> The subject that has acquired the learning set has not merely learned to discriminate stimuli; he has learned a strategy or principle of responding that will result in a greater frequency of reward. The principle is: correct on trial one, stay with that stimulus; incorrect on trial one, switch to the other stimulus. In terms of evolutionary significance, the ability to develop an abstract principle from past learning would seem to have a higher survival value than the ability merely to learn about the motivational consequences of particular responses to particular stimuli. Learning set provides the organism with the means for acquiring new skills rapidly in new situations with minimal failure. Without a learning set each problem must be learned anew. An animal that can apply a principle to a new situation would seem to have considerable advantage over one that could not. (Hodos, 1970, p. 34)

The extent to which an animal learns to learn or acquires a learning set has been used as a measure of animal intelligence and as a basis for comparing the intelligence of different species (Hodos). In such comparisons the rat and other "lower" species have fared rather poorly compared to the performance of primates and carnivores. The vast majority of such studies have used visual stimuli as discriminative cues. However, as discussed above, animals are not equally attentive to all stimuli and it is possible that the perceptual salience of a cue may be as important as phylogenetic status in determining whether a complex task such as a learning set is acquired. As described below, this hypothesis has received strong support from learning set experiments in which odors are used as discriminative cues.

In an initial study, Nigrosh et al. tested animals on a sequential discrimination reversal task (SDR). Animals were first trained on an olfactory, visual, or auditory discrimination. In the next session the significance of the stimuli was reversed; the former positive stimulus was now negative and the former negative stimulus was now positive. When criterion performance was achieved the significance of the stimuli was again reversed. This procedure was continued for 10 reversals. The results, shown in Fig. 8, illustrate three important outcomes. First, as measured by errors to criterion, the initial olfactory, visual, and auditory discriminations were approximately equal in difficulty. Second, on the first reversal the visual and auditory groups made many more errors than they did in initial learning (negative transfer). This is a common finding in reversal studies and is due to initial perseverative responding on the prior $S^+$ stimulus. In sharp contrast, the odor group made fewer errors on the first reversal (positive transfer). Finally, the number of errors made by the odor group rapidly decreased over the test series. All rats made fewer errors on reversal two than they did in original learning and after reversal six most

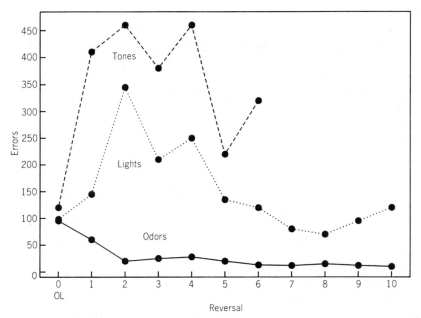

**FIGURE 8.** Median number of errors on original learning (OL) and 10 successive discrimination reversals for rats trained with odors (*n* = 4), lights (*n* = 2), or tones (*n* = 4). Note that only rats in the odor group showed positive transfer on the first reversal and acquired a reversal learning set. Three of the four rats in the odor group made only 2–5 errors in each of the last 5 reversal problems. Figure based on data from Nigrosh, Slotnick, and Nevin (1975).

rats made only one to three errors on each subsequent problem. These rats showed consistent interproblem transfer and rapid acquisition of a reversal learning set. The asymptotic performance of those trained on visual stimuli approached or slightly exceeded that achieved in initial training. (The animals trained on auditory stimuli continued to make many errors and their training was discontinued.) The rapid acquisition of an olfactory SDR set has been replicated in other studies and appears to be a reliable and robust phenomenon (Slotnick, 1984; Slotnick & Kaneko, 1981). The acquisition of an olfactory reversal learning set is not simply a function of stimulus discriminability: It occurs with odors that are qualitatively similar (Slotnick, 1984) and even when the stimuli are different concentrations of the same odor (B. M. Slotnick, unpublished results). In contrast, the performance of rats trained on highly discriminable nonolfactory stimuli such as lights versus tones (Slotnick, 1984) or sucrose versus sodium chloride tastants (Slotnick & Brosvic, 1987) is much worse than those trained with odors.

The rapid acquisition of an olfactory learning set also occurs in tests using a series of novel two-odor discrimination problems. Initial evidence for positive interproblem transfer over a series of olfactory discrimination problems was reported by Jennings and Keefer (1969) and Langworthy and Jennings (1972). These results were extended by Slotnick and Katz (1974) who trained rats to discriminate 16 different pairs of floral odors. For eight rats the odors within each pair were qualitatively similar (e.g., two variants of rose) and for four rats the odors of a pair were qualitatively dissimilar (e.g., rose versus lavender). Human observers could reliably discriminate between two floral types but performed largely at or near chance in discriminating between two variants of the same floral type.

In the first problem rats made 30–90 errors and in successive problems made fewer errors (Fig. 9). By the last problem most rats achieved criterion performance with less than five errors and many of these animals made 0–2 errors in the last 5 problems of the series. One of these animals was trained on an additional 40 problems. It made an average of 2 errors per problem and 0 or 1 error in 11 of the last 20 problems. Rats trained on the qualitatively similar odors performed almost as well as those trained on the qualitatively different odor pairs. These results are notable not only because they demonstrated that rats can acquire a learning set (adopt a "win–stay, lose–shift" rule) comparable to those achieved by primates (trained with visual stimuli) but also because of the rapidity with which the learning occurred. In subsequent studies other investigators have also

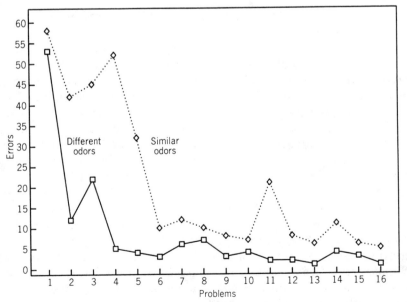

**FIGURE 9.** Mean errors for each of 16 sequential 2-odor discrimination tasks for 4 rats trained on qualitatively similar and 4 rats qualitatively different floral odors. A new and unique pair of odors were used for each problem. Three of the rats trained on qualitatively different odors made only 0–4 errors in the last 10 problems of the series.

reported that rats show rapid interproblem transfer over a series of two-odor discrimination tasks (Eichenbaum, Fagan, & Cohen, 1986; Staubli, Fraser, Kessler, & Lynch, 1986; Staubli, Ivy, & Lynch, 1984).

An even more impressive demonstration of complex olfactory learning and odor memory is provided by the results of a recent study of Kufera, Slotnick, and Risser (1988). Six rats were trained on nine sequential eight-odor discrimination tests. The stimuli for each problem were obtained by random selection (without replacement) from a stock of 100 food, floral, and pure chemical odorants. On each problem the eight odors (four arbitrarily designated as positive and four as negative stimuli) were presented in mixed order in two separate 160-trial sessions (each odor presented 40 times). After completing this training two memory tests were given. The first consisted of repeating the fourth problem of the series. In the second test the odors used in problem 6 were presented but the significance of each odor was reversed. The rats showed a rapid and dramatic improvement in performance over the nine eight-odor discrimination tasks (Fig. 10). In the first only one rat reached criterion performance of 90% correct responding on each stimulus, but in problems 6–9 all rats achieved criterion. In the latter part of the test series rats required only a few exposures to each odor for learning to occur. Thus, as shown in Fig. 10, in the last few problems response accuracy was higher than 70% in the first 40 trials (five exposures to each odor on average) and was higher than 90% in the second set of 40 trials (after approximately 10 exposures to each odor). The retention tests demonstrated that rats remembered odors they had been exposed to in prior sessions. Thus, they made few errors when odors from problem 4 were given again but many errors on the reversal of problem 6 (Fig. 10). Obviously, the rat has a considerable ability to rapidly acquire a rule for discriminative responding and to associate numerous odors with reinforcement or nonreinforcement. The rapid within-session learning of eight odors is impressive considering that odors are sampled for only about 0.5 sec on each trial and that there is ample opportunity for short-term interference effects among odors during these brief (approximately one hour) test sessions. Equally impressive is the evidence from the reversal test that individual odors are remembered in spite of the considerable potential for anterograde and retrograde interference from exposure to other odors in prior test sessions.

A relatively long-term memory for odors in the guinea pig has been reported by Beauchamp and Wellington (1984). Because male guinea pigs readily investigate urine samples from female conspecifics but quickly habituate (show less olfactory investigation) to repeated stimuli from the same female, habituation provides a measure of odor memory. After several exposures (approximately 90 sec of stimulus sampling) to odors from one set of females, a memory for these odors was evident when they were presented again after five weeks.

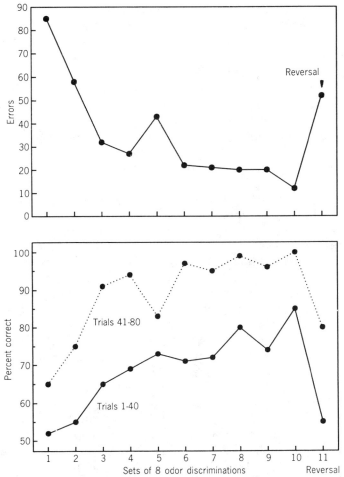

**FIGURE 10.** *Top panel:* Mean errors for 6 rats trained on a series of 11 eight-odor discrimination problems. Problems 10 and 11 were tests for memory. Problem 10 was a repetition of problem 4. Problem 11 was a repetition of problem 6 but with the significance of all stimuli reversed. *Bottom panel:* Mean percent correct performance in the first 40 and second 40 trials. Each odor is presented approximately 5 times in each set of 40 trials. Data from Kufera, Slotnick, and Risser, 1988.

## D. Olfactory Learning in the Pigeon

Why do rats perform simple and complex discriminations so well when odors are provided as cues? Their facility to use odors as discriminative stimuli may simply be an extension of the fact that they, like other macrosmatic animals, have a well-developed olfactory system, use odors in almost all aspects of their daily life, and are biologically prepared to attend to odors. It is also possible that there are properties unique to odors that will allow for exceptionally good stimulus control even in microsmatic species like birds and primates. This question was addressed by Duncan and Slotnick (1985) in a study that repeated in the pigeon the major features of the Nigrosh et al. (1975) report on olfactory stimulus control in the rat. Pigeons were trained in an operant chamber similar to that used by Henton, Smith, and Tucker (1966) to discriminate visual stimuli (red and green lights) or olfactory stimuli (vapors of isoamyl acetate and isopropyl acetate). The test procedures were similar to those described by Nigrosh et al. for the rat but modified for use with the pigeon. After achieving criterion performance, both groups were trained on series of discrimination reversals (reversal test). In a second experiment animals from both groups were trained using compound odor and visual stimuli and then tested on one of the components of the compound (transfer test). Those birds that were originally trained on odors were tested on the visual components of the compound and those trained on visual stimuli were tested on the odor components. In the reversal experiment the visual and odor groups acquired the initial discrimination at the same rate. However, when the significance of the stimuli was reversed, the pigeons trained on odors made many errors and failed to acquire a reversal learning set. Those trained on visual stimuli had positive transfer on the first reversal and improved performance on each subsequent reversal (Fig. 11). On the transfer test pigeons that were originally trained with odors acquired the visual discrimination during compound stimuus training. Those originally trained with visual stimuli did not acquire the olfactory discrimination during compound stimulus training (Fig. 12).

The extent of stimulus control exerted by visual and olfactory stimuli in the reversal set and the transfer test in this study with pigeons was almost precisely opposite that obtained in the corresponding study with rats (Nigrosh et al., 1975). These data strongly suggest that the advantage gained by odors in discrimination learning is species-dependent and is not due to a unique quality of odor stimuli per se. Whatever the basis for this species difference these results provide additional evidence that stimulus modality may be at least as important as task variables and phylogenetic status in determining performance on a learning task. These outcomes also indicate that the ability of a species to acquire a learning set is probably less a measure of intelligence than it is a measure of the perceptual salience of the stimulus.

## E. Stimulus Sampling

Stimulus filtering mechanisms insure that only a small percentage of passively received but otherwise adequate stimuli will influence behavior. Active sampling increases the probability that a stimulus will be perceived and it provides the opportunity to measure attentiveness to a stimulus

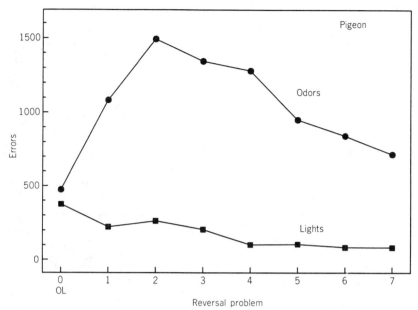

**FIGURE 11.** Mean errors made by 4 pigeons trained on an odor reversal set task and 4 pigeons trained on a visual reversal set task. OL, original learning. Compare with results from reversal study with rats shown in Fig. 8. Data from Duncan and Slotnick (1985).

domain. Visual fixation, tactile manipulation, and pinna erection are easily observable motor acts associated with sampling of visual, tactile, and auditory stimuli, respectively. Sniffing, the active sampling of olfactory stimuli can be measured by assessing increased volume and rate of airflow through the nasal passages. Studies with human subjects have examined parameters of sniffing and optimal sniffing strategies for maximizing detection of odors (e.g., Laing, 1982, 1983, 1985, 1986; Rehn, 1978; Stuiver, 1958; Teghtsoonian, Teghtsoonian, Berglund & Berglund, 1978). Although a consideration of these reports is beyond the scope of the present review, it is interesting to note that presentation of an odor for only 0.18 sec is sufficient for odor detection (Stuiver, 1958) and that a single sniff of only slightly longer duration (0.4 sec) is sufficient to identify an odor even when its intensity is just above recognition threshold (Laing, 1986).

Several investigators have described aspects of sniffing in behaving animals (Marshall & Moulton, 1977; W. H. Teichner, Price, & Nalwalk, 1967; Welker, 1964) but only Youngentob, Mozell, Sheehe, and Hornung (1987) have performed a systematic and detailed analysis of sniffing behavior in an animal trained to detect odors. Rats were trained using operant conditioning to detect the odor of amyl acetate or pyridine. The

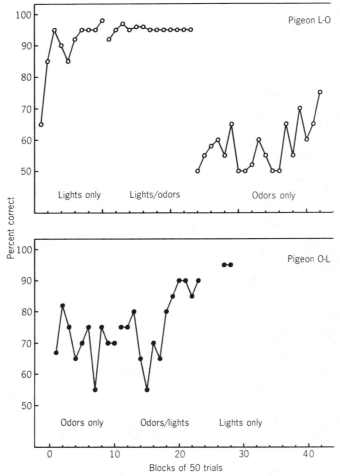

**FIGURE 12.** Percent correct responding for two birds. *Top:* Pigeon L–O was first trained to discriminate visual stimuli (lights only), then given additional training with redundant odor cues and then tested on odors only. *Bottom:* Pigeon O–L was first trained on odor stimuli (odors only), then given additional training with redundant visual cues, and then tested on visual cues. Compare with the results of a similar experiment with rats shown in Fig. 7. See text for additional details. Data from Duncan and Slotnick (1985).

stimuli were sampled from an enclosed port fitted with a pneumotacho-graph and differential pressure transducer. These served to continuously monitor flow rate and total flow volume during active sniffing. Of 52 direct and derived measures used to characterize sniffing behavior (e.g., sniff bout duration, number of sniff bouts per trial, duration of expiratory and inspiratory sniffs, sniff volume, peak flow rate of sniffs, decay slope of sniffs), 12 measures were found to discriminate reliably between air and odor trials in a detection task. Basically, these measures were descriptors of volume, duration, average and peak flow rate, and sniff number. Generally, there was one primary bout of sniffing during a trial. Within a bout there were more inspiratory than expiratory sniffs and the number of sniffs and sniff duration varied with odor concentration: As concentration decreased there was an increase in inspiratory and expiratory sniffs, total inspiratory volume, peak flow rate, and duration of the sniff bout. The average duration of a sniffing bout in different experiments was approximately 0.6 to 0.7 sec for high concentration of the stimulus and increased by 0.1 to 0.2 sec in tests with low (near threshold) concentrations. Interestingly, sniff bout duration was significantly longer on air trials (0.79 sec) than on odor trials (0.66 sec). As described below, similar results have been obtained in our studies on sample duration.

This brief summary highlights only a few of the interesting outcomes of Youngentob's study. The results provide a basic description of active, purposeful odor sampling and evidence that some parameters of sniffing may be used to measure attentive behavior to an olfactory stimulus.

The pneumotachograph method for recording sniff behavior is, unfortunately, not compatible with the dynamic positive pressure airflow odor generator used in most automated olfactometers. In an effort to obtain a simple measure of stimulus sampling we have employed the vertical stimulus sampling port described above. Rats quickly learn to sample odors from this port. Sample duration is defined by the amount of time a photocell at the neck of the port is blocked by the animal's snout during the stimulus-on period of a trial.

While duration is a relatively gross measure of stimulus sampling it does change in a meaningful way as a function of problem difficulty and parameters of the stimulus. Sample duration has not been examined systematically but some interesting outcomes have emerged in the course of testing animals on detection and discrimination tasks (B. M. Slotnick, unpublished results).

First, it is clear that the minimum sample duration required to detect stimuli that are at least several orders of magnitude above threshold is probably less than the 0.2 sec required by our operant schedule. Thus, a number of rats tested in the absence of the 0.2-sec-sample requirement accurately detected odors and discriminated among odors after sampling for only 0.15–0.2 sec. Nevertheless, and in agreement with the results of Youngentob, most animals sampled the stimulus for about 0.4–0.8 sec before responding.

Second, sampling duration changes as a function of a number of variables including which stimulus is presented ($S^+$ or $S^-$), amount of training, problem difficulty, and novelty of the stimulus. Under most conditions animals sample longer on $S^-$ trials then on $S^+$ trials. This is in agreement with Youngentob's finding that sniff bout duration was reliably longer on $S^-$ than on $S^+$ trials. Sampling duration on $S^-$ trials generally decreases as training proceeds but still tends to remain about 0.1–0.2 sec longer than sample duration on $S^+$ trials (Fig. 13).

Introduction of a novel stimulus results in a sharp increase in sample duration. A new stimulus presented at any time during a training session is almost always sampled for approximately 1 to 2 sec (Fig. 13). For example, in a recent unpublished study 19 rats were first trained on a detection task ($S^+$, 0.5% amyl acetate; $S^-$, air) and then trained to discriminate between amyl acetate and 2% butanol ($S^-$). The butanol stimulus was approximately equal in perceived intensity to that of the amyl acetate stimulus. The mean sample duration on the very first trial in which butanol was presented was 1.5 sec (range: 0.7–2.0) and each rat sampled the butanol stimulus longer than it did the $S^+$ (amyl acetate) stimulus. Further, differences in sample duration within this group were related or predictive of whether they responded to the novel stimulus on its first presentation. The mean sample duration of 8 rats that responded was 0.9 sec or, on average, 0.4 sec longer than they sampled the $S^+$ stimulus. The 11 rats that did not respond sampled the stimulus for a mean of 1.9 sec or, on average, 1.3 sec longer than they sampled the $S^+$ stimulus.

Why do rats sample the stimulus longer on $S^-$ trials? One explanation is that reinforcement is not available on these trials while, on $S^+$ trials, short

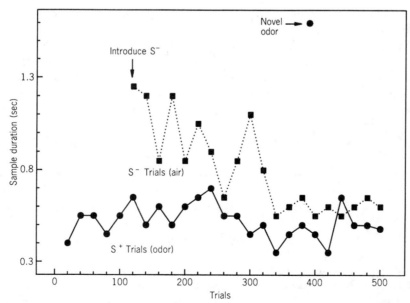

**FIGURE 13.** Mean sample duration on $S^+$ (odor) and $S^-$ (air) trials for one rat. Note the sharp increase in sampling time when the novel odor was presented in one trial. Mean performance accuracy over all blocks of 20 trials was 97% (range, 85%–100%). See text for additional details (B. M. Slotnick, unpublished results).

samples minimize the delay of reinforcement. However, several observations suggest that this cannot be the only factor governing the difference in sampling duration on $S^+$ and $S^-$ trials. First, a novel stimulus will almost always be sampled for a longer time than training stimuli. A second observation relates to stimulus detectability. As stimulus concentration decreases, sample duration on $S^+$ trials increases and may exceed that on $S^-$ trials. This is illustrated in Fig. 14 which shows the change in sampling duration for three rats tested for absolute threshold on propionic acid. Note that sample duration on $S^+$ trials increased as stimulus concentration was lowered but sample duration on $S^-$ trials remained the same or increased only slightly. By the fifth problem of the series ($S^+$, 0.001% propionic acid; $S^-$, air) each rat sampled longer on most $S^+$ trials than on $S^-$ trials (Fig. 14). While one might expect sampling to increase as stimulus concentration decreases, it is less clear why sampling duration should be longer on $S^+$ than on $S^-$ trials. That is, if only a short sample is required to determine that no stimulus is present, why should the animal sample longer to determine that a stimulus is present? One possible explanation is that even in the absence of differential odor training, the animal samples

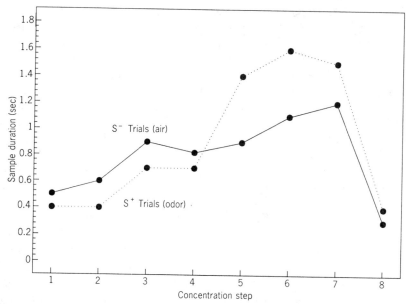

**FIGURE 14.** Mean sample duration on $S^+$ (propionic acid vapor) and $S^-$ (air) trials for three rats as a function of odor concentration. Concentration steps 1–8 were 0.5, 0.1, 0.01, 0.005, 0.001, 0.0005, 0.0001, and 0.00001% (of vapor saturation), respectively. Each data point is based on 90 trials. Performance accuracy at concentration steps 1–7 was 80–100%. Concentration step 8 was below detection threshold and performance accuracy was 55% (B. M. Slotnick, unpublished results).

not only to detect the stimulus but also to recognize it. At low concentration a longer sample may be required for recognition than that needed to ascertain that no stimulus is present. This explanation is admittedly speculative but it is true that even in a detection task recognition probably plays a role in governing sampling and responding: Rats vigorously sample but often do not respond to novel stimuli introduced at irregular times during a session and sample duration, even on simple detection problems, far exceeds that needed to detect the stimulus. The important point is that factors governing sampling duration may be quite complex and not determined entirely by reinforcement contingencies.

Sampling duration may also provide a relatively sensitive measure of an animal's ability to detect or discriminate odors. In the acquisition of a difficult detection or discrimination task in which accuracy is initially at chance levels, sampling duration may predict improvement in performance. An example of this is shown in Fig. 15. The rat whose performance is illustrated was trained on a relatively difficult task in which the positive stimuli were two concentrations of propionic acid ($S^+$ stimuli) and the negative stimulus was a mixture of propionic acid and acetic acid ($S^-$ stimulus). The perceived intensity of the mixture ($S^-$) was intermediate between intensities of the two $S^+$ stimuli. As shown, performance accuracy was at chance levels for the first 240 trials and then gradually improved. For the first 100 trials there were no consistent differences in sample duration on $S^+$ and $S^-$ trials. However, after trial 100 sampling duration increased on $S^-$ trials and, thereafter, performance accuracy improved. In general, a difference in sampling duration on $S^+$ and $S^-$ trials at a time when accuracy is low is predictive of a subsequent improvement in performance. In our experience, a clear change in sampling between $S^+$ and $S^-$ trials is generally followed by an increase in accuracy within 100–200 trials.

In tests of absolute threshold using a method of limits procedure, an interesting pattern of sampling behavior occurs. In well-trained animals there may be little difference in $S^+$ and $S^-$ sampling at high concentrations of the stimulus. As described above, as the stimulus concentration is lowered, sampling time on $S^+$ trials increases. At concentrations that are below threshold, sampling duration decreases and becomes equally low on $S^+$ and $S^-$ trials (Fig. 14). The animal responds on every trial and many short sample trials occur (trials in which it responds after sampling the stimulus for less than 0.2 sec). In short, the animal is no longer under stimulus control; it has stopped attending to the stimuli. With extended training under these conditions it may learn not to pay attention to the stimulus. The evidence for this is that after discriminable stimuli are reintroduced the animal will continue to perform at chance for many trials before coming under stimulus control. In this case also, a gradual increase in $S^-$ sampling duration precedes an increase in performance accuracy.

In summary, it is evident that parameters of sniffing behavior provide

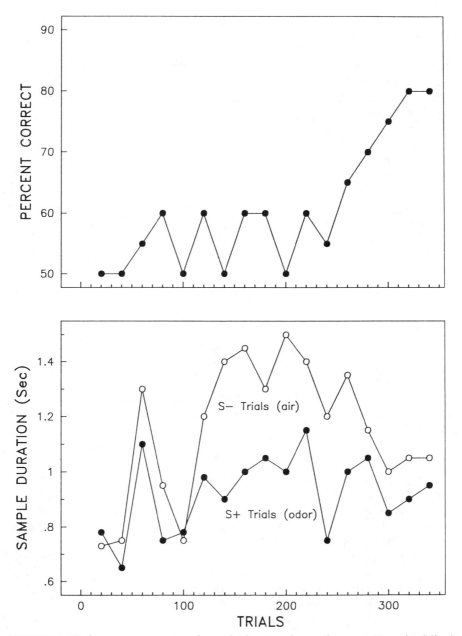

**FIGURE 15.** Performance accuracy and sample duration during the acquisition of a difficult odor mixture discrimination task for one rat. Note that a large difference in sample duration between S$^+$ and S$^-$ trials precedes an increase in response accuracy (B. M. Slotnick, unpublished results). See text for additional details.

information about the detection and discrimination of odors that is not available from the more traditional behavioral measure of response accuracy. The time the animal keeps its snout in an odor stream is a relatively crude and indirect measure of odor sampling but it is easy to obtain and provides an index of attention to the stimulus. These results strongly argue in favor of including some measure of stimulus sampling in future studies on odor detection and discrimination.

These data on stimulus sampling are also relevant to a methodological issue. In some studies animals have been trained to hold their head in the stimulus stream for one or more seconds to signal detection of the correct ($S^+$) stimulus (e.g., Eichenbaum, Kuperstein, Fagan, & Nagode, 1987; Moulton & Marshall, 1976). However, this method confounds stimulus sampling with operant responding. Also, because rats tend to sample longer on $S^-$ trials, requiring the animal to keep its nose in the stimulus stream to signal detection of the positive stimulus may unnecessarily increase the difficulty of the task.

## IV. THE NEURAL BASIS OF OLFACTORY PERCEPTION

### A. Connections of the Olfactory Cortex

Prior to the advent of contemporary anatomical studies it was believed that the olfactory system did not have a thalamic–neocortical representation and that its higher-order connections were primarily to the limbic system and hypothalamus. Indeed, it was common for neuroanatomists to include these structures within a rhinencephalon (smell brain) and assume that processing of olfactory impulses occurred primarily in the subcortical components of the limbic system. Recent studies have identified four areas of the brain which receive input from cells of the olfactory cortex: the limbic system (amygdala and entorhinal cortex), the hypothalamus, the ventral pallidum, and the thalamus. These areas, presumably, constitute higher-order processing centers for olfactory information. In addition, but not considered further in this review, are the centrifugal projections of the olfactory cortex upon the olfactory bulb.

As described below, the results of studies on olfactory discrimination suggest that the various central projections of the olfactory system subserve different functions. However, this area of investigation is in its infancy and, in some respects, the most interesting outcomes of many lesion studies are their failure to demonstrate a specific olfactory dysfunction.

### B. Lesions of the Olfactory Cortex

The lateral olfactory tract (LOT) is the primary efferent pathway connecting the olfactory bulb with the forebrain and thus one might expect that transection of the LOT would produce severe deficits in olfactory perception. While several studies reported that transection of the LOT results in anosmia (Fleming & Rosenblatt, 1974; Long & Tapp, 1970; Sapolsky & Eichenbaum, 1980; Thompson, 1980), these outcomes were probably due to inadequate behavioral tests (Slotnick & Berman, 1980). In our own experiments (Slotnick & Berman, 1980; B. M. Slotnick & F. W. Schoonover, unpublished results) rats with lesions were tested for a preoperatively learned odor detection task (0.5% amyl acetate versus air) and on a two-odor discrimination task (0.1% propyl acetate versus 0.1% ethyl acetate). In some rats horseradish peroxidase (HRP) was injected into the olfactory bulbs prior to sacrifice and the brain was processed according to the method of Mesulam (1978). Brain sections were examined for anterograde transport of HRP to sites of termination of olfactory mitral and tufted cells.

Completely bulbectomized rats had no retention of the odor detection task and performed at chance in 2000 retraining trials but performed well

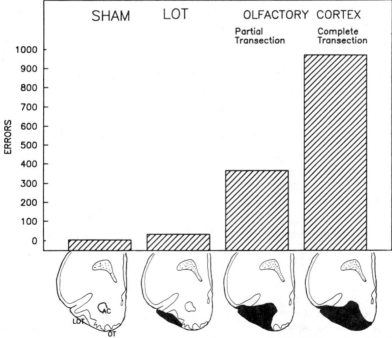

**FIGURE 16.** Mean errors made on a simple odor detection task (S⁺, 0.5% amyl acetate; S⁻, air) after lesions of the lateral olfactory tract ($n = 7$), partial transection of the olfactory cortex ($n = 6$), or complete transection of the olfactory cortex ($n = 4$). Rats in the complete transection group performed at or near chance in 2000 trials. All other achieved criterion performance. Abbreviations: AC, anterior commissure; LOT, lateral olfactory tract; OT, olfactory tubercle. From B. M. Slotnick and F. W. Schoonover (unpublished results).

on a preoperatively learned visual discrimination. Surprisingly, rats with discrete transection of the LOT performed nearly as well as nonlesioned controls. This was true even if the tract was interrupted just as it distributes collaterals to the olfactory cortex. Rats with larger lesions of the olfactory cortex had significant deficits in detection and discrimination. The magnitude of these deficits was clearly related to the extent of the transection in the frontal plane. Thus, lesions that extended from the lateral aspect of the piriform cortex to the lateral edge of the olfactory tubercle (transecting the LOT, anterior commissure, and piriform cortex) produced a severe deficit in detection and discrimination (Fig. 16). However, these rats were not anosmic; they were able to reacquire the odor detection task and most performed above chance in the two-odor discrimination test. Anosmia, defined as a failure to reacquire the odor detection task in 2000 trials, was produced only in rats with complete transection through the olfactory peduncle or the rostral olfactory cortex (and in those with complete removal of the olfactory bulbs). If the transection spared a small ventral or medial segment of the olfactory cortex (or anterior olfactory nucleus, in rats with more rostral lesions) the rat was able to detect and, in some cases, discriminate odors. No critical site within the olfactory cortex for odor detection or discrimination was found: Discrete lesions of the olfactory tubercle, the lateral olfactory tract, the anterior commissure, or piriform cortex produced only small deficits.

The savings in olfaction in rats with large lesions of the olfactory peduncle or cortex can best be understood with reference to recent anatomical studies showing that a considerable number of fibers from the olfactory bulb do not travel in the main body of the LOT but project to the medial half of the olfactory cortex via medial and deep regions of the olfactory peduncle (Broadwell, 1975; Devor, 1976; Scalia, 1966). The exact course of these projections are not known but their termination in olfactory cortex could be demonstrated in our anatomical studies of lesioned animals: Anterograde transported HRP from the olfactory bulb of rats with transections through the lateral half or two-thirds of the olfactory cortex was found in layer 1 of the olfactory tubercle several millimeters posterior to the level of the transection. These projections probably supported odor detection in animals with lesions that spared only the medial segment of the olfactory peduncle or olfactory cortex.

It is clear from these experiments that rats are able to perceive odors even after rather massive insult or deafferentation of the olfactory cortex. If frank anosmia is to be achieved by lesions of the CNS it is necessary to either remove the olfactory bulbs or essentially separate the bulbs from the forebrain.

The finding that rats with more discrete lesions (including those with transection of the LOT, the major efferent pathway from the olfactory bulb) performed as well or nearly as well as controls was unexpected. These behavioral savings may reflect the fact that the detection and discrimination tasks were relatively simple and, hence, insensitive to small deficits, or, alternatively, that there is considerable redundancy in the projection of the olfactory bulb onto the olfactory cortex. The later hypothesis is, in fact, well supported by studies showing that the olfactory bulb projections onto olfactory cortex are not topographically organized in a point-to-point fashion but are rather broad and overlapping. These anatomical results are consistent with Haberly's suggestion that olfactory information is stored as an ensemble code in the cortex such that any one odor would activate many thousands of cells and each cell might participate in the coding of many different odors (Haberly, 1985). Our results are compatible with this hypothesis although the experiments were not specifically designed to test it. However, it is evident that behavioral tests of olfactory detection and discrimination in combination with the brain lesion method could provide a powerful tool for assessing this and other hypotheses of the functional organization of the olfactory cortex.

## C. Interruption of Olfactory Afferents to the Limbic System

Axons of the LOT project to periamygdaloid cortex and, via a synapse in the lateral entorhinal cortex, to the dentate gyrus and fields CA1 and CA3 of the hippocampus proper. Evidence that the projections to the amygdala are somehow involved in olfactory perception stems from human clinical studies of olfaction after amygdalectomy (Andy, Jurko, & Hughes, 1975) and of psychomotor epilepsy associated with morphological or electroencephalographic abnormalities of the periamygdaloid cortex or uncus. The auras that preceded these fits included strong olfactory sensations. These were frequently described as highly disagreeable and often were accompanied by taste and gastrointestinal sensations (Jasper, 1958). Interestingly, sites in the amygdala in which electrical stimulation arouses sniffing and olfactory investigation are adjacent to or overlap sites involved with ingestive behavior (licking, chewing, salivation, and swallowing) or fear and other signs of autonomic arousal (Ursin & Kaada, 1960).

However, experimental attempts to produce olfactory discrimination deficits by lesions of the amygdala or posterior pyriform lobe have been singularly unsuccessful. Thus, Swann (1934) reported that rats with large temporal lobe lesions had no deficits in following an odor cue in a maze. Likewise, studies on primates with temporal lobe lesions failed to find deficits in simple odor discrimination tasks or changes which were specific to olfaction (Brown, 1963; Brown, Rosvold, & Mishkin, 1963; Santibanez & Pinto Hamuy, 1957; Shuckman, Kling, & Orbach, 1969).

The olfactory tasks used in these early studies were relatively simple, provided little control of stimulus parameters, and were probably insensitive to small changes in discrimination capacity. To reexamine this problem with more stringent and better controlled tests, rats with posterior piriform cortical lesions were tested on simple (Slotnick, 1985) and complex

(Slotnick & Kaneko, 1981) discrimination tasks. In most rats the lesions transected the piriform cortex and LOT just rostral to the anterior amygdala, thus eliminating all direct olfactory projections to the amygdala and entorhinal cortex. Surprisingly, experimental animals performed as well as controls in tests of retention, acquisition of two-odor discrimination problems, reversal learning, and in a test for olfactory intensity-difference threshold. Eichenbaum et al. (1986) have also reported no deficits in the acquisition of successive two-odor discriminations or learning an olfactory reversal task in rats with amygdala lesions. These results are in agreement with those of earlier studies on the amygdala and suggest that olfactory projections to the limbic system do not play a significant role in simple or complex olfactory learning or in olfactory sensitivity (as measured by a test for intensity-difference threshold).

In contrast, other investigators have reported that amygdala lesions or posterior transection of the LOT eliminated or produced marked changes in mating of male hamsters (Devor, 1973, 1975; Macrides, Firl, Schneider, Bartke, & Stein, 1976; Marques, O'Connell, Benimoff, & Macrides, 1982). Olfactory function was not directly assessed in these experiments but male sexual behavior in the hamster is known to be strongly dependent upon smell (Murphy, 1976) and it is reasonable to assume that the behavior deficits were caused by interruption of olfactory afferents to the amygdala and entorhinal cortex. This assumption is also supported by the finding that posterior LOT transection in the hamster also disrupts other olfactory-dependent behaviors (scent marking, hoarding, and nesting; Marques et al., 1982) and that disruption of mating can be produced in animals with unilateral posterior LOT transection if the contralateral olfactory bulb is removed (Devor, 1975).

The different outcomes of these two sets of experiments indicate that a deficit in an olfactory-dependent behavior may not be revealed by tests of olfactory learning or discrimination. Of course, it is conceivable that animals with posterior transection of the LOT may have specific anosmias for pheromonal cues. However, it seems more likely that the deficits in species-specific behaviors resulted from disconnection of olfactory input from brain areas involved in the arousal or execution of these acts rather than a disruption of sensory capacity per se (see also Devor, 1973). Certainly, there are important limitations in using only a learning task or only an olfactory-dependent, species-specific behavior to assess the functional role of an olfactory projection.

The olfactory projections to the entorhinal cortex are of considerable interest because this cortex provides a major input to the hippocampus, a structure that appears to be important for memory. There is electrophysiological evidence that olfactory impulses alter hippocampal activity (Komisaruk & Bayer, 1972; J. L. Price, 1985) and Eichenbaum et al. (1987) have described cells in the hippocampus that respond to different motor components of an operant olfactory discrimination task. Of particular interest were cells (cue-sampling cells) that were active when the animal sampled the discriminative stimulus but not during exploratory sniffing in the chamber. As discussed by Eichenbaum et al. (1987), these cells may be involved in evaluating the stimulus or be part of a neural mechanism for matching the new input with a retrieved neocortical image or representation.

Evidence for hippocampal involvement in olfactory memory has been reported by Staubli et al. (1984). Rats were trained in a maze on a series of 2-odor discrimination tasks and then given lesions of the lateral entorhinal cortex (which receives an input from the LOT and, in turn, projects directly to the hippocampus) or the dorsal entorhinal cortex (which does not receive an olfactory input). In postoperative tests the animals were trained on new odors. The intertrial interval was 0–2 min for some problems and 3–10 min for others. The performance of control animals (those with lesions of dorsal entorhinal cortex) was unaffected by the intertrial interval manipulation. Those with lateral entorhinal cortical lesions performed as well as controls when the intertrial interval was short but performed more poorly when the intertrial interval was 3–10 min. In a second test both groups were trained using a short intertrial interval, returned to their home cage for 1 hr, and then retested but with the significance of the olfactory cues reversed. In the reversal test, controls showed negative transfer (responded to the prior positive stimulus) but experimental rats had no such preference and rapidly acquired the new task. In a later study Staubli et al. (1986) replicated this reversal effect but also found that if rats learned the original task preoperatively then entorhinal lesions had little effect on the reversal test. In general, the results of these studies suggest that olfactory deafferentation of the hippocampus can interfere with some aspect of short-term memory for odors. Thus, experimental animals were apparently unable to retain an association between odor and reward for more than a few minutes (results of the intertrial interval manipulation) or retain a learned odor discrimination for more than 1 hr (results of the reversal manipulation). However, there was no evidence that lesions disrupted long-term memories for odors.

These findings are supported by results of Eichenbaum et al. (1986). Rats with lesions of either the ventral hippocampal commissure (which interrupt projections of the hippocampal fornix) or of the amygdala were trained using operant conditioning on three sequential two-odor discrimination tasks and on a reversal task. Differences among groups were obtained only for the reversal task. Rats with hippocampal damage acquired this reversal more rapidly than did controls or those with lesions of the amygdala. As in the study by Staubli et al. (1986), an analysis of errors indicated that rats with hippocampal damage showed less negative transfer than did controls and, hence, made fewer errors in learning the reversal task. While there were many important procedural differences between these two studies, the results of both suggest that interference with hippocampal function does not disrupt simple olfactory discrimination learning but may interfere with olfactory memory. These studies are the first to provide evidence for hippocampal involvement in odor memory. However, both used an indirect method to test memory and the basis of the disruption is unclear. The deficit may have been due to forgetting or proactive interference. Other factors, such as increased distraction during long intertrial intervals or those unique to reversal learning, must also be considered. In any case it seems clear that the olfactory projections to the hippocampus may play an important but, as yet, not easily specified role in olfactory learning and memory.

## D. The Olfactory Thalamocortical System

Certainly one of the more notable advances in our understanding of the organization of the olfactory system was the discovery that olfactory cortex projected to the mediodorsal nucleus of the thalamus (Powell, Cowan, & Raisman, 1965). This thalamic projection has now been well established in studies using axonal transport methods and neurophysiology (e.g., Benjamin, Jackson, Golden, & West, 1982; J. L. Price & Slotnick, 1983) and exists in marsupials, lagomorphs, rodents, and primates. In the rat the projection originates from cells deep to the piriform cortex and olfactory tubercle. These cells probably receive their olfactory input from pyramidal cells in layer II of olfactory cortex and not directly from the LOT. In addition, cells from a more restricted part of the olfactory cortex project to the submedial nucleus of the ventromedial thalamus. Both thalamic nuclei project to the prefrontal cortex, specifically the ventral agranular insular and lateral orbital areas on the dorsal bank of the rhinal sulcus (J. L. Price & Slotnick, 1983). The existence of an olfactory thalamocortical system raises the interesting possibility that it, like other sensory thalamocortical systems, may subserve higher integrative and discriminative functions. However, such speculation must be tempered by anatomical findings which suggest that the olfactory thalamocortical system differs in important ways from other, more "typical" sensory systems which relay information through the thalamus to the neocortex. Thus, this neocortical area also receives olfactory input from the piriform cortex and this projection appears to be more dense and more direct than that from the thalamus. There also may be important species differences. In some species an olfactory connection to the ventral agranular insular part of this cortex comes directly from the olfactory bulb. This direct bulbar-neocortical projection has been found in the mouse (Shipley & Adamek, 1984) and opossum (Meyer, 1981) but has not been identified in the rat (Scott, 1986). In the monkey, there appear to be two separate neocortical olfactory areas. One, in the central posterior part of the orbital cortex, receives a direct input from the mediodorsal nucleus. The second, in the lateral posterior orbital cortex, does not appear to receive a thalamic input but is connected to the olfactory system via the basal forebrain and amygdala (Takagi, 1984). Finally, of course, the thalamic inputs into olfactory neocortex originate from a cortical structure which, although more simple in organization than neocortex, provides the basis for considerable processing of olfactory information.

What role does the olfactory thalamocortical system have in processing olfactory information? The results of two studies on olfactory learning set suggest that these projections may be essential for complex olfactory learning. In the first, controls and rats with lesions of the mediodorsal nucleus or with posterior transection of the LOT were tested on an odor reversal learning set task (Slotnick & Kaneko, 1981). Animals were trained preoperatively on a visual discrimination and, after surgery, tested for retention, trained to discriminate the odor of propyl acetate ($S^+$) from ethyl acetate ($S^-$) and then tested on a series of 6 discrimination reversals. There were no significant differences in error scores among groups on retention of the visual discrimination or acquisition of the initial olfactory discrimination. Controls had rapid acquisition of the reversal set. Most had improved scores on the first reversal (positive transfer) and made few or no errors in the last 2–3 reversals. However, rats with MD lesions performed much more poorly. Those with complete lesions of MD made many more errors on the first reversal than in original learning (negative transfer). Their performance gradually improved but none demonstrated acquisition of the learning set. Rats with partial damage to MD performed somewhat better and those with thalamic lesions that did not damage MD had no deficits. Three rats with MD lesions were also tested on a visual discrimination reversal task. Normal rats generally perform poorly on a visual reversal task (Nigrosh et al., 1975). Rats with MD lesions performed in a similar manner and were not different from controls.

In a second study Lu and Slotnick (1988) tested rats with MD lesions on a series of two-odor discrimination problems. If the rat did not achieve criterion performance within 400 trials training on that problem was terminated and the next problem was given. Lesioned and control groups had excellent retention of a preoperatively learned odor detection task and their performance on the first three of the two-odor discrimination problems was similar. In subsequent problems control rats demonstrated acquisition of a learning set. In contrast, the performance of rats with MD lesions was variable and none showed a consistent improvement over the test series (Fig. 17). Further, five of the seven experimental rats but none of the controls failed to reach criterion in one or two of the discriminations. Discrimination failures occurred on different problems for different rats and no one problem appeared to be difficult (resulted in high error scores) for all or even a majority of experimental animals.

Studies by Eichenbaum and his associates have also demonstrated that rats with lesions of the mediodorsal nucleus or of the olfactory neocortex make more errors than do controls in learning an odor discrimination.

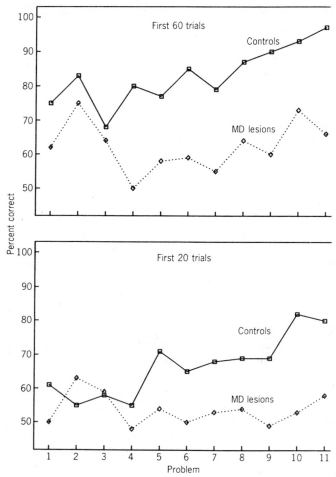

**FIGURE 17.** Mean performance of 5 sham lesioned control rats and 5 rats with destruction of the thalamic mediodorsal nucleus on the first 60 (top) or first 20 trials (bottom) of a series of 11 sequential two-odor discrimination problems. Note that controls rats demonstrate acquisition of a learning set but rats with thalamic lesions do not. See text for additional details. Data from Lu and Slotnick (1988).

Eichenbaum, Shedlack, and Eckmann (1980) tested rats on a preoperatively learned olfactory discrimination and for acquisition of new problems. Complex effects were observed but, overall, rats with thalamic lesions could perform as well as controls on some tasks but made approximately twice as many errors as controls in other tasks, particularly those that used novel odors or that could be characterized as difficult discriminations. Rats with lesions of the rhinal sulcal cortex (which included the cortical projection site for the central or olfactory part of MD) performed more poorly than did those with MD lesions. The scores of control rats that had lesions of frontal cortical areas that receive a projection from the lateral (nonolfactory) part of MD were similar to that of sham lesioned controls. Eichenbaum, Clegg, and Feeley (1983) replicated these effects of olfactory neocortical lesions and, in addition, demonstrated that they were specific to an olfactory task and occurred whether a symmetrical or asymmetrical reinforcement procedure was used.

The possibility that the thalamocortical system might play a significant role in memory storage was tested by B. M. Slotnick and J. M. Risser (unpublished study). Prior to surgery rats were trained to discriminate among eight odors. Four were designated positive and four negative and the stimuli were presented in mixed order within a session. The animals were then rested for 14 days and then retested under conditions of extinction (preoperative memory test). They were then retrained, operated on, and, 14 days later, they were again tested under extinction (postoperative memory test). Because the extinction test provides no feedback for correct or incorrect responses, performance accuracy during this test provides a reasonably pure measure of memory.

All rats showed nearly perfect retention in both the preoperative and postoperative retention tests. The thalamocortical system does not appear to play a significant role in long-term memory for odors.

In summary, lesions that interrupt either the thalamic or cortical component of the thalamocortical system can produce severe deficits in some forms of olfactory learning and these are much greater than those observed after olfactory deafferentation of the limbic system. Eichenbaum et al. (1980) suggested that these deficits are related to stimulus novelty or discrimination difficulty. The latter possibility appears to be more likely since in our studies most rats with MD lesions had little or no learning deficits when presented with novel, albeit simple, odor discrimination tasks. The effects of lesions on difficult olfactory discriminations has not been examined systematically but this hypothesis is consistent with the neurophysiological results of Tanabe, Iino, and Takagi (1975) that cells in the monkey orbital cortex have narrower receptive fields than do those in the olfactory cortex or amygdala.

Our own studies support the notion that an intact thalmocortical system is important for the interproblem transfer that underlies the acquisition of a learning set. At present it is uncertain whether this deficit in the acquisition of a learning set is specific to olfaction. This is difficult to determine because rats do not readily acquire a learning set if olfactory cues are not used (Slotnick & Brosvic, 1987). However, it is likely that the deficits in olfactory discrimination in rats with cortical lesions (Eichenbaum, Clegg, & Feeley, 1983; Eichenbaum et al., 1980) are specifically related to olfaction. In these studies discrimination deficits were obtained with lesions of olfactory neocortex (orbital frontal cortex) but not with lesions of (nonolfactory) anteromedial frontal cortex. In nonolfactory tasks (e.g., operant responding on a partial reinforcement schedule, spatial delayed alternation, spatial reversal) the opposite pattern of results is generally obtained: Performance deficits are associated with lesions of anteromedial prefrontal cortex and, generally, not with lesions of orbital frontal cortex (review by Rosenkilde, 1983).

## V. OLFACTORY CODING

A better appreciation of olfactory perception will require an understanding of how the olfactory system is able to detect and discriminate different odors, that is, how olfactory information is coded by the nervous system. The problem of olfactory coding has been a central issue of physiological and anatomical studies, but until recently has not been addressed by behavioral neuroscientists.

Recent investigations have concentrated on mechanisms that may be used at different stages of olfactory processing, from the receptor to the olfactory cortex. Examples include studies on the molecular basis of receptor activation (e.g., Anholt, 1987; Lancet & Pace, 1987), the extent to which channels of information are segregated or preserved from the receptor sheet to the forebrain (Saucier & Astic, 1986; Shepherd, 1985), circuit or synaptic analysis of information processing within the olfactory bulb (Shepherd, 1972), pattern analysis of electrophysiological events (Freeman, 1978, 1982; Meredith, 1986), and the search for potential primary odors (Amoore, 1977). Behavioral studies that have been specifically designed to test some of the hypotheses derived from this research are reviewed in this section. Studies of olfactory perception provide unique tests for the functional significance of such hypotheses and it is likely that they will play an increasingly important role in this area of investigation.

### A. Odor Classification and Stimulus Generalization

The earliest such behavioral study provided a test of Amoore's stereochemical theory of odor classification. Briefly, considerations of the shape and other attributes of odorous molecules led Amoore to propose several classes (or primary types) of odors (i.e., camphoraceous, ethereal, floral, musky, pepperminty, pungent, putrid) and, on the basis of molecular structure, to predict into which class a specific odor would fall. According to this theory odors within a class should be perceptually more similar than are odors from different classes. This aspect of the theory was tested by Braun and Marcus (1969) in the only published account of olfactory stimulus generalization in animals. This study is also one of the earliest to use olfactometric methods and operant conditioning with rats. Rats were trained to discriminate between an ethereal and a floral, an ethereal and a camphoraceous, or a floral and a camphoraceous odor. Odors were presented for 1-min periods and responses in the presence of the positive odor were reinforced on a variable-interval 30-sec schedule. During training the concentration of the odors was varied to insure that animals would discriminate on the basis of odor quality and not on the intensity differences between stimuli. In the stimulus generalization test each animal was presented with other odors from the same class as the training odors. Response rate in the presence of the test odor was used as a measure of the perceptual similarity of that odor to the training odor. For example, if an animal was trained to discriminate between an $S^+$ of acetone (ethereal odor) and an $S^-$ of geraniol (floral odor), it was tested with a different ethereal and floral odor. Strictly speaking, this does not constitute a test of stimulus generalization because few test odors were used and there was no basis for defining a physical gradient among odors. The test is perhaps better described as one of odor naming or odor classification. In any case the results demonstrated that rats judged odors within a class to be more similar than those from different classes. Thus, an animal trained on an ethereal $S^+$ odor and a floral $S^-$ odor made more responses to a novel ethereal odor than it did to a novel floral odor. These results support, in part, those from a human odor classification study of Amoore and Venstrom (1967) and provide what is perhaps the most objective behavioral assessment from a presumably unbiased observer of an odor classification scheme. However, it should be noted that the outcome of Braun and Marcus study does not provide evidence for any specific mechanism of coding and the results may be compatible with other concepts of coding which are quite different from that proposed by Amoore (e.g., S. Price, 1984). Perhaps a more important contribution of the Braun and Marcus study is its demonstration that animals can be used to classify or rank order odors on the basis of perceived similarity. A stimulus generalization or odor classification test in combination with neurobiological manipulations could provide a powerful tool for assessing concepts of odor coding that are now based primarily on studies of odorant–receptor interactions, organization of olfactory projections to the brain, and metabolic studies.

## B. Spatial Coding of Olfactory Information

The possibility that odors may be coded topographically has received considerable attention. Briefly, and in its simplest form, the suggestion is that different classes of odors excite different sets of sensory receptors which, in turn, project to separate loci on the olfactory bulb and the olfactory cortex. The topographic organization might be based on a finite set of "primary odors" with fine discriminations made in a manner analogous to that used by the visual system for coding wavelength. In any case, overlap in the sensitivity of receptors for different odors and in the projection fields of these receptors is not incompatible with a topographic coding mechanism. In other sensory systems (e.g., vision, audition, and somesthesis) there exists a rather precise point-to-point projection of the receptor surface onto the relevant cortical surface and there is little doubt that this spatial organization is one basis for coding sensory quality or localization.

However, there are special difficulties in applying this model to olfaction. The anatomical organization of the olfactory system is somewhat unique because there is little evidence for a topographically organized projection of olfactory bulb mitral and tufted cells onto the olfactory cortex. Numerous studies using degeneration and axonal transport methods have failed to uncover any marked order in the projection of the olfactory bulb onto the olfactory cortex (Luskin & Price, 1982; J. L. Price, 1973). In fact, it appears that the axons of mitral and tufted cells are highly branched and terminate in widely divergent parts within the cortex (Ojima, Mori, & Kishi, 1984). The marked difference between the organization of projections in this sensory system and those of the well-ordered somatosensory, visual, and auditory systems has frustrated attempts to understand how qualitatively different odors are recognized and discriminated within the olfactory cortex.

## C. Anatomical Evidence for Spatial Coding in the Olfactory Bulb

Attempts to find sources of order in the projections of the receptor sheet onto the olfactory bulb have been somewhat more successful. While the apparent absence of topographic organization in the projection of the bulb upon olfactory cortex remains a problem, several lines of evidence have revived the possibility that the olfactory bulb might use a spatial mechanism to code its afferent input. Some of these results and related behavioral studies are described below.

The anatomical investigations of Land (1973; Land & Shepherd, 1974) demonstrating order in the projection of receptor neurons to the olfactory bulb have recently been extended by Saucier and Astic (1986) and Clancy, Schoenfeld, and Macrides (1985) using retrogradely transported horseradish peroxidase or fluorescent dyes injected into discrete areas of the olfactory bulb. Although the outcome of these later studies differ in certain details, they are in basic agreement that the mediolateral and dorsoventral axes of the receptor sheet are represented in the same axes in the bulb but that the organization in the rostral–caudal axis is more complex.

These studies demonstrate more order in the projection of the epithelium upon the bulb than had hitherto been suspected, but the pattern revealed is regional rather than point to point. This may simply reflect the limits of the method; the use of more discrete injections or more sensitive techniques may reveal a more precise spatial organization.

A second line of evidence for topographic specificity in odor coding stems from metabolic studies using 2-deoxyglucose (e.g., Bell & Laing, 1984; Jourdan, 1982; Skeen, 1977; Stewart, Kauer, & Shepherd, 1979; M. H.

Teicher, Stewart, Kauer, & Shepherd, 1980). In these studies animals treated with 2-deoxyglucose (2-DG) were exposed to an odor for 45 or more minutes and autoradiographs of histological sections from the olfactory bulbs were examined for patterns of metabolic activity. Exposure to different odors, including complex natural odors and vapors of pure chemicals, did not produce diffuse activity but, rather, specific foci or "domains" of activity within the glomerular layer of the olfactory bulb. The size of the foci and the intensity of the metabolic activity varied with the concentration of the odorant, while the distribution or pattern of foci within the glomerular layer varied as a function of the odor quality. Unfortunately, relatively few odors have been tested and even fewer mapped in detail.

Experiments on the effects of prolonged exposure to odors also support the hypothesis of spatial coding. Pinching and Doving (1974), in an extension of an initial study of prolonged exposure to an odor (Doving & Pinching, 1973), exposed separate groups of rats to 44 different odors for two or more weeks. Subsequent histological analysis indicated that in each group the exposure resulted in transneuronal-like degenerative changes in the bulb mitral cell layer and, further, that a somewhat different and specific region of the mitral cell layer was affected in response to each of the odors. Unfortunately, quantitative methods were not used in determining the patterns of these degenerative-like changes and, as described below, it has proven difficult to replicate some of these findings.

## D. Behavioral Studies Related to Spatial Coding

One interpretation of the Pinching and Doving results is that different regions of the olfactory bulb mitral cell layer mediate different odors and that the integrity of these cells can be compromised by prolonged exposure to that odor. This was first tested by Laing and Panhuber (1978) who assessed the ability of rats exposed for several months to deodorized air or to one of two odors to detect these odors. The behavioral results did not reveal any marked differences among groups. Rats had no deficit in detecting an odor to which they had received prolonged exposure but, interestingly, rats from each exposure group did not perform quite as well when tested with a novel odor. Dalland and Doving (1981) and Cunzeman and Slotnick (1984) have also reported that rats given prolonged exposure to an odor have no deficits in detecting the odor or in their absolute threshold for the odor.

One interpretation of these behavioral studies is that the anatomical changes produced by prolonged exposure to an odor result not from overstimulation of mitral cells but from lack of stimulation. According to this view, exposure to a particular odor serves to maintain mitral cells that normally respond to that odor. If this is so then early odor deprivation should result in olfactory deficits. This hypothesis was tested by determin-

ing the absolute olfactory threshold to cyclohexanone of separate groups of
rats exposed for two months to different concentrations of the odor or only
to deodorized air (Laing & Panhuber, 1980). Rats raised in a normal rat
colony served as controls. The results provided only modest support for
the hypothesis: Rats raised in deodorized air performed at least as well as
other groups when tested on suprathreshold concentrations of the odor
but were slightly (but significantly) worse than other groups when tested
on the lowest concentrations of the odor. In any case, there appeared to be
no differences among groups in detection threshold. While there was little
evidence for a deficit in rats raised in deodorized air, the outcomes are in
good agreement with the reports of Laing and Panhuber (1978) and
Cunzeman and Slotnick (1984) that prolonged exposure to an odor does
not alter an animal's sensitivity to that odor.

   A factor that complicates the interpretation of these behavioral results is
that the anatomical findings in these studies did not closely correspond to
those reported by Pinching and Doving. Cunzeman and Slotnick failed to
find any differences among experimental and control rats in cell number or
cell size in any of the regions of the mitral cell layer that were examined,
but their histological procedures may have been less sensitive to such
changes than those used by Pinching and Doving. Dalland and Doving did
find areas of the mitral cell layer that had degenerative-like cells in treated
rats but there was considerable variability in the patterns of these changes.
The anatomical results of Laing and Panhuber (1978) corresponded more
closely to those described by Pinching and Doving but the patterns of
degenerative-like cells were similar in the two different odor exposure
groups. Other investigators have reported histological effects of prolonged
exposure to an odor (As, Smit, & Koster, 1980; C. A. Greer, personal
communication; Jourdan, Holley, Glaso-Olsen, Thommesen & Doving,
1980) but, thus far, none have found patterns of altered and normal cells
identical to those described by Pinching and Doving.

   The significance of these findings for theories of odor coding is, at
present, unclear. Behavioral tests have not revealed any marked changes
in olfaction as a result of prolonged exposure to an odor or to deodorized
air and it has proven difficult to replicate the initial and promising
anatomical results of Pinching and Doving. It appears that the odor
exposure effect is far more complex than was initially appreciated and, as
described by Laing (1984) and Panhuber (1986), the anatomical results of
prolonged exposure may vary considerably with such factors as airflow
rate, age, duration of exposure, type of odor, histological procedure, and
criteria established for judging or classifying cells as having degenerative-
like changes. Recently, Laing and his associates have used an objective
computer-assisted method to quantify changes in mitral cells in response
to prolonged odor exposure (Panhuber, Laing, Willcox, Eagleson, &
Pittman, 1985). Their results revealed an interesting interaction between
age of exposure and changes in cell size: In adult rats prolonged exposure

caused significant shrinkage in mitral cell size in most areas of the bulb, but in rats exposed from day of birth the results were quite different; there was cell shrinkage in some areas but significant increases in cell size in other areas of the bulb (Panhuber & Laing, 1987; Panhuber, Mackay-Sim, & Laing, 1987). These outcomes may help resolve some of the disparities in the anatomical results of earlier reports and the method provides a rigorous standard for evaluating the morphological results of any future study of prolonged exposure to odors.

Studies using 2-DG have demonstrated that exposure to different odors produces heightened metabolic activity in different regions of the glomerular area of the bulb and these results provide more satisfactory evidence for spatial patterning. The functional significance of these 2-DG results could be assessed by testing animals in which lesions have been made in parts of the olfactory bulb known to be activated by a specific odor. However, many of the odors studied with 2-DG produce foci of activity in several parts of the glomerular layer and this makes it difficult or impractical to destroy these regions without also damaging widespread areas of the bulb. Two separate studies have reported that exposure to odors produce quite discrete foci of heightened metabolic activity in the bulb and the significance of these results have been examined using behavior.

M. H. Teicher et al. (1980) found that neonatal rats treated with 2-DG and exposed to the ventrum of a lactating dam had high levels of metabolic activity localized in a discrete cluster of glomeruli in the posterior medial part of the olfactory bulb (the modified glomerular complex or MGC; Greer, Stewart, Teicher, & Shepherd, 1982). This focus of activity was not present in pups exposed to clean air or to amyl acetate vapor. M. H. Teicher et al. (1980) suggested that this glomerular complex might mediate the suckling pheromone which was known to play an important role in nipple attachment behavior in neonatal rats. That these glomeruli might be of special significance to the neonate is also supported by developmental studies demonstrating that they are among the earliest glomeruli to develop ontogenetically (Friedman & Price, 1984; Greer et al., 1982). In behavioral tests olfactory bulbectomized neonatal rats fail to show nipple attachment and those with large but subtotal removal of the bulbs have severe deficits in tests of attachment and suckling (Risser & Slotnick, 1987a). However, pups with discrete lesions of this glomerular complex and those with lesions of other areas of the bulb had no deficits in nipple attachment or suckling (Risser & Slotnick, 1987b). Hudson and Distal (1986) also found that destruction of the modified glomerular complex in neonatal rabbits were without effect on suckling behavior even when lesions were produced within two hours of birth.

In one respect, these behavioral results do not provide a very satisfactory test for the functional significance of 2-DG findings in the olfactory bulb. As yet, there is only indirect evidence that the MGC mediates pheromonal cues from the dam and, likewise, tests of nipple attachment and suckling are only indirect measures of olfaction. Future research on this topic could be improved by incorporating measures of odor detection such as those developed by Alberts and May (1980) for the behaving neonatal rat.

A more stringent test of spatial coding has been made possible by the recent finding that exposure of rtas to propionic acid vapor produces a particularly discrete focus of 2-DG activity in the olfactory bulb. The heightened activity occurred in a small group of glomeruli in the dorsomedial sector of the rat olfactory bulb (Fig. 18). Minor foci were generally located on the medial side within 2 mm of the major focus (Bell & Laing, 1984). The potential significance of this focus was enhanced by the finding that vapor mixtures of limonene and propionic acid which greatly reduce the perceived intensity of propionic acid odor for humans also reduce the intensity of the 2-DG from propionic acid in the rat (Bell, Laing, & Panhuber, 1987).

Two recent studies have examined the functional significance of this discrete focus of bulbar activity. In an initial behavioral study, rats with lesions that removed this part of the bulb were tested for their acquisition of a propionic acid detection task and for their threshold for propionic acid (Slotnick, Graham, Laing, & Bell, 1987). The performance of experimental animals was indistinguishable from that of controls. In a second and more extensive study rats with both discrete lesions of the propionic acid focal area and those with larger lesions of the rostral one-third of the bulb were tested on a variety of tasks (B. M. Slotnick, D. G. Laing, H. Panhuber, & G. A. Bell, unpublished study). These included tests of absolute detection threshold, intensity difference threshold, discrimination, and tests using

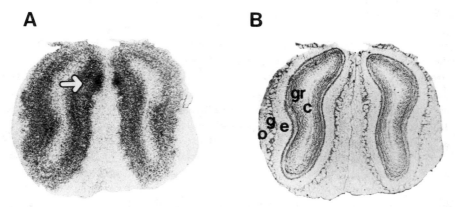

**FIGURE 18.** (*A*) Autoradiograph showing a major foci of accumulation of 2-DG (arrows) in the dorsomedial glomeruli of each olfactory bulb in a rat exposed to propionic acid vapor. (*B*) Nissl-stained section from the same region of the olfactory bulbs. Abbreviations: c, bulbar core; e, external plexiform layer; g, glomerular layer; o, olfactory nerve layer. Adapted from Slotnick, Graham, Laing, and Bell (1987).

odor mixtures. In 2-odor discrimination tests the concentration of odors was varied to insure that stimulus intensity did not provide a cue for correct responding. In general, these tests were specifically designed to provide reasonably sensitive measures of the ability of rats to detect propionic acid at low concentrations, to detect small differences in the intensity of two propionic acid stimuli, and to detect propionic acid in a mixture with a very similar odor (acetic acid). In all tests experimental rats with discrete lesions performed as well as controls. The results are in agreement with those of the initial study (Slotnick et al., 1987) and demonstrate that glomeruli of the olfactory bulb that are specifically activated by an odor can be removed without producing a significant deficit in the animal's ability to use that odor in a variety of tasks.

A recent study on the effects of subtotal olfactory bulb lesions is also relevant to this question of odor coding (McBride, Slotnick, Graham, & Graziadei, 1985). Rats with lesions of 20 to 90% of the olfactory bulbs were tested on a variety of odor detection tasks. Deficits in performance were related to the amount of bulbar tissue removed but not the region of the bulb that was destroyed, although these two variables were confounded in animals with very large lesions. The other interesting outcome of this study was that severe deficits were obtained only in rats with more than 80% of the bulb destroyed. These rats could detect a variety of odors but their performance was variable and they required more training than rats with smaller lesions.

In summary, behavioral tests of spatial patterning have yielded consistently negative results and failed to support the notion that specific areas of the olfactory bulb are essential for an animal to perceive an odor. The contrast between the null outcomes of behavioral studies and the anatomical evidence supporting the idea of spatial patterning is puzzling. It is unlikely that the topographical organization revealed in numerous physiological and anatomical studies (particularly those using 2-DG) is an epiphenomenon unrelated to coding. Rather, one might question whether the available behavioral studies have provided a critical or even adequate test of this evidence. In the experiments with propionic acid it is possible that more sensitive psychophysical tests might have revealed sensory deficits in brain-lesioned animals. On the other hand, the areas of the olfactory bulb identified with 2-DG may play an important role in mediating olfactory functions that were not assessed in these studies (e.g., following an odor gradient, selective attention to an odor, arousal, odor adaptation).

It might also be noted that the anatomical and physiological evidence for spatial patterning is far from perfect. The available methods have important limitations and, except by analogy with other sensory systems, many of the results do not provide specific and testable functional hypotheses. Some of the problems with the prolonged exposure method are mentioned above and have been discussed in detail by Laing (1984) and Panhuber

(1986). The evidence from 2-DG studies is complicated by the fact that the relatively long exposure to an odor used to generate maps would also produce significant olfactory adaptation and, as an odor stimulus, departs significantly from that normally used by behaving animals. In addition, the method as currently practiced does not allow identification of activity at the cellular level and may not be sufficiently sensitive to identify all glomeruli whose metabolic activity is increased by exposure to an odor. While foci of relatively high activity have been given most attention in such studies, other areas that have lower activity and are more variable in location are often observed but their significance remains unknown. Future studies using high-resolution 2-DG autoradiography (Lancet, Greer, Kauer, & Shepherd, 1982) may help resolve some of these problems.

## VI. COMMENTS AND CONCLUSIONS

Studies of pheromonally guided behavior demonstrate that the sense of smell plays a unique and, in many cases, critical role in the behavior of rodents and, presumably, other macrosmatic species (e.g., Doty, 1976). This is dramatically illustrated by the major disruption produced in reproductive, aggressive, and other social behaviors by experimentally induced anosmia and by the influence olfactory stimuli have on neuroendocrine function (Macrides, 1976; Murphy, 1976). Such studies show clearly that odors play a major and, perhaps, dominant role in the perceptual life of rodents. An important contribution of the studies on learning is the demonstration that effective odors are not confined to the narrow band of stimuli identified as pheromones nor are the responses to such stimuli limited to those of social and other species-specific behaviors. Functionally, olfaction is not a primitive sensory modality. The evidence from stimulus sampling behavior and learning overwhelmingly supports the conclusion that rats selectively attend to odors, use odors in preference to other sensory cues in many instrumental learning tasks, and learn quickly to associate odors with response consequences. Indeed, as shown by studies of interproblem transfer and memory, when provided with odors as discriminative cues the rodent has an exceptional ability to learn, one that has not or perhaps cannot be duplicated when nonolfactory cues are used. Further, rats acquire olfactory tasks very rapidly. Under appropriate training conditions, simple two-odor discriminations are learned virtually without errors and a learning set is acquired after training on only a few (5 to 8) problems even when each requires discriminating among many stimuli. By any criterion these tasks are easy for the rat; other, more demanding tests will be needed to determine the animal's full capacity to remember, classify, and use odors in its instrumental behavior.

The available studies have examined only very few of the factors that can influence olfactory perception. Many basic sensory and perceptual processes have not been tested. For example, odors are generally emitted at point sources and sampling behavior and response to an odor may be significantly affected by perception of an odor gradient. Although there have been several studies of rats following odor trails in mazes (review by Doty, 1975), there appears to be none of the response of rats to odor gradients in an open field. Another important factor governing odor perception is adaptation, that is, the relatively rapid increase in odor threshold after even brief exposures to an odor. Adaptation in the rat has been demonstrated (Bennett, 1968), but it has only been examined systematically in humans. Adaptation degrades the stimulus in a psychophysical sense and can alter perception of odor quality. For example, consider the effects produced by cross-adaptation (the change induced in the perception of one odor by exposure to a different odor) and odor mixtures. Studies with human subjects (see Engen, 1982, for a general review) have demonstrated that the effects upon perception of two sequentially presented simple odors may be quite complex. These may include no effect, reciprocal or nonreciprocal cross-adaptation between the odors, or, less common, facilitation. The effects of odor mixtures may be different. A mixture of two qualitatively different vapors can produce synthesis (a different quality is perceived), reciprocal or nonreciprocal masking, or an analytic result in which the components are detected and recognized.

In exerting maximal control of the stimulus, most learning experiments have eliminated these factors. However, in their natural environment animals are seldom exposed to bursts of pure odors. Instead, they receive a continuously changing stream of complex odor stimuli, the perception of which must be influenced by the effects of odor mixture and adaptation. However, the influence of these variables in animals are virtually unknown. Their effects may be similar or perhaps quite different from those obtained with human subjects. Quantitative differences in sensitivity to odors between microsmatic and macrosmatic species is well established (Marshall, Blumer, & Moulton, 1981; Marshall & Moulton, 1981; Slotnick & Ptak, 1977). Might there also be qualitative differences in perception? Or, in more operational terms, might the effects of cross-adaptation or odor mixture for the rat differ significantly from that in humans? Such questions can be attacked experimentally. The methods that have proved successful to assess an animal's ability to discriminate between pure odorants can be used, for example, to determine whether the perception of an odor mixture is synthetic or analytic or the extent to which an odor mixture masks detection of a component of that mixture. Recently, there has been an increased interest in odor adaptation and mixtures in human subjects (Laing, Ache, Cain, & McBride, in preparation) and the generality of many hypotheses generated by this work could be tested using an animal model.

The studies on the effects of brain lesions reviewed here should be viewed as initial attempts to determine the role of different anatomical projections in olfactory perception. The abilities tested are those of stimulus discrimination, learning, and memory. The results of some of these studies suggest some potential disassociations between lesion site and function. Thus, acquisition of a learning set is severely disrupted by lesions of the thalamocortical system but is unaffected by olfactory deafferentation of the limbic system (Slotnick & Kaneko, 1981), while short-term memory for odors is disrupted by lesions of the hippocampus but not by lesions of the amygdala (Eichenbaum et al., 1986).

The results of these studies suggest that different parts of the olfactory pathway may have separate functions in the perception of odors. However, the data base is far too meager and the anatomy too complicated to identify specific functions with different parts of the olfactory forebrain. At the very least, more systematic studies comparing different lesions on a standard series of tasks assessing sensory function, learning, and memory are needed. Further, some of the assumptions underlying these experiments may be questioned. There are difficulties in knowing what capacities are revealed by different behavioral tests, or whether lesion-induced changes reflect deficits in a specific (and sometimes arbitrarily defined) category of behavior. There has also been a tendency to overinterpret behavioral results or to simplify the results of anatomical studies. An implicit assumption in many of these studies is that distinctly different functions will be subserved by different olfactory forebrain projections. There may be some heuristic value in emphasizing the independence of these pathways but this ignores the complexity of the system. For example, because there are interconnections among the amygdala, orbital cortex, and mediodorsal thalamic nucleus, it is unlikely that the olfactory thalamocortical system and the olfactory limbic system can be considered independent of one another either anatomically or functionally.

There also may be more parsimonious explanations for some lesion-induced deficits. For example, rats with lesions of the thalamic mediodorsal nucleus often make more errors than do controls in discriminating odors and, in the studies of Slotnick and Kaneko (1981) and Lu and Slotnick (1988), experimental rats failed to acquire an olfactory learning set. This may reveal a deficit in the ability of lesioned rats to employ a cognitive strategy. However, in the case of reversal learning, the performance of lesioned rats trained on odor cues was, in fact, very similar to that of normal rats trained with visual cues. This suggests that the lesion-induced deficit may stem from changes in attentiveness or in the salience of odors. The possibility that lesion-induced deficits in performance might be due to changes in perception could be evaluated in future studies using measures of stimulus sampling, cue competition, and other tests of stimulus salience such as those described by Nigrosh et al. (1975).

While the behavioral studies of olfactory coding are subject to many of the same pitfalls and criticisms, they have one important advantage: The experiments can be designed to test specific predictions based on physiological or anatomical evidence for a coding mechanism. Indeed, the precision with which such predictions can be made is one measure of the adequacy of a theory of odor coding. The fact that behavior studies have, thus far, failed to provide evidence for the concept of spatial patterning does not mean spatial patterning does not exist. It clearly does and has been demonstrated in numerous anatomical and physiological studies. The behavioral experiments serve to question what use is made of spatially patterned inputs. The finding that animals with discrete lesions of a topographically well-ordered part of a sensory system may have no detectable deficits in sensation is hardly unique in the history of neuro-psychological experimentation. Such outcomes often force investigators to reevaluate the evidence for a theory or generate more specific hypotheses regarding function. They may also indicate that a theory is too simple, that it does not adequately account for the complexities of behavior and the different ways animals can use sensory inputs, or that the theory has overemphasized certain lines of evidence. For example, in evaluating the results of experiments using 2-DG, there is tendency to stress the evidence provided for localization of function and to ignore or discount other, perhaps less dramatic, evidence for widespread distribution of receptor axons to olfactory bulb glomeruli (e.g., Kauer, 1981). However, the failure to produce any appreciable deficits from lesions of parts of the bulb identified with 2-DG may require a reconsideration of these data and of the functional significance of the observed metabolic results.

The broad range of topics covered in this chapter precludes any simple summary statement. One theme that unites the different areas of research is the contribution behavioral analyses can make to our understanding of both simple and complex aspects of olfaction. The success of many of these studies is due, in part, to the fact that they combine proven and efficient conditioning procedures with reasonably precise control of discriminative stimuli. These studies have demonstrated the usefulness of the behaving animal model for examining aspects of olfactory perception. In the future we can anticipate further advances in our understanding of olfactory behavior. In addition, it is likely that the rapid advances in anatomical, physiological, neurochemical, and molecular biological studies of olfaction will stimulate new areas of behavioral investigation and, probably, require the development of better and more sophisticated tests of olfactory perception.

## ACKNOWLEDGMENTS

Preparation of this chapter was supported, in part, by NIH Grant NS22043, NSF Grant BNS-8319872, an NIH Fogarty Senior International Fellowship (FO6 TWO1215), and the excellent facilities of the Australian CSIRO Division of Food Research. I thank David Laing, Helmut Panhuber, Graham Bell, Len Fisher, Margaret Wigney, and Catherine Sheehan for their many useful comments during the writing of this review.

## REFERENCES

Alberts, J. R., & May, B. (1980). Ontogeny of olfaction: development of the rats' sensitivity to urine and amyl acetate. *Physiology and Behavior, 24,* 965–970.

Amoore, J. E. (1977). Specific anosmia and the concept of primary odors. *Chemical Senses & Flavour, 2,* 267.

Amoore, J. E., & Venstrom, D. (1967). Correlations between stereochemical assessments and organoleptic analysis of odorous compounds. In T. Hayashi (Ed.), *Olfaction and taste II* (pp. 3–17). Oxford: Pergamon.

Andy, O. J., Jurko, M. F., & Hughes, J. R. (1975). The amygdala in relation to olfaction. *Confinia Neurologica, 37,* 215–222.

Anholt, R. R. H. (1987). Primary events in olfactory reception. *Trends in Biochemical Sciences, 12,* 58–62.

As, W. van, Smit, K. G. J., & Koster, E. P. (1980). Effects of long-term odour exposure on mitral cells of the olfactory bulb in rats. In H. van der Starre (Ed.), *Olfaction and taste VII* (p. 296). London: IRL Press.

Beauchamp, G. K., & Wellington, J. L. (1984). Habituation to individual odors occurs following brief, widely-spaced presentations. *Physiology and Behavior, 32,* 511–514.

Bell, G. A., & Laing, D. G. (1984). Activity patterns revealed by [$^3$H] 2-deoxyglucose in the main olfactory bulb of the rat after stimulation with odor molecules of varying polarity. *Society for Neuroscience Abstracts, 10,* 657.

Bell, G. A., Laing, D. G., & Panhuber, H. (1987). Odour mixture suppression: Evidence for a peripheral mechanism in human and rat. *Brain Research, 426,* 8–18.

Benignus, V. A., & Prah, J. D. (1980). A computer-controlled vapor-dilution olfactometer. *Behavior Research Instrumentation, 12,* 535–540.

Benjamin, R. M., Jackson, J. C., Golden, G. T., & West, C. H. D. (1982). Sources of olfactory input to opossum mediodorsal nucleus identified by horseradish peroxidase and autoradiographic methods. *Journal of Comparative Neurology, 207,* 358–368.

Bennett, M. H. (1968). The role of the anterior limb of the anterior commissure in olfaction. *Physiology & Behavior, 3,* 507–515.

Braun, J. J., & Marcus, J. (1969). Stimulus generalization among odorants by rats. *Physiology and Behavior, 4,* 245–248.

Broadwell, R. D. (1975). Olfactory relationships of the telencephalon and diencephalon in the rabbit. I. An autoradiographic study of the efferent connections of the main and accessory olfactory bulbs. *Journal of Comparative Neurology, 163,* 329–346.

Brown, T. S. (1963). Olfactory and visual discrimination in the monkey after selective lesions of the temporal lobe. *Journal of Comparative and Physiological Psychology, 56,* 764–768.

Brown, T. S., Rosvold, H. E., & Mishkin, M. (1963). Olfactory discrimination after temporal lobe lesions in monkeys. *Journal of Comparative and Physiological Psychology, 56,* 190–195.

Bruce, H. M. (1960). A block to pregnancy in the mouse caused by proximity of strange males. *Journal of Reproduction and Fertility, 1,* 96–103.

Clancy, A. N., Schoenfeld, T. A., & Macrides, F. (1985). Topographic organization of peripheral input to the hamster olfactory bulb. *Chemical Senses, 10,* 399.

Cunzeman, P., & Slotnick, B. M. (1984). Prolonged exposure to odors in the rat: Effects on odor detection and on mitral cells. *Chemical Senses, 9,* 229–239.

Dalland, T., & Doving, K. B. (1981). Reaction to olfactory stimuli in odor-exposed rats. *Behavioral and Neural Biology, 32,* 79–88.

Davis, R. G., & Tapp, J. T. (1972). Odor artifacts of an olfactometer evoke a CER in the rat. *Perceptual and Motor Skills, 35,* 931–936.

Devor, M. (1973). Components of mating dissociated by lateral olfactory tract transection in male hamsters. *Brain Research, 64,* 437–441.

Devor, M. (1975). Neuroplasticity in the sparing of deterioration of function after early olfactory tract lesions. *Science, 190,* 998–1000.

Devor, M. (1976). Fiber trajectories of olfactory bulb efferents in the hamster. *Journal of Comparative Neurology, 166,* 31–48.

Doty, R. L. (1975). Determination of odour preferences in rodents: Methodological review. In D. G. Moulton, A. Turk, & J. W. Johnston, Jr. (Eds.), *Methods in olfactory research* (pp. 395–406). New York: Academic Press.

Doty, R. L. (1976). *Mammalian olfaction, reproductive processes, and behavior.* New York: Academic Press.

Doty, R. L., Brugger, W. E., Jurs, P. C., Orndorff, M. A., Snyder, P. J., & Lowry, L. D. (1978). Intranasal trigeminal stimulation from odorous volatiles: Psychometric responses from anosmic and normal humans. *Physiology and Behavior, 20,* 175–185.

Doving, K. B., & Pinching, A. J. (1973). Selective degeneration of neurons in the olfactory bulb following prolonged odour exposure. *Brain Research, 52,* 115–129.

Dravnieks, A. (1975). Instrumental aspects of olfactometry. In D. G. Moulton, A. Turk, & J. W. Johnston, Jr. (Eds.), *Methods in olfactory research* (pp. 1–61). New York: Academic Press.

Duncan, H. J., & Slotnick, B. M. (1985). Comparison of visual and olfactory stimuli in reversal learning with pigeons. *Chemical Senses, 10,* 410.

Eayrs, J. T., & Moulton, D. G. (1960). Studies in olfactory acuity. 1. Measurement of olfactory thresholds in the rat. *Quarterly Journal of Experimental Psychology, 12,* 90–98.

Eichenbaum, H., Clegg, R. A., & Feeley, A. (1983). Rexamination of functional subdivisions of the rodent prefrontal cortex. *Experimental Neurology, 79,* 434–451.

Eichenbaum, H., Fagan, A., & Cohen, N. (1986). Normal olfactory discrimination learning set and facilitation of reversal learning after medial-temporal damage in rats: Implications for an account of preserved learning abilities in amnesia. *Journal of Neuroscience, 6,* 1876–1884.

Eichenbaum, H., Kuperstein, M., Fagan, A., & Nagode, J. (1987). Cue-sampling and goal-approach correlates of hippocampal unit activity in rats performing an odor-discrimination task. *Journal of Neuroscience, 7,* 716–732.

Eichenbaum, H., Shedlack, K. J., & Eckmann, K. W. (1980). Thalmo-cortical mechanisms in odor guided behavior. I. Effects of lesions of the mediodorsal thalamic nucleus and frontal cortex on olfactory discriminations in the rat. *Brain, Behavior and Evolution,* **17,** 255–275.

Engen, T. (1973). The sense of smell. *Annual Review of Psychology,* **24,** 187–206.

Engen, T. (1982). *The perception of odors.* New York: Academic Press.

Field, B. F., & Slotnick, B. M. (1987). A multi-purpose photocell and touch circuit amplifier. *Physiology and Behavior,* **40,** 127–129.

Fleming, A. S., & Rosenblatt, J. S. (1974). Olfactory regulation of maternal behavior in rats. II. Effects of peripherally induced anosmia and lesions of the lateral olfactory tract in pup-induced virgins. *Journal of Comparative and Physiological Psychology,* **86,** 233–246.

Freeman, W. J. (1978). Spatial properties of an EEG event in the olfactory bulb and cortex. *Electroencephalography and Clinical Neurophysiology,* **44,** 586–605.

Freeman, W. J. (1982). Changes in spatial patterns of rabbit olfactory EEG with conditioning to odors. *Psychophysiology,* **19,** 44–56.

Friedman, B., & Price, J. L. (1984). Fiber systems in the olfactory bulb and cortex: A study in adult and developing rats using the Timm method with the light and electron microscope. *Journal of Comparative Neurology,* **223,** 88–109.

Greer, C. A., Stewart, W. B., Teicher, M. H., & Shepherd, G. M. (1982). Functional development of the olfactory bulb and a unique glomerular complex in the neonatal rat. *Journal of Neuroscience,* **2,** 1744–1759.

Haberly, L. B. (1985). Neuronal circuitry in olfactory cortex: Anatomy and functional implications. *Chemical Senses,* **10,** 219–238.

Henton, W. W., Smith, J. C., & Tucker, D. (1966). Odor discrimination in pigeons. *Science,* **153,** 1138–1139.

Hodos, W. (1970). Evolutionary interpretation of neural and behavioral studies of living vertebrates. In F. O. Schmitt (Ed.), *The neurosciences: Second study program* (pp. 26–39). New York: Rockefeller University Press.

Hudson, R., & Distal, H. (1986). Regional autonomy in the peripheral processing of nipple-search odors. *Society for Neuroscience Abstracts,* **12,** 1181.

Jasper, H. H. (1958). Functional subdivision of the temporal region in relation to seizure patterns and subcortical connections. In M. Baldwin & P. Bailey (Eds.), *Temporal lobe epilepsy.* Springfield, IL: Thomas.

Jennings, J. W., & Keefer, L. H. (1969). Olfactory learning set in two varieties of domestic rat. *Psychological Reports,* **24,** 3–15.

Jourdan, F. (1982). Spatial dimension in olfactory coding: a representation of the 2-deoxyglucose patterns of glomerular labeling in the olfactory bulb. *Brain Research,* **240,** 341–344.

Jourdan, F., Holley, A., Glaso-Olsen, G., Thommesen, G., & Doving, K. B. (1980). Comparison between the patterns of selective degeneration and marking with $^{14}C$ 2-deoxy-glucose in the rat olfactory bulb. In H. van der Starre (Ed.), *Olfaction and taste VII* (p. 295). London: IRL Press.

Kauer, J. S. (1981). Olfactory receptor cell staining using horseradish peroxidase. *Anatomical Record,* **200,** 331–336.

Komisaruk, B. R., & Bayer, C. (1972). Response of diencephalic neurons to olfactory bulb stimulation, odor, and arousal. *Brain Research,* **36,** 153–170.

Kufera, A. M., Slotnick, B. M., & Risser, J. M. (1988). *Rats learn to label lots of odors: A remarkable demonstration of learning-set and odor memory in the rat* [Abstract Book] Tenth annual meeting of the Association for Chemoreception Sciences, Abstract No. 57.

Laing, D. G. (1982). Characterization of human behaviour during odour perception. *Perception*, **11**, 221–230.

Laing, D. G. (1983). Natural sniffing gives optimum odour perception for humans. *Perception*, **12**, 99–117.

Laing, D. G. (1984). The effect of environmental odours on the sense of small. In N. W. Bond (Ed.), *Animal models in psychopathology* (pp. 59–98). Sydney, Australia: Academic Press.

Laing, D. G. (1985). Optimum perception of odor intensity by humans. *Physiology and Behavior*, **34**, 569–574.

Laing, D. G. (1986). Identification of single dissimilar odors is achieved by humans with a single sniff. *Physiology and Behavior*, **37**, 163–170.

Laing, D. G., Ache, B., Cain, W. S., & McBride, R. L. (in preparation). *Perception of complex smells and tastes*. Sydney, Australia: Academic Press.

Laing, D. G., Murray, K. E., King, M. G., & Cairncross, K. D. (1974). A study of olfactory discrimination in the rat with the aid of a new odor delivery technique. *Chemical Senses & Flavour*, **1**, 197–212.

Laing, D., & Panhuber, H. (1978). Neural and behavioral changes in the rat following continuous exposure to an odor. *Journal of Comparative Physiology*, **124**, 259–265.

Laing, D. G., & Panhuber, H. (1980). Olfactory sensitivity of rats reared in an odorous or deodorized environment. *Physiology and Behavior*, **25**, 555–558.

Lancet, D., Greer, C. A., Kauer, J. S., & Shepherd, G. M. (1982). Mapping of odor-related neuronal activity in the olfactory bulb by high-resolution 2-deoxyglucose autoradiography. *Proceedings of the National Academy of Sciences of the U.S.A.*, **79**, 670–674.

Lancet, D., & Pace, U. (1987). The molecular basis of odor recognition. *Trends in Biochemical Sciences*, **12**, 63–66.

Land, L. J. (1973). Localized projection of olfactory nerves to rabbit olfactory bulb. *Brain Research*, **63**, 153–166.

Land, L. J., & Shepherd, G. W. (1974). Autoradiographic analysis of olfactory receptor projections in the rabbit. *Brain Research*, **70**, 506–510.

Langworthy, R. A., & Jennings, J. W. (1972). Odd ball, abstract, olfactory learning in laboratory rats. *Psychological Record*, **22**, 487–490.

Long, C. J., & Tapp, J. T. (1970). Significance of olfactory tracts in mediating response to odors in the rat. *Journal of Comparative and Physiological Psychology*, **72**, 435–443.

Lu, X. I., & Slotnick, B. M. (1988). *MD lesions and olfactory learning* [Abstract Book], Tenth annual meeting of the Association for Chemoreception Sciences, Abstract No. 56.

Ludvigson, H. W., & Sytsma, D. (1967). The sweet smell of success: Apparent double alternation in the rat. *Psychonomic Science*, **9**, 283–284.

Luskin, M. B., & Price, J. L. (1982). The distribution of axonal collaterals from the olfactory bulb and the nucleus of the horizontal limb of the diagonal band to the olfactory cortex demonstrated by double retrograde labelling techniques. *Journal of Comparative Neurology*, **209**, 249–263.

Macrides, F. (1976). Olfactory influences on neuroendocrine function in mammals. In R. L. Doty (Ed.), *Mammalian olfaction, reproductive processes, and behavior* (pp. 29–65). New York: Academic Press.

Macrides, F., Firl, A. C., Jr., Schneider, S. P., Bartke, A. & Stein, D. G. (1976). Effects of one-stage or serial transections of the lateral olfactory tract on behavior and plasma testosterone levels in male hamsters. *Brain Research*, **109**, 97–109.

Marques, D. M., O'Connell, R. J., Benimoff, N., & Macrides, F. (1982). Delayed deficits in behavior after transection of the olfactory tracts in hamsters. *Physiology and Behavior*, **28**, 353–365.

Marshall, D. A., Blumer, L., & Moulton, D. G. (1981). Odor detection curves for *n*-pentanoic acid in dog and man. *Chemical Senses*, **6**, 445, 453.

Marshall, D. A., Doty, R. L., Lucero, D. P., & Slotnick, B. M. (1981). Odor detection thresholds in the rat for the vapors of three related perfluorocarbons and ethylene glycol dinitrate. *Chemical Senses*, **6**, 421, 433.

Marshall, D. A., & Moulton, D. G. (1977). Quantification of nasal air flow patterns in dogs performing an odor detection task. In J. LeMagnen & P. MacLeod (Eds.), *Olfaction and taste VI* (p. 197). London: IRL Press.

Marshall, D. A., & Moulton, D. G. (1981). Olfactory sensitivity to alpha-ionone in humans and dogs. *Chemical Senses*, **6**, 53–61.

McBride, S., Slotnick, B. M., Graham, S., & Graziadei, P. P. C. (1985). Failure to find specific anosmias in rats with olfactory bulb lesions. *Chemical Senses*, **10**, 410.

Meredith, M. (1986). Patterned response to odor in mammalian olfactory bulb: The influence of intensity. *Journal of Neurophysiology*, **56**, 572–597.

Mesulam, M. M. (1978). Tetramethyl benzidine for horseradish peroxidase neuro-histochemistry: A non-carcinogenic blue reaction-product with superior sensitivity for visualizing neural afferents and efferents. *Journal of Histochemical Cytochemistry*, **26**, 106–117.

Meyer, R. P. (1981). Central connections of the olfactory bulb in the American opossum (*Didelphys virginiana*): A light microscopic degeneration study. *Anatomical Record*, **201**, 141–156.

Moulton, D. G. (1973). The use of animals in olfactory research. In W. I. Gray (Ed.), *Methods of animal experimentation: Vol. 4. Environment and the special senses* (pp. 143–223). New York: Academic Press.

Moulton, D. G., Celebi, G., & Fink, R. P. (1970). Olfaction in mammals–two aspects: Proliferation of cells in the olfactory epithelium and sensitivity to odours'. In G. E. W. Wolstenholme & J. Knight (Eds.), *Ciba foundation symposium on taste and smell in vertebrates* (pp. 227–250). London: Churchill.

Moulton, D. G., & Marshall, D. A. (1976). The performance of dogs in detecting a-ionone in the vapor phase. *Journal of Comparative Physiology*, **110**, 287–306.

Murphy, M. R. (1976). Olfactory impairment, olfactory bulb removal, and mammalian reproduction. In R. L. Doty (Ed.), *Mammalian olfaction, reproductive processes, and behavior* (pp. 96–117). New York: Academic Press.

Nigrosh, B. J., Slotnick, B. M., & Nevin, J. A. (1975). Olfactory discrimination, reversal learning, and stimulus control in rats. *Journal of Comparative and Physiological Psychology*, **89**, 285–294.

Ojima, H., Mori, K., & Kishi, K. (1984). The trajectory of mitral cell axons in the rabbit olfactory cortex revealed by intracellular HRP injection. *Journal of Comparative Neurology*, **230**, 77–87.

Palmerino, C. C., Rusiniak, K. W., & Garcia, J. (1980). Flavor-illness: The peculiar roles of odor and taste in memory for poison. *Science*, **208**, 753–755.

Panhuber, H. (1982). Effect of odor quality and intensity on conditioned odor aversion learning in the rat. *Physiology and Behavior*, **28**, 149–154.

Panhuber, H. (1986). The effect of long duration postnatal odour exposure on the development of the rat olfactory bulb. In W. Breipohl (Ed.), *Ontogeny of olfaction* (pp. 127–141). Springer-Verlag: Berlin.

Panhuber, H., & Laing, D. G. (1987). The size of mitral cells is altered when rats are exposed to an odor from their day of birth. *Developmental Brain Research*, **34**, 133–140.

Panhuber, H., Laing, D. G., Willcox, M. E., Eagleson, G. K., & Pittman, E. A. (1985). The distribution of the size and number of mitral cells in the olfactory bulb of the rat. *Journal of Anatomy*, **140**, 297–308.

Panhuber, H., Mackay-Sim, A., & Laing, D. G. (1987). Prolonged odor exposure causes severe cell shrinkage in the adult rat olfactory bulb. *Developmental Brain Research*, **31**, 307–311.

Passe, D. H., & Walker, J. C. (1985). Olfactory psychophysics in vertebrates. *Neuroscience and Biobehavioral Reviews*, **9**, 431–467.

Phillips, D. S. (1968). Olfactory cues in visual discrimination problems. *Physiology and Behavior*, **3**, 683–685.

Pierson, S. C. (1974). Conditioned suppression to odorous stimuli in the rat', *Journal of Comparative and Physiological Psychology*, **86**, 708–717.

Pietras, R. J., & Moulton, D. G. (1974). Hormonal influences on odor detection in rats: Changes associated with estrous cycle, pseudopregnancy, ovariectomy, and administration of testosterone propionate. *Physiology and Behavior*, **12**, 475–491.

Pinching, A. J., & Doving, K. B. (1974). Selective degeneration in the rat olfactory bulb following exposure to different odours. *Brain Research*, **82**, 195–204.

Powell, T. P. S., Cowan, W. M., & Raisman, G. (1965). The central olfactory connections. *Journal of Anatomy*, **99**, 791–793.

Price, J. L. (1973). An autoradiographic study of complementary laminar patterns of termination of afferent fibers to the olfactory cortex. *Journal of Comparative Neurology*, **150**, 87–108.

Price, J. L. (1985). Beyond the primary olfactory cortex: Olfactory-related areas in the neocortex, thalamus, and hypothalamus. *Chemical Senses*, **10**, 239–258.

Price, J. L., & Slotnick, B. M. (1983). Dual olfactory representation in the rat thalamus: An anatomical and electrophysiological study. *Journal of Comparative Neurology*, **214**, 63–77.

Price, S. (1984). Mechanisms of stimulation of olfactory neurons: an essay. *Chemical Senses*, **8**, 341–354.

Rehn, T. (1978). Perceived odor intensity as a function of air flow through the nose. *Sensory Processes, 2,* 198–205.

Risser, J. M., & Slotnick, B. M. (1987a). Nipple attachment and survival in neonatal olfactory bulbectomized rats. *Physiology and Behavior, 40,* 545–549.

Risser, J. M., & Slotnick, B. M. (1987gb). Suckling behavior in rats pups with lesions which destroy the modified glomerular complex. *Brain Research Bulletin, 19,* 273–281.

Rosenkilde, C. E. (1983). Functions of the prefrontal cortex. *Acta Physiological Scandinavica, Supplementum, 514,* 6–58.

Santibanez, G., & Pinto Hamuy, T. (1957). Olfactory discrimination deficits in monkeys with temporal lobe ablations. *Journal of Comparative and Physiological Psychology, 50,* 472–474.

Sapolsky, R. M., & Eichenbaum, H. (1980). Thalamocortical mechanisms in odor-guided behavior. II. Effects of lesions of the mediodorsal thalamic nucleus and frontal cortex on odor preferences and sexual behavior in the hamster. *Brain, Behavior and Evolution, 17,* 276–290.

Saucier, D., & Astic, L. (1986). Analysis of the topographical organization of the olfactory epithelium projections in the rat. *Brain Research Bulletin, 16,* 455–462.

Scalia, F. (1966). Some olfactory pathways in the rabbit brain. *Journal of Comparative Neurology, 126,* 285–310.

Scott, J. W. (1986). The olfactory bulb and central pathways. *Experientia, 42,* 223–232.

Shepherd, G. M. (1972). Synaptic organization of the mammalian olfactory bulb. *Physiological Reviews, 52,* 864–917.

Shepherd, G. M. (1985). Are there labeled lines in the olfactory pathway? In D. W. Pfaff (Ed.), *Taste, olfaction, and the central nervous system* (pp. 307–321). New York: Rockefeller University Press.

Shettleworth, S. J. (1972). Constraints on learning. In D. S. Lehrman, R. A. Hinde, & E. Shaw (Eds.), *Advances in the study of behavior* (Vol. 4, pp. 1–68). New York: Appleton-Century-Crofts.

Shipley, M. T., & Adamek, G. D. (1984). The connections of the mouse olfactory bulb: A study using orthograde and retrograde transport of wheat germ agglutinin conjugated to horseradish peroxidase. *Brain Research Bulletin, 12,* 669–688.

Shuckman, H., Kling, A., & Orbach, J. (1969). Olfactory discrimination in monkeys with lesions in the amygdala. *Journal of Comparative and Physiological Psychology, 67,* 212–215.

Silver, W. L., & Moulton, D. G. (1982). Chemosensitivity of rat nasal trigeminal receptors. *Physiology and Behavior, 28,* 927–931.

Skeen, L. C. (1977). Odor-induced patterns of deoxyglucose consumption in the olfactory bulb of the tree shrew, *Tupaia glis. Brain Research, 124,* 147–153.

Slotnick, B. M. (1984). Olfactory stimulus control in the rat. *Chemical Senses, 9,* 157–165.

Slotnick, B. M. (1985). Olfactory discrimination in rats with anterior amygdala lesions. *Behavioral Neuroscience, 99,* 956–963.

Slotnick, B. M., & Berman, E. J. (1980). Transection of the lateral olfactory tract does not produce anosmia. *Brain Research Bulletin*, **5**, 441–445.

Slotnick, B. M., & Brosvic, G. M. (1987). Failure of rats to acquire a taste reversal learning set. *Chemical Senses*, **12**, 333–339.

Slotnick, B. M., Graham, S., Laing, D. G., & Bell, G. A. (1987). Detection of propionic acid vapor by rats with lesions of olfactory bulb areas associated with high 2-DG uptake. *Brain Research*, **417**, 343–346.

Slotnick, B. M., & Kaneko, N. (1981). Role of mediodorsal thalamic nucleus in olfactory discrimination learning. *Science*, **214**, 91–92.

Slotnick, B. M., & Katz, H. (1974). Olfactory learning-set formation in rats. *Science*, **185**, 796–798.

Slotnick, B. M., & Ptak, J. E. (1977). Olfactory intensity-difference thresholds in rats and humans. *Physiology and Behavior*, **19**, 795–802.

Slotnick, B. M., & Schoonover, F. W. (1984). Olfactory thresholds in normal and unilaterally bulbectomized rats. *Chemical Senses*, **9**, 325–340.

Southall, P. F., & Long, C. J. (1971). Odor stimuli, training procedures, and performance in a T-maze. *Psychonomic Science*, **24**, 4–6.

Staubli, U., Fraser, D., Kessler, M., & Lynch, G. (1986). Studies on retrograde and anterograde amnesia of olfactory memory after denervation of the hippocampus by entorhinal cortex lesions. *Behavioral and Neural Biology*, **46**, 432–444.

Staubli, U., Ivy, G., & Lynch, G. (1984). Hippocampal denervation causes rapid forgetting of olfactory information in rats. *Proceedings of the National Academy of Science of the U.S.A.*, **81**, 5885–5887.

Stevens, D. A. (1975). Laboratory methods for obtaining olfactory discrimination in rodents. In D. G. Moulton, A. Turk, & J. W. Johnston, Jr. (Eds.), *Methods in olfactory research* (pp. 375–394). New York: Academic Press.

Stewart, W. B., Kauer, J. S., & Shepherd, G. M. (1979). Functional organization of rat olfactory bulb analysed by the 2-Deoxyglucose method. *Journal of Comparative Neurology*, **185**, 715–734.

Stone, H., Pryor, G., & Steinmetz, G. (1969). The design and operation of an improved olfactometer for behavior and physiological investigation. *Behavior Research Instrumentation*, **1**, 153–156.

Stuiver, M. (1958). *Biophysics of the sense of smell*. Doctoral thesis, Gröningen University, Biology Department, Gröningen, The Netherlands.

Swann, H. G. (1934). The function of the brain in olfaction. II. The results of destruction of olfactory and other nervous structures upon the discrimination of odors. *Journal of Comparative Neurology*, **59**, 175–201.

Takagi, S. F. (1984). The olfactory nervous system of the old world monkey. *Japanese Journal of Physiology*, **34**, 561–573.

Tanabe, T., Iino, M., & Takagi, S. F. (1975). Discrimination of odors in olfactory bulb, pyriform-amygdaloid areas, and orbitofrontal cortex of the monkey. *Journal of Neurophysiology*, **38**, 1284–1296.

Teghtsoonian, R., Teghtsoonian, M., Berglund, B., & Berglund, U. (1978). Invariance of odor strength with sniff vigor: An olfactory analogue to size constancy. *Journal of Experimental Psychology, Human Perception and Performance*, **4**, 144–152.

Teicher, M. H., Stewart, W. B., Kauer, J. S., & Shepherd, G. M. (1980). Suckling pheromone stimulation of a modified glomerular region in the developing rat olfactory bulb revealed by the 2-deoxyglucose method. *Brain Research, 194,* 530–535.

Teichner, W. H., Price, L. M., & Nalwalk, T. (1967). Suprathreshold olfactory responses of the rat measured by sniffing. *Journal of Psychology, 66,* 63–75.

Terrace, H. S. (1963). Discrimination learning with and without "errors." *Journal of the Experimental Analysis of Behavior, 6,* 1–27.

Terrace, H. S. (1966). Stimulus control. In W. K. Honig (Ed.), *Operant behavior: Areas of research and application* (pp. 271–344). New York: Appleton-Century-Crofts.

Thompson, R. (1980). Some subcortical regions critical for retention of an odor discrimination in albino rats. *Physiology and Behavior, 24,* 915–921.

Thorne, B. M., & O'Brien, A. L. (1971). The use of olfactory cues in solving a visual discrimination task. *Behavioral Research Methods and Instrumentation, 3,* 240.

Tucker, D. (1963). Physical variables in the olfactory stimulation process. *Journal of General Physiology, 46,* 453–489.

Tucker, D. (1971). Nonolfactory responses from the nasal cavity: Jacobson's organ and the trigeminal system. In L. M. Beidler (Ed.), *Handbook of sensory physiology* (Vol. 4, Part 1, pp. 151–181). New York: Springer-Verlag.

Ursin, H., & Kaada, B. R. (1960). Functional localization within the amygdaloid complex in the cat. *Electroencephalography and Clinical Neurophysiology, 12,* 1–20.

Welker, W. F. (1964). Analysis of sniffing of the albino rat. *Behaviour, 22,* 223–244.

Youngentob, S. L., Mozell, M. M., Sheehe, P. R., & Hornung, D. E. (1987). A quantitative analysis of sniffing strategies in rats performing odor detection tasks. *Physiology and Behavior, 41,* 59–69.

# 7

# THE SENSE OF FLUTTER–VIBRATION IN MONKEYS AND HUMANS

*Robert H. LaMotte*

*Department of Anesthesiology, Yale University School of Medicine, New Haven, Connecticut*

The primate uses the hand both to feel and to manipulate objects in the environment. Both humans and nonhuman primates such as monkeys use tactile information to guide movements of the hand, for example, to pick up an object, as well as to identify the structural properties of the object such as its overall size, shape, texture, and compliance. During the process of active touch, the brain may integrate information about intended movements with sensory input not only from tactile cutaneous receptors but also from receptors in joints, muscles, and tendons.

The studies I will summarize in this chapter were concerned only with the processing of tactile information derived from cutaneous receptors in the hand. They asked the following questions: First, what are the relative sensory capacities of monkeys and/or humans to detect and discriminate between tactile stimuli delivered to the glabrous (smooth) skin of the hand? Second, how are these stimuli coded in the responses of cutaneous mechanoreceptors in the monkey's hand? Finally, how is the coded information processed in the somatosensory cortex?

The tactile stimuli produced during tactile exploration of an object are normally variable and complex. They are variable because they are generated by the movement of the hand, which is under control of the observer and not the experimenter. The stimuli are complex because the local variations in the object's shape, texture, and so on, result in spatial as well as temporal patterns of skin indentations that shift in locus on the skin as the hand moves over the object. Therefore, to simplify the task of carrying out combined psychophysical and neurophysiological studies of tactile sensation it was thought useful to simplify the tactile stimulus and make it more reproducible by producing a pattern of indentations at only one spot on the skin of the restrained hand. Passive stimulus delivery had the additional advantage that the same stimuli could be delivered in psychophysical studies in the awake subject as well as in neurophysiological experiments in the anesthetized monkey. The stimuli, which consisted of mechanical sinusoids delivered to the skin with a coil-driven probe, were simple in the sense that they were spatially synchronous and determined by only two physical parameters—sine-wave frequency and sine-wave amplitude.

It is known that the quality of the tactile sensation elicited by sinusoidal mechanical stimulation of the skin differs depending on sine-wave frequency (Verrillo, 1968). Frequencies below 40 Hz evoke the sense of flutter, while higher frequencies of 50–400 Hz elicit the sense of vibration. For this reason the vibratory sense is considered to be dual and has been called the sense of flutter–vibration (Talbot, Darian-Smith, Kornhuber, & Mountcastle, 1968).

In the late 1960s, Vernon Mountcastle and his colleagues published several papers on the peripheral and cortical neural mechanisms contributing to vibratory sensation (Mountcastle, Talbot, Sakata, & Hyvarinen, 1969; Talbot et al., 1968). Psychophysical measures of human sensation were compared with the response properties of primary afferent and cortical neurons in the monkey. Two types of rapidly adapting mechanoreceptive fibers were studied during recordings from the peripheral nerve in the monkey. One type (which I will call RAs) was presumed to terminate in Meissner corpuscles in the papillary ridges in the glabrous skin; these responded best (i.e., at lowest sine-wave amplitudes) to frequencies of 5–40 Hz, which evoked, in humans, a sensation of flutter. A second set of afferents, those terminating in Pacinian corpuscles beneath the skin (and called PCs), responded best to higher frequencies that evoked the sensation of vibration. It was also found that every fiber had two amplitude thresholds. The lowest was that evoking minimal activity (one or a few impulses—the "absolute threshold"); a higher one was that amplitude required to evoke a phase-locked entrainment of one impulse per cycle of the sine wave (the "tuning threshold") (Talbot et al., 1968).

From recordings of activity in single neurons in the primary somatosensory cortex in the monkey two sets of neurons were identified that served the sense of flutter–vibration (Mountcastle et al., 1969). One set was linked to the RA mechanoreceptive afferent fibers and responded best to sinusoidal frequencies within the flutter range whereas the other set was linked to the PC primary afferents and responded best to higher frequencies. The PC cortical neurons in SI, unlike the PC afferents, exhibited no phase-locked activity (although those in the second somatic area, SII, apparently do, at least in the cat; e.g., Bennett, Ferrington, & Rowe, 1980; Ferrington & Rowe, 1980). However, the RA neurons in SI did exhibit phase-locked discharge albeit with more jitter in the number and temporal position of impulses per cycle in comparison with the tuned activity of the primary afferents. It was further shown that both the strength of this periodic neuronal signal and the capacities of humans to discriminate between different sinusoidal frequencies within the flutter range depended on sine-wave amplitude. The variability in the periodic responses of the RA cortical neurons (standard deviations of cycle histograms) was greatest near the detection thresholds of humans where frequency discrimination was poorest. The neuronal variability decreased, and the sensory capacity to discriminate frequency greatly improved, when sine-wave amplitude was increased to 7 dB above detection threshold. This range between detection threshold and 7 dB above, within which detection was possible but frequency discrimination and pitch recognition were poor, was called the "atonal interval" (Mountcastle et al., 1969).

Based on the results obtained by Mountcastle's group in the late 1960s two neuronal mechanisms emerge as candidates for encoding vibratory frequency and serving as a basis for pitch recognition and frequency discrimination. One is a "place" mechanism that mediates gross discriminations between high versus low frequencies that activate in varying degrees the two populations of PC versus RA afferent fiber and their cortical counterparts. The second coding mechanism serves the discrimination between different frequencies of flutter and is based on the temporal order or periodicity of neuronal responses in RAs and the cortical neurons to which they are linked in SI. A hypothetical cortical mechanism, presumably in SI, would operate upon differences in the dominant period of the phase-locked activity in the "RA" cortical neurons set up by each of the frequencies to be discriminated.

When I joined Mountcastle's laborator there was one important piece of information missing in the foregoing hypotheses and that was whether or not monkeys and humans had similar sensory capacities to detect and discriminate between mechanical sinusoids of different frequencies. That is, do neurophysiological mechanisms observed in monkey serve as a valid model for those mediating the sense of flutter vibration in humans? A scond question of interest to us was the role of the somatosensory cortex in the detection, discrimination, and identification of vibratory stimuli. Specifically, to what extent is the sense of flutter–vibration impaired after lesions in the primary somatosensory cortex and adjoining areas in the parietal lobe?

## METHODS

### Experimental Design

The following is a general overview of the experiments performed.

1. Monkeys and humans were first trained to detect sinusoids delivered to the glabrous skin of the restrained hand as a function of sine-wave amplitude and frequency.

2. The same stimuli were delivered to the receptive fields of single RA and PC afferent fibers innervating the glabrous skin of the monkey hand (Mountcastle, LaMotte, & Carli, 1972). Despite the earlier work of Talbot et al. (1968), it was considered necessary to use exactly the same stimuli in both sensory and neuronal studies in the monkey and in particular to search for fibers with response thresholds as low as those obtained in our new psychophysical study (the sensory thresholds in the newer study were lower than those published in Talbot et al. (1968).

3. Monkeys and humans were further trained to discriminate between 30 Hz sine waves of different amplitudes; discrimination thresholds and the Weber fraction were obtained for different amplitude standards (LaMotte & Mountcastle, 1975).

4. Test frequencies of varying amplitudes were occasionally presented during the tests of amplitude discrimination without feedback to the subject in order to obtain matches in subjective amplitude between those frequencies to be used in subsequent tests of frequency discrimination (LaMotte & Mountcastle, 1975).

5. Another group of monkeys and humans was trained to discriminate between sine waves of different frequencies (but the same subjective intensities) within the flutter range. One of the monkeys was further trained to make identifications of 4 different classes of sine-wave frequency (LaMotte & Mountcastle, 1975).

6. The capacities of monkeys and humans to make frequency discriminations were measured under different levels of amplitude to measure the sensory atonal interval. These data were then compared with the neuronal atonal interval that lies between the absolute and tuning thresholds of the RA afferent fibers. These data provided a test of the hypothesis that frequency discrimination between flutter frequencies depended on differences in the tuned discharges of these fibers (LaMotte & Mountcastle, 1975).

7. Animals trained to detect, discriminate, or identify sine-wave stimuli were tested before and after various lesions of the parietal lobe (LaMotte & Mountcastle, 1978, 1979).

## Subjects and Apparatus

Seven monkeys, *Macaca mulatta,* weighing from 3 to 4 kg were trained and tested in restraining chairs. Eight human subjects were also tested. Each monkey or human was tested in only some of the experiments described below. Both monkeys and humans were tested with the stimulated hand immobilized, palm-up, in plasticene by adhesive tape, with the glabrous skin of either the fingertip or the thenar eminence exposed to the 3-mm-diameter probe tip of a mechanical stimulator (Talbot et al., 1968). The stimulator was an electromechanical device that indented the skin vertically to a displacement that was then maintained under servocontrol at the desired value via a displacement transducer and electronic feedback circuitry. Each sensory test consisted of a series of trials each initiated by a sustained key response, after which the mechanical probe indented the skin by 575 $\mu$m and one or more sinusoidal stimuli were deliverd superimposed on the step (Fig. 1). For human subjects, the other hand operated a telegraph key during detection and additional keys during discrimination. In most experiments, monkeys had both hands restrained so that either could be stimulated. A response key was operated with the forefinger of one hand (see LaMotte & Mountcastle, 1979, for further details). If desired, as in the case of tests after certain cortical lesions, the response could be made by the stimulated hand with the idea of confining cortical input and output to the contralateral hemisphere.

## Outline of Each Psychophysical Task

**1. Absolute Detection.** The subject was required to release the key during the presentation of a mechanical sine wave delivered to the hand after a variable delay (Fig. 1A). On each trial, after a warning tone, the subject pressed the response key by flexing the forefinger, resulting in the step indentation. If the key response was sustained for the required time (0.8 to 6.4 sec), a sinusoid (0.8 sec) was delivered. Key release during this period, except within the first 200 msec, was a correct response ("hit" or correct detection) and was followed by onset of a panel light and apple juice reward to monkeys. Key release before onset of the sinusoid or during its first 200 msec was an anticipatory error. Failure to release the key during the sinusoid was a "miss." Anticipations and misses were not rewarded and resulted in a brief prolongation of the 1-sec intertrial interval. For each sine-wave frequency tested, typically 7 different amplitudes were presented, centered near the anticipated detection threshold (method of constant stimuli). The percentage of trials on which a hit occurred was plotted as a function of sine-wave amplitude for each frequency tested. Detection threshold was defined as the amplitude that corresponded to a hit rate of 50%.

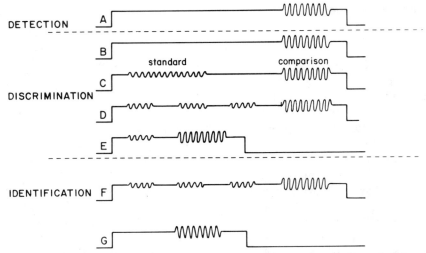

**FIGURE 1.** Outline of stimulus events occurring during each sensory task. On each trial a step indentation into the skin (upward deflection) was followed by one or more sine waves superimposed on the step (see Methods). The standard and comparison are symbolized as small and large amplitudes, respectively. (*A*) The detection task in which the sine wave to be detected is preceded by a variable waiting period. (*B*) First stage of training for multiple-standard (MS) discrimination task in which only a single comparison stimulus is presented on each trial. (*C*) Second stage of training in the MS task wherein the comparison is preceded by a single standard of variable duration. (*D*) Final version of the MS task in which the comparison is preceded by a variable number of presentations of the standard. (*E*) The single-standard (SS) discrimination task in which the comparison is always preceded by a single standard. (*F*) A version of the identification task in which the sine wave to be identified is preceded by a variable number of presentations of a standard. This task was used as a training exercise. (*G*) The version of the identification task used during testing in which only a single stimulus was presented on each trial.

**2. Two-Alternative, Forced-Choice Discrimination: The "Single Standard" (SS) Task.** On each trial during a sustained key press, two sine waves were delivered—a "standard" (of fixed amplitude and frequency) and a "comparison" (always either greater or lesser than the standard in frequency or amplitude, depending on which attribute was being discriminated) (Fig. 1*E*). The subject was required to release the key during the comparison (a correct "detection") and then "tap" the key once if the conparison was greater than the standard or not tap it and merely wait for an interval if the comparison were lesser. A correct choice was followed by a drop of apple juice for the monkey and a blink of either a red panel light (for a correct "lesser" response) or a green light for a correct "greater" response). Two monkeys and 3 humans were trained and tested on an amplitude discrimination task in which the comparison was always the same frequency as the standard but differed in amplitude. Other subjects, 6 monkeys and 5 humans, were tested in a frequency discrimination task in which the standard and comparison had the same subjective amplitude but differed in frequency.

Typically there were three or more different comparisons above and below the value of the standard with one selected pseudorandomly on each trial (the method of constant stimuli). The percentage of trials on which the comparison was called greater than the standard was plotted as a function of the magnitude of the comparison (amplitude or frequency). The discrimination or "difference" threshold was defined as half of the difference between the amplitudes (frequencies) of comparisons corresponding to 75 and 25% respectively.

**3. Incremental–Decremental Detection and Forced-Choice Discrimination: The "Multiple Standard" (MS) Task.** This task was similar to the SS task except that the comparison was preceded on each trial by a variable number of presentations of the standard (as opposed to just one presentation) (Fig. 1D). During a sustained key press from 1 to 6 presentations of the standard occurred followed by the comparison. The comparison had to be detected by a key release within a short interval after its onset in order to have the opportunity of making the forced-choice discrimination as described above for the SS task. A correct detection was signaled by the onset of a yellow panel light but the apple joice reward was given only after a correct discrimination was made. Premature key releases ("anticipations") aborted the trial and were followed by a brief delay of the next trial. A failure to detect the comparison (a "miss") did not result in such a delay. For any given test there were typically 6 comparisons that were greatr in value than the standard (increments) and 6 that were lesser (decrements), with selection carried out pseudorandomly. The percentage of trials on which the comparison was detected was plotted as a function of the value of the comparison (amplitude or frequency). The increment or decrement corresponding to a 50% detection was defined respectively as the incremental or decremental detection threshold. The difference threshold was also obtained based on the forced-choice discrimination responses as described for the SS task.

The MS paradigm was used primarily as a technique to force the subject to maintain a high level of vigilance and to attend to small differences in the relevant attribute (amplitude or frequency). Since the monkey was rewarded only for correct discriminatory responses and not for correct detections and since the probability of detecting correctly by chance was low ($\leq 1$ in 6) it was easier for the subject to be correct by paying attention than by guessing. Although this was obviously less so in the SS paradigm wherein the chance of guessing correctly was 1 in 2 on each trial, the SS task allowed a much more rapid rate of data collection and generated all of our published data on amplitude and frequency discrimination. The MS paradigm was used primarily as a training exercise each day prior to tests with the SS task.

**4. Identification.** One monkey trained to discriminate frequency was further trained to identify frequency by pushing one of 4 panel switches for each of 4 categories of sine-wave frequencies. The monkey's stimulated hand was restrained while the other hand operated a telegraph key to make the detection response and the panel switches to make the identification. There were two versions of this task. Under one version, a standard that was midway between the categories of frequencies to be identified was presented on each trial from 1 to 6 time prior to the "comparison" (Fig. 1F). The subject was required to hold the telegraph key down until delivery of the comparison, at which time the key had to be released (exactly as described for the MS discrimination task) and a correct identification of the comparison made by selecting the appropriate panel switch. This version of the task was used primarily as a training exercise each day. The second version of the identification task was the same as the first except that only a single sine wave, the one to be identified, was presented after a fixed interval of time on each trial (Fig. 1G). The monkey was rewarded by apple juice and illumination of the selected panel switch only after a correct identification.

### Training Procedures

Monkeys were first trained on the variable-delay detection task, then on the SS and MS discrimination tasks. For each task, training proceeded in stages of gradually increasing complexity under programmed control of a minicomputer. The nature of the task at any given stage of training was governed by two lists: One was a set of parameters organized by task difficulty into numbered classes, while the other was the sequence of class numbers to be chosen. During training the numbers were arranged in order of increasing difficulty. A selected class number specified the parameters to be used. These included the *stimulus values* to be presented (e.g., sine-wave amplitudes and frequencies), *stimulus timing* (number of stimuli, their durations, interstimulus intervals), *response contingencies* (the kinds of responses, required, types of feedback, such as panel light cues, probabilities of reward, durations of time out following errors, etc.), and a *performance criterion*, the number of errors permitted under a given class of parameters over a specified number of trials before the next class of parameters was automatically selected (a higher class number, if the criterion was met, or a lower number if it was not).

Training on detection began by teaching the animal to maintain a key press for a fixed interval of time. This interval was gradually increased and eventually made variable. The amplitude of the sine wave to be detected was well above detection threshold and the interval within which the key had to be released was initially long but eventually reduced to 0.8 sec. Once the animal could detect a sine wave of fixed amplitude under the variable-delay paradigm, the amplitude was varied and selected pseudorandomly on each trial from a matrix that was centered at the anticipated threshold amplitude.

Discrimination training began by presenting a single stimulus on each trial, 1–5 sec after onst of the step indentation (Fig. 1*B*). The stimulus was either of two possible "comparison" stimuli readily discriminable by trained subjects. One comparison required a response of "greater," as cued by onset of a green panel light after a correct detection, and the other stimulus required a response of "lesser," as cued by onset of a red panel light after detection. During further training the cue lights were gradually faded out so that discrimination was then based entirely on tactile and not visual cues. A standard of variable length was then introduced (its amplitude gradually increased and its separation in time from the comparison gradually reduced) prior to the comparison (Fig. 1*C*). The value of the standard was midway between the values (amplitude or frequency) of the two comparisons. Next the standard was fragmented into from 1 to 6 parts, the durations and separations of which were gradually adjusted to fixed values of 0.8 sec which resulted in the final version of the MS task (Fig. 1*D*). Next, the separation in magnitude between the two comparisons was gradually reduced and other comparisons with values in-between introduced. Finally, tests were initiated in which the standard was presented only one time on each trial (the SS task) (Fig. 1*E*).

Training on frequency identification began with a variable number of presentations of the standard (38 Hz) followed by a comparison of either 30 or 70 Hz. (Fig. 1*F*). Only the second and fourth panel switches were available; the others were covered. After this task had been learned the same standard was presented followed by a comparison of either 10 or 50 Hz, with only the first and third panel switches available. Eventually, all four panel switches were available and all four stimuli used, each selected pseudorandomly. After performance exceeded 90% for each comparison, the comparison frequencies were gradually changed to values closer together until no further change was possible without a significant deterioration in performance. Tests were then introduced in which no standard preceded the stimulus to be identified (Fig. 1*G*). The latter version of the identification task was then used for data collection.

## RESULTS AND DISCUSSION

### Peripheral Neural Mechanisms Contributing to the Sense of Flutter–Vibration

**Sensory Detection Threshold As a Function of Sine-Wave Frequency.** Detection thresholds in monkeys and humans were measured for test frequencies of 2, 5, 10, 20, 30, 40, 60, 100, 200, 300, and 400 Hz. During a test with each frequency, typically 8 different sine-wave amplitudes were delivered 16 or 32 times each. The range of amplitudes was centered on the anticipated detection threshold, the lowest value equal to zero and the highest a value detected on 90% of the trials. For each monkey and each frequency tested, detection threshold was obtained from psychometric functions (percentage correct versus amplitude) formed by pooling the results from several runs. The thresholds were then averaged for all 6 monkeys and plotted for each frequency in Fig. 2 (solid line in each panel). Detection thresholds obtained in the same manner from 6 humans were averaged and plotted in Fig. 2 (dashed line in each panel). The characteristic U-shaped function is similar to that obtained by previous investigator; thresholds were lowest to the higher frequencies and increased with decreasing sine-wave frequency. Although threshold can be influenced by a host of factors such as skin temperature (Verrillo & Bolanowski, 1986), or the area of mechanical contact (Verrillo, 1985), the important point for us was that the monkey and human thresholds were virtually identical when measured under the same stimulus conditions.

The same vibratory stimuli used in the psychophysical experiments were delivered to the receptive fields of the rapidly adapting (Meissner corpuscle) and Pacinian mechanoreceptive afferent fibers (RAs and PCs

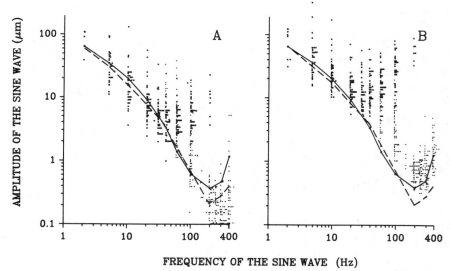

**FIGURE 2.** Frequency–threshold curves for monkeys and humans compared with the absolute and tuning thresholds of RA and PC mechanoreceptive peripheral nerve fibers from the median nerve of the monkey. —, The mean sensory detection threshold for 6 monkeys; ---, the mean detection threshold for 5 human subjects. (*A*) Absolute thresholds of 36 PA afferent fibers (■), and 43 PC fibers (●) are plotted for each frequency tested. (*B*) Tuning thresholds for the same RA (■) and PC (●) fibers. Data replotted from Fig. 5 in Mountcastle, LaMotte, and Carli (1972).

respectively) innervating the glabrous skin of the monkey hand. Each fiber's responses were recorded to each sine-wave frequency at amplitudes that ranged from below to above the absolute and tuning thresholds. The absolute threshold (minimal amplitude eliciting at least one impulse) is plotted in Fig. 2A as a function of sine-wave frequency for each RA fiber (large solid squares). Absolute thresholds for each PC fiber are plotted in Fig. 2A as small closed circles. Tuning thresholds (minimal amplitude eliciting tuning, i.e., a discharge of 1 impulse per cycle of the sine wave at a probability of 0.9) are plotted in Fig. 2B as large squares and small circles for the same RAs and PCs, respectively. The tuning and absolute thresholds differ by 7 dB at 30 Hz for RAs and by about 2 dB at 200 Hz for PCs. Regardless of which type of fiber threshold is chosen for comparison, it is clear that the RA thresholds overlap the detection thresholds for monkeys and humans within the flutter range of 2 to 40 Hz but not the detection thresholds at higher frequencies. Conversely, PC thresholds overlap sensory detection thresholds for higher frequencies of 50–400 Hz (those that elicit the sense of vibration) but not the detection thresholds for the lower frequencies. The idea that two different mechanoreceptor systems contribute to the perception of vibratory stimuli was predicted from the "duplex" theory of Verrillo and colleagues based on their psychophysical studies in humans (Verrillo, 1968).

Evidence against a contribution to the sense of flutter–vibration from slowly adapting (SA) mechanoreceptive afferent fibers was provided in the experiments of Talbot et al. (1968). Further evidence that the RAs and PCs and not the SAs provide "labeled lines" for flutter–vibration comes from percutaneous recording and electrical stimulation of single mechanoreceptive afferent fibers in awake humans (Ochoa & Torebjork, 1983). Human subjects reported that the quality of sensation when a single RA was a electrically stimulated through a microelectrode inserted into the median nerve was an intermittant tapping or flutter (or less often, a vibration) which changed pitch with changes in the frequency of electrical stimulation. Similar results were found for PCs except that the sensation was primarily that of vibration or tickle. In contrast a single SA (type I) evoked a sense of continuous pressure which changed in subjective magnitude with changes in the frequency of electrical stimulation. These data support the notion that the central nervous system interprets incoming impulses in RA afferents in terms of a temporal code, whereas impulses from SAs are decoded in terms of intensive information.

**Sine-Wave Amplitude Discrimination.** The capacity of 1 monkey and 1 human to detect a change in amplitude and then discriminate the direction of change (the MS task) is shown in Fig. 3. The capacities of these and other subjects to make amplitude discriminations with only a single presentation of the standard on each trial (the SS task) are shown in Fig. 4. The standard was 30 Hz, 102 $\mu$m for each of these tasks and the amplitudes of the 30 Hz

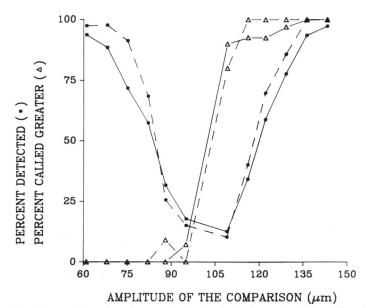

**FIGURE 3.** Psychometric functions for amplitude discrimination made by 1 monkey (—) and 1 human (---) under the MS discrimination task. Each U-shaped function is the percentage of trials on which each comparison amplitude was detected following 1 to 6 presentations of the standard (102 μm, 30 Hz). Each S-shaped function is the percentage of detected comparisons chosen as greater than the standard. Each symbol is the weighted average of data from 150–200 trials.

comparison stimuli varied from 61 to 143 μm. Threshold measurements for those subjects tested on both tasks are given in Table 1. Note that the difference thresholds for the SS task ($DT_{SS}$) were roughly half the size of the increment and decrement thresholds. This might be explained as follows. In the SS task the subjects were evidently not really comparing each comparison with the standard but instead were making a two-category identification of "greaters" versus "lessers." The evidence for this was that the performance of monkeys and humans on the SS task changed little or not at all when the standard was omitted and only the comparison was presented on each trial.

Also evident from Table 1 is that the difference thresholds for the MS task are lower than those for the SS task. The reason for this is that the difference thresholds from the SS task were based on all trials because all comparisons were detected. In the MS task, data from trials on which the subject was uncertain and did not detect the comparison are excluded from determinations of the difference thresholds.

A difference threshold (DT) for each monkey and human tested under the SS task was obtained for three different amplitude standards: 36, 103,

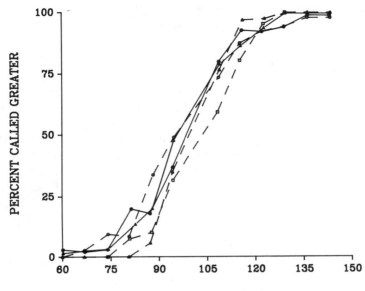

**AMPLITUDE OF THE COMPARISON ($\mu$m)**

**FIGURE 4.** Psychometric functions for amplitude discrimination made by 2 monkeys (—) and 3 humans (---) under the SS discrimination task. Each function is the percentage of trials on which each comparison was chosen as greater than the standard. The standard was 102 $\mu$m, 30 Hz and was presented only once on each trial. Each symbol is the weighted average of data from 150–200 trials. Data obtained from Fig. 3 in LaMotte and Mountcastle (1975).

**TABLE 1. Discrimination Thresholds[a]**

|                          | Subject | DT   | IT   | $DT_{MS}$ | $DT_{SS}$ |
|--------------------------|---------|------|------|-----------|-----------|
| Amplitude discrimination | M54     | 21.5 | 20.2 | 5.4       | 9.6       |
|                          | M33     | 18.6 | 17.7 | 4.3       | 10.2      |
|                          | H1      | 18.7 | 18.4 | 5.6       | 10.8      |
|                          | H2      | 17.6 | 15.4 | 4.2       | 7.1       |
| Frequency discrimination | M38     | 5.2  | 4.6  | 1.4       | 1.9       |
|                          | M41     | 4.3  | 4.4  | 1.0       | 2.0       |
|                          | M52     | 3.8  | 4.4  | 1.1       | 1.8       |
|                          | H3      | 4.5  | 4.1  | 1.0       | 1.7       |
|                          | H4      | 5.7  | 5.6  | 1.3       | 1.8       |

[a] DT, Decrement threshold; IT, increment; threshold; $DT_{MS}$ and $DT_{SS}$ are difference thresholds for the multiple- and single-standard tasks, respectively. Thresholds are in micrometers and hertz for amplitudes and frequency discriminations, respectively. Monkey and human identification numbers are preceded by M and H.

and 154 $\mu$m. The Weber function (DT/standard) was flat for each subject and difference thresholds averaged about 10% of each standard for both monkey and human subjects.

**Matching Different Sine-Wave Frequencies for Subjective Intensity.** From the U-shaped threshold–frequency function in Fig. 2 it is clear that a higher frequency needs less amplitude to be judged equal in subjective magnitude with a lower frequency, at least at threshold. This is also true for suprathreshold stimuli, as shown by the equal-intensity contours obtained by Stevens (1968) for different vibratory frequencies delivered to human skin. Thus, before training monkeys to discriminate frequency we had to be certain that the frequencies were matched for subjective amplitude. Otherwise, monkeys might attend to small differences in subjective magnitude as a cue for frequency discrimination, particularly during training when the frequencies were far apart.

Frequencies ranging from 20 to 40 Hz were matched for subjective intensity with a 30-Hz standard having an amplitude of either 48 or 129 $\mu$m. Matches were first obtained for human subjects using both a direct and an indirect procedure. With the direct method, the 30-Hz standard was followed on each trial by a comparison whose frequency and amplitude were varied. Subjects were told to ignore pitch and judge whether the comparison was greater or lesser in subjective intensity. For each comparison frequency, the percentage of trials on which the comparison was judged greater than the standard was plotted as a function of the amplitude of the comparison. The amplitude corresponding to 50% (the "point of subjective equality") on the psychometric function was chosen as the subjective match with the amplitude of the standard.

The indirect procedure, given to monkeys as well as to humans was similar to the direct procedure except that comparison frequencies were introduced only occasionally (on about 20% of the trials) to subjects being tested on an amplitude discrimination task in which comparisons of 30 Hz were given on most trials. When a comparison frequency other than 30 Hz was given there was no feedback (panel light or liquid reward) after the choice response. The points of subjective equality were obtained as described above and are plotted in Fig. 5 along with those from the direct procedure as a function of sine-wave frequency. The results obtained under the indirect method were the same for monkeys and humans and similar to those obtained for humans using the direct method. This suggests that both monkeys and humans were attending to differences in amplitude and not in pitch. Further, human subjects, when questioned at the end of the indirect procedure, said that they were not aware of any changes in pitch.

The peripheral and central neuronal mechanism for encoding the intensity of a vibrating stimulus within the flutter frequency range was hypothesized to be based on the size of the active population of RA

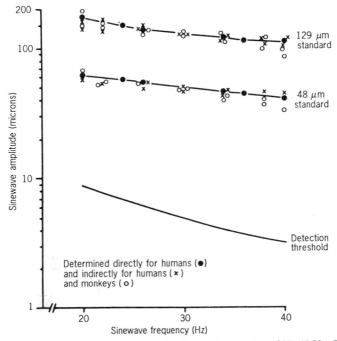

**FIGURE 5.** Equal subjective intensity curves for flutter frequencies of 20–40 Hz. Each symbol is the match (point of subjective equality) between the amplitude of each test frequency and the amplitude of a 30-Hz standard (which was either 48 or 129 $\mu$m). The solid circles connected by a line are the subjective amplitude matches made by humans using a direct procedure. The open circles (monkeys) and crosses (humans) are matches obtained by an indirect procedure (see main text). Reproduced from Fig. 6 in LaMotte and Mountcastle (1975), with permission.

afferent fibers and associated cortical neurons activated by the stimulus and by the total amount of evoked activity in these neurons (Johnson, 1974; Mountcastle et al., 1969; Talbot et al., 1968). If similar neural mechanisms operate in humans, then monkeys and humans should make similar amplitude discriminations and similar subjective judgments of amplitude. The evidence presented above suggests that monkeys and humans do indeed possess the same capacities to detect small differences in amplitude, to discriminate between different categories of amplitude, and to match different frequencies for subjective intensity.

**Sine-Wave Frequency Discrimination.** Typical psychometric functions for the MS task are shown in Fig. 6 for 1 monkey and 1 human subject. The standard was 30 Hz and 129 $\mu$m; comparisons ranged from 20 to 40 Hz and were matched with the standard for subjective intensity. Threshold measurements are given in Table 1. Difference thresholds from these same

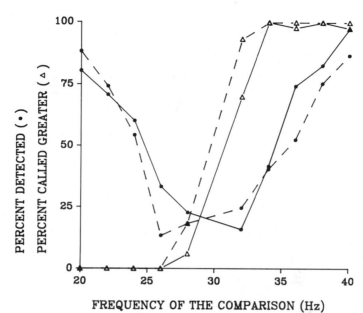

**FIGURE 6.** Psychometric functions for frequency discrimination made by 1 monkey (—) and 1 human (---) under the MS task. Same format as in Fig. 3 except that the comparisons were equal in subjective intensity to the standard but different in frequency. The standard was 30 Hz, 129 μm.

subjects for the SS discrimination task are also given. These thresholds are greater than those for the MS task, but less than the incremental and decremental thresholds. Explanations for these differences in threshold are the same as those given above for amplitude discrimination with the MS versus SS tasks.

Psychometric functions for frequency discrimination with the SS task are shown in Fig. 7 for the same and additional subjects (a total of 6 monkeys and 5 humans). Comparison frequencies ranged from 24 to 36 Hz. The data in each figure were fitted by a logistic function (solid line). This function resembles the cumulative normal distribution and is given by $P = 1/(1 + e^{-(a + Bx)})$. Further details on the curve-fitting procedure are given elsewhere (LaMotte & Mountcastle, 1975). The difference thresholds for the SS task varied from 1.9 to 2.9 Hz (a mean of 2.7 Hz) for monkeys and from 1.2 to 2.9 Hz (a mean of 1.8 Hz) for humans. Thus, while thresholds from the MS task were nearly the same for monkeys and humans, those from the SS task were slightly higher for monkeys than for humans. It is not clear whether the latter discrepancy is due to a genuine difference in sensory capacity between species or to differences in performance (e.g., tendencies to be distracted and exhibit guessing behavior).

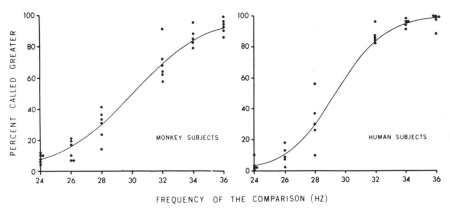

**FIGURE 7.** Psychometric functions for frequency discrimination made by 6 monkeys (left panel) and 5 humans (right panel) under the SS task. The function for each individual subject is plotted as individual points and the solid line is a logistic function that was fitted to the data. The equations for the solid lines are $P = 1/(1 - e^{-(0.414x-12.4)})$ for monkeys and $P = 1/1 - e^{-(0.62x-18.3)})$ for humans. Reproduced with permission from Fig. 7 of LaMotte and Mountcastle (1975).

**Frequency Identification.** A wide range of frequencies (from flutter to vibratory) were presented to the monkey trained to identify 4 different categories of sine-wave frequency. The monkey was first trained to identify 10, 30, 50, and 70 Hz with an accuracy of 90–95% correct for each response. Then the number of frequencies within each category was increased from 1 to 3. Each frequency within the same category required the same category of response. Each stimulus was presented 40 times in each of 3 runs and the results were averaged. The stimulus frequencies and the animal's accuracy of responding are shown in Fig. 8. Performance was excellent with only slight confusions made between adjacent categories. Standard deviations ranged from 0 to 10.2%.

These data were then used to select four new stimuli, 20, 34, 46, and 60 Hz (one stimulus per category). These were closer together along the frequency continuum than were the stimuli used in training. Each stimulus was presented 78 times in each of 3 runs and the results averaged. The monkey's capacity to identify them is shown in the left panel of Fig. 12. Performance was 80% correct or better. Standard deviations ranged from 0 to 3%. Several attempts to reduce the frequency range even further resulted in a marked deterioration in performance. Two human subjects were able to perform this task only slightly better (90 to 100% correct) after considerable training.

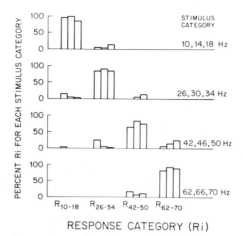

RESPONSE CATEGORY (Ri)

**FIGURE 8.** Capacities of a monkey to identify 4 different categories of sinewave frequency. The monkey was given a single sine wave on each trial and required to identify it (categorize its frequency) by pressing the correct panel switch (1 of 4). The mean percentage of trials on which each of the 4 response categories ($R_i$) was chosen in response to each stimulus category ($S_i$) is presented. Different frequencies were presented with equal probability for each stimulus category (a total of 12 frequencies). Thus, a correct identification was made in response to either 10, 14, or 18 Hz by pressing panel switch 1. A correct response to either 26, 30, or 34 Hz was made by pressing switch 2 and so on. Each stimulus was presented 40 times in each of 3 runs and the percent correct averaged and plotted.

**The Peripheral Neural Code for Flutter Frequency Discrimination: Comparisons between Sensory and Neuronal Atonal Intervals.** Overall, the above results support the hypothesis that both monkeys and humans possess remarkably good capacities to identify different categories of flutter and vibratory frequencies and to discriminate very small differences between frequencies within the flutter range. Our next goal was to test the hypothesis that flutter frequency discrimination required the presence of tuned discharges in RA afferent fibers. If this were so then sensory discrimination should deteriorate if the amplitudes of the stimuli should fall below tuning threshold, that is, within the neuronal atonal interval between absolute and tuning thresholds of the RAs.

We tested the capacities of 4 monkeys and 5 humans to make frequency discriminations between a 30-Hz standard and comparisons of 24–36 Hz under different amplitude levels where the amplitude of the standard (matched with those of the comparison frequencies) ranged from 4 to 20 dB above each subject's detection threshold. Under each amplitude condition a percent correct score was obtained. This score was defined as the difference between the percentage of trials on which 36 Hz was judged higher in frequency than 30 Hz and the percentage of trials in which 24 Hz was judged higher than 30 Hz. The percent correct was averaged and plotted for monkeys and humans as a function of sine-wave amplitude relative to detection threshold (Fig. 9). Threshold discrimination was defined as the amplitude corresponding to 50% correct—the latter situated halfway between the percentile for chance performance (0%) and that for perfect performance (100%).

Psychometric functions for detection, averaged separately for the same monkey and human subjects, are also given in Fig. 9. Detection threshold is also the amplitude at 50% correct. The left side of the shaded area is the

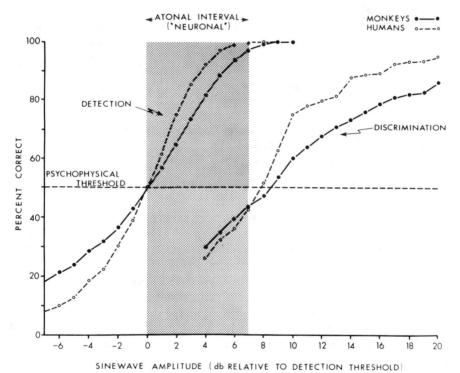

**FIGURE 9.** The sensory and neuronal atonal interval for the sense of flutter. The S-shaped curves are psychometric functions based on averaged data for monkeys (—) and humans (---). Those on the left are the percent correct for detecting a 30-Hz sine wave as a function of sine-wave amplitude. The curves on the right are the percent correct for discriminating between a 30-Hz standard and comparisons of 24 and 36 Hz as a function of the amplitude of the standard to which the comparisons were matched for subjective intensity. For both detection and discrimination tasks, sine-wave amplitude is given in decibels relative to detection threshold for 30 Hz. Threshold for either task is the amplitude at 50% correct; the difference in amplitude thresholds for detection and discrimination (about 8 dB) is the sensory atonal interval. The left edge of the shaded area is the median absolute threshold for the RA afferent fibers (data from Fig. 2A) which is arbitrarily centered on 0 dB. The right edge of the shaded area is the median tuning threshold for these same fibers. The width of the shaded region is the neuronal atonal interval (about 7 dB). Reproduced from Fig. 11 of LaMotte and Mountcastle (1975), with permission.

median absolute threshold for the RAs (data from left panel of Fig. 2A). This value is 5.5 μm and is arbitrarily placed at 0 dB because it is very close to the average detection threshold of the monkey subjects (4.8 μm). The right side of the shaded area is the median tuning threshold for the same fibers (obtained from Fig. 2B) expressed in decibels relative to absolute neural threshold.

Two conclusions can be made from the data in Fig. 9. First, there exists for the sense of flutter an atonal interval extending from detection threshold to an amplitude of about 8 dB above threshold within which sine-wave stimuli can be detected but frequencies cannot be discriminated nor, by inference, pitch recognized. Second, this sensory atonal interval is close to the neuronal atonal interval of 7 dB between absolute and tuning thresholds of RA afferent fibers. These results are consistent with the hypothesis that flutter frequency discrimination requires tuned discharge within the RA population, whereas absolute detection can be based on the presence of a few impulses in a small number of these fibers.

### Cortical Mechanisms for Detecting and Discriminating Sine-Wave Frequencies within the Flutter Range

Mountcastle and colleagues recorded the responses of single neurons in the primary somatosensory cortex (SI) to vibratory stimuli delivered to the glabrous skin of the monkey hand. These neurons responded to tactile stimulation of the contralateral but not the ipsilateral hand. The recordings were made first in awake paralyzed monkeys (Mountcastle et al., 1969) and subsequently from awake monkeys simultaneously performing the detection task (Carli, La Motte, & Mountcastle, 1971). The neurons linked to RA primary afferents (via the lemniscal system) exhibited threshold responses at sine-wave amplitudes near the average monkey sensory detection threshold and close to the absolute thresholds of the RA primary afferents. At threshold, a minimal degree of phase locking was present in neuronal responses. The strength of this periodic response increased as sine-wave amplitude was increased from 0 to 7 dB or greater relative to threshold, as revealed in several response measures. First, a Fourier analysis of the cycle histogams (impulses per time bin within the period of each sine wave) obtained from the RA cortical neurons in the behaving monkey indicated that the amplitudes and energy fractions of the first harmonic increased with sine-wave amplitude in parallel with the increasing probability of sensory detection by the monkey and reached maximal values at the top of the sensory atonal interval (where frequency discrimination is first possible for monkeys and humans—see Fig. 9). A second measure of the strength of the period signal in the cortical neurons was the standard deviation of the cycle histogram. This variability measure was highest near the absolute thresholds of the cortical neurons and decreased with increases in sine-wave amplitude to an asymptote near the top of the sensory atonal interval (about 7 dB above threshold). These and related findings led to the hypothesis that flutter-frequency discrimination depended on a sufficiently strong representation of the periods of the sine waves to be discriminated in the phase-locked discharges of the RA cortical neurons. The data did not support two alternative hypotheses for neuronal coding of flutter-frequency discrimination such as a place code (e.g., relative amount of activity in RA versus PC driven cortical activity) or an intensity code (e.g., total number of impulses in cortical neurons elicited by different frequencies of flutter) as discussed in the paper by Mountcastle et al. (1969).

**The Sense of Flutter–Vibration after Lesions of the Parietal Lobe.** If certain neurons in SI are important for processing information about vibratory stimuli delivered to the contralateral hand then lesions of the parietal lobe that include part or all of SI should impair the animal's capacity to detect, discriminate, and identify such stimuli on the contralateral but not the ipsilateral hand. The results of the next series of experiments demonstrate that this is, in fact, what happens. Monkeys trained on the sensory tasks of detection, discrimination (SS paradigm), or identifcation were tested before and after a partial or total unilateral removal of the parietal lobe. No deficit on any task was found on the hand ipsilateral to any lesion; sensory impairments were entirely limited to the hand contralateral to the cortical lesion. Detection thresholds were elevated to 2 to 7 times preoperative values after removal of SI alone or all of the postcentral gyrus. Detection thresholds were not significantly impaired after removal of SII (second somatosensory cortex) alone or the posterior parietal cortex (Brodmann's areas 5 and 7). Extensive practice over a 2- to 8-week postoperative period resulted in a lowering of detection thresholds back to or close to preoperative values (LaMotte & Mountcastle, 1978). Similar effects were obtained on the capacities of monkeys to make amplitude discriminations; a parietal lobectomy (that, in one case, included the adjacent motor cortex) or removal of the postcentral gyrus elevated difference thresholds to 2 to 5 times preoperative values (LaMotte & Mountcastle, 1978). Removal of the posterior parietal areas alone had no effect on amplitude difference thresholds.

The capacities of monkeys to discriminate between different frequencies within the flutter range (using comparisons of 24 to 36 Hz) were permanently lost on the hand contralateral to a lesion that included the entire parietal lobe. Postoperative tests were administered for 42 to 119 days after the lesion. A similar loss followed removal of the postcentral gyrus although one monkey exhibited some recovery of function by the third postoperative week of testing. Psychometric functions for one monkey (Fig. 10) were obtained before and after parietal lobectomy under conditions of varying probability of reward for a correct "lesser" or "greater." Most animals had little bias toward choosing one response category over

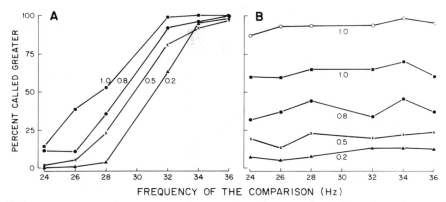

**FIGURE 10.** Flutter frequency discrimination by a monkey before and after unilateral removal of the parietal lobe contralateral to the tested hand. ( A ) Preoperative psychometric functions obtained under conditions in which the probability of reward for a correct choice of "greater" was varied, as indicated for each value beside the appropriate curve. The probability of reward for a correct "lesser" was always 0.8. Each symbol is the weighted average of 5 runs of 120 trials each. The standard was 30 Hz, 129 μm. ( B ) Postoperative results obtained under similar conditions as described for panel A.

the other but this monkey had, preoperatively, a bias toward "greater" when the probabilities of reward were equal for the two responses. By varying these probabilities so that a correct lesser was more likely to be rewarded than a correct greater one could shift the animal's point of subjective equality (the stimulus value corresponding to 50% called greater) to the right along the abscissa. This shift had little effect on the slopes of the psychometric functions and thus little effect on the difference thresholds. A similar effect of varying response bias occurred postoperatively when the animal was unable to discriminate. Results such as these indicated that response bias, which in any case is minimal for a two-alternative, forced-choice task had little influence on our measurements of the effects of cortical lesions on discriminatory performance.

Frequency discrimination between a 30-Hz standard and widely disparate comparison frequencies that extended from 10 to 50 Hz was impaired (threshold elevations of 4 to 5 times preoperative values) but not eliminated after removal of the parietal lobe contralateral to the tested hand (Fig. 11). Human subjects reported that discrimination between these stimuli was based to a large extent on the identification of whether a comparison belonged to the category of flutter (10–25 Hz) or vibration (45 and 50 Hz). We hypothesized that the capacity to make frequency discriminations within the flutter range (based on a neuronal periodicity code) was lost after parietal lobectomy but that the capacity to discriminate flutter from vibratory frequencies (based on a "place" code, i.e., relative

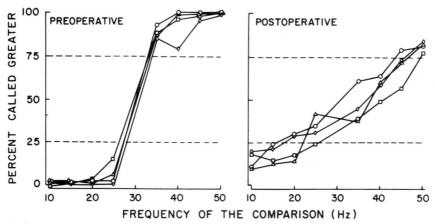

**FIGURE 11.** Frequency discriminations by 4 monkeys before and after removal of the parietal lobe contralateral to the tested hand. The standard was 30 Hz and comparisons extended from flutter to vibratory frequencies. Each function was a weighted average of data obtained from 3 to 6 runs of 150 to 200 trials each. Reproduced with permission from Fig. 10 of LaMotte and Mountcastle (1979).

activation of RA and PC systems) remained but was impaired. This hypothesis was also supported by the effects of parietal lobectomy on frequency identification as described next.

**Frequency Identification before and after Parietal Lobectomy.** Both pre- and postoperative performance of the monkey trained to identify sine-wave frequencies are given in Fig. 12. Frequency identification was greatly impaired by removal of the parietal lobe contralateral to the tested hand (Fig. 12*B*). Postoperatively, the monkey exhibited little confusion between a flutter frequency (20 or 34 Hz) and the vibratory frequency of 60 Hz but could not distinguish frequencies within the flutter range (i.e., 20 versus 34 Hz). As a result, the number of categories that could be reliably identified was reduced from 4 to 2.

For reasons unrelated to the study, the lesion experiments were brought to conclusion before we had the opportunity of making lesions confined to SI, that is, exclusive of areas 5 and SII. However, the evidence available to us suggested a predominant role of SI and possibly SII as well for the sense of flutter–vibration, particularly for flutter frequency discrimination. SI was thought to play a particularly important role in tasks requiring a neural code based on periodicity or temporal order; SI plays a less important role in sensory tasks based on neural codes of intensity or place. Detection or amplitude discrimination is believed to be based on the amount of activity and number of active neurons within central neurons driven by activity in

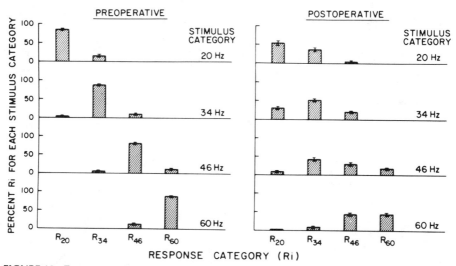

**FIGURE 12.** Frequency identification by a monkey before and after a unilateral removal of the parietal lobe contralateral to the tested hand. Only a single stimulus was presented on each trial. The monkey chose 1 of 4 response categories ($R_i$) by pressing the correct panel switch, $R_{20}$, $R_{34}$, $R_{46}$, or $R_{60}$ in response to a stimulus of 20, 34, 46, or 60 Hz, respectively. The percentage ($\pm 1$ SD) of trials on which each response was made to each stimulus was averaged for 78 presentations of each stimulus and shown for preoperative and postoperative performance. Reproduced with permission from Fig. 11 of LaMotte and Mountcastle (1979).

RAs or PCs. Similarly, the discrimination between flutter and vibratory frequencies is presumed to be based on the relative amount of activity and number of active neurons in each of two different populations of neurons, namely those linked to RAs versus those driven by PCs.

## Advantages of Present Methods for Future Psychophysical Studies of Tactile Sensation in Animals

The methods used in these studies of flutter–vibration may prove useful in other studies of tactile sensation in animals. The passive delivery of stimuli to the restrained hand allows for more precise control over stimulus parameters than would be possible during active touch. Consequently, exactly the same stimuli can be given to both monkeys and humans and in both psychophysical and neurophysiological experiments. This facilitates making correlations between species and between measurements of sensory and neuronal events. Also, comparisons of sources of response bias and response variability for humans and monkeys are facilitated when both species respond behaviorally in the same way to the stimuli.

An advantage of the detection, identification, and the MS discrimination tasks was that they discouraged guessing behavior by making the probability of guessing correctly relatively low. In the variable-delay detection paradigm, the stimulus duration (and the interval of time within which a detection response was correct) was typically short in relation to the duration of the variable foreperiod which preceded it. In the identification task the chance of guessing correctly was only 1 in 4. During discrimination with the MS tasks, the opportunity of making a correct discrimination of "lesser" or "greater" was contingent upon correctly detecting the presence of the comparison stimulus, the latter being preceded by 1 to 6 presentations of the standard. Thus, the chance of correctly guessing the change from the standard to the comparison and then correctly guessing in the two-choice discrimination would be no better than 1 in 12.

A disadvantage of the MS discrimination paradigm was that the rate of data collection was slow due to the length of time it took for multiple presentations of the standard. In contrast, the SS discrimination task in which the standard was presented only once on each trial provided a more rapid means of collecting data. However the chance of guessing correctly was relatively high, 1 in 2. Both humans and well-trained animals apparently view the decision outcomes in such a two-alternative, forced-choice task as equal in value, resulting in measures of sensitivity that are relatively free of decision bias (Bloom & Berkley, 1977; Green & Swets, 1966). But in our experience, it was not unusual during the training of animals for a strong response bias to develop, that is, a preference for one or the other discrimination responses. This is why we found it advantageous to use the MS discrimination paradigm as a training procedure and as a daily warm-up exercise.

In our psychophysical studies of flutter–vibration, we obtained measurements of sensory thresholds using the procedures classically used with data collected by the method of constant stimuli. It is now possible to obtain independent measures of sensitivity and response bias for our variable-delay detection task using a model derived from decision theory (W. J. Williams and R. H. LaMotte, unpublished observation). In this model it is assumed that the foreperiod consists of a variable number of intervals or "slots" of time, the last one of which contains the stimulus to be detected. It was shown that the task can be used to force the subject's response criterion to change with time during each trial such that correct detections (hits) and anticipations (false alarms) increase as a function of time during the foreperiod on each trial. A model of the ideal observer was derived and it was shown how receiver-operating characteristic curves could be obtained.

Recently a decision theory model was also developed for the MS discrimination task (K. O. Johnson, unpublished observations). In this model it is assumed that for each stimulus presented during each trial the subject must make one of three decisions—that the stimulus is less than, the same as, or greater than the previous stimulus. The decision task is represented in terms of three normal distributions with two decision

boundaries separating three decision regions; it was shown how independent measures of sensitivity and response bias can be calculated.

### Combined Neurophysiological and Psychophysical Studies of Spatially Organized Tactile Stimuli

The vibratory stimuli used in our studies of flutter–vibration produce
synchronous indentations within and surrounding the area of contact
between the vibratory probe and the skin. Such "pure" vibratory stimuli
can be encountered with contact between the hand and a vibrating object
during everyday use of the hand. But more typically, a pattern of skin
indentation varies spatially as well as temporally and is produced not by an
object vibrating against a passive skin but by the lateral movement of the
hand over the surface of the object. The physical properties of the object, in
contact with the skin, contribute to the pattern of skin indentation. The
surfaces of objects are typically not flat and smooth but have contour,
pattern, texture and compliance. When stroked with the hand, such
surfaces will prduce asynchronous indentations at different loci within
each contact area at each point in time.

Such cutaneous events cannot be mimicked by experimentally varying
the waveforms of vertical indentations with a mechanical probe placed
normal to the stationary skin. Consequently, I am interested in studying
how certain features of the surfaces of objects are encoded in the responses
of primary mechanoreceptive afferent fibers when the surfaces are indented and stroked laterally across the stationary skin of the fingerpad.
During these experiments, the hand is restrained palm up and a robotic
device selects a surface and applies it to the fingerpad, indenting the skin
and then stroking the object across the skin under servocontrolled force
and velocity (LaMotte, Whitehouse, Robinson, & Davis, 1983). The object
is transparent so that the movement of skin beneath its surface can be
viewed through a microscope.

One question we have asked is how the perception of relative motion
between an object and the skin is encoded in the discharges of primary
mechanoreceptive afferent fibers (LaMotte, Whitehouse, & Srinivasan,
1986). It was shown that the perception of ongoing motion requires the
existence of detectable surface features and cannot occur if the surface is
perfectly smooth.

A related problem is how small a surface feature on a perfectly smooth
surface can be tactually detected by humans (LaMotte & Whitehouse,
1986). The capacity of humans to detect the presence of a single raised dot
(diameter about 0.5 mm) on a smooth plate stroked across the passive
fingerpad was compared with the response sensitivities of mechanoreceptive afferent nerve fibers in monkeys as a function of dot height. Humans
were able to detect a dot as low as 2 $\mu$m and it was shown that this
detection depended solely on activity in the RA mechanoreceptive fibers.

Only the RAs had response thresholds to dot heights within the range of human detection thresholds; the thresholds for SAs and PCs were significantly higher.

In a related experiment, it was found that the lowest height of a texture (consisting of an array of small dots each 50 $\mu$m diameter with 100-$\mu$m center-to-center spacing) that could be detected by humans during lateral stroking was less than 1 $\mu$m and that detection of texture was based on activity exclusively in the PC afferents (LaMotte et al., 1986). SAs and RAs did not respond to such textures and required greater dot spacings and/or dot heights for activation. Thus, it is interesting that the same two fiber types (RAs and PCs) that could be selectively activated by appropriate frequencies of near-threshold vibratory stimuli could be selectively activated by the choice of appropriate experimental surface features.

In other laboratories, spatially patterned stimuli such as gratings and Braille or other dot arrays have been applied to monkeys' glabrous skin (Darian-Smith, Davidson, & Johnson, 1980; Darian-Smith & Oke, 1980; Goodwin & Morley, 1987a, & 1987b; Johnson & Lamb, 1981; Johnson & Phillips, 1981; Lamb, 1983; Morley & Goodwin, 1987; Phillips & Johnson, 1981). These studies have provided new information about the peripheral neural mechanisms contributing to the tactile perception of surface roughness and spatial pattern. For example, one important finding was that only the responses of SA mechanoreceptive afferent fibers could account for the tactile spatial resolution achieved by humans. The spatial detail in the responses of RA and particularly PC fibers was poorer than that in SA responses (Johnson & Phillips, 1981; Phillips & Johnson, 1981).

In the above studies the spatial parameters that were varied were the sizes and spacings of raised elements on a surface but not the shapes of the elements. Yet the shapes of local variations in surface contour might well be expected to contribute importantly to the perception of surface pattern or texture. Recently we investigated how a small shape was represented in the responses of SA and RA mechanoreceptive afferents in the monkey (LaMotte & Srinivasan, 1987a, 1987b; Srinivasan & LaMotte, 1987). The stimuli consisted of flat plates each containing a single step, the cross-sectional shape of which approximated a half-cycle of a sinusoid. Step height was fixed at 0.5 mm while the step width (i.e., the half-cycle wavelength) was varied from 0.45 to 3.13 mm, resulting in step shapes that varied in steepness and curvature. The various features of the spatial response profiles of SAs and RAs as recorded from monkey peripheral nerve were correlated with the capacities of humans to discriminate between step shapes. It was found that the RAs provided more accurate "intensive" information than SAs about the sharpness of the steps but primarily only when the steps were stroked across the skin. In contrast the SAs and not the RAs conveyed spatial information about step shape, for example, the curvature profile, under either stimulus conditions of statically indenting the step into the skin or stroking it across the skin.

Further, it was shown that all the major characteristics of SA responses could be interpreted as due to SA sensitivity to the amount and rate of change in skin curvature.

In most of these recent studies in which spatial tactile stimuli have been used, recordings of activity in mechanoreceptive afferent fibers in the monkey were correlated with psychophysical measures of tactile sensation in humans. None of these studies have obtained psychophysical measurements in the monkey. Detection and discrimination tasks such as those used in our studies of flutter–vibration could be used with spatially patterned stimuli. The identification task should prove particularly useful in studies of spatial pattern recognition. Finally, combining psychophysical observations with simultaneous electrophysiological recordings of neuronal activity in the awake monkey will certainly be fruitful in studies of how tactile spatial information is processed in the central nervous system.

## REFERENCES

Bennett, R. E., Ferrington, D. G., & Rowe, M. (1980). Tactile neuron classes within second somatosensory area (SII) of cat cerebral cortex. *Journal of Neurophysiology*, **43**, 292–309.

Bloom, M., & Berkley, M. A. (1977). Visual acuity and the near point of accommodation in cats. *Vision Research*, **17**, 723–730.

Carli, G., LaMotte, R. H., & Mountcastle, V. B. (1971). *A comparison of sensory behavior and the activity of postcentral cortical neurons observed simultaneously, elicited by oscillatory mechanical stimuli delivered to the contralateral hand in monkey.* Munich: International Union of Physiological Sciences.

Darian-Smith, I., Davidson, I., & Johnson, K. O. (1980). Peripheral neural representation of spatial dimensions of a textured surface moving across the monkey's fingerpad. *Journal of Physiology (London)*, **309**, 135–146.

Darian-Smith, I., & Oke, L. E. (1980). Peripheral neural representation of the spatial frequency of a grating moving across the monkey's fingerpad. *Journal of Physiology (London)*, **309**, 117–133.

Ferrington, D. G., & Rowe, M. J. (1980). Differential contributions to coding of cutaneous vibratory information of cortical somatosensory areas I and II. *Journal of Neurophysiology*, **43**, 310–331.

Goodwin, A. W., & Morley, J. W. (1987a). Sinusoidal movement of a grating across the monkey's fingerpad: Representation of grating and movement features in afferent fiber responses. *Journal of Neuroscience*, **7**, 2168–2180.

Goodwin, A. W., & Morley, J. W. (1987b). Sinusoidal movement of a grating across the monkey's fingerpad: Effect of contact angle and force of the grating on afferent fiber responses. *Journal of Neuroscience*, **7**, 2191–2202.

Green, D. M., & Swetz, J. A. (1966). *Signal detection theory and psychophysics.* Wiley, New York.

Johnson, K. O. (1974). Reconstruction of population response to a vibratory stimulus in quickly adapting mechanoreceptive afferent fiber population innervating glabrous skin of the monkey. *Journal of Neurophysiology, 37,* 48–72.

Johnson, K. O., & Phillips, J. R. (1981). Tactile spatial resolution. I. Two point discrimination, gap detection, grating resolution and letter of recognition. *Journal of Neurophysiology, 46,* 1177–1191.

Lamb, G. (1983). Tactile discrimination of textured surfaces. Peripheral neural coding in the monkey. *Journal of Physiology (London), 338,* 567–587.

LaMotte, R. H., & Mountcastle, V. B. (1975). Capacities of humans and monkeys to discriminate between vibratory stimuli of different frequency and amplitude: A correlation between neural events and psychophysical measurements. *Journal of Neurophysiology, 38,* 539–559.

LaMotte, R. H., & Mountcastle, V. B. (1978). Neural processing of temporally-ordered somesthetic input: Remaining capacity in monkeys following lesions of the parietal lobe. In G. Gordon (Ed.), *Active touch: The mechanism of recognition of objects by manipulation,* (pp. 73–78). Oxford: Pergamon.

LaMotte, R. H., & Mountcastle, V. B. (1979). Disorders in somesthesis following lesions of parietal lobe. *Journal of Neurophysiology, 42,* 400–419.

LaMotte, R. H., & Srinivasan, M. A. (1987a). Tactile discrimination of shape: Responses of slowly adapting mechanoreceptive afferents to a step stroked across the monkey's fingerpad. *Journal of Neuroscience, 7,* 1655–1671.

LaMotte, R. H., & Srinivasan, M. A. (1987b). Tactile discrimination of shape: Responses of rapidly adapting mechanoreceptive afferents to a step stroked across the monkey's fingerpad. *Journal of Neuroscience, 7,* 1672–1681.

LaMotte, R. H., & Whitehouse, J. (1986). Tactile detection of a dot on a smooth surface: Peripheral neural events. *Journal of Neurophysiology, 56,* 1109–1128.

LaMotte, R. H., Whitehouse, G. M., Robinson, C. J., & Davis, F. (1983). A tactile stimulator for controlled movements of textured surfaces across the skin. *Journal of Electrophysiological Techniques, 10,* 1–17.

LaMotte, R. H., Whitehouse, J. M., & Srinivasan, M. A. (1986). Responses of cutaneous receptors to smooth and finely embossed surfaces stroked across the primate fingerpad. *Society for Neuroscience Abstracts, 12,* 335.

Morley, J. W., & Goodwin, A. W. (1987). Sinusoidal movement of a grating across the monkey's fingerpad: Temporal patterns of afferent fiber responses. *Journal of Neuroscience, 7,* 2181–2191.

Mountcastle, V. B., LaMotte, R. H., & Carli, G. (1972). Detection thresholds for vibratory stimuli in humans and monkeys: Comparison with threshold events in mechanoreceptive afferent nerve fibers innervating the monkey hand. *Journal of Neurophysiology, 35,* 122–136.

Mountcastle, V. B., Talbot, W. H., Sakata, H., & Hyvarinen, J. (1969). Cortical neuronal mechanisms in flutter-vibration studied in unanesthetized monkeys. Neuronal periodicity and frequency discrimination. *Journal of neurophysiology, 32,* 452–484.

Ochoa, J., & Torebjork, E. (1983). Sensations evoked by intra-neural microstimulation of single mechanoreceptor units innervating the human hand. *Journal of Physiology (London), 342,* 633–654.

Phillips, J. R., & Johnson, K. O. (1981). Tactile spatial resolution. II. Neural representation of bars, edges, and gratings in monkey primary afferents. *Journal of Neurophysiology, 46,* 1192–1203.

Srinivasan, M. A., & LaMotte, R. H. (1987). Tactile discrimination of shape: Responses of slowly and rapidly adapting mechanoreceptive afferents to a step indented into the monkey's fingerpad. *Journal of Neuroscience, 7,* 1682–1697.

Stevens, S. S. (1968). Tactile vibration: Change of exponent with frequency. *Perception and Psychophysics, 3,* 223–228.

Talbot, W. H., Darian-Smith, I., Kornhuber, H. H., & Mountcastle, V. B. (1968). The sense of flutter-vibration: Comparison of the human capacity with response patterns of mechanoreceptive afferents from the monkey hand. *Journal of Neurophysiology, 31,* 301–334.

Verrillo, R. T. (1968). A duplex mechanism of mechanoreception. In D. R. Kenshalo & C. Thomas (Eds.), *The skin senses,* (pp. 139–159). Springfield, IL: Thomas.

Verrillo, R. T. (1985). Psychophysics of vibrotactile stimulation. *Journal of the Acoustical Society of America, 77,* 225–232.

Verrillo, R. T., & Bolanowski, S. J., Jr. (1986). The effects of skin temperature on the psychophysical responses to vibration of glabrous and hairy skin. *Journal of the Acoustical Society of America, 80,* 528–532.

# LOCALIZATION

# 8

# SOUND LOCALIZATION AND BINAURAL PROCESSES

**Charles H. Brown**

*Department of Psychology, University of South Alabama, Mobile, Alabama*

**Bradford J. May**

*Department of Biomedical Engineering, The Johns Hopkins University, Baltimore, Maryland*

Since the early detection of the proximity of other organisms is often acoustically conveyed, it is likely that an evolutionary premium has been placed on the ability of animals to locate the origin of sounds swiftly and accurately. Sound localization and acoustic orientation may be preparatory for successful encounters with both prey and predators, as well as with kin and competitors. In many instances the determination of the origin of a signal may be as important or even more important than its recognition. The biological significance of hearing and sound-localization capabilities is underscored by the fact that though there are a variety of cases of vertebrates that lost the capacity for vision after having radiated into a lightless habitat, there are no comparable instances of species that have lost the capacity for hearing (Masterton, 1974). Evidently there may be no silent habitat where the ability to hear has no biological significance. Furthermore, though different representative species have been shown to vary in the acuity of their sound-localization capabilities (Erulkar, 1972; H. E. Heffner & Heffner, 1984; R. S. Heffner & Masterton, this volume, Ch. 9; Masterton, 1974; Masterton & Diamond, 1973; Park, Okanoya, & Dooling, 1987; Ravizza & Diamond, 1974), there are no known cases of warm-blooded vertebrates that lack the ability to orient to sound (though localization abilities are nearly vestigial in pocket gophers, a trait linked to its underground habitat, R. S. Heffner, Richard, & Heffner, 1987). Localization abilities are also present in many species of fish (Moulton & Dixon, 1967; Schuijf & Buwalda, 1980) and amphibians (Feng, Gerhardt, & Capranica, 1976; Rheinlaender, Gerhardt, Tager, & Capranica, 1979), and are likely to be found in most reptiles as well.

The primal significance of sound localization is also represented developmentally. Directional hearing may be expressed at or near birth in the behavior of a variety of avian and mammalian species including laughing gull chicks (Beer, 1969, 1970), peeking ducklings (Gottleib, 1965), infant cats (Clements & Kelly, 1978a), rats (Potash & Kelly, 1980), guinea pigs (Clements & Kelly, 1978b), and humans (Muir & Field, 1979; Wetheimer, 1961). Thus, the evidence suggests that organisms that are nearly helpless at birth, as well as many precocial species, are able to localize sound or orient toward the origin of sound nearly as soon as they can hear. In fact, for many species, the perceptual ability to localize sound may be intact prior to birth, and its expression may be obscured until the organism has achieved a level of physical maturation sufficient to permit a behavioral response to sound position. Though sound-localization abilities are widely evident at the earliest stages of development, these abilities may not be preserved through the later stages of development. The directional hearing ability of rats declines with age (Brown, 1984), and this decline appears to be associated with degeneration in the central auditory system (Casey & Feldman, 1982; Keithley & Feldman, 1979).

In certain species sound localization may play a critical role in orchestrating shifts of attention and acoustically guiding the visual apparatus to the site of potentially "interesting" events. Animals with evolutionarily modern binocular and stereoscopic visual systems (and consequently limited hemispheric or peripheral vision) such as humans and the advanced primates may be particularly dependent upon sound localization for the rapid orientation of the eyes to the location of action (Harrison & Irving, 1966). Furthermore, there appears to be a direct correlation between the acuity of sound localization in the horizontal plane and the width of the binocular visual field (R. S. Heffner & Heffner, 1985). An appreciation of the natural correspondence between sight and sound is expressed early in the development of human infants (Mendelson & Haith, 1976), and the artificial spatial separation of visual and acoustic events is distressing to young infants (Aronson & Rosenbloom, 1971). Harrison and his associates have shown that the correspondence between visual and auditory space is important for location discriminations in animals. Both rats (Beecher & Harrison, 1971) and monkeys (Harrison, Downey, Segal, & Howe, 1971) almost instantly learn to respond on levers adjacent to the location of an acoustic cue, but are slow to learn to respond on levers that are spatially distant from the position of the discriminative acoustic stimulus. In general, organisms appear to be quick to learn to direct responses toward the origin of sounds, but slow to learn to direct responses to locations that violate the contiguity of visual and acoustic space.

Though sound-localization mechanisms evolved because of the resulting survival value conferred upon organisms in their natural habits, directional hearing has nearly always been studied in idealized quiet, echo-free environments, with synthetic clicks, tones, or noises (Waser, 1977). The acoustically simplified approach to the study of directional hearing obscures the chief impediments and obstacles that auditory systems must over come to function in nature, and this in turn obscures the remarkable acoustic capabilities organisms have acquired. Thus, the central theme of this chapter is to focus on sound localization within a natural context, and the impediments to localization imposed by the natural ecology. We first review the physical cues for directional hearing available to terrestrial vertebrates. A brief description follows of the more common behavioral methodologies used to measure sound-localization abilities in animals. We then survey the acuity of directional hearing in selected birds and mammals. The remainder of this chapter is focused on sound localization in the natural environment, and ecological constraints on directional hearing.

## I. LOCALIZATION CUES

The location of sound in space may be expressed in reference to its distance, horizontal position (or azimuth), and vertical position (or elevation) relative to the location of the listener. In contrast to the case of terrestrial vertebrates, the auditory receptor organs found in a large number of insects respond differentially themselves to changes in the location of sound direction (Pumphery, 1940). These ears, or transducers, act as displacement detectors, and because particle displacement is a vector quantity, the direction of propagation of the sound wave may be coded by the individual receptor organ. In contrast, the terrestrial vertebrate ear is widely recognized as being a pressure transducer (Henson, 1974; Manley, 1973; Shaw, 1974). Sound pressure is a scalar and not a vector quantity, and information regarding the direction of origin of a sound wave is unavailable to a pressure transducer. As a consequence, in terrestrial vertebrates directional hearing is dependent upon the development of structures in the central nervous system dedicated to the triangulation of sound location through the comparison of differences in the pressure wave incident at each ear.

### A. Resolving the Horizontal Coordinate of Sound Direction

In most vertebrates the accurate localization of sound is dependent upon the comparative analysis of the signal incident at both ears. Whenever a sound originates off to one side of a listener, the pressure wave will necessarily be received by the "near" ear before it reaches the "far" ear

(Fig. 1). Thus, interaural differences in the *time of arrival* of the signal at each ear serve as one of the principal cues for directional hearing. Time-of-arrival cues occur both for the leading edge of the wave front, and for ongoing time or phase differences in the waveforms. The magnitude of time-of-arrival cues is dependent on the azimuth of the signal and the radius of the organism's head: Time-of-arrival cues are at a maximum when the signal is 90° to the right or left of midline, and larger organisms will enjoy larger interaural differences in the time of arrival of the right- and left-ear waveforms.

Time-of-arrival cues are frequency dependent as well as head-size and azimuth dependent. The effective acoustic radius of the head for low-frequency signals is larger than the skull perimeter, but the acoustic radius of the head is equal to the skull perimeter when high-frequency signals are presented (Kuhn, 1977). The acoustic radius for the leading edge of a signal is also equal to the skull parameter (Kuhn, 1977). Signals with a gradual onset (slow risetime) lack a crisp leading edge, and as a result, time-domain localization would likely be restricted to the comparison of interaural differences in the phase of the fine structure of the signal, or to the comparison of time-of-arrival differences in the envelope of complex signals (Henning, 1974; McFadden & Pasanen, 1978). That is, when a signal begins imperceptibly, the onset of the wave front cannot cue localization. However, in spectrally and temporally complex signals the envelope, or the amplitude contour of the waveform, will be modulated,

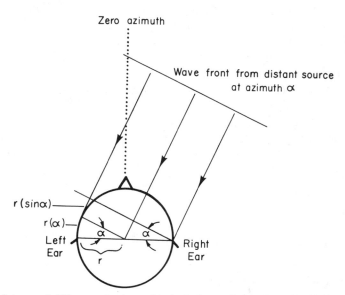

**FIGURE 1.** Interaural difference in time of arrival of waveforms from sources positioned off the subject's midline. Adopted from Woodworth (1938).

and the envelope of these fluctuations at the near and far ear will differ in time of arrival. These interaural differences in signal envelope are localizable by human listeners (Henning, 1974; McFadden & Pasanen, 1978), and are likely to play a significant role in sound localization in nonhuman primates and other mammals as well (Brown, Beecher, Moody, & Stebbins, 1980). Furthermore, the auditory system may detect interaural differences in the phase of the fine structure (each cycle of the waveform) of low-frequency signals, and the processing of these phase differences has been measured behaviorally and physiologically in a variety of vertebrates. Physiological constraints on the vertebrate auditory system limit the highest frequency for which differences in signal phase may be resolved. Therefore, localization based on ongoing interaural differences in phase are restricted to signals less than some critical value between approximately 1 and 5 kHz, with different species exhibiting some variation regarding the highest frequency that may be resolved by this mechanism (Anderson, 1973; Brown, Beecher, Moody, & Stebbins, 1978a; Johnson, 1980; Kiang, Watanabe, Thomas, & Clark, 1965; Klump & Eady, 1956; Rose, Brugge, Anderson, Hind, 1967).

Interaural differences in *signal amplitude* serve as the second principal cue for directional hearing. When signal frequency is high enough that the organism's head and pinnae cast an "acoustic shadow" the amplitude of the waveform incident at the near ear is likely to exceed that heard by the far ear. In general, interaural amplitude diferences occur only for sounds with wavelengths equal to or less than the diameter of the listener's head, and lower-frequency, long-wavelength sounds will not give rise to interaural amplitude differences. Furthermore, because the magnitude of interaural amplitude differences is frequency dependent, complex high-frequency signals will yield different amplitude differences for the various frequency components in the signal. This phenomenon will give rise to an interaural intensity-difference spectrum which may subserve the localization of complex high-frequency signals (see Fig. 2).

While the resolution of the azimuth of sound in mammals is likely due to the presence of mechanisms dedicated to the analysis of interaural differences in the temporal and spectral features of sound waves (see R. S. Heffner & Masterton, this volume, Ch. 9), other mechanisms are likely to play a significant role for directional hearing in nonmammalian vertebrates. In fish the lateral line system is likely to contribute to directional hearing in the acoustic near field (Bleckmann, 1980, 1988; for a brief review, see Coombs & Janssen, this book, Vol. II, Ch. 3). Furthermore, in birds, amphibians, and some other vertebrates, air chambers may connect the middle ear cavities on both sides of the skull (Henson, 1974; Smith, 1904; Stellbogen, 1930; Wada, 1923). The eustachian tubes and the anterior air spaces connect the left and right middle ear cavities in many of these species. Under this arrangement sound waves may pass through the interaural pathway, and sound incident at the tympanic membrane from

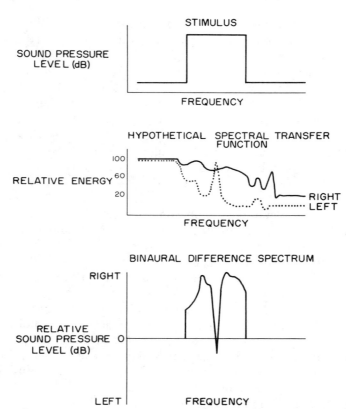

**FIGURE 2.** Hypothetical interaural level differences as a function of signal frequency and the azimuth of the source. Negative interaural level differences result from cases in which the sound-pressure level is greater at the ear farther from the source than that at the ear nearer to the source. Adopted from Brown, Beecher, Moody, and Stebbins (1978a).

the external ear may interact with the sound wave that passed from one ear to the other via the interaural pathway (Coles, Lewis, Hill, Hutchings, & Gowee, 1980; Hill, Lewis, Hutchings, & Coles, 1980; Rosowski & Saunders, 1980; Wever & Vernon, 1957). Depending on the wavelengths of the sound stimulus relative to the dimensions of the listener's head, the azimuth of the sound source, and so forth, the interaural and external waveforms may vary from being in phase to being 180° out of phase and the resulting constructive and destructive interference may markedly alter the net stimulation of the tympanic membrane as a function of changes in sound direction. Thus, the resolution of sound azimuth may be based on different cues in different vertebrate groups.

## B. Resolving the Vertical Coordinate of Sound Direction

The discrimination of variations in the height or elevation of sound sources may be particularly important for nonterrestrial organisms, such as marine organisms and the arboreal primates. In marine and rain forest habitats, for example, the sounds associated with biologically critical events may emanate from above or below, as well as from the right or from the left. If the external ears of listeners are immobile and bilaterally symmetrical, vertical sound localization would not involve binaural processing, unless of course listeners strategically cock their heads (Menzel, 1980). In barn owls it has been shown that the left and right external ears are not bilaterally symmetrical, and this distinct asymmetry is implicated in vertical sound localization (Norberg, 1977; Payne, 1962). The external ears of human listeners are not bilaterally identical (Shaw, 1974), and vertical sound localization shows some evidence of binaural processing (Butler, 1969; Gardner, 1973). Nevertheless, in human listeners (and presumably in many other mammals) vertical localization is attributed to the fact that the transformation function of the external ear is elevation dependent. With high-frequency, broad bandwidth stimuli, the spectral content of stimuli changes with changes in the elevation of the source (Butler, 1969; Gardner, 1973; Hebrank & Wright, 1974; Kuhn, 1979). For the change in spectrum content to serve as a cue for elevation the signal must be relatively fixed in frequency composition, and the subject must be familiar with the signal. The requirement of high-frequency signals for vertical localization is due to the fact that the wavelength must be small relative to the dimensions of the convolutions of the pinnae for the geometry of the external ear to effectively "filter" the signal in an elevation-dependent fashion (Kuhn, 1979; Shaw, 1974). However, with lower-frequency signals elevation-dependent reflections of the signal off the organism's torso may provide a basis for vertical localization (Brown, Schessler, Moody, & Stebbins, 1982; Kuhn, 1979). It is possible that the retention of the general mammalian attribute of high-frequency hearing (see Heffner and Masterton, this volume) may be more critical for accurate vertical localization than horizontal localization, though we know of no direct tests of this possibility.

As implied above, pinnae mobility may also play a role in vertical localization. Most mammals have relatively mobile pinnae, and asymmetrical changes in pinnae shape or orientation (Searle, Braida, Cuddy, & Davis, 1975) may augment vertical localization just as ear canal asymmetry augments vertical localization in the barn owl (Norberg, 1977; Payne, 1962). We know of no direct tests of this possibility. Because the wavelength of sound in water is approximately 4.3 times its value in air, marine environments present special problems for vertical sound localization. Nevertheless, some marine species exhibit remarkable vertical localization abilities, and the cues underlying this capability are not well understood.

## C. Resolving the Distance Coordinate of Sound

Visual depth and distance perception is attributed to a variety of monocular cues including size constancy (retinal image size varies with changes in object distance), interposition (near objects are in front of or partially obscure more distant objects), linear perspective (parallel lines appear to converge at the horizon), textural perspective (the density of items per retinal area increase with distance; that is, more people are visible in a crowd at progressively greater distances), aerial perspective (very distant objects lose their color saturation and appear to be tinged with blue), relative brightness (objects at greater distances from a light source, such as a street lamp, have less luminance than do objects closer to the source), relative motion parallax (the image of distant objects is shifted less by a change in the position of the viewer than are the images of closer objects), and the binocular cue of convergence (the relative inward turn of the eyes; Kaufman, 1979). Given the richness of our appreciation of visual depth and distance perception it is surprising how little is known about auditory distance-perception capabilities of various species, and the physical cues underlying these abilities. Also surprising is the fact that kinetic cues, those involving relative motion between source and receiver, are better understood than are static cues. In some echo-locating bats the Doppler shift in the apparent "pitch" of the echo-location signal yields exquisite information on target distance (Simmons, Howell, & Suga, 1975) and target texture (Schmidt, 1988), and Doppler shifts may be used by human listeners to cue the approach or departure of trains, aircraft, and the like (for a recent review of the localization of dynamic cues by humans, see Perrott, 1982; Perrott & Tucker, 1988). Relative sound amplitude also yields some indication of distance, but studies of sound transmission have shown that sound amplitude in natural environments may fluctuate 20 dB or more in short periods of time (Waser & Brown, 1986; Wiley & Richards, 1978). Thus, amplitude per se may be a poor index of transmission distance. Broadcast sounds are degraded by the natural habitat, and habitat-induced changes in sound quality may cue transmission distance. The relative frequency content of complex signals may change as a function of propagation distance. The attenuation rates of sound in the natural habitat differ as a function of broadcast frequency, and also differ between habitats (Waser & Brown, 1986). Furthermore, the temporal patterning of signals may change as a function of broadcast distance. Reverberation may act to "smear" the temporal patterning of signals by overlaying a reflected (and hence delayed) indirect wave on the direct wave. Reverberation is frequency dependent, and reverberation is a more serious problem for organisms resident in forested habitats than for those resident in savanna or grassland habitats (Waser & Brown, 1986). Because of these acoustical complexities various signals may differ in respect to how they are degraded by the natural habitat, and in respect to how well they reveal the distance parameter. In fact, some signals may be relatively immune to habitat-induced degradation, while other signals may degrade systematically with propagation distance.

Brown and Waser (1988) have shown that different primate calls are degraded differently by the acoustics of the natural environment. Signal degradation has been measured both in respect to the frequency composition of the utterance and in respect to the temporal patterning of the call. Figure 3 shows the waveform of a blue monkey's *trill* vocalization (averaged for 6 repetitions) at propagation distances of 1, 12.5, and 100 m recorded in the monkey's natural habitat. To provide a numeric characterization of the change in the temporal pattern of waveforms, Brown and Waser (1988) submitted the averaged signal at each propagation distance and the source signal to a normalized cross-correlation program. If the temporal pattern of the signal did not change as a function of propagation in the natural habitat the correlation value would be 1.0; alternatively, progressively smaller values would indicate greater degradation of the waveform by the acoustics of the habitat. Table 1 shows the results of this analysis for 6 blue monkey calls. The results show that the waveform of some signals (e.g., the *chirp* call) are more susceptible to degradation than are those of other vocalizations (e.g., the *boom* call).

Brown and Waser (1988) also measured degradation of the frequency content of broadcast calls. The spectrum of the source signal was compared

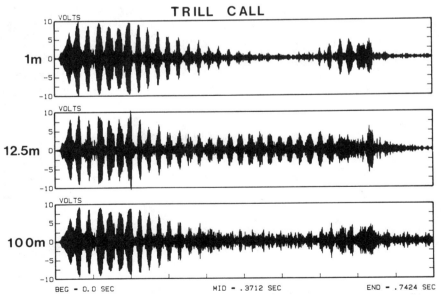

**FIGURE 3.** Averaged waveforms for six repetitions of the blue monkey's (*Cercopithecus mitis*) *trill* vocalization broadcast in the Kakamega forest (Kenya) at transmission distances of 1, 12.5, and 100 m. Adopted from Brown and Waser (1988).

**TABLE 1. Time-Domain Change Index**[a]

| Signal | 12.5 m | 100 m |
|--------|--------|-------|
| Grunt | 0.9667 | 0.8925 |
| Pyow | 0.9542 | 0.8571 |
| Trill | 0.3033 | 0.8063 |
| Boom | 0.9584 | 0.9413 |
| Ka-train | 0.9153 | 0.9101 |
| Chirp | 0.9348 | 0.7637 |

SOURCE: Adopted from Brown and Waser (1988).

[a] Normalized cross-correlation of the averaged signal at 1 m with the averaged signal at 12.5 or 100 m. The greater the difference between signals, the smaller the coefficient.

with the spectrum of the signal at each propagation distance, and a difference spectrum was computed to indicate the change in the frequency composition of the utterance (Fig. 4). A numeric index of the magnitude of frequency degradation was computed by dividing the power of the difference spectrum by the power of the source signal, and the results of this analysis for 6 monkey calls are shown in Table 2. In this analysis the greater the frequency degradation, the larger the value of the change index. The data in Table 2 show that some vocalizations (e.g., the *boom* call) are relatively unchanged by the acoustics of the natural habitat, while other calls (e.g., the *chirp* or *pyow* calls) are susceptible to degradation by the natural habitat. The pattern of these degradation scores shows that different utterances are degraded in different ways by the acoustics of the natural habitat. Interestingly, the blue monkey's *boom* and *pyow* calls are both long-distance signals (Brown, 1989), yet the two calls differ in respect to their susceptibility to habitat-induced degradation. These findings would suggest that the *pyow* call would be good for revealing the

**TABLE 2. Frequency-Domain Change Index**[a]

| Signal | 12.5 m | 100 m |
|--------|--------|-------|
| Grunt | 0.26 | 0.24 |
| Pyow | 0.20 | 0.35 |
| Trill | 0.34 | 0.11 |
| Boom | 0.10 | 0.06 |
| Ka-train | 0.34 | 0.14 |
| Chirp | 0.40 | 0.35 |

SOURCE: Adopted from Brown and Waser (1988).

[a] Ratio of the difference in area of the power density of the averaged signal at 1 m and the averaged signal at 12.5 or 100 m. The greater the difference between signals, the larger the coefficient.

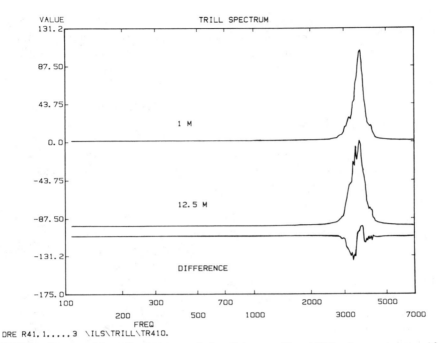

VALUE
131. 2

87. 50

43. 75

0. 0

−43. 75

−87. 50

−131. 2

−175. 0

TRILL SPECTRUM

1 M

12. 5 M

DIFFERENCE

100        300        700        2000        5000
   200        500    1000        3000        7000
FREQ

DRE R41. 1.....3  \ILS\TRILL\TR410.

**FIGURE 4.** Spectra of the *trill* call at transmission distances of 1 and 12.5 m in comparison with the difference spectrum of the two signals. Adopted from Brown and Waser (1988).

distance dimension of sound location, while the *boom* utterance would not. Presumably the presence or absence of distance information in the sound quality of vocalizations is relevant for the social function of different vocalizations, though we know of no investigations addressing this idea. However, it is clear that habitat-induced degradation of vocal signals is used by both birds and primates to monitor possible intrusions into their territories.

Whitehead (1987) has shown that experimental manipulation of the degradation of bark vocalizations (mimicking degradation that occurs in the natural habitat) influences the perception of distance by howler monkeys. Furthermore, playback experiments suggest that acoustic degradation of song is used by great tits to judge distance to the source, but these judgments require that the listener be familiar with the song to gauge distance (and the possibility of intrusion into one's territory: McGregor & Krebs, 1984; McGregor, Krebs, & Ratcliffe, 1983; also see Nelson & Marler, this book, Vol. II, Ch. 14). These studies show that the qualitative degradation of signals is an important cue for the localization of distance, and that the distance coordinate of sound location is a parameter of substantial biological interest. The sound-degradation cue, like the textural cue or the aerial-perspective cue of visual distance perception, is a unilateral and not a bilateral cue (monaural or monocular). Though binocular vision is important for distance and depth perception, the available evidence suggests that binaural hearing is not critical for auditory distance judgments.

## II. LOCALIZATION METHODOLOGY: EGOCENTRIC AND OTOCENTRIC STUDIES

Many organisms will approach the origin of biologically interesting sounds, and it is possible to score the accuracy of their approach. This method has been used to study localization in frogs (Feng et al., 1976), monkeys (Waser, 1977) and many other species. Food or some other reward has been employed to motivate cats (Casseday & Neff, 1973) and a variety of other species to perform the task, but the fundamental feature of these methods is that localization is measured, according to the language of Walls (1951), in an *egocentric* fashion: that is, in reference to the subject's physial orientation in spaces.* In these procedures the accuracy of localization is dependent upon the ability of the perceptual system to detect an acoustic change due to a change in the location of sound, as well as upon the performance of the spatial/motor systems of the organism to accurately guide the subject to move toward the perceived location of the source. Differences in acuity of localization may be due to differences in the accuracy of perceptual systems or due to differences in the integrity of spatial/motor systems.

Orientation paradigms, like sound approach methods, have been used to score the acuity of localization (Brown, 1982a, 1982b; Knudsen & Konishi, 1978). With these methods a head turn or orientation response is used to index the perception of sound direction. Orientation paradigms are an egocentric method requiring the integration of sensory and spatial/motor capabilities.

Behavioral tasks in which listeners operate levers, disks, or keys to signal the detection of a change in sound quality induced by a change in sound location have been used with both human (Mills, 1958) and animal

---

* Working in the visual system Walls (1951) distinguished between egocentric localization and oculocentric localization: that which was based on the spatial resolving capability of the retina itself. We introduce the concept of otocentric localization to make a parallel distinction in the auditory system, and to emphasize the fact that the discrimination of two stimuli that differ in spatial location does not necessarily mean that the listener has any idea where either of the two stimuli originated in reference to the receiver's body. This distinction is not widely appreciated, and it is particularly important for the investigation of the neural mechanisms underlying localization. Experimental destruction of neural tissue may impair the sensory processing of stimuli (otocentric impairments), or alternatively it may impair the integration of an undisturbed sensory input with spatial/motor output (egocentric impairments). Furthermore, clinically observed deficits in human spatial abilities may exhibit this dichotomization.

subjects (Brown et al., 1978a). These later procedures are ear centered or *otocentric*. Few investigators have compared performance between egocentric and otocentric methods, though R. S. Heffner and Masterton (this volume, Ch. 1) report nearly identical performance levels in normal cats when tested with either otocentric or egocentric methods.

Within each of the general categories of egocentric and otocentric methodologies great variation is possible in procedural details, and differences in signal duration, signal amplitude, the noise and reverberation level of test environments, and so forth, are all likely to impact on sound localization performance. Given the potential for variation due to procedural variables, can it be shown that species differences in localization performance are sufficiently consistent within a species and sufficiently robust between species to be accurately measured?

The data show that in addition to the consistency in results between egocentric and otocentric methods reported within a laboratory by R. S. Heffner and Masterton (this volume, Ch. 1), good agreement between laboratories has been observed with otocentric methods (Brown et al., 1978a; Houben & Gourevitch, 1979). Figure 5 shows the comparison of interaural time-difference thresholds directly measured with earphones in macaque monkey (Houben & Gourevitch, 1979) and human listeners (Klump & Eady, 1956) with interaural time-difference thresholds estimated from free field measurements of macaque monkey minimum audible angles (Brown, Beecher, Moody, & Stebbins, 1978). Also displayed in Fig. 5 are interaural phase-difference contours representing interaural phase differences of 11 and 2.5°, corresponding with monkey and human localization, respectively. The results show high agreement between different laboratories (using different procedures) for the macaque monkey data, and marked differences between human and monkey subjects. Furthermore, the results suggest that the localization of low-frequency tones in both human and macaque monkey subjects is determined by a mechanism sensitive to interaural differences in the phase of tonal stimuli, though human listeners possess a more sensitive mechanism. The uniformity, precision, and consistency of results both within and between laboratories make it possible to rigorously compare localization abilities of different species.

## III. THE COMPARATIVE ACUITY OF DIRECTIONAL HEARING

Tables 3 and 4 present measurements of localization acuity in some representative vertebrates when tested with highly locatable stimuli. Perhaps because humans tend to focus on terrestrial events, researchers have more intensely studied localization in the horizontal plane (Table 3) than in the vertical plane (Table 4). Many organisms are capable of detecting changes in auditory space of only a few degrees, an achievement

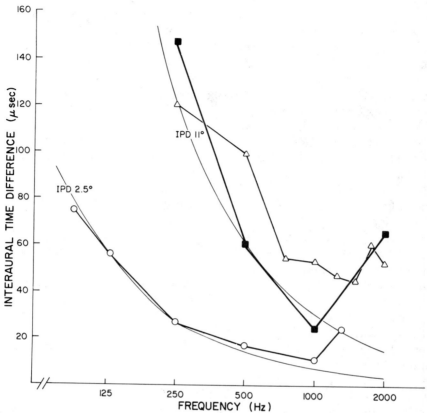

**FIGURE 5.** Interaural time-difference thresholds. The ordinate is threshold in microseconds and the abscissa is signal frequency. ■, Old World monkey thresholds transposed from localization data obtained in a free field (Brown et al., 1978a); ▽, monkey interaural time-difference thresholds measured with earphones (Houben & Gourevitch, 1979); ○, human interaural time-difference thresholds measured with earphones (Klumpp and Eady, 1956). IPD 2.5° and IPD 11° denote interaural phase-difference contours corresponding to the human and monkey thresholds. Adopted from Brown, Beecher, Moody, and Stebbins (1978a).

not equaled by any artificial listening device. Human (Mills, 1958) and nonhuman primates (Brown et al., 1978a), parakeets (Park et al., 1987) and barn owls (Knudsen, Blasdel, & Konishi, 1979), opossums (Ravizza & Masterton, 1972), cats (Casseday & Neff, 1973), dogs (H. Heffner, 1977), sea lions (Moore & Au, 1975), seals (Mohl, 1964), porpoises (McDonald-Renaud, 1974), and elephants (R. S. Heffner & Heffner, 1982) all exhibit excellent spatial-resolving power in the horizontal plane with appropriate stimuli (Table 3). On the other hand, some other organisms are decidedly less sensitive to changes in acoustic space. Some species of birds, and rodents, as well as horses (see Table 3) are surprisingly insensitive to

**TABLE 3.  Localization Acuity in Representative Vertebrates in the Horizontal Plane**[a]

| Group | Species | Acuity (degrees) | Source |
|---|---|---|---|
| Fish | Lemon shark | 19 | Nelson, 1967 |
| | Cod & haddock | 20 | Chapman & Johnstone, 1974 |
| Amphibian | Green treefrog | 30 | Feng et al., 1976 |
| Aves | Parakeet | 5 | Park et al., 1987 |
| | Canary | 15 | Park et al., 1987 |
| | Zebra finch | 15 | Park et al., 1987 |
| | Great tit | 16 | Klump et al., 1986 |
| | Bull finch | 24 | Schwartzkopff, 1950 |
| | Pine grosbeak | 20 | Granit, 1941 |
| | European sparrowhawk | 15 | Kretschmar, 1982 |
| | Barn owl | 2 | Knudsen et al., 1979 |
| | Chicken | 4 | Engelmann, 1928 |
| | Quail | 60 | Gatehouse & Shelton, 1978 |
| | Red-tailed hawk, great horned owl | 52 | Brown, 1982a |
| Marsupialia | Opossum | 5 | Ravizza & Masterton, 1972 |
| Rodentia | Kangaroo rat | 20 | H. Heffner & Masterton, 1980 |
| | Wood rat | 19 | H. Heffner, 1977 |
| | White rat | 10 | Kelly, 1980 |
| Carnivora | Least weasel | 12 | R.S. Heffner & Heffner, 1987 |
| | Cat | 5 | Casseday & Neff, 1973 |
| | | 4 | Martin & Webster, 1987 |
| | Dog | 8 | H. Heffner, 1977; cited in H.E. Heffner & Heffner, 1984 |
| Primates | Rhesus/pig-tailed monkey | 3 | Brown et al., 1978a |
| | Grey-cheeked mangabey | 6 | Waser, 1977 |
| | Human | 1 | Mills, 1958 |
| Pinnipedia | Sea lion | 4 | Moore & Au, 1975 |
| | Harbor seal | 2 | Mohl, 1964 |
| Cetactea | Harbor porpoise | 3 | Andersen, 1970 |
| | Bottlenose porpoise | 1 | McDonald-Renaud, 1974 |
| Perissodactyla | Horse | 22 | H.E. Heffner & Heffner, 1984 |
| Proboscidea | Elephant | 1 | R.S. Heffner & Heffner, 1982 |

[a] The data summarized in this table are rounded to the nearest integer and are for the best signal tested. In some cases the test signal was a pure tone. In most cases, however, the test signals were bands of noise, clicks, or species-specific vocalizations.

**TABLE 4. Localization Acuity in Representative Vertebrates in the Vertical Plane[a]**

| Group | Species | Acuity (degrees) | Source |
|---|---|---|---|
| Fish | Cod | 16 | Hawkins & Sand, 1977 |
| Amphibian | Reed frog | 37 | Passmore et al., 1984 |
| Aves | Barn owl | 2 | Knudsen et al., 1979 |
| Marsupialia | Opossum | 13 | Ravizza & Masterton, 1972 |
| Carnivora | Cat | 4 | Martin & Webster, 1987 |
| Primate | Rhesus/pig-tail | 3 | Brown et al., 1982 |
|  | Human | 3 | Wettschreck, 1973 |
| Cetactea | Bottlenose porpoise | 2 | McDonald-Renaud, 1974 |

[a] The data summarized in this table are rounded to the nearest integer and are for the best signal tested. In some cases the test signal was a pure tone. In most cases, however, the test signals were bands of noise, clicks, or species-specific vocalizations.

changes in sound location. Though Masterton and his associates have made major advances in examining the physiological substrate for differences in localization abilities in mammals (see R. S. Heffner & Masterton, this volume, ch. 1), at present there are no general theories that adequately account for the variation in precision of localization exhibited by different vertebrates (Tables 3 and 4). For example, there is no anatomical, physiological, or theoretical factor that would lead one to expect quail directional hearing to be an order of magnitude less sensitive than that of parakeets.

It has been widely noted that animals with large heads generate large binaural localization cues both in terms of interaural differences in time-of-arrival cues and the frequency–intensity-spectrum cues (Brown et al., 1978a, 1980; H. Heffner & Masterton, 1980; R. S. Heffner & Heffner, 1982; Masterton & Diamond, 1973; Masterton, Heffner, & Ravizza, 1969). Physically more robust cues should in turn permit the nervous system to ascertain sound position with greater precision, and large-headed animals should, as a consequence, exhibit superior acuity. Though the head-size/robust-cue hypothesis was generally successful in accounting for variations in the acuity of terrestrial mammal directional hearing (R. S. Heffner & Heffner, 1982; R. S. Heffner & Masterton, this volume, Ch. 1), it remains unsuccessful in explaining variations in the sound localization abilities in birds and marine mammals. According to this hypothesis, organisms with small heads should have relatively poor localization acuity, and that which they do possess should be dependent upon high-frequency hearing and the presentation of sounds that capitalize upon this high-frequency hearing and the exploitation of the frequency–intensity spectrum. Birds with heads less than half the size of those of rats (Klump, Windt, & Curio, 1986; Park et al., 1987) exhibit localization abilities as good as or better than rat sound localization (H. Heffner & Masterton, 1980), and small birds have impoverished high-frequency hearing (Dooling, 1980).

Because the wavelength of sound in water is about 4.3 times its value in air, the effective head size of many marine mammals is about half that of rats, yet the acuity of underwater localization may be superb (McDonald-Renaud, 1974). Though rats have good high-frequency hearing (Kelly & Masterton, 1977) their sound localization may not be dependent upon the presentation of high-frequency sounds. C. H. Brown (unpublished observation) found no difference in localization performance of rats presented with noise bands 20 Hz–40 kHz in bandwidth or 20 Hz–1 kHz in bandwidth. At lease at certain angles, localization in rats was not dependent upon high-frequency hearing and the utilization of the frequency–intensity spectrum. Furthermore, research with large mammals (H. E. Heffner & Heffner, 1984) indicates that the availability of large interaural differences does not insure high-precision sound localization. Though the interaural pathway and the pressure gradient system (if functional in some vertebrate groups and not in others) may account for some of these differences, it is unlikely to account for all of these species differences.

These observations indicate that species differences in the acuity of directional hearing are not simply a passive consequence of physical differences in head size and the resulting magnitude of the interaural cues. Some species appear to have evolved specializations for high-acuity sound localization and others have not. A number of investigators have argued that the function of sound localization is to direct the eyes to the site of sound (Hafter & DeMaio, 1975; Harrison & Irving, 1966; R. S. Heffner & Heffner, 1983, 1985; Pumphery, 1950), and differences in the spatial fields of other sensory modalities may underlie species differences in the precision of sound localization, as may differences in life history strategies, type of predation threat, sources of food, and so forth (R. S. Heffner & Heffner, 1987). A greater understanding of localization mechanisms is likely dependent upon the examination of life history strategies, the ecology of different organisms, and the role of hearing, sound localization, and binaural processes in the natural environment.

## IV. SOUND LOCALIZATION AND THE NATURAL ENVIRONMENT

Classically, sound-localization abilities have been studied with simple tonal stimuli (see Fig. 5), yet the auditory system has evolved to localize vocal signals and the noises that cue the movements, presence, and activities of other organisms.

### A. Contact–Seeking Vocal Signals

Vocal signals are issued to further a variety of biological objectives including position marking, the localization and attraction of kin and mates, as well as territorial delineation and defense. The localization of the greatest variety of vocal signals has been studied in Old World monkeys. Macaque monkeys produce extensive acoustic variation in their utterances, and their calls may be initially grouped into two classes: harsh calls, a group of noisy strident sounds; and clear calls, a group of tonal, musical sounds. Figure 6 displays localization thresholds (top panel) and sound spectrograms for two clear calls and two harsh calls (bottom panel), respectively. From the sound spectrograms it can be seen that the clear calls are harmonically structured with the dominant frequency band centered at about 0.8 kHz and overtones ranging to about 4 kHz. The harsh

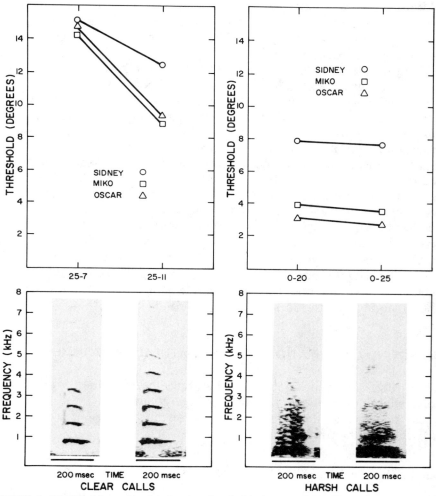

**FIGURE 6.** Old World monkey localization thresholds for two clear calls and two harsh calls (top panel). Sound spectrograms for the four test signals are shown in the bottom panel. Adopted from Brown, Beecher, Moody, and Stebbins (1979).

calls are broad bandwidth signals, often four or more octaves wide, ranging from approximately 100 Hz to several thousand Hertz. In this frequency region localization acuity is near asymptotic levels for sounds greater than one octave in bandwidth (Brown et al., 1980), thus the acoustic variation between different harsh calls has little effect on sound localization (Brown, Beecher, Moody, & Stebbins, 1979). The acuity of localization of macaque clear calls, however, is quite variable, and this variation is presumably due to subtle differences in their acoustic composition. The presence or absence of the acoustical detail contained in the harmonic elements of clear calls has been shown to have no effect on sound localization (Brown et al., 1979), but the magnitude of frequency modulation (or the effective bandwidth) of the dominant frequency band of various clear calls strongly influences sound localization (Brown et al., 1978b). Figure 7 displays the relation between the effective bandwidth of the dominant frequency band of macaque clear calls and the locatability of these signals. Frequency-modulated clear calls are located four or five times more accurately than are unmodulated calls. Thus, frequency modulation may be employed in vocal production to promote sound localization. Studies of vocal behavior in Japanese macaques (Green, 1975; May, Moody, & Stebbins, 1988) have shown that the magnitude and form of frequency modulation is a communicatively relevant dimension. Clear calls that are not frequency modulated and presumably hard to localize are given in social contexts at the low end of the arousal continuum, while frequency-modulated clear calls and many harsh calls may be issued in high-arousal contexts. The acoustic structure of these signals would make them easy to localize. This correspondence favors the idea that the magnitude of contact seeking registered by different calls is communicatively underscored by their locatability. Interestingly, heightened frequency-modulation contours appear to play a role in human language acquisition. Fernald (personal communication) has noted that human infant care takers exaggerate the frequency-modulation contours of their speech when attempting to "capture" an infant's gaze and attention.

Frequency modulation is not a one-dimensional parameter. Signals may differ in respect to the duration, bandwidth, and direction of frequency change, and these parameters individually as well as combined may be critical for localization. Furthermore, certain modulation parameters may have the effect of disrupting sound-localization behavior, while others may enhance it (Brown, 1982b). Because natural utterances vary in amplitude and duration as well as in frequency-modulation characteristics, before the significance of a particular frequency-modulation parameter can be studied, it first must be isolated through computer synthesis techniques. Such techniques allow each parameter of a complex acoustic stimulus to be precisely specified and independently manipulated. For example, May and his associates (1988) have created synthetic sounds that resemble macaque clear calls. The stimuli shown in Fig. 8a are linear frequency

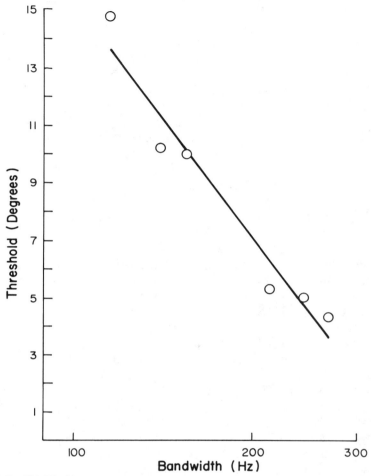

**FIGURE 7.** Old World monkey localization thresholds of six clear calls as a function of frequency modulation bandwidth) of the dominant band of the call. The correlation coefficient is −0.98. Adopted from Brown, Beecher, Moody, and Stebbins (1978b).

sweeps and were designed to reflect frequency changes like those observed in the lower harmonies of natural utterances.

How nonhuman primates perceive highly abstracted versions of their vocal signals can be measured with operant testing paradigms (see Moody et al., this book, Vol. II, ch. 10). To determine the significance of frequency modulation in the sound-localization behaviors of Old World Monkeys, the operant task used by May, Moody, Stebbins, and Norat (1986) required monkeys to release a metallic disk to signal the detection of a change in the location of a FM stimuli. The sound-localization threshold, or minimum audible angle (MAA), was defined as the angle of speaker separation that produced 50% correct releases.

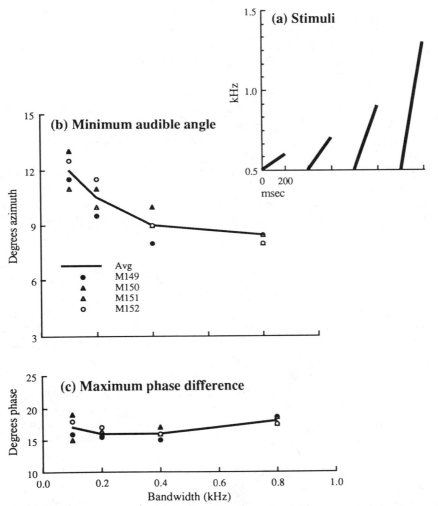

**FIGURE 8.** Frequency-modulated tones (upsweeps) with a fixed 0.5-kHz starting frequency and 200-ms duration (*a*). Localization acuity (minimum audible angles, MAA) as a function of the bandwidth of the frequency-modulated tones (*b*). Interaural phase differences corresponding to the panel (*b*) thresholds as a function of the bandwidth of the modulated tones (panel *c*). No significant difference was observed for interaural phase differences. Adopted from May, Moody, Stebbins, and Norat (1986).

The results of these tests are shown in Fig. 8*b*. As the upper frequency limit and, consequently, the bandwidth of the frequency sweep increased, the minimum audible angle obtained with the stimulus decreased. The inverse relation of stimulus bandwidth to the sound-localization threshold suggests that the FM components of macaque clear calls may facilitate the localization of the signaller, which in this case is a conspecific animal seeking social contact. How the acoustical features of vocal signals are designed to promote or impede sound localization in other social situations is discussed in greater detail later in this chapter.

As previously noted for simple tones, behavioral methods are also capable of revealing the mechanisms that underlie FM sound-localization processes. Figure 8c displays the interaural phase-difference (IPD) contour calculated for the four frequency sweeps shown in panel ( a ). For each of these complex stimuli, the IPD contour is based on the interaural time-of-arrival difference at the corresponding MAA and the fastest rate of phase change in the sweep, that is, the frequency that yields the maximum phase difference at the minimum detectable change in localization. The constant phase disparity at the changing minimum audible angle suggests that variations in localization thresholds for FM signals are governed by a mechanism maximally sensitive to a specific phase difference. This finding for frequency-modulated signals parallels that seen for pure tone localization (Fig. 5).

## B. Ventroliquial and Locatable Signals

Acoustical features of vocal signals may have been selected to promote sound localization in certain cases and impair sound localization in others. When threatened by predators many communal birds respond in one of two contrasting modes. In the attack mode, even very small birds may chase and mob potential predators. These aggressive attacks are accompanied by raucous and incessant mobbing calls which act as an attractant, recruiting other small birds to join in the chase. The contrasting response is the flight mode. In this mode only a faint alarm call may be issued; congeners appear to be repelled by the call, and the call evokes evasive maneuvers rather than mobbing responses. Marler (1955, 1957) proposed that the acoustic structure of the hawk alarm call may impair sound localization by the predator, and thus reduce the risk of exposing oneself to predation which may be incurred by calling. On the other hand, the mobbing call should be locatable in keeping with its role as an attractant and recruitment signal.

Sound spectrograms of the two utterances are presented in Fig. 9. The hawk alarm call (Fig. 9a ), or the seeet call, is a narrow-bandwidth whistle 7–8 kHz in frequency which is near the upper limit of hearing in most birds (Dooling, 1980). The mobbing call (Fig. 9b ) is frequency-modulated, broad bandwidth (ranging 2–6 kHz), and presented reiteratively. Figure 10 displays the mean error of orientation measured by video taping orientation responses of hawks and owls under seminaturalistic conditions (Brown, 1982a). At call onset (the column labeled 1 for both calls) the subject's orientation relative to the location of hidden speakers was about 90°, the value expected by chance. However, at call offset (the column labeled 2 for both calls) the mean error of orientation was 51.5° for the mobbing call, but 124.5° for the alarm call. Hawks and owls were able to locate the position of the mobbing calls, but were disoriented by the

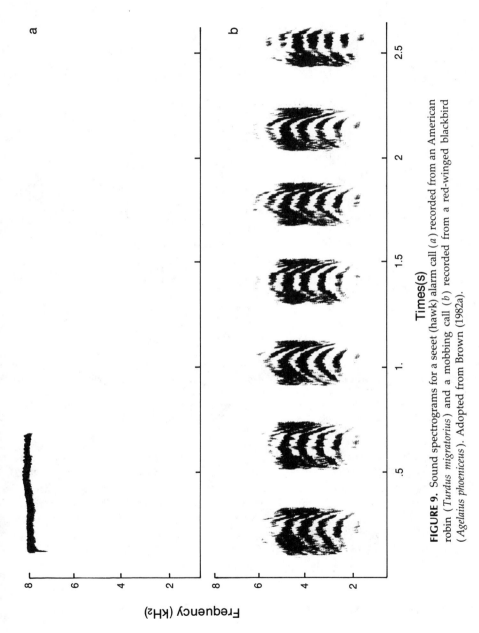

**FIGURE 9.** Sound spectrograms for a seeet (hawk) alarm call (*a*) recorded from an American robin (*Turdus migratorius*) and a mobbing call (*b*) recorded from a red-winged blackbird (*Agelaius phoeniceus*). Adopted from Brown (1982a).

269

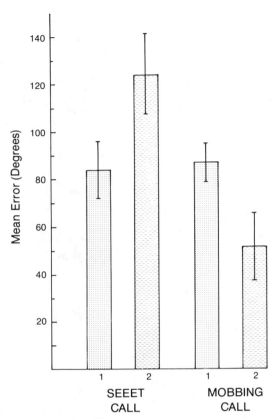

**FIGURE 10.** Mean error of orientation scores for seeet calls and mobbing calls. The error of orientation at trial onset is indicated by column 1, and the error of orientation at trail termination is indicated by column 2. Adopted from Brown (1982a).

presentation of the alarm call. This finding suggests that under field conditions, the alarm call may be perceived by the auditory system of some predatory birds to originate from locations other than its actual position. Thus, the call may be acoustically ventroloquial. A mechanistic explanation of acoustical ventriloquism is still speculative. Nevertheless, these observations emphasize that vocalizations may have acquired structural features that actively hinder or promote sound localization, and localization may strongly influence the function of different signals in nature.

## V. ECOLOGICAL CONSTRAINTS ON LOCALIZATION

Under laboratory conditions sound-localization abilities are tested almost without exception in extraordinarily quiet and echo-free environments; yet under natural conditions organisms must localize sounds under reverberant, noisy conditions.

## A. Masking and Localization

Biologically significant sounds are often embedded in an on-going fluctuating background noise, and signals are degraded by the action of frequency-specific absorption, reflection, and diffraction of the waveform with environmental surfaces (Brown & Waser, 1988). Remarkably only limited attention has been directed to sound localization capabilities of organisms under typical natural conditions. Localization thresholds for narrow-band and broadband signals simulating the bandwidth parameters of macaque vocalizations are shown in Fig. 11 (Brown, 1982b). In the quiet a broadband signal acoustically similar to a harsh call (1 kHz in bandwidth) was more accurately localized than were two narrow-band signals (resembling clear calls). Masking noise (presented at a level at which all signals were still clearly audible) impaired localization for all signals, but the

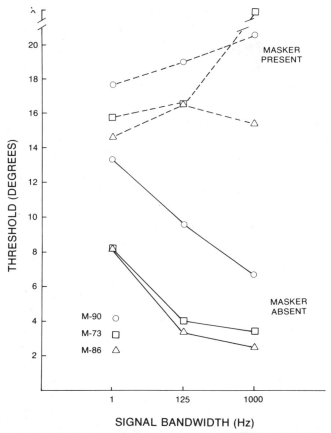

**FIGURE 11.** Localization acuity as a function of signal bandwidth for Old World monkeys in the presence and absence of a white-noise masker. The stimuli were bands of noise geometrically centered at 800 Hz presented at 30 dB SPL. The white-noise masker was 55 dB. Adopted from Brown (1982b).

impairment was markedly less for the narrow-band signals. These results do not show that the impairment in localization was independent of a decrease in audibility per se. Nevertheless, the results do show that localization acuity will be influenced by fluctuations in ambient noise level even though the fluctuations in level do not mask the actual detection of the signal.

## B. Reverberation and Localization

In anechoic environments the waveform of the to-be-localized signal is undisturbed by reflections of the signal with environmental surfaces, and the auditory system is only confronted with interaural differences in the direct wave of the signal. Under natural conditions environmental reflections introduce a variety of complications. As illustrated in Fig. 12 the path length of the direct wave is shorter than that of the reflected wave, and this difference in path length will give rise to constructive or destructive interference of the waveform at the point where it is received by a microphone or a listener's ear. Because the subject's two ears sample the interference characteristics of the direct and reflected waveforms at two

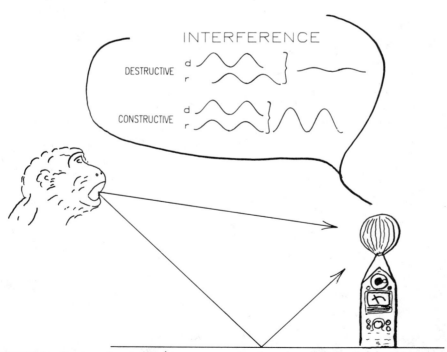

**FIGURE 12.** Environmental reflections and destructive and constructive interference between the reflected and direct waves at the point of measurement.

different locations in space, interaural differences in the waveform incident at each ear will be a consequence of the reverberation characteristics of the environment as well as the geometrical relation between the orientation of the listener with respect to the location of the signal's source. In the case of complex signals the interference characteristics of the signal at the point of measurement would likely differ for the various frequency components. As a result, the relative representation of each frequency component would be determined by the direct-wave frequency composition of the signal and the interference characteristic of each frequency component of the signal at that point in the environment.

May and his associates (B. May, D. B. Moody, W. C. Stebbins, & M. A. Norat, unpublished observations) explored the effects of reverberation on the sound-localization behaviors of Old World monkeys by comparing the minimum audible angles obtained for tones, frequency sweeps, and noise under anechoic and echoic conditions. The echoic testing environment was created in an anechoic chamber by placing two sound-reflecting panels near the subject. The panels were arranged so that the path-length difference between the direct and reflected route to the subject's head was short (approximately 0.3 m). Thus, under echoic conditions, perturbations of the waveforms produced by destructive and constructive interference would begin within 1–2 msec. Hence, interference would begin well before the 5-msec signal onset was completed.

Although the localization of sweeps and noise was not significantly influenced by the presence of reverberation, when monkeys were required to detect changes in the location of tones, the introduction of the reflectant surfaces was found to produce an increase in the behavioral threshold at certain frequencies (e.g., 6 kHz), but to aid localization at others (e.g., 500 Hz; see Table 5). These results were surprising and indicate that complex perturbations of the waveform produced by interference do not necessarily confound primate sound-localization mechanisms. Instead, the primate auditory system may be designed to utilize these naturally occurring perturbations to promote sound localization and other critical auditory processes. The actual characteristics of the signals incident at the subject's ears, and the mechanisms by which different signal features are processed are not yet known. However, these findings emphasize the importance of exploring environmental degradation and perturbation of signals, and the mechanisms by which the auditory system processes acoustic signals in real-world situations. In this context it is interesting to note that reverberation should seriously distort waveforms in many experimental situations in which sound localization is reportedly described as occurring naturally and effortlessly (Beecher & Harrison, 1971; Harrison et al., 1971; Mendelson & Haith, 1976; Muir & Field, 1979; Wetheimer, 1961). In contrast, many animals are slow to learn to respond to changes in sound position that occur in the controlled, acoustically simple environments created in the laboratory (Brown, 1984; Brown et al., 1978a).

**TABLE 5. Localization Acuity for Old World Monkeys in Anechoic and Echoic Environments**

| Stimulus | Subject | Anechoic (degrees) | Echoic (degrees) |
|----------|---------|--------------------|------------------|
| 500 Hz | 142 | 6.6 | 2.6 |
| | 147 | 4.9 | 2.5 |
| | 149 | 13.7 | 4.3 |
| | 150 | 8.7 | 6.9 |
| | 151 | 10.6 | 7.4 |
| | 168 | 5.8 | 2.9 |
| 6 kHz | 142 | 7.9 | 13.2 |
| | 147 | 4.9 | 5.7 |
| | 149 | 8.7 | 15.5 |
| | 150 | 5.9 | 9.2 |
| | 151 | 5.8 | 6.5 |
| | 168 | 11.1 | 10.3 |

From B. May, D.B. Moody, W.C. Stebbins, and M.A. Norat, unpublished observations.

## C. Localization and Release from Masking

Though the evolutionary advantage of sound localization has likely been a persistent source of selection for improvements in the processing of subtle interaural differences, binaural processing is also critical for the detection of signals in noise, and in some cases selection for improved sound-detection mechanisms may have been of greater importance. When the origin of the source of a masking noise and the origin of a signal are separated in space, binaural processes may reduce the masking effect of the noise. The magnitude of the release from masking produced by the spatial separation of the signal and the masker differs for various kinds of signal and masker parameters in human subjects. Under certain conditions the release from masking effect with human listeners may exceed 25 dB (Durlach & Colburn, 1978). However, it is unclear if reductions of this magnitude are due to testing with stimuli that possess characteristics that make them particularly susceptible to masking, or alternatively if the effect is due to testing with stimuli that capitalize on the spatial-processing abilities of the binaural auditory system. Though some investigators have used loudspeakers and naturally produced binaural differences (Bronkhorst & Plomp, 1988; Dirks & Wilson, 1969; Plomp, 1976; Plomp & Mimpen, 1981), most release from masking studies have used earphones and synthetic shifts of the position of the masker or the signal through the employment of computers or delay lines, or more simply through reversing the polarity of signal or noise inputs to the earphones (Buus, 1985; Carhart, Tillman, & Johnson, 1967, 1968; Carhart, Tillman, & Greetis,

1969a, 1969b; Durlach & Colburn, 1978; Hall, 1986; Hall, Cokely, & Grose, 1988; Levitt & Rabiner, 1967; Schooneveldt & Moore, 1987, 1988). Synthetiticly generated stimulus conditions are frequently out of bounds of the stimulus parameters that could occur naturally. Thus, under real-world conditions the release from masking effect may not reach the 25-dB level. Using a white-noise masker located at 0 or 90° and tonal signals (250 Hz, 1 kHz, or 4 kHz) positioned at 0° C. H. Brown (unpublished observations) has compared release from masking in Old World monkeys and human listeners. As shown in Fig. 13 the release from masking effect varied from −2.3 to 8.1 dB for the human subjects, and from 2.6 to 4.8 dB for monkey listeners. Individual variation in the size of the effect was marked for both species; however, the monkeys experienced a smaller average change in audibility produced by changing the location of the masking noise relative to the location of the signal than did human subjects.

The reduced release from masking effect found for monkeys may be due to the monkey's smaller head size, and the resultingly more modest interaural differences in the time or spectral domains of the waveforms incident at each ear. However, other kinds of signals may result in greater effects: Preliminary data show a 9.0-dB release from masking for the blue monkey pyow vocalization, and other vocal signals may exhibit yet an even greater release from masking. Studies using speech sounds presented to human listeners have reported a greater release from masking with spondee words compared to some other speech sounds (Pb words). Though investigators have argued that the low-frequency content of the spondee words is responsible for their greater masking level difference (Dirks & Wilson, 1969; Schubert & Schultz, 1962), the frequency trend seen in Fig. 13 suggests that low-frequency signals do not enjoy a superior release from masking effect. Clearly, more observations are warranted here, and studies should be conducted with both acoustically complex naturally occurring signals as well as with more simple synthetic ones.

The release from masking phenomenon may have strong implications for the audible range of signals in nature. The attenuation rate of vocal signals in nature is dependent upon the acoustical ecology of the habitat, and on the dominant frequency band of the vocalization (Brown, 1989; Waser & Brown, 1986). Nevertheless, the audible range of many monkey vocalizations would be nearly doubled by a release from masking factor of 6 dB, and signals at certain frequencies would gain an even greater propagation advantage (Brown, 1989; Brown & Waser, 1986). However, the significance of scattering and reverberation on the directivity of both signals and background noise in different habitats is an unknown factor pertinent for the role of release from masking in nature. To our knowledge there are no studies of the directional characteristics of ambient noise and animal vocalizations in nature, yet these habitat and signal characteristics may have influenced the evolution of binaural mechanisms in the mammalian auditory system.

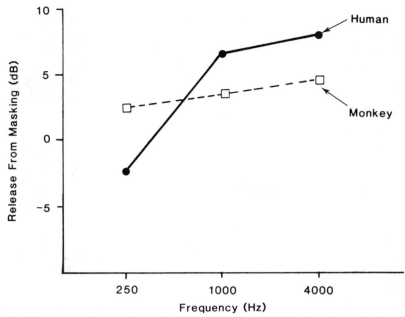

**FIGURE 13.** Release from masking (threshold audibility of the signal positioned at 0°, masker 90°; threshold audibility of the signal positioned 0°, masker 0°) as a function of tone frequency. The data are averaged from 3 Old World monkeys and 6 human subjects (C. H. Brown, unpublished observations).

Sound localization has been one of the most persistently studied topics in auditory perception, and the mechanisms subserving directional hearing have been shown to be important for the detection of signals in noise as well as for sound localization per se. However, as noted here, most research has focused on the processing of acoustically simple synthetic signals in quiet, echo-free environments, and this research bias has limited our understanding of the full capabilities of the directional-hearing components of the auditory system. Fortunately, our current technology may be adopted to investigate localization mechanisms and capabilities at strategic junctures across a continuum from the quiet, acoustically simplified laboratory environment to the reverberant, acoustically cluttered natural world. Carefully engineered laboratory studies designed to mimic the acoustical complexities of both natural environments and natural signals may be expected to play a progressively more prominent role in localization research in the future. As has been surveyed here, many problems in directional hearing have yet to be solved, and it is hoped that this chapter will help stimulate investigators to develop new research approaches directed toward the resolution of these problems.

## ACKNOWLEDGMENTS

Portions of the research reviewed in this chapter were supported by NIH Grants R01 NS16632-08 and K04 NS00880-04.

## REFERENCES

Andersen, S. (1970). Directional hearing in the habor porpoise (*Phocoena phocoena*). In G. Pilleri (Ed.), *Investigations on Cetacea* (Vol. 2, pp. 260-263). Benteli.

Anderson, D. J. (1973). Quantitative model for the effects of stimulus frequency upon synchronization of auditory nerve discharges. *Journal of the Acoustical Society of America,* **54,** 361–364.

Aronson, E., & Rosenbloom, S. (1971). Space perception in early infancy: Perception within a common auditory-visual space. *Science,* **172,** 1161–1163.

Beecher, M. D., & Harrison, J. M. (1971). Rapid acquisition of an auditory location discrimination by rats. *Journal of the Experimental Analysis of Behavior,* **16,** 193–199.

Beer, C. G. (1969). Laughing gull chicks: Recognition of their parent's voices. *Science,* **166,** 1030–1032.

Beer, C. G. (1970). Individual recognition of voice in social behavior of birds. In D. S. Lehrman, R. A. Hinde, & E. Shaw Eds.), *Advances in the study of behavior* (Vol. 3). New York: Academic Press.

Bleckmann, H. (1980). Reaction time and stimulus frequency in prey localization in the surface feeding fish *Aplocheilus lineatus. Journal of Comparative Physiology,* **140,** 163–172.

Bleckmann, H. (1988). Prey identification and prey localization in surface feeding fish and fishing spiders. In J. Atema, R. R. Fay, A. N. Popper, & W. N. Tavolga (Eds.), *Sensory biology of aquatic animals* (pp. 619–641). New York: Springer-Verlag.

Bronkhorst, A. W., & Plomp, R. (1988). The effect of head-induced interaural time and level differences on speech intelligibility in noise. *Journal of the Acoustical Society of America,* **83,** 1508–1516.

Brown, C. H. (1982a). Ventroloquial and locatable vocalizations in birds. *Zeitschrift fuer Tierpsychologie,* **59,** 338–350.

Brown, C. H. (1982b). Auditory localization and primate vocal behavior. In C. T. Snowdon, C. H. Brown, & M. R. Petersen (Eds.), *Primate communication.* London & New York: Cambridge University Press.

Brown, C. H. (1984). Directional hearing in aging rats. *Experimental Aging Research,* **10,** 35–38.

Brown, C. H. (1989). The active space of blue monkey and grey-cheeked mangabey vocalizations. *Animal Behaviour,* **37,** 1023–1034.

Brown, C. H., Beecher, M. D., Moody, D. B., & Stebbins, W. C. (1978a). Localization of pure tones in Old World monkeys. *Journal of the Acoustical Society of America,* **63,** 1484–1494.

Brown, C. H., Beecher, M. D., Moody, D. B., & Stebbins, W. C. (1978b). Localization of primate cells by Old World monkeys. *Science, 201,* 753–754.

Brown, C. H., Beecher, M. D., Moody, D. B., & Stebbins, W. C. (1979). Locatability of vocal signals in Old World monkeys: Design features for the communication of position. *Journal of Comparative and Physiological Psychology, 93,* 806–819.

Brown, C. H., Beecher, M. D., Moody, D. B., & Stebbins, W. C. (1980). Localization of noise bands by Old World monkeys. *Journal of the Acoustical Society of America, 68,* 127–132.

Brown, C. H., Schessler, T., Moody, D. B., & Stebbins, W. C. (1982). Vertical and horizontal sound localization in primates. *Journal of the Acoustical Society of America, 68,* 1804–1811.

Brown, C. H., & Waser, P. M. (1988). Environmental influences on the structure of primate vocalizations. In D. Todt, P. Geodeking, & D. Symmes (Eds.), *Primate vocal communication.* (pp. 51–66). Berlin: Springer-Verlag.

Butler, R. A. (1969). Monaural and binaural localization of noise bursts vertically in the median sagittal plane. *Journal of Auditory Research, 9,* 230–235.

Buus, S. (1985). Release from masking caused by envelope fluctuations. *Journal of the Acoustical Society of America, 78,* 1958–1965.

Carhart, R., Tillman, T. W., & Johnson, K. R. (1967). Release of masking for speech through interaural time delay. *Journal of the Acoustical Society of America, 42,* 124–138.

Carhart, R., Tillman, T. W., & Johnson, K. R. (1968). Effects of interaural time delays on masking by two competing signals. *Journal of the Acoustical Society of America, 43* 1223–1230.

Carhart, R., Tillman, T. W., & Greetis, E. S. (1969a). Release from multiple maskers: Effects of interaural time disparities. *Journal of the Acoustical Society of America, 45,* 411–418.

Carhart, R., Tillman, T. W., & Greetis, E. S. (1969b). Perceptual masking in multiple sound backgrounds. *Journal of the Acoustical Society of America, 45,* 694–703.

Casey, M. A., & Feldman, M. L. (1982). Aging in the rat medial nucleus of the trapezoid body. *Neurobiology of Aging, 3,* 187–195.

Casseday, J. H., & Neff, W. D. (1973). Localization of pure tones. *Journal of the Acoustical Society of America, 54,* 365–372.

Chapman, C. J., & Johnstone, A. D. F. (1974). Some auditory discrimination experiments on marine fish. *Journal of Experimental Biology, 61,* 521–528.

Clements, M., & Kelly, J. B. (1978a). Directional responses by kittens to an auditory stimulus. *Developmental Psychobiology, 11,* 505–511.

Clements, M., & Kelly, J. B. (1978b). Auditory spatial responses of young guinea pigs (*Cavia porcellus*) during and after ear blocking. *Journal of Comparative and Physiological Psychology, 92,* 34–44.

Coles, R. B., Lewis, D. B., Hill, K. G., Hutchings, M. E., & Gower, D. M. (1980). Directional hearing in the Japanese quail (*Coturnix coturnix japonica*). II. Cochlear physiology. *Journal of Experimental Biology, 86,* 153–170.

Dirks, D. P., & Wilson, R. H. (1969). The effect of spatially separated sound sources on speech intelligibility. *Journal of Speech and Hearing Research, 12,* 5.

Dooling, R. J. (1980). Behavior and psychophysics of hearing in birds. In A. N. Popper, & R. R. Fay (Eds.), *Comparative studies of hearing in vertebrates.* Berlin: Springer-Verlag.

Durlach, N. I., & Colburn, H. S. (1978). Binaural phenomena. In E. C. Carterette & M. Friedman (Eds.), *Handbook of Perception: Vol. 4. Hearing* (pp. 365–466). New York: Academic Press.

Engelmann, W. (1928). Untersuchungen uber die Schallokalisation bei Tieren. *Zeitschrift fuer Psychologie,* **105,** 317–370.

Erulkar, S. D. (1972). Comparative aspects of spatial sound localization. *Physiological Review,* **52,** 237–359.

Feng, A. S., Gerhardt, H. C., & Capranica, R. R. (1976). Sound localization behavior of the green tree frog (*Hyla cinerea*) and the barking tree frog (*H. gratiosa*). *Journal of Comparative Physiology,* **107,** 241–252.

Gardner, M. B. (1973). Some monaural and binaural facets of median plane localization. *Journal of the Acoustical Society of America,* **54,** 1489–1495.

Gatehouse, R. W., & Shelton, B. R. (1978). Sound localization in bob white quail (*Colinus virginianus*). *Behavioral Biology,* **22,** 533–540.

Gottlieb, G. (1965). Imprinting in relation to parental and species identification by avian neonates. *Journal of Comparative and Physiological Psychology,* **59,** 345–356.

Granit, O. (1941). Beitrage zur Kenntnis des Gehorsinns der Vogel. *Ornis Fennica,* **18,** 49–71.

Green, S. (1975). Variation of vocal pattern with social situation in the Japanese monkey (*Macaca fuscata*): A field study. In L. A. Rosenblum (Ed.), *Primate behavior* (Vol. 4). New York: Academic Press.

Hafter, E. R., & DeMaio, J. (1975). Difference thresholds for interaural delay. *Journal of the Acoustical Society of America,* **57,** 181–187.

Hall, J. W. (1986). The effect of across-frequency differences in masking level on spectro-temporal pattern analysis. *Journal of the Acoustical Society of America,* **79,** 781–787.

Hall, J. W., Cokely, J. A., & Grose, J. H. (1988). Combined monaural and binaural masking release. *Journal of the Acoustical Society of America,* **83,** 1839–1845.

Harrison, J. M., Downey, P., Segal, M., & Howe, M. (1971). Control of responding by location of auditory stimuli: Rapid acquisition in monkey and rat. *Journal of the Experimental Analysis of Behavior,* **15,** 379–386.

Harrison, J. M., & Irving, R. (1966). Visual and non-visual auditory systems in mammals. *Science,* **154,** 738–743.

Hawkins, A. D., & Sand, O. (1977). Directional hearing in the median plane by the cod. *Journal of Comparative Physiology,* **122,** 1–8.

Hebrank, J., & Wright, D. (1974). Spectral cues used in the localization of sound sources in the median plane. *Journal of the Acoustical Society of America,* **56,** 1829–1834.

Heffner, H. (1977). Hearing and sound localization in the kangaroo rat (*Dipodomys merriami*). *Journal of the Acoustical Society of America,* **61,** S59.

Heffner, H., & Masterton, R. B. (1980). Hearing in glires: Domestic rabbit, cotton rat, feral house mouse, and kangaroo rat. *Journal of the Acoustical Society of America,* **68,** 1584–1599.

Heffner, H. E., & Heffner, R. S. (1984). Sound localization in large mammals: Localization of complex sounds by horses. *Behavioral Neuroscience, 98,* 541–555.

Heffner, R. S., & Heffner, H. E. (1982). Hearing in the elephant (*Elephas maximus*): Absolute sensitivity, frequency discrimination, and sound localization. *Journal of Comparative and Physiological Psychology, 96,* 926–944.

Heffner, R. S., & Heffner, H. E. (1983). Hearing in large mammals: Horses (*Equus caballus*) and cattle (*Bos taurus*). *Behavioral Neuroscience, 97,* 299–309.

Heffner, R. S., & Heffner, H. E. (1985). Auditory localization and visual fields in mammals. *Neuroscience Abstracts, 11,* 547.

Heffner, R. S., & Heffner, H. E. (1987). Localization of noise, use of binaural cues, and a description of the superior olivary complex in the smallest carnivore, the least weasel (*Mustela nivalis*). *Behavioral Neuroscience, 101,* 701–708.

Heffner, R. S., Richard, M. M., & Heffner, H. E. (1987). Hearing and the auditory brainstem in a fossorial mammal, the pocket gopher. *Neuroscience Abstracts, 13,* 546.

Henning, G. B. (1974). Detectability of interaural delay in high-frequency complex waveforms. *Journal of the Acoustical Society of America, 55,* 84–90.

Henson, O. W. (1974). Comparative anatomy of the middle ear. In W. D. Keidel & W. D. Neff (Eds.), *Handbook of sensory physiology* (Vol. 5, pp. 39–110). New York: Springer-Verlag.

Hill, K. G., Lewis, D. B., Hutchings, M. E., & Coles, R. B. (1980). Directional hearing in Japanese quail (*Coturnix coturnix japonica*). I. Acoustic properties of the auditory system. *Journal of Experimental Biology, 86,* 135–151.

Houben, D., & Gourevitch, G. (1979). Auditory lateralization in monkeys: An examination of two cues serving directional hearing. *Journal of the Acoustical Society of America, 66,* 1057–1063.

Johnson, D. H. (1980). The relationship between spike rate and synchrony in responses of auditory nerve fibers to single tones. *Journal of the Acoustical Society of America, 68,* 1115–1122.

Kaufman, L. (1979). *Perception: The world transformed.* New York: Oxford University Press.

Keithley, E. M., & Feldman, M. L. (1979). Spiral ganglion cell counts in an age-graded series of rat cochleas. *Journal of Comparative Neurology, 188,* 429–442.

Kelly, J. B. (1980). Effects of auditory cortical lesions on sound localization in the rat. *Journal of Neurophysiology, 44,* 1161–1174.

Kelly, J. B., & Masterton, R. B. (1977). Auditory sensitivity of the albino rat. *Journal of Comparative and Physiological Psychology, 91,* 930–936.

Kiang, N. Y. S., Watanabe, T., Thomas, E. C., & Clark, L. F. (1965). *Discharge patterns of single fibers in the cat's auditory nerve.* Cambridge, MA: MIT Press.

Klump, R. G., & Eady, H. R. (1956). Some measurements of interaural time difference thresholds. *Journal of the Acoustical Society of America, 28,* 859–860.

Klump, R. G., Windt, W., and Curio, E. (1986). The great tit's (*Parus major*) auditory resolution in azimuth. *Journal of Comparative Physiology, 158,* 383–390.

Knudsen, E. I., Blasdel, G. G., & Konishi, M. (1979). Sound localization by the barn owl measured with the search coil technique. *Journal of Comparative Physiology, 133,* 1–11.

Knudsen, E. I., & Konishi, M. (1978). A neural map of auditory space in the owl. *Science,* **200,** 795–797.

Kretchmar, E. (1982). *Wie hort ein Sperber (Accipiter nisus L.) Alarmrufe seiner Beutevogel?* Thesis, Ruhr-University Bochum.

Kuhn, G. F. (1977). Model for the interaural time differences in the azimuthal plane. *Journal of the Acoustical Society of America,* **62,** 157–167.

Kuhn, G. F. (1979). The effect of the human torso, head, and pinna on the azimuthal directivity and on the median plane vertical directivity. *Journal of the Acoustical Society of America,* **65,** (S1), S8(A).

Levitt, H., & Rabiner, L. R. (1967). Binaural release from masking for speech and gain in intelligibility. *Journal of the Acoustical Society of America,* **42,** 601–608.

Manley, G. A. (1973). A review of some of current concepts of the functional evolution of the ear in terrestrial vertebrates. *Evolution (Lawrence, Kansas).* **26,** 608–621.

Marler, P. (1955). Characteristics of some animal calls. *Nature (London).* **176,** 6–8.

Marler, P. (1957). Specific distinctiveness in the communication of some birds. *Behaviour,* **11,** 13–39.

Martin, R. L., & Webster, W. R. (1987). The auditory spatial acuity of the domestic cat in the interaural horizontal and median vertical planes. *Hearing Research,* **30,** 239–252.

Masterton, R. B. (1974). Adaption for sound localization in the ear and brainstem of mammals. *Federation Proceedings,* **133,** 1904–1910.

Masterton, R. B., & Diamond, I. T. (1973). Hearing: Central neural mechanisms. In E. C. Carterette & M. P. Friedman (Eds.), *Handbook of perception* (Vol. 3, pp. 408–448). New York: Academic Press.

Masterton, R. B., Heffner, H. E., & Ravizza, R. (1969). The evolution of human hearing. *Journal of the Acoustical Society of America,* **45,** 966–985.

May, B., Moody, D. B., Stebbins, W. C., & Norat, M. A. (1986). Sound localization of frequency-modulated sinusoids by Old World monkeys. *Journal of the Acoustical Society of America,* **80,** 776–782.

May, B., Moody, D. B., & Stebbins, W. C. (1988). The significant features of Japanese macaque coo sounds: A psychophysical study. *Animal Behaviour.* **36,** 1432–1444.

McDonald-Renaud, D. (1974). *Sound localization in the bottlenose porpoise (Tursiops truncatus) (Montagu).* Doctoral dissertation, University of Hawaii, Department of Zoology, Honolulu.

McFadden, D., & Pasanen, E. G. (1978). Lateralization at high frequencies based on interaural time differences. *Journal of the Acoustical Society of America,* **59,** 634–639.

McGregor, P. K., & Krebs, J. R. (1984). Sound degradation as a distance cue in great tit (*Parus major*) song. *Behavioral Ecology and Sociobiology,* **16,** 49–56.

McGregor, P. K., Krebs, J. R., & Ratcliffe, L. M. (1983). The reaction of great tits (*Parus major*) to playback of degraded and undegraded songs: The effect of familiarity with the stimulus song type. *Auk,* **100,** 898–906.

Mendelson, M. J., & Haith, M. M. (1976). The relation between audition and vision in the human newborn. *Monographs of the Society for Research in Child Development,* **41,** (Serial No. 167).

Menzel, C. R. (1980). Head cocking and visual perception in primates. *Animal Behaviour,* **28,** 151–159.

Mills, A. W. (1958). On the minimum audible angle. *Journal of the Acoustical Society of America,* **30,** 237–246.

Mohl, B. (1964). Preliminary studies on hearing in seals. *Videnskabelige Meddelelser Fra Dansk Naturhistorisk Forening i Kjobenhavn,* **127,** 283–294.

Moore, P. W. B., & Au, W. W. L. (1975). Underwater localization of pulsed tones by the California sea lion (*Zalophus californianus*). *Journal of the Acoustical Society of America,* **58,** 721–727.

Moulton, J. M., & Dixon, R. H. (1967). Directional hearing in fishes. In W. N. Tavolga (Ed.), *Marine bioacoustics* (Vol. 2, pp. 187–232). Oxford: Pergamon.

Muir, D., & Field, J. (1979). Newborn infants orient to sounds. *Child Development,* **50,** 431–436.

Nelson, D. R. (1967). Hearing thresholds, frequency discrimination and acoustic orientation in the lemon shark, *Negaprio brevirostris* (*Poey*). *Bulletin of Marine Science,* **17,** 741–768.

Norberg, R. A. (1977). Occurrence and independent evolution of bilateral ear symmetry in owls and implications on owl taxonomy. *Philosophical Transactions of the Royal Society of London, Seris B,* **282,** 375–408.

Park, T., Okanoya, K., & Dooling, R. (1987). Sound localization in the budgerigar and the interaural pathways. *Journal of the Acoustical Society of America,* **81,** S28.

Passmore, N. I., Capranica, R. R., Telford, S. R., & Bishop, P. J. (1984). Phonotaxis in the painted reed frog (*Hyperolius marmoratus*): The localization of elevated sources. *Journal of Comparative Physiology,* **154,** 189–197.

Payne, R. S. (1962). How the barn owl locates its prey by hearing. *Living Bird,* **1,** 151–159.

Perrott, D. (1982). Dynamic factors in sound localization. In W. Gatehouse (Ed.), *Localization of sound: Theory and application.* Groton, CT: Amphora Press.

Perrott, D., & Tucker, J. (1988). Minimum audible movement angle as a function of signal frequency and the velocity of the source. *Journal of the Acoustical Society of America,* **83,** 1522–1527.

Plomp, R. (1976). Binaural and monaural speech intelligibility of connected discourse in reverberation as a function of azimuth of a single competing sound source (speech or noise). *Acoustica,* **34,** 200–211.

Plomp, R., & Mimpen, A. M. (1981). Effect of the orientation of the speaker's head and the azimuth of a noise source on the speech reception threshold for sentences. *Acoustica,* **48,** 325–328.

Potash, M., & Kelly, J. (1980). Development of directional responses to sounds in the infant rat (*Rattus norvegicus*). *Journal of Comparative and Physiological Psychology,* **94,** 864–877.

Pumphery, R. J. (1940). Hearing in insects. *Biological Review of the Cambridge Philosophical Society,* **15,** 107.

Pumphery, R. J. (1950). Hearing. *Symposium of the Society for Experimental Biology,* **4,** 1–18.

Ravizza, R. J., & Diamond, I. T. (1974). Role of auditory cortex in sound localization: A comparative ablation study of hedgehog and bushbaby. *Federation Proceedings*, **33**, 1917–1919.

Ravizza, R. J., & Masterton, R. B. (1972). Contribution of neocortex to sound localization in opossum (*Didelphis virginiana*). *Journal of Neurophysiology*, **35**, 344–356.

Rheinlaender, J., Gerhardt, H. C., Yager, D. D., & Capranica, R. R. (1979). Accuracy of phonotaxis by the geen treefrog (*Hyla cinerea*). *Journal of Comparative Physiology*, **133**, 247–255.

Rose, J. E., Brugge, J. F., Anderson, D. J., & Hind, J. E. (1967). Phase-locked responses to low frequency tones in single auditory nerve fibers of the squirrel monkey. *Journal of Neurophysiology*, **30**, 769–793.

Rosowski, J. J., & Saunders, J. C. (1980). Sound transmission through the avian interaural pathways. *Journal of Comparative Physiology*, **136**, 183–190.

Schmidt, S. (1988). Evidence for a spectral basis of texture perception in bat sonar. *Nature* (*London*), **331**, 617–618.

Schooneveldt, G. P., & Moore, B. C. J. (1987). Comodulation masking release (CMR): Effects of signal frequency, flanking-band frequency, masker bandwidth, flanking band level, and monotic verses dichotic presentation of the flanking band. *Journal of the Acoustical Society of America*, **82**, 1944–1956.

Schooneveldt, G. P., & Moore, B. C. J. (1988). Failure to obtain comodulation masking release with frequency-modulated maskers. *Journal of the Acoustical Society of America*, **83**, 2290–2292.

Schubert, E. D., & Schultz, M. C. (1962). Some aspects of binaural signal selection. *Journal of the Acoustical Society of America*, **34**, 844–849.

Schwartzkopff, J. (1950). Beitrag zum Problem des Richtungshorens bei Vogeln. *Zeitschrift Vergleichende für Physiologie*, **32**, 319–327.

Searle, C. L., Braida, L. D., Cuddy, D. R., & Davis, M. F. (1975). Binaural pinna disparity: Another auditory localization cue. *Journal of the Acoustical Society of America*, **57**, 448–455.

Shaw, E. A. G. (1974). The external ear. In W. D. Keidel & W. D. Neff (Eds.), *Handbook of sensory physiology* (Vol. I, pp. 240–249). Berlin: Springer-Verlag.

Simmons, J. A., Howell, D. J., & Suga, N. (1975). Information content of bat sonar signals. *American Scientist*, **63**, 204–215.

Smith, G. (1904). The middle ear and columella of birds. *Ovart. Journal Microscopic Science*, **48**, 11–22.

Stellbogen, E. (1930). Uber das aussere und mittlere Ohr des Waldkrauzes. *Zeitschrift Morphologie und fuer Oekologie der Tiere*, **19**, 686–731.

Wada, Y. (1923). Beitrage zur vergleichenden Physiologie des Gehororgans. *Pfluegers Archiv fuer die Gesamte Physiologie des Menschen und der Tiere*, **202**, 46–69.

Walls, G. L. (1951). The problem of visual direction. Part II. The tangible basis for nativism. *American Journal of Optometry*, **28**, 115–146.

Waser, P. M. (1977). Sound localization by monkeys: A field experiment. *Behavioral Ecology and Sociobiology*, **2**, 427–431.

Waser, P. M., & Brown, C. H. (1986). Habitat acoustics and primate communication. *American Journal of Primatology*, **10**, 135–154.

Wetheimer, M. (1961). Psychomotor coordination of auditory and visual space at birth. *Science,* **134**, 1692.

Wever, E. G., and Vernon, J. A. (1957). Auditory responses in the spectacled caiman. *Journal of Cellular and Comparative Physiology*, **50**, 333–339.

Whitehead, J. M. (1987). Vocally mediated reciprocity between neighboring groups of mantled howling monkeys, *Aloutta palliata palliata*. *Animal Behaviour*, **35**, 1615–1627.

Wiley, R. H., & Richards, D. G. (1978). Physical constraints on acoustic communication in the atmosphere: Implications for the evolution of animal vocalization. *Behavioral Ecolology and Sociobiology*, **3**, 69–94.

Woodworth, R. S. (1938). *Experimental psychology*. New York: Holt.

# 9

# SOUND LOCALIZATION IN MAMMALS: BRAIN-STEM MECHANISMS

*Rickye S. Heffner*

*Department of Psychology, University of Toledo, Toledo, Ohio*

*R. Bruce Masterton*

*Department of Psychology, Florida State University, Tallahassee, Florida*

Due to technological advances in animal psychophysics and to a persistent interest in the evolution of hearing and its physiological substrates, the hearing abilities of a widening array of animals are becoming better known. There are now good behavioral audiograms on more than 45 species of mammals, representing at least 10 orders and 2 subclasses (e.g., see R. S. Heffner & Heffner, 1983a, 1985a). There are also sound-azimuth thresholds for at least 19 species in 9 orders of mammals (R. S. Heffner & Heffner, 1988a; Table 2 below). Although these totals are large and growing, they remain a very small fraction of the entire population of extant mammals (which contains more than 900 genera in 18 orders and 3 subclasses). And regardless of their numerical size, the two samples remain far from random. In both samples there are too many Primates and too few Rodents, Marsupials (pouched mammals), Chiroptera (bats), Insectivora (shrews, moles, hedgehogs, etc.), and Artiodactyla (deer, antelope, cattle, etc.). Nevertheless, the range of hearing and sound localization abilities, and the morphological specializations accompanying a few unusual ecologies can now be glimpsed.

The increase in the available sample of species now permits a tentative comparative analysis of sound localization—an analysis that often allows a kind of insight into a sensory system's contribution that is not possible

285

through physiological experimentation alone. To be sure, physiological experimentation cannot be replaced as the ultimate test for a specific structure–function hypothesis. However, the hypotheses themselves are often best generated by the wider vista provided by comparative methods. The present review is directed toward that general goal.

Broad and detailed reviews of the psychophysics and the anatomical or physiological mechanisms for sound localization can be found elsewhere and only inescapable references to this broad literature will be made here (e.g., Aitkin, 1986; Irvine, 1986; Masterton & Imig, 1984; Phillips & Brugge, 1985). Instead, this review focuses on some of the comparative data surrounding sound localization and the neural mechanisms subserving it in the hope that it can suggest directions for further research both into structure–function relations and into the evolution and adaptation of the auditory system. Because little is yet known about the anatomy, physiology, or psychophysics of the sound-localization dimensions of elevation or distance, the following remarks are confined to azimuth and azimuthal localization.

Any discussion of the evolution of hearing abilities of mammals and the brainstem mechanisms on which they depend is based on a foundation of psychophysical data that allows us to compare the abilities of a wide variety of mammals. Therefore, it is appropriate to devote at least a small amount of space to a discussion of how those data are obtained and how the methods of comparative psychophysics have themselves evolved to a point where comparisons can be made between species that differ greatly in their life-styles and their motor and intellectual capacities.

## ANIMAL PSYCHOACOUSTICS

All modern psychophysical methods designed for animal testing have in common the ability to control motivation and to provide quick and reliable consequences not only for "hits" and "correct rejections" (both correct responses), but also to provide clear negative consequences for errors in the form of "misses" and "false alarms." Thus, it is now possible to reduce errors and maximize correct responses to obtain very reliable discriminations. With the use of a microcomputer both types of errors and correct responses can be routinely detected and measured by any of a variety of electronic devices. Raw data are easily recorded, statistics calculated, and both the stimuli and consequences modified during the testing session. For example, a hungry animal licking pureed food extruded from a small tube can be quickly trained to indicate a change in its sound field if the change is followed by an avoidable shock. In this case, the cessation of licking (sensed by an electronic contact circuit connected between the lick tube and the cage floor) serves as an indicator of detection (R. S. Heffner & Heffner, in press). The microcomputer continuously records and calculates false alarm rate, hits, and misses, and changes the stimulus parameters. If desired, it can even track the animal's threshold by incrementing the stimulus after each miss and decrementing it after each hit (R. S. Heffner & Heffner, 1983a; Stebbins, 1970). The availability of this exquisitely detailed information on the behavior of the subject allows the use of strict statistical definitions of threshold including signal-detection analyses of performance (e.g., R. S. Heffner & Heffner, 1988a).

Another important feature of modern psychoacoustic methods is that they allow the animal to maintain its head and ears in a fixed location relative to the sound source. As a result, a uniform sound field can be maintained around the ears and the direction of sound sources relative to the animal's head and ears can be determined precisely. Most conditioned-suppression and avoidance methods and several "go/no-go" techniques accomplish head and ear stabilization by requiring the animal to place its mouth on a water spout or lick plate where it can be detected by a contact circuit (e.g., R. S. Heffner & Heffner, 1983b, 1987; Ravizza, Heffner, & Masterton, 1969). The two-choice procedures can accomplish this same degree of head stabilization by requiring an orienting or stabilizing response for the acoustic stimulus (and a chance for reward) to be presented (e.g., H. Heffner & Masterton, 1980; R. S. Heffner & Heffner, 1982, 1988c, in press).

These new procedures, depending largely on electronic circuitry and the slavish dependability of the microcomputer, stand in sharp contrast to previously available techniques which required constant vigilance from the experimenter to observe and judge responses, resolve ambiguities, and record and calculate performance throughout a session while the animal moved about within a relatively large and heterogenous sound field. However, it is the ability to control precisely the location of the subjects' ears relative to a sound source that has probably contributed more than any other factor to the reliability of animal psychoacoustical data.

The results obtained by the modern methods have led to a high degree of agreement in results between different methods within a laboratory and between different laboratories. For example, Fig. 1 shows the close agreement both in asymptomatic performance and in sound-localization thresholds of cats using two different behavioral methods. These thresholds also agree with those obtained for cats in other laboratories (R. S. Heffner & Heffner, 1988e). Such empirical support for both the precision and the validity of animal psychoacoustical measures suggests that most modern data provide an acceptable foundation for the study of the natural variation in hearing and sound localization and their relation to ecology on one hand and to neurological and otological structures on the other.

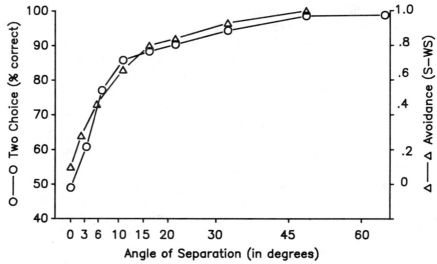

**FIGURE 1.** Sound-localization performance of 2 groups of domestic cats measured by two different behavioral methods approximately 2 years apart. In the two-choice method the cat touches a center "orienting" switch to hear a sound, then touches a left or right touch switch (corresponding to left or right sound source) with its nose to receive reward. In the avoidance method a thirsty cat indicates a shift in sound source location by cessation of drinking in order to avoid a mild electric shock. The close similarity of the results suggests that both methods are probably valid indicators of the cat's ability. From R. S. Heffner & Heffner (1988e).

## COMPARATIVE ANATOMY OF SOUND-LOCALIZATION MECHANISMS

Although an interesting array of relations are present between behavioral performance on one hand, and gross morphological or ecological variables on the other hand, the comparative investigation of sound-localization ability has now gone far beyond these to include neuroanatomical, neurophysiological, and neurochemical correlates as well. To make sense of this large variety of data, it is necessary first to review briefly what is known about the neural mechanisms of sound localization.

The second-order anatomical projection known to subserve the fundamental elements of hearing and azimuthal sound localization consists of the ventral acoustic stria and trapezoid body of the medulla. Figure 2 shows the second-order auditory fibers from cochlear nucleus reaching the right superior olivary complex. Figure 2 also shows that there is a major intrinsic pathway within the superior olivary complex itself, one arising from the cells of the medial nucleus of the trapezoid body (MTB) and ending in the lateral superior olive (LSO). The presence of this pathway, together with the extrinsic pathways, brings two major binaural nuclei, the LSO and MSO, into relatively immediate neural contact with both cochlear

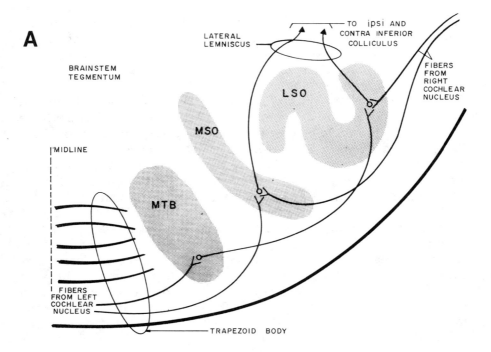

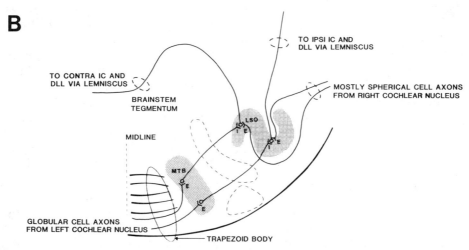

**FIGURE 2.** (*A*) Major sound-localization nuclei in superior olivary complex of cat. Nine periolivary nuclei are not shown. Note intrinsic pathway from MTB to LSO. (*B*) LSO–MTB system for analyzing binaural spectrum differences. E, Excitatory synapse; I, inhibitory synapse.

nuclei and, therfore, with both ears. it should be noted that there are direct connections by which convergence of binaural input can occur in the cochlear nuclei (Cant & Gaston, 1982). However, the time delays over this pathway are too long (>1 msec in cat) for neural interaction to occur during most forms of sound localization (Mast, 1970), leaving the second-order projection target as the primary locus of binaural interaction. Discovered by Stotler (1953) and by Rasmussen (1946) more than 35 years ago, these systems have proven to be the neuroanatomical substrate serving azimuthal sound localization in the sense that damage anywhere along their route results in severe and unrecoverable deficits in azimuthal sound-localization ability (e.g., Casseday & Smoak, 1981; Jenkins & Masterton, 1982; Masterton, Glendenning, & Nudo, 1981; Masterton, Jane, & Diamond, 1967).

The pathways shown in Fig. 2 are well suited to their task of analyzing binaural locus cues. They are very fast and reliable; the synapses of the ventral cochlear nucleus and MTB are perhaps the shortest-latency synapses found anywhere in the nervous system (e.g., Guinan, Guinan, & Norris, 1972; Guinan, Norris, & Guinan, 1972; Li & Guinan, 1971). At both the ventral cochlear nucleus and the MTB, the incoming fibers encapsulate the postsynaptic cells in a way that virtually guarantees a minimum time delay in synaptic action (e.g., Morest, 1968; Tolbert, Morest, & Yurgelun-Todd, 1982). Therefore, the difference in the two pathways terminating in the superior olives is small enough (<1 msec in cat) for the impulses evoked by stimulation of one ear to reach the MSO and LSO in time to interact neurophysiologically with the impulses evoked at the other ear (e.g., Rosenzweig & Amon, 1955). This neuroanatomical convergence and neurophysiological interaction is the reason that students of sound localization have focused many of their physiological and comparative investigatioans on the superior olivary complex or on the inferior colliculus which receives most of the efferent projections from the superior olivary complex (Fig. 2).

A second, perhaps more satisfying line of evidence, which leads to the conclusion that it is the superior olivary complex that is the key to azimuthal sound localization, is shown in Fig. 3. The figure summarizes a number of ablation–behavior experiments in which sound-localization ability was tested before and after section of one or another of the pathways shown in Fig. 2 (Jenkins & Masterton, 1982; Thompson & Masterton, 1978). Figure 3 shows that there is a sharp difference in the behavioral deficits resulting from monaural deafness (M in Fig. 3) or section of the trapezoid body (T in Fig. 3), which contains the input to the superior olives, as opposed to section of one lateral lemniscus (L in Fig. 3), which contains the output from the superior olives. Damage to the input results in a broad sound-localization deficit encompassing both left and right auditory hemifields. In sharp contrast, damage to one lateral lemniscus results in a virtually total loss of sound-localization ability in the

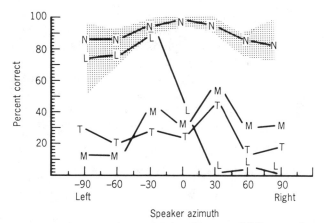

**FIGURE 3.** Behavioral performance of normal (N) and monaural (M) cats and cats with section of either the trapezoid body (T) or left lateral lemniscus (L). Gray area represents range of 10 normal cats. Damage of pathway to the superior olives (M or T) results in poor performance throughout auditory field. However, unilateral section of output of superior olives results in sharp deficit confined to contralateral hemifield. Adapted from Jenkins & Masterton (1982). Analogy with the effects of section before or after optic chiasm gives rise to the notion of an acoustic chiasm, a process localizable to the superior olives.

hemifield of auditory space contralateral to the damage, while the same ability in the ipsilateral hemifield seems to remain normal. Because the lateral lemniscus provides the output of LSO and MSO while the trapezoid body provides their input (Fig. 2), it follows that the difference between the two deficits is the result of the integrative and distributive activity taking place in the superior olivary complex itself.

It should also be noted that the differences in the deficits resulting from section of the trapezoid body (T in Fig. 3) as opposed to section of one lateral lemniscus (L in Fig. 3) is analogous to the difference in deficits seen in the visual system with sections before and after the optic chiasm. It is this similarity in deficits (auditory nerve, ventral acoustic stria, or trapezoid body section equivalent to optic nerve section; lateral lemniscus section equivalent to optic tract section) that has given rise to the notion that the superior olivary complex serves mammals as an acoustic chiasm (Glendenning, Hutson, Nudo, & Masterton, 1985; Glendenning & Masterton, 1983).

The analogy between the *function* of the acoustic, optic, and even the somatosensory chiasms does not hold for their *anatomy*. In the optic or somatosensory systems the sensory fields are represented point for point on the receptor surface. Therefore, the chiasmatic function of their respective decussations contralateralizes the neural activity representing one sensory hemifield merely by sorting and distributing the ascending fibers to the appropriate side of the brain—no subtle interactions or reintegrations are necessary. In the auditory system, however, the auditory hemifield is *not* represented in the chochlea. Therefore, a mere sorting and redistribution of the ascending fibers cannot accomplish the same task. It follows that while the optic or somatosensory chiasms are structures, the acoustic chiasm must be a process. Nevertheless, it is a process that performs the same function and it is localizable to the superior olivary complex. The following discussion turns on variations in the SOC among mammals and the associated variations in auditory abilities.

## VARIATIONS OF THE SUPERIOR OLIVARY COMPLEX

The superior olivary complex in the cat (stripped of its 9 periolivary nuclei) is shown in Fig. 2. Although often accepted as the standard configuration for mammals, the same set of structures in other mammals takes on markedly different forms. This variation has stimulated several analyses of the relation of the form of the SOC to phyletic lineage, to other morphological variables, and to auditory abilities. An example of a phyletic analysis is given in Fig. 4 which shows the form of the superior olivary complex in seven mammals with sequential kinship to humans. Both the absolute and relative size of the constituent structures varies markedly. On one hand, MSO varies from large in macaque and human to very small in the opossum and none at all in the hedgehog. Although the MSO clearly increased in size in the human lineage, this change is not unique in the human line. It can be found in any set of species ordered according to increasing size. Of particular relevance to audition, there seems to be an even stronger relation between size of the MSO and low-frequency auditory sensitivity since even very small species with good low-frequency hearing, such as gerbils, kangaroo rats, and least weasels, have a well-developed MSO (for a review, see R. S. Heffner & Heffner, 1985a). In this instance, comparative psychophysics has supported conclusions derived from electrophysiological and anatomical studies that the MSO is primarily concerned with localization of sound sources emitting low frequencies.

Table 1 shows the number of neurons in the MSO in 25 animals. The data reveal that MSO is larger in large land mammals and smaller or nonexistent in either small or marine mammals: In general, the size of the medial superior olive is loosely related to the size of the animal. However, statistical analysis shows that it is not body size itself, or even head size, that is the closest correlate of the number of cells in the MSO (note dolphin in Table 1). Instead, the closest correlate ($r = 0.79$) is the functional distance between the ears (that is, interaural distance divided by the speed of sound; Masterton, Heffner, & Ravizza, 1969). In land mammals with large interaural distances, MSO is large; in most mammals with small interaural distances or in marine mammals even with large absolute but small functional interaural distances, MSO is either small or nonexistent

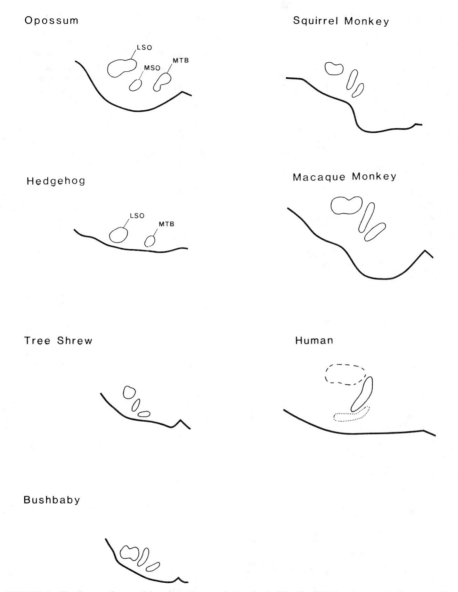

**FIGURE 4.** Outlines of sound-localization nuclei in the left half of the brainstem in 7 mammals with increasing kinship with humans, not drawn to scale. Note variation in relative size of MSO (and its absence in hedgehog) and in the size and shape of LSO. The broken lines representing the MTB and LSO in humans indicates that these nuclei are only marginally present.

**TABLE 1. Number of Neurons in the Medial Superior Olivary Nucleus in 25 Mammals**

| Animal | Number of Cells in MSO | Source[a] |
|---|---|---|
| Hedgehog | 0 | 2 |
| Bat, *Carollia* | 0 | 2 |
| Bat, *Phyllostomus* | 0 | 2 |
|  | 100 | 1 |
| Bat, *Myotis* | 0 | 1,2 |
| Tree shrew | 1269 | 3 |
| Loris | 2723 | 3 |
| Galago | 2781 | 3 |
| Marmoset | 2423 | 3 |
| Squirrel monkey | 3279 | 2 |
| Owl monkey | 4263 | 3 |
| Spider monkey | 2570 | 1 |
|  | 4270 | 3 |
| Vervet monkey | 5480 | 3 |
| Macaque monkey | 3420 | 1 |
|  | 5253 | 2 |
|  | 5285 | 3 |
| Human | 4260 | 2 |
| Rat | 690 | 1 |
|  | 616 | 2 |
| Mouse | 210 | 1 |
|  | 128 | 2 |
| Gerbil | 1100 | 1 |
| Hamster | 300 | 1 |
| Guinea pig | 2360 | 1 |
|  | 2547 | 2 |
| Chinchilla | 3090 | 1 |
|  | 3157 | 2 |
| Squirrel | 1426 | 2 |
| Ground squirrel |  |  |
|    *Citellus tridecemlineatus* | 1240 | 1 |
|    *Citellus beechii* | 1010 | 1 |
| Cat | 4200 | 1 |
|  | 5895 | 2 |
|  | 5795 | 3 |
| Dolphin | 0 | 2 |

[a] 1, Harrison and Feldman (1970); 2, Harrison and Irving (1966); 3, Moore and Moore (1971).

(Table 1). However, this sample is heavily weighted with primates, which tend to be large animals with good low-frequency sensitivity, and does not include MSO cell counts for some very small species with large MSOs and good low-frequency sensitivity, such as kangaroo rats and least weasels. Therefore, it is still possible, as discussed below, that the number of cells in the MSO is a function of the low-frequency hearing ability of the species and the correlation with interaural distance will decrease as the sample becomes more representative of mammals.

Similarly, the LSO varies in size and shape within the human lineage and to an even greater degree when other species are included. In some mammals, such as cats and rats, the LSO has a characteristic horizontal S shape, but its orientation varies and it even appears inverted in rats relative to cats. In other mammals, the LSO is U, M, or W shaped, or an irregular oval. Figure 5 shows the form of LSO in a variety of mammals together with their high- and low-frequency hearing limits. Also illustrated is a hypothetical composite, or "prototype" LSO. The modifications of the prototype LSO indicated by shading in Fig. 5 give rise to each of the forms of LSO presently known. It can be concluded that the form of LSO has been established by the addition or subtraction of similar elements or modules which have been added or lost medially as high-frequency hearing increased or decreased in the evolution of a species. Because the LSO is primarily a high-frequency nucleus that rarely contains cells with characteristic frequencies near an animal's low-frequency hearing limit, it is not surprising that there are not similar additions to the lateral limb of the LSO as low-frequency hearing extends below approximately 0.1 kHz.

In summary, because of their separate relations to high- and low-frequency hearing, the variation in size and differentiation of LSO is almost inverse to that of MSO (cf. Fig. 5 and Table 1). LSO reaches its most complex form in animals with the smallest functional interaural distance, the same animals whose hearing extends into the highest frequency range (such as bats and dolphins) whereas LSO is smallest and least well-differentiated in animals with a large functional interaural distance and poorest high-frequency hearing (such as humans and elephants).

Therefore, it has been concluded that in the absence of relatively large interaural time-difference cues, small animals have exploited the second interaural cue for sound localization, the intensity-difference or, more precisely, the spectrum-difference cue. Since it is high-frequency sounds that are best shadowed by head and pinna and therefore produce the greatest interaural spectrum differences, these same animals have an extended high-frequency hearing range. This inverse relation between functional interaural distance and high-frequency hearing limit was first noted nearly 20 years ago (Masterton et al., 1969) and has remained strong despite a doubling of the sample of species—including animals such as mouse and elephant which were chosen specifically to test the application of the relation to the extremes of small and large species. Figure 6

# Right LSO Prototype

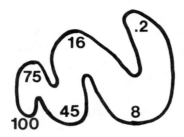

### Right LSOs with Upper and Lower Hearing Limits

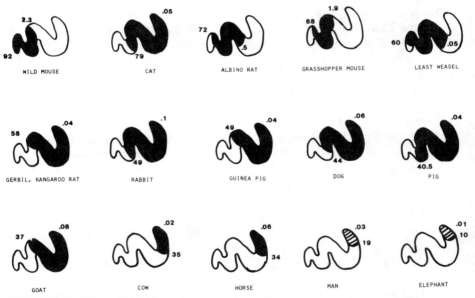

**FIGURE 5.** Outlines of (right) lateral superior olive and upper and lower limits of hearing in 15 mammals together with imaginary composite or prototype above. Outlines are drawn from coronal sections through largest part of each LSO and are not to the same scale. Shaded area indicates shape actually found in each species. Because LSO is strictly tonotopic, its form whether S, M, U, or W shaped roughly indicates the animal's frequency range of hearing. Thus, horizontal S-shaped LSOs in rat and cat are not really inversions of each other but instead are left–right transpositions along a sequential tonotopic map reflecting their differences in range of hearing.

illustrates this relation as it now stands with 45 species. All subgroups of mammals contribute to the correlation of $-0.85$ ($p < 0.001$, solid line in Fig. 6). The correlations for rodents alone (dashed line in Fig. 6) and primates alone (dotted line in Fig. 6) are $-0.83$ and $-0.81$, respectively.

Figure 6 also notes the existence of a dramatic exception to the relation in the form of the pocket gopher represented in Fig. 6 by G (R. S. Heffner, Richard, & Heffner, 1987). In these animals azimuthal sound-localization ability is vestigial possibly as a result of their extreme fossorial habits (see Table 2). The absence of high-frequency sensitivity in a small species with vestigial sound localization is further evidence for the belief that the chief selective pressure for high-frequency hearing derives from pressures for accurate sound localization.

In contrast to the LSO–MTB system that is well developed in mammals with high-frequency hearing, the MSO is best developed in large animals with large functional interaural distances which maximize the range of interaural time differences. Animals with large MSOs tend to have extended low-frequency hearing ranges (see H. E. Heffner & Heffner, 1984; R. S. Heffner & Heffner, 1982, 1983a, 1985a; Masterton et al., 1969). It should be noted that these two systems are not incompatible and that some

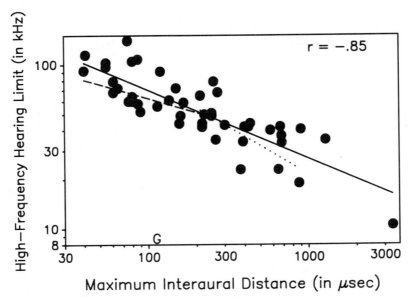

**FIGURE 6.** Relation between maximum functional interaural distance (interaural distance/speed of sound) and the highest audible frequency (at 60 db SPL) for the more than 40 mammals with complete behavioral audiograms. The correlation of $-0.85$ (—) does not vary with order as shown by the dashed line for Rodents and the dotted line for Primates. Nor does it vary with most ecologies, whether marine or echolocation. Only the highly fossorial pocket gopher, G, is exceptional.

**TABLE 2. Azimuth Thresholds for 19 Species of Mammals**

| Animal | Stimulus | Threshold[a] (deg) | Source |
|---|---|---|---|
| Human | Click | 0.8 | H. E. Heffner & Heffner (1984) |
| Dolphin | Click | 0.9 | Renaud & Popper (1975) |
| Elephant | Noise | 1.2 | R. S. Heffner & Heffner (1982) |
| Seal | Click | 3.2 | Terhune (1974) |
| Macaque | Noise | 4 | Brown, Beecher, Moody, & Stebbins (1980) |
| Pig | Noise | 4.5 | R. S. Heffner & Heffner (1989) |
| Opossum | Noise | 4.6 | Ravizza & Masterton (1972) |
| Cat | Noise | 5 | R. S. Heffner & Heffner (1988e) |
| Dog | Click | 8 | H. Heffner (unpublished) |
| Albino rat | Noise | 10 | Kelly (1980) |
| Least weasel | Noise | 12 | R. S. Heffner & Heffner (1987) |
| Norway rat, wild | Noise | 12.8 | H. E. Heffner & Heffner (1985) |
| Hedgehog | Click | 19 | Chambers (1971) |
| Wood rat | Noise | 19 | R. S. Heffner & Heffner (1988a) |
| Grasshopper mouse | Noise | 19.3 | R. S. Heffner & Heffner (1988a) |
| Horse | Noise | 22 | H. E. Heffner & Heffner (1984) |
| | Click | 30 | H. E. Heffner & Heffner (1984) |
| Kangaroo rat | Click | 27 | H. Heffner & Masterton (1980) |
| Gerbil | Noise | 27 | R. S. Heffner & Heffner (1988c) |
| Pocket gopher | Noise | >60 | R. S. Heffner, Richard, & Heffner (1987) |

[a] Threshold is 75% correct for two-choice procedures and 0.50 performance level for conditioned suppression and avoidance procedures.

species have developed both to a high degree of refinement. The best examples of this are found in the Carnivora, particularly the domestic cat and least weasel, which possess large and well-developed MSOs and LSOs and have hearing ranges that extend far into both the low and high frequencies.

The presence of two binaural cues for azimuthal localization, two types of adaptation of hearing range to these cues, and two sets of anatomical structures for analyzing these cues suggests that the discussion might profitably turn to each alone.

## THE HIGH-FREQUENCY, LSO–MTB SYSTEM

Through electrophysiological experimentation, the LSO–MTB system has been shown to be an interaural spectrum-difference analyzer (Boudreau & Tsuchitani, 1968, 1970). Figure 7 shows that a sound reaching the far ear of an animal has a different spectrum from that reaching the near ear due to the unequal attenuation of high frequencies over the somewhat longer

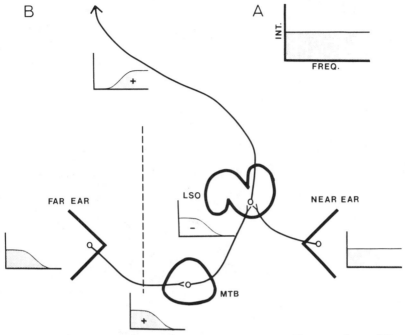

**FIGURE 7.** Binaural spectrum subtraction performed by LSO–MTB system for a white-noise sound source to the right of midline (---). (*A*) Graph of sound spectrum (frequency-by-intensity) used in (*B*) to show spectral differences present at the ears and represented at each nucleus in LSO–MTB system; + and − indicate excitatory and inhibitory synaptic activity, respectively. Spectral subtraction at LSO depends on change in transmitter substance from excitatory to inhibitory only in contralateral pathway. Spectrum represented in output of LSO is spectrum at right ear minus spectrum at left ear.

distance traveled and the sound shadow provided by the head and pinnae. After the two different spectra are faithfully encoded by each cochlea, they are relayed via the cochlear nuclei to the lateral superior olives. However, as the neural activity representing the contralateral spectrum is transmitted to the LSO, the MTB changes it from excitatory to inhibitory in its effect, so that when it reaches the LSO the neural activity evoked by the spectrum from the far ear is subtracted from the neural activity evoked by the spectrum of the near ear (Fig. 7).

This simple neurochemical trick of changing the transmitter substance from excitatory to inhibitory along the contralateral pathway but not along the ipsilateral pathway means that the output of LSO represents the interaural spectrum difference, frequency by frequency, exactly as might be expected from psychophysical experiments into interaural intensity-difference phenomena (Boudreau & Tsuchitani, 1970; Hutson, Glendenning, & Masterton, 1987).

**A**

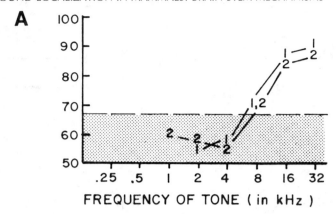

FREQUENCY OF TONE ( in kHz )

**B**

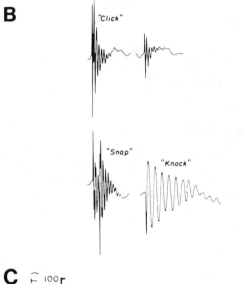

"Click"

"Snap"   "Knock"

**C**

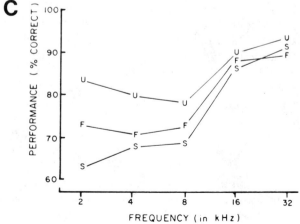

PERFORMANCE ( % CORRECT )

FREQUENCY ( in kHz )

In animals that have only an LSO–MTB system and little or no MSO system, this apparatus serves well enough for sound localization, certainly for sounds whose content includes high frequencies. However, as the frequency of a single pure tone is lowered, the sound shadows provided by head and pinnae become less effective and such animals have more and more difficulty localizing the source of the sound. Figure 8A shows the poor low-frequency sound-localization ability of hedgehogs, which have an MTB–LSO system but no MSO system at all (see Fig. 4 and Table 1). It can be seen that a hedgehog is a relatively accurate localizer of high-frequency tones while it cannot localize low-frequency tones at all, despite the fact that it can hear and respond to such tones in other ways (Masterton, Thompson, Brunso-Behtold, & RoBards, 1975). Fortunately for the hedgehog and similar mammals, however, most natural environmental sounds as well as all brief or transient sounds have sharp onsets which always contain high frequencies (Fig. 8B). Therefore, even animals with a pure MTB–LSO system, such as hedgehogs, dolphins, and most bats, have no difficulty localizing natural sounds despite their difficulty with low-frequency tones (Fig. 8C).

In general, sound-localization mechanisms depending only on MTB–LSO processing provide a reasonably good system for most small animals. Certainly it permits easy localization of a sound to at least one quadrant of space and to within plus or minus 15° or less with 50% accuracy (Table 2).

However, evidence of the limitations of this interaural spectrum-difference system is now obvious—small species tend to be among the least accurate localizers (Table 2). Several reasons for this limitation can be noted. For example, interaural intensity differences for pure-tone stimuli are unreliable and even occasionally reversed at some frequencies (Aitkin, 1986; Harrison & Downey, 1970; Irvine, 1986). Yet natural sounds usually contain a range of frequencies permitting a more reliable interaural spectral analysis ($\Delta f_i$) as opposed to a simple interaural intensity comparison for only one frequency ($\Delta I$). For very large animals with large head and pinna shadows, another disadvantage arises. The interaural intensity difference can become so great that many sounds are entirely inaudible in the far ear, thus stimulating only the near ear and indicating only the hemifield of its source (R. S. Heffner & Heffner, 1982). It now appears that some very large mammals (e.g., horses, pigs, and elephants) have given up the ability to

---

**FIGURE 8.** (A) Hedgehog sound-localization performance for pure tones at 60° speaker separation. The hedgehog's pure LSO–MTB system provides only interaural intensity-difference cue and this cue begins to disappear below 10 kHz. (B) Waveforms of three natural sounds. Note rapid, nonsinusoidal onset in addition to fundamental resonance sinusoid. (C) Same as (A) except tone onsets were either unkeyed (U), fast-keyed (F), or slow-keyed (S). Note improvement in performance at low-frequencies when onset time of tone is shortened and transient high frequencies intrude. From Masterton, Thompson, Brunso-Bechtold, & RoBards, (1975).

use interaural intensity differences to localize sound over some or all of their high-frequency hearing range just as the hedgehog has given up the use of interaural phase over its low-frequency range (Fig. 9). Instead, these animals have turned to an MSO system which takes advantage of the large interaural time differences produced by their large heads (R. S. Heffner & Heffner, 1982, 1986a, 1986b, 1988b). Monaural pinna cues may also contribute to sound localization in these species as a supplement to or substitute for the binaural cues. Indeed, the horse, which lacks the ability to use interaural intensity differences, requires high frequencies to localize sound in the lateral fields using monaural pinna cues and cannot locate sounds on the cone of confusion if the sound does not contain high frequencies (R. S. Heffner & Heffner, 1983b).

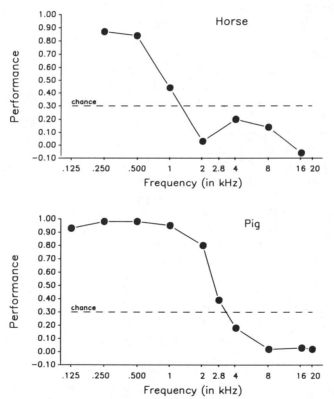

**FIGURE 9.** Average sound-localization performance for pure tones at 60° speaker separation for two horses and three pigs. Performance is good for low-frequency pure tones which provide an interaural phase-difference cue, but falls to chance at high frequencies which provide only an interaural intensity-difference cue. This pattern of performance suggests a pure MSO, or non-LSO, system which may be the case for the horse. However, it is not the case for the pig which has a prominent LSO as well as MSO. After R. S. Heffner & Heffner, 1986a, 1986b, 1988b, 1989.

Before leaving the discussion of the LSO–MTB system, we should note that there is no simple relation between the size and complexity of the LSO and the ability to use the interaural spectrum difference cue to localize sound (R. S. Heffner & Heffner, 1986b). Most species have a well-developed LSO and good ability to use the cue (e.g., R. S. Heffner & Heffner, 1987, 1988c; Masterton et al., 1975). But species exist in which the size of the LSO is seemingly unrelated to the ability to localize using the interaural intensity cue. For example, humans have an undistinguished LSO but retain accurate intensity-difference analysis (Mills, 1958; Moore, 1987). Conversely, the pig has lost nearly all ability to use the cue yet retains a well-developed LSO (R. S. Heffner & Heffner, 1989). Therefore, it can be concluded that the LSO has functions beyond sound localization and these functions may also be differentially represented in different species. In rodents, for example, LSO has the additional property of being a major source of olivocochlear efferents (Altschuler, Parakkal, & Fex, 1983; White & Warr, 1983). Therefore, even in the absence of participation in sound localization, rodents would probably retain an LSO. Therefore, the presence of a large or well-differentiated LSO is neither a necessary nor sufficient condition to predict capability for interaural spectrum-difference analysis. A clearer understanding of the range of functions of the LSO must await additional comparative behavioral and electrophysiological data on species that differ in their use of the binaural locus cues and in their central auditory anatomy.

With the several limitations on the use of interaural spectrum differences, it is not surprising that animals large enough to carry a large head with widely spaced ears (and therefore an interaural distance that provides large time differences between the two ears) invariably add a second system to their LSO–MTB system. As noted, this time-analysis system seems to be a major function of the MSO.

## THE LOW-FREQUENCY MSO SYSTEM

The possibility that the medial superior olive might be analyzing interaural time differences was first suggested by the neuroanatomical work of Stotler (1953) and then demonstrated with a neurophysiological recording by Galambos, Schwartzkopff, and Rupert (1959). As already noted, Stotler described the convergence of second-order neurons on MSO as shown in Fig. 2. He then pointed out that the synapses on the dendrites of the MSO cells might be exactly what was needed to explain the psychophysics of azimuthal localization based on interaural time differences (Stotler, 1953).

Galambos and his colleagues managed to record the responses of a single MSO neuron in 1959. They showed that the cell's probability of response was a reliable function of the interaural time difference. Although they managed to record from only one cell, their discovery encouraged others to investigate the phenomenon and more cells with the same property were soon found despite the technical difficulty involved (e.g., Chan & Yin, 1984; Goldberg & Brown, 1968; Guinan, Norris, & Guinan, 1972; Hall, 1965). The MSO units that change their response over the physiological range for sound localization have a characteristic delay, that is, a particular interaural time difference to which each is most sensitive (Rose, Gross, Geisler, & Hind, 1966). If one envisions a population of such cells, each cell with a different characteristic delay, the response of the entire population could serve as a high-resolution time-difference encoder. Since there are about 5000 neurons available in each MSO of a cat (although not all are sensitive to interaural time differences), such a mechanism is not improbable.

Figure 10 shows the way that the MSO is thought to analyze binaural time differences. In general, the idea is that the time delay of the sound reaching the far ear is made up for by time delays in conduction and transmission over the neural pathways from each ear to the MSO contralateral to the sound source. In reality this system can accommodate an even larger range of interaural time delays than might be expected from the schematic in Fig. 10. That is, because of intensity–latency trades in the cochlea itself and at each synapse of the pathway, the near ear is favored and the far ear disfavored beyond that due to the difference in neural conduction distances alone (Masterton et al., 1967). Therefore, it can be expected that a population of near-coincidence detecting units in MSO, something like those envisioned by von Bekesy (1930) and Jeffress (1958) on psychophysical grounds alone, may be very close to the way MSO actually functions (Chan & Yin, 1984; Yin & Kuwada, 1983).

Although there are cells in MSO that do not seem concerned with interaural time differences and there are cells in other structures that might be time-sensitive, it now seems relatively safe to conclude that high resolution of interaural time differences depends on the MSO system. However, there is one further detail to the MSO interaural time-difference hypothesis provided by comparative data.

The MSO time-sensitive cells can be used for analyzing interaural *phase* differences at frequencies low enough for first- and second-order neurons to phase-lock to the waves of the tone. However, interaural phase differences cannot be analyzed for azimuth if the interaural distance is greater than one-half the wavelength of the stimulating tone. This physical limit is present because at higher frequencies (or at longer interaural distances) the MSO cannot tell which ear is leading and which ear is lagging in phase. Further, there is a neurophysiological upper limit to the frequency that can be followed by phase-locking in order for the two monaural phases to be faithfully represented in the neural activity reaching the MSO. In the mammals studied so far, phase-locking cannot be detected much above 2 or 3 kHz (R. S. Heffner & Heffner, 1987; Johnson, 1980). Therefore, it was a surprise when it was discovered that some bats with

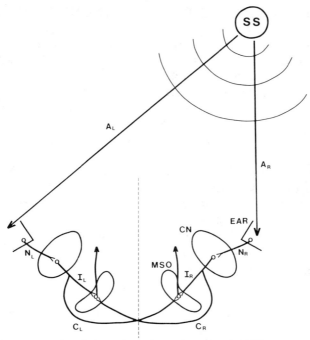

**FIGURE 10.** Schematic of still hypothetical mechanism of MSO system for analyzing interaural time-differences. Difference in lengths of neural pathways (e.g., IL versus CR) presumably offset the difference in length of the sound path to the two ears (AR versus AL). For a sound source to the right, neural activity from both ears would reach the left MSO at (nearly) the same time. Neurons in MSO are known to be sensitive to small time-of-arrival differences.

interaural distances too large to analyze the phase differences of the very high frequencies in their own echo-locating chirp (not to mention the unlikelihood that their neurons could phase-lock at such high frequencies) nevertheless possessed an MSO (see Zook & Casseday, 1982). However, this apparent contradiction to the general rule seems to be resolved since it has been shown that even in humans, the interaural time delay in the *envelope* of a complex sound (as opposed to the phase difference of the frequencies constituting the sound itself) can also be used for sound-localization purposes (McFadden & Posanen, 1976). That is, even though a bat cannot analyze the interaural phase differences within its own chirp, the envelope of the chirp's echo provides a time-of-arrival and phase differences at the two ears and these differences are related to the azimuth of the echo source. Apparently, some bats make use of this cue with an MSO system in addition to the usual MTB–LSO system for analyzing spectrum difference cues.

## SELECTIVE PRESSURES AFFECTING SOUND LOCALIZATION

Despite much progress toward understanding the neural mechanisms of sound localization, there is still no generally accepted explanation as to why sound localization thresholds vary as widely as they do among the species shown in Table 2. In other words, we do not yet understand the selective pressures acting on sound-localization acuity and the relative reliance on the binaural locus cues. Even less is known about the contribution of monaural localization and the advantage or disadvantage conferred by a mobile pinna.

From the foregoing discussion concerning the brainstem nuclei involved in sound localization, it would seem that a mechanistic explanation for the range of acuity might be evident. Although such an explanation would tell us how some species are capable of more accurate localization than others and not why, interest in such explanations is high and they merit a brief examination. By perusing the data in Table 1 (number of cells in MSOs), Fig. 5 (configuration of LSOs), and Table 2 (localization acuities of mammals), one can see that possession of a large LSO or MSO is no guarantee of good localization acuity. For examples, horses have a large MSO but are poor localizers; gerbils and kangaroo rats have large MSOs and LSOs but are poor localizers. On the other hand, species that are good localizers tend to have at least one of the olivary nuclei well developed (the MSO in humans, monkeys, and elephants; the LSO in dolphins); sometimes both are well developed (seal, pig, cat). Thus, it is possible that among mammals good development of at least one of the main olivary nuclei may be necessary to support accurate sound localization.

A second kind of explanation for the variation in sound-localization acuity is based on interaural distance. For many years it seemed reasonable to accept that all mammals are under strong and equal selective pressure to localize as accurately as possible and that the source of variation in acuity is the difference in the magnitude of the physical cues available to them. Since the magnitude of the locus cues are, in turn, mostly determined by interaural distance, this idea remained uncontradicted by the limited data available at the time: Humans with their large interaural distances were the most accurate localizers, monkeys and cats with intermediate interaural distances were somewhat less accurate, and rats with their small interaural distances were least accurate of all (Table 2). As more species were examined, however, it became apparent that a large interaural distance does not automatically result in good localization acuity—as exemplified by the poor acuity of some large mammals such as horses and cattle (H. E. Heffner & Heffner, 1984; R. S. Heffner & Heffner, 1986b). Nor is a very small interaural distance always accompanied by poor localization acuity, as demonstrated by the ability of the least weasel and grasshopper mouse to localize more accurately than many other species with the same or larger interaural distances (R. S. Heffner & Heffner, 1987, 1988a).

The relation between interaural distance and sound-localization acuity for the 18 species already tested is illustrated in Fig. 11. The correlation is statistically reliable ($r = -0.59$). However, it accounts for only 35% of the variance in acuity and the presence of markedly deviant animals suggests the presence of at least one other factor. The search for other plausible factors has taken the form of two related questions: First, are particular life-styles associated with particular localization abilities (e.g., predators versus prey, underground versus above-ground habitat)? Second, can the overall variation in sound localization be related to a single unifying factor which might in turn lead to an explanation of the role of localization acuity in the life of all mammals? Limited evidence has begun to accumulate that bears on each of these questions.

It has been noted that intermediate-sized predators (cats and dogs) seem to localize sound more accurately than prey species whether large or small (e.g., hoofed mammals, rats). This observation suggests that trophic level, that is the degree to which an animal is a predator or prey, might be an important factor in acuity. To determine the generality of this observation, several species were selected for testing. First, the only family of the Artiodactyla containing predatory species, Suidae, was examined. The domestic pig was found to be a very accurate localizer compared to mammals in general and particularly when compared to the prey species of

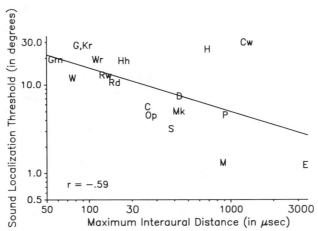

**FIGURE 11.** Relation between interaural distance and sound-localization threshold among 18 species of mammals. Although the correlation is reliable ($p < 0.01$), the presence of deviant species, specifically the large prey animals and the small predatory animals, suggests that additional factors have influence. (C, cat; Cw, cow; D, dog; E, elephant; G, gerbil; Gm, grasshopper mouse; H, horse; Hh, hedgehog; Kr, kangaroo rat; M, man; Mk, rhesus monkey; Op, opossum; P, pig; Rd, domestic Norway rat; Rw, wild Norway rat; S, seal; W, least weasel; Wr, wood rat.

that order (R. S. Heffner & Heffner, 1989). Second, two small predatory species were selected to determine whether a predatory life-style might overcome some of the disadvantage resulting from a small interaural distance. The species were the least weasel, a carnivore but the smallest member of that order, and the grasshopper mouse, again the only predatory member of an order comprising prey species. The mouse-sized carnivore (the least weasel) although not as accurate as much larger carnivores nevertheless was found to localize more accurately than any prey (R. S. Heffner & Heffner, 1987). The grasshopper mouse, which is even smaller, localizes more accurately than other rodents of similar size (R. S. Heffner & Heffner, 1988a). Thus, interaural distance may be a factor that limits the localization acuity of very small mammals, but a predatory life-style does appear to be associated with increased sound-localization acuity, and prey seem to be capable of less acuity than their interaural distances can support.

Despite the interest that this hypothesis holds, it is limited in that trophic level is not easily quantified. Many mammalian species occupy intermediate trophic levels (some primates and rodents), others are scavengers (opossum), and others are neither predator nor prey (elephant). Thus, even if, after additional species are examined, strong predators remain more accurate localizers than exclusively prey species, a more general explanatory factor that would apply readily to all species would remain desirable.

The second group of mammals that may be under different selective pressure for localization are those living in a one-dimensional space (in tunnels below the ground) as opposed to those living in two- or three-dimensional space (on the ground or in the air). This factor is plausible because both sound propagation and the directional responses available to animals are greatly restricted in tunnels. Several of the species listed in Table 2 live underground but still forage on the surface (least weasel, grasshopper mouse, kangaroo rat, and gerbil), thus remaining subject to the selective pressures common to other species that live exclusively on the surface. However, the pocket gopher (*Geomys bursarius*), the most specialized North American rodent for underground living, and the Old World mole rats rarely if ever venture above ground. Recently completed auditory tests with the pocket gopher reveal that they have the most restricted frequency range and least sensitive hearing of any mammals yet tested. In addition, the pocket gopher is also unable to localize single brief (100 msec or less) noise bursts emitted from loudspeakers 180° apart (R. S. Heffner, et al., 1987). Thus the pocket gopher has very unusual auditory characteristics which may be attributable to its underground habitat. Only one other strongly fossorial mammal has been examined, the mole rat, *Spalax*, and electrophysiological recordings from its auditory system indicate that it may also have a very restricted hearing range (Bruns, Muller, Hofer, Heth, & Nevo, 1988). Thus, it seems likely that hearing is affected by unusual environmental adaptations including underground hearing and it will be useful to examine this idea in additional species that vary in their degree of fossorial specialization.

Although it seems clear that factors such as interaural distance, lifestyle, and habitat have contributed to the differences in sound-localization acuity among different mammalian species, the question remains as to whether there is some more fundamental factor that alone accounts for the variation. The possibility exists that there may be a common factor that explains why very different species should possess similar localization acuity, such as humans and elephants, or horses and gerbils. In searching for such a factor we have noted, as have others, that localizing a sound source is closely tied to localizing it visually (e.g., Pumphrey, 1950). That is, a principal function of sound localization seems to be to direct the eyes toward the source of a sound so that it can be identified visually. It is this functional relation that may be the basis for the heretofore puzzling correlation between the number of neurons in the MSO and the number of neurons in the abducens nucleus (Harrison & Irving, 1966). Although anatomical studies since that time have revealed no direct neural connection between the MSO and the abducens nerve nucleus (or other nuclei in the eye-position system), the abducens nucleus is an "eye-azimuth" motor nucleus and is involved in visually locating objects in space just as the MSO is involved in locating them acoustically. These observations led to a search for a visual parameter that might be a unifying factor explaining the differences in sound-localization acuity among all mammals.

In searching for visual correlates of sound-localization acuity, it can be noted that animals with large overlapping, or binocular, visual fields (mostly primates and carnivores and perhaps predatory species in general) tend to have good sound-localization acuity, whereas those with less overlapping, or small binocular, fields (usually hoofed animals and rodents and perhaps prey species in general) have poorer acuity (R. S. Heffner & Heffner, 1985b). Because the size of the visual fields is quantifiable, a correlation coefficient can be computed. At a statistically reliable value of $r = 0.70$, this factor accounts for almost 50% of the variation in sound-localization acuity, but it lacks intuitive explanatory value and contains several deviant points (i.e., species) which suggest other related visual parameters should be examined.

One such parameter that seems promising is a measure of the horizontal width of the subfield of most acute vision. That is, a very narrow field of best vision, such as the foveal field in humans, may place demands on sound-localization accuracy in order to place the narrow fovea directly on the sound source. Animals with their most acute vision located in a broad horizontal streak (horses, cattle, and some rodents) may, however, have less demands on accurate eye direction. Since retinal ganglion-cell density maps can be used to derive a measure of the width of the area of best vision that can be applied to all species, this measure of vision is quantifiable and allows comparison with sound-localization acuity. Among eight species the correlation between sound-localization threshold and the width of the area of best vision is strikingly high and reliable ($r = 0.96$, $p < 0.01$; R. S. Heffner & Heffner, 1988d). If this relation remains strong with a larger number and variety of species, it will reinforce the notion that a primary function of the azimuthal sound-localization system is to allow an animal to direct its visual system for scrutinizing an object or event more closely. That this function might have been the single most influential factor in the evolution of sound localization among mammals has some intuitive appeal (Pumphrey, 1950).

## REFERENCES

Aitkin, L. (1986). *The auditory midbrain*. Contemporary neuroscience. Clifton,NJ: Humana Press.

Altschuler, R. A., Parakkal, M. H., & Fex, J. (1983). Localization of enkephalin-like immunoreactivity in acetylcholinesterase-positive cells in the guinea-pig lateral superior olivary complex that project to the cochlea. *Neuroscience, 9*, 621–630.

Boudreau, J. C., & Tsuchitani, C. (1968). Binaural interaction in the cat superior olive S-segment. *Journal of Neurophysiology, 31*, 442–454.

Boudreau, J. C., & Tsuchitani, C. (1970). Superior olive S-segment cell discharge to tonal stimulation. In W. D. Neff (Ed.), *Contributions to Sensory Physiology* (Vol. 4, pp. 144–213). New York: Academic Press.

Brown, C. H., Beecher, M. D., Moody, D. B., & Stebbins, W. C. (1980). Localization of noise bands of Old World monkeys. *Journal of the Acoustical Society of America, 68*, 127–132.

Bruns, V., Muller, M., Hofer, W., Heth, G., & Nevo, E. (1988). Inner ear structure and electrophysiological audiograms of the subterranean mole rat, *Spalax ehrenbergi. Hearing Research, 33*, 1–10.

Cant, N. B., & Gaston, K. C. (1982). Pathways connecting the right and left cochlear nuclei. *Journal of Comparative Neurology, 212*, 313–326.

Casseday, J. H., & Neff, W. D. (1973). Localization of puretones. *Journal of the Acoustical Society of America, 54*, 365–372.

Casseday, J. H., & Smoak, H. A. (1981). Effects of unilateral ablation of anteroventral cochlear nucleus on localization of sound in space. In J. Syka & L. Aitkin (Eds.), *Symposium on neuronal mechanisms of hearing* (pp. 277–282). New York: Plenum Press.

Chambers, R. E. (1971). *Sound localization in the hedgehog (Paraechinus hypomelas)*. Unpublished master's thesis, Florida State University, Tallahassee.

Chan, J. C. K., & Yin, T. C. T. (1984). Interaural time sensitivity in the medial superior olive of the cat: Comparison with the inferior colliculus. *Neuroscience Abstracts, 10*, 844.

Galambos, R., Schwartzkopff, J., & Rupert, A. (1959). Microelectrode studies of superior olivary nuclei. *American Journal of Physiology, 197*, 527–536.

Glendenning, K. K., Hutson, K. A., Nudo, R. J., & Masterton, R. B. (1985).

Acoustic chiasm, II. Anatomical basis of binaurality in lateral superior olive of cat. *Journal of Comparative Neurology, 232,* 261–285.

Glendenning, K. K., & Masterton, R. B. (1983). Acoustic chiasm: Efferent projections of the lateral superior olive. *Journal of Neuroscience, 3,* 1521–1537.

Goldberg, J. M., & Brown, P. B. (1968). Functional organization of dog superior olivary complex: An anatomical and electrophysiological study. *Journal of Neurophysiology, 31,* 639–656.

Guinan, J. J., Jr., Guinan, S. S., & Norris, B. E. (1972). Single auditory units in the superior olivary complex. I. Responses to sounds and classification based on physiological properties. *International Journal of Neuroscience, 4,* 101–120.

Guinan, J. J., Jr., Norris, B. E., & Guinan, S. S. (1972). Single auditory units in the superior olivary complex. II. Locations of unit categories and tonotopic organization. *International Journal of Neuroscience, 4,* 147–166.

Hall, J. L. (1965). Binaural interaction in the accessory superior olivary nucleus of the cat. *Journal of the Acoustical Society of America, 37,* 814–823.

Harrison, J. M., & Downey, P. (1970). Intensity changes at the ear as a function of the azimuth of a tone source: A comparative study. *Journal of the Acoustical Society of America, 47,* 1509–1518.

Harrison, J. M., & Feldman, M. L. (1970). Anatomical aspects of the cochlear nucleus and superior olivary complex. In W. D. Neff (Ed.), *Contributions to Sensory Physiology* (Vol. 4, pp. 95–142). New York: Academic Press.

Harrison, J. M., & Irving R. (1966). Visual and nonvisual auditory systems in mammals. *Science, 154,* 738–743.

Heffner, H., & Masterton, R. B. (1980). Hearing in Glires: Domestic rabbit, cotton rat, feral house mouse, and kangaroo rat. *Journal of the Acoustical Society of America, 68,* 1584–1599.

Heffner, H. E., & Heffner, R. S. (1984). Sound localization in large mammals: Localization of complex sounds by horses. *Behavioral Neuroscience, 98,* 541–555.

Heffner, H. E., & Heffner, R. S. (1985). Sound localization in wild Norway rats (*Rattus norvegicus*). *Hearing Research, 19,* 151–555.

Heffner, R. S., & Heffner, H. E. (1982). Hearing in the elephant (*Elephas maximus*): Absolute sensitivity, frequency discrimination, and sound localization. *Journal of Comparative and Physiological Psychology, 96,* 926–944.

Heffner, R. S., & Heffner, H. E. (1983a). Hearing in large mammals: Horses (*Equus caballus*) and cattle (*Bos taurus*). *Behavioral Neuroscience, 97,* 299–309.

Heffner, R. S., & Heffner, H. E. (1983b). Sound localization and high-frequency hearing in horses. *Journal of the Acoustical Society of America, 73,* S42.

Heffner, R. S., & Heffner, H. E. (1985a). Hearing in mammals: The least weasel. *Journal of Mammalogy, 66,* 745–755.

Heffner, R. S., & Heffner, H. E. (1985b). Auditory localization and visual fields in mammals. *Neuroscience Abstracts, 11,* 547.

Heffner, R. S., & Heffner, H. E. (1986a). Localization of tones by horses: Use of binaural cues and the role of the superior olivary complex. *Behavioral Neuroscience, 100,* 93–103.

Heffner, R. S., & Heffner, H. E. (1986b). *Variation in the use of binaural localization cues among mammals.* Abstracts of the ninth midwinter research meeting of the Association for Research in Otolaryngology, p. 108.

Heffner, R. S., & Heffner, H. E. (1987). Localization of noise, use of binaural cues, and a description of the superior olivary complex in the smallest carnivore, the least weasel (*Mustela nivalis*). *Behavioral Neuroscience,* **101,** 701–708.

Heffner, R. S., & Heffner, H. E. (1988a). Sound localization in a predatory rodent, the northern grasshopper mouse (*Onychomys leucogaster*). *Journal of Comparative Psychology,* **102,** 66–71.

Heffner, R. S., & Heffner, H. E. (1988b). *Interaural phase and intensity discrimination in the horse using dichotically presented stimuli.* Abstracts of the eleventh midwinter research meeting of the Association for Research in Otolaryngology, p. 233.

Heffner, R. S., & Heffner, H. E. (1988c). Sound localization and use of binaural cues by the gerbil (*Meriones unguiculatus*). *Behavioral Neuroscience,* **102,** 422–428.

Heffner, R. S., & Heffner, H. E. (1988d). The relation between vision and sound localization acuity in mammals. *Society for Neuroscience Abstracts,* **14.**

Heffner, R. S., & Heffner, H. E. (1988e). Sound localization acuity in the cat: Effect of azimuth, signal duration, and test procedure. *Hearing Research,* **36,** 221–232.

Heffner, R. S., & Heffner, H. E. (1989). Sound localization, use of binaural cues, and the superior olivary complex in pigs. *Brain, Behaviour and Evolution,* **33,** 248–258.

Heffner, R. S., Richard, M. M., & Heffner, H. E. (1987). Hearing and the auditory brainstem in a fossorial mammal, the pocket gopher. *Society for Neuroscience Abstracts,* **13,** 546.

Hutson, K. A., Glendenning, K. K., & Masterton, R. B. (1987). Biochemical basis for the acoustic chiasm? *Society for Neuroscience Abstracts,* **13,** 548.

Irvine, D. R. F. (1986). The auditory midbrain. *Progress in Sensory Physiology,* **7,** pp. 1–279.

Jeffress, L. A. (1958). Medial geniculate body: A disavowal. *Journal of the Acoustical Society of America,* **30,** 802–803.

Jenkins, W. M., & Masterton, R. B. (1982). Sound localization: The effects of unilateral lesions in the central auditory system. *Journal of Neurophysiology,* **47,** 987–1016.

Johnson, D. H. (1980). The relationship between spike rate and synchrony in responses of auditory-nerve fibers to single tones. *Journal of the Acoustical Society of America,* **68,** 1115–1122.

Kelly, J. B. (1980). Effects of auditory cortical lesions on sound localizations of the rat. *Journal of Neurophysiology,* **44,** 1161–1174.

Li, R. Y.-S., & Guinan, J. J. (1971). Antidromic and orthodromic stimulation of neurons receiving calyces of Held. *M.I.T. Quarterly Progress Report,* No. 100, pp. 227–234.

Mast, T. E. (1970). Binaural interaction and contralateral inhibition in the dorsal cochlear nucleus of the chinchilla. *Journal of Neurophysiology,* **33,** 108–115.

Masterton, R. B., Glendenning, K. K., & Nudo, R. J. (1981). Anatomical-behavioral analyses of hindbrain sound localization mechanisms. In J. Syka & L. Aitkin (Eds.), *Symposium on neuronal mechanisms of hearing* (pp. 263–275). New York: Plenum Press.

Masterton, R. B., Heffner, H. E., & Ravizza, R. J. (1969). The evolution of human hearing. *Journal of the Acoustical Society of America,* **45,** 966–985.

Masterton, R. B., & Imig, T. (1984). Neural mechanisms for sound localization. *Annual Review of Physiology, 46,* 275–287.

Masterton, R. B., Jane, J. A., and Diamond, I. T. (1967). The role of brainstem auditory structures in sound localization. I. Trapezoid body, superior olive and lateral lemniscus. *Journal of Neurophysiology, 30,* 341–359.

Masterton, R. B., Thompson, G. C., Brunso-Bechtold, J. K., & RoBards, M. J. (1975). Neuroanatomical basis of binaural phase-difference analysis for sound localization: A comparative study. *Journal of Comparative and Physiological Psychology, 89,* 379–386.

McFadden, D., & Posanen, E. G. (1976). Lateralization at high frequencies based on interaural time differences. *Journal of the Acoustical Society of America, 59,* 634–639.

Mills, A. W. (1958). On the minimum audible angle. *Journal of the Acoustical Society of America, 30,* 237–246.

Moore, J. K. (1987). The human auditory brain stem: A comparative view. *Hearing Research, 29,* 1–32.

Moore, J. K., & Moore, R. Y. (1971). A comparative study of the superior olivary complex in the primate brain. *Folia Primatologica, 16,* 35–51.

Morest, D. K. (1968). The collateral system of the medial nucleus of the trapezoid body of the cat, its neuronal architecture and relation to the olivocochlear bundle. *Brain Research, 9,* 288–311.

Phillips, D. P., & Brugge, J. F. (1985). Progress in neurophysiology of sound localization. *Annual Review of Psychology, 36,* 245–274.

Pumphrey, R. J. (1950). Hearing. *Symposium of the Society for Experimental Biology, 4,* 1–18.

Rasmussen, G. L. (1946). The olivary peduncle and other fiber projections of the superior olivary complex. *Journal of Comparative Neurology, 84,* 141–219.

Ravizza, R. J., Heffner, H., & Masterton, R. B. (1969). Hearing in primitive mammals. I. Opossum (*Didelphis virginiana*). *Journal of Auditory Research, 9,* 1–7.

Ravizza, R. J., & Masterton, R. B. (1972). Contribution of neocortex to sound localization in opossum (*Didelphis virginiana*). *Journal of Neurophysiology, 35,* 344–356.

Renaud, D. L., & Popper, A. N. (1975). Sound localization by the bottlenose porpoise *Tursiops truncatus. Journal of Experimental Biology, 63,* 569–585.

Rose, J. E., Gross, N. B., Geisler, C. D., & Hind, J. E. (1966). Some neural mechanisms in the inferior colliculus of the cat which may be relevant to localization of a sound source. *Journal of Neurophysiology, 29,* 288–314.

Rosenzweig, M. R., & Amon, A. H. (1955). Binaural interaction in the medulla of the cat. *Experientia, 11,* 498–500.

Stebbins, W. C. (1970). Studies of hearing and hearing loss in the monkey. In W. C. Stebbins (Ed.), *Animal psychophysics* (pp. 41–66). New York: Appleton-Century-Crofts.

Stotler, W. W. (1953). An experimental study of the cells and connections of the superior olivary complex of the cat. *Journal of Comparative Neurology, 98,* 401–432.

Terhune, J. M. (1974). Directional hearing of the harbor seal in air and water. *Journal of the Acoustical Society of America,* **56,** 1862–1865.

Thompson, G. C., & Masterton, R. B. (1978). Brain stem auditory pathways involved in reflexive head orientation to sound. *Journal of Neurophysiology,* **41,** 1183–1202.

Tolbert, L. P., Morest, D. K., & Yurgelun-Todd, D. A. (1982). The neuronal architecture of the anteroventral cochlear nucleus of the cat in the region of the cochlear nerve root: Horseradish peroxidase labelling of identified cell types. *Neuroscience,* **7,** 3031–3052.

von Bekesy, G. (1930). Ueber das Richtunghoren bei einer Zeitdifferenz oder Lautstarkenungleichheit der beiderseitigen Schalleinwerkungen. *Physikalische Zeitschrift,***31,** 857–868.

White, J. S., & Warr, W. B. (1983). The dual origins of the olivocochlear bundle in the albino rat. *Journal of Comparative Neurology,* **219,** 203–214.

Yin, T. C. T., & Kuwada, S. (1983). Binaural interaction in low-frequency neurons in inferior colliculus of the cat. III. Effects of changing frequency. *Journal of Neurophysiology,* **50,** 1020–1042.

Zook, J. M., & Casseday, J. H. (1982). Cytoarchitecture of auditory system in lower brainstem of the mustache bat, *Pteronotus parnellii. Journal of Comparative Neurology,***207,** 1–13.

# 10

# ECHOLOCATION IN DOLPHINS

*Whitlow W. L. Au*

*Naval Ocean Systems Center, Kailua, Hawaii*

The term *echolocation* was first proposed by Griffin (1944) to describe an acoustic method of orientation by animals based on the projection of ultrasonic pulses and the reception of echoes reflected from objects. One of the most effective methods for an animal to probe an underwater environment for the purposes of navigation, obstacle and predator avoidance, and prey detection is by echolocation. Many odontocetes emit sounds and analyze returning echoes to detect and recognize objects underwater. Acoustic energy propagates in water more efficiently than almost any other form of energy, so the use of echolocation and passive acoustics (listening) by dolphins is ideal. Electromagnetic, thermal, light, and other forms of energy are severely attenuated in water. Often the natural habitat of shallow bays, inlets, coastal waters, swamps, marshlands, and rivers of certain dolphin species are so murky or turbid that vision is severely limited. These animals must then rely almost exclusively on their auditory perception, including echolocation, for survival.

Most dolphin auditory research has been performed with the Atlantic bottlenose dolphin (*Tursiops truncatus*), the most common dolphin found in oceanariums, zoos, and research facilities. In 1947, Arthur McBride, the first curator of Marine Studios in Florida, presented evidence in his personal notes that *Tursiops* may detect objects underwater by echolocation (McBride, 1956). McBride died in 1950 and it was only later that his notes were discovered and published with an introductory note by Schevill (1956). When Kellogg and Kohler (1952) found that *Tursiops* could hear sounds at frequencies greater than 50 kHz, they also proposed that dolphins may be able to echolocate. However, it was not until 1960 that Norris, Prescott, Asa-Dorian, and Perkins (1961) performed the first unequivocal demonstration of echolocation in dolphins. They trained a blindfolded *Tursiops* to perform a variety of behavioral tasks. The blind-folded dolphin emitted click signals which were interpreted to be echolocation signals, and went about performing the tasks as if it could see. Since then many experiments have shown that dolphins possess highly sophisticated and adaptive echolocation systems.

Dolphins echolocate by emitting high-intensity broadband acoustic pulses in a directional beam and listening to echoes reflected from objects in their environment. By scanning their echolocation beam across objects and by analyzing the characteristics of the echoes, dolphins can obtain considerable information about their environment. The presence, size, structure, material composition, and shape of objects can be determined. The relative distance of objects can also be determined by estimating the time between the transmission of a pulse and the reception of echoes. The distance over which a dolphin's echolocation system can operate depends on the size of the objects and the ambient background noise of the environment. Distances of several hundred meters are well within the range of the animal's echolocation system.

This chapter will concentrate mainly on research performed at the Naval Ocean Systems Center (NOSC) laboratory in Hawaii, although significant findings by others will be noted. Most of the research results discussed here will pertain to *Tursiops truncatus*. The goals of dolphin echolocation research performed at NOSC are (a) to determine the echolocation capabilities of different species, (b) to quantify the parameters of the echolocation system, (c) to understand possible signal-processing mechanisms used by dolphins to detect targets in noisy and reverberant environments, (d) to understand possible signal-processing signal-processing mechanisms used by dolphins to make fine discrimination between different targets, and (e) to construct mathematical and physical models of the dolphin's echolocation system.

In all of the auditory perception experiments performed at NOSC—Hawaii, operant conditioning techniques with positive reinforcement and correct-response feedback are used to achieve and maintain stimulus control. The positive reinforcement usually consists of a secondary bridge tone in the form of an airborne whistle sound or an underwater tone pulse to provide immediate correct-response feedback, followed by the primary reinforcement, a fish reward. An animal is trained by dividing a complex task into many simpler components that can be chained together and eventually lead to the establishment of stimulus control which indicates that a change in the stimulus will result in a measurable change in behavior. Once stimulus control is established, the dolphin's acuity is generally determined by using standard psychophysical testing techniques such as the method of up/down staircase or the tracking procedure, the method of constant stimuli, and the probe technique. Two methods of response are usually used: the go/no-go procedure, and the two-alternate, forced-choice procedure. In the go/no-go procedure, the dolphin indicates a positive response by leaving station (a specific, defined location) to strike a paddle, or remains at station to indicate a negative response. In the two-alternative, forced-choice, the animal must strike one of two paddles, each paddle indicating a particular response concerning the condition of the stimulus or stimuli. Schusterman (1980) presented an excellent description of some of the behavioral methodology utilized in echolocation research at NOSC and elsewhere.

One unique feature of the research performed at NOSC–Hawaii is that echolocation experiments are performed in a natural bay (Kaneohe Bay) with animals housed in floating pens secured to pier structures. An example of a floating-pen structure with a dolphin stationing in a hoop performing an echolocation experiment is shown in Fig. 1. The bottom of the bay consists of mud and silt, which absorbs acoustic energy and minimizes bottom reflections and reverberation. Therefore, the dolphins are exposed to a natural, open and spacious environment instead of a concrete or redwood tank. A tank environment may have an adverse influence on a dolphin's echolocation capabilities. The close proximity of tank walls, which are good acoustic reflectors, would tend to discourage a dolphin from emitting high-intensity acoustic signals as it would in open waters. This could introduce the possibility that unnatural and suboptimal signals may actually be used in tanks. The reverberant environment of tanks can also complicate the measurement and analysis of echo-location signals.

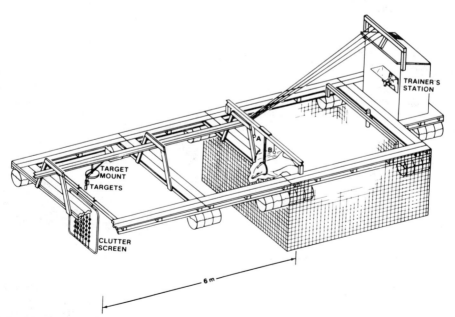

**FIGURE 1.** An example of a floating pen structure enclosed with a wire mesh and supported with 50-gallon oil drums. The diagram also shows a dolphin stationed in a hoop.

## I. ACOUSTIC CHARACTERISTICS OF THE ECHOLOCATION SYSTEM

### A. Transmission System

The dolphin has a bistatic sonar system with the upper portion of its head used as the transmitter and the lower portion as the receiver. Although there has been a long-standing controversy on whether sounds are produced in the larynx or in the nasal system of odontocetes (Morris, 1986; Norris, 1986), almost all experimental data with *Tursiops* indicate that sounds are produced in the nasal system in the vicinity of the nasal plugs. Evidence from acoustic measurements (Diercks, Trachta, Greenlaw, & Evans, 1971), x-ray measurements (Dormer, 1979; Hollien, Hollien, Caldwell, & Caldwell, 1976; Norris, Dormer, Pegg, & Liese, 1971;), muscle activity and pressure measurements (Ridgway et al., 1980), and ultrasonic Doppler motion detection measurements (Mackay & Liaw, 1981) all implicated the nasal system in the vicinity of the nasal plugs in the generation of sounds. Proponents of the theory that the larynx is the site of sound production (Purves, 1967; Purves & Pilleri, 1983) offer only anatomical arguments and results from laboratory simulation with dead animals, without any supportive experimental evidence from live sound-producing specimens. Figure 2 is a schematic of a bottlenose dolphin head

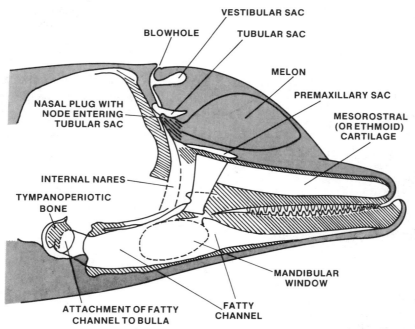

**FIGURE 2.** Section of a dolphin's head showing various structures associated with sound reception and production. Adapted from Norris (1968).

depicting various structures associated with sound production and reception.

The melon in front of the nasal plug may play a role in focusing sounds in the water. Norris and Harvey (1974) found a low-velocity core extending from just below the anterior surface toward the right nasal plug, and a grated outer shell of high-velocity tissue. Such a velocity gradient would focus signals originating at the nasal plug region in both the vertical and horizontal planes.

The ability of dolphins to perform difficult echolocation detection, recognition, and discrimination tasks should depend heavily on the kinds of signals emitted. The signals must have sufficient information-carrying capacity so that important features of a target can be coded in the echoes. Extensive measurements of echolocation signals utilized by the bottlenose dolphin performing a variety of echolocation tasks have been made at our facility (Au, 1980; Au, Floyd, & Haun, 1978; Au, Floyd, Penner, & Murchison, 1974; Au, Moore, & Pawloski, 1986; Au, Penner, & Kadane, 1982). Typical signals are short, broadband, transient-like clicks with durations between 50 and 70 $\mu$sec. An example of a typical echolocation click train emitted by a bottlenose dolphin performing a target detection task in Kaneohe Bay is shown in Fig. 3. The frequency spectra plotted as a function of time are shown on the left and the individual click waveform is displayed on the right. The peak-to-peak sound-pressure level (SPL) is expressed in decibels (dB) referenced to a SPL of one micropascal ($\mu$Pa) used in underwater acoustics. This can be compared with the standard reference pressure level of 0.0002 dyne/cm$^2$ or 20 $\mu$Pa used with in-air acoustics. The signals in a typical click train tend to be highly repetitive. Their shape in the time domain resembles exponentially damped sine waves of 6 to 10 cycles. The frequency spectrum usually rises to a peak between 110 and 130 kHz, with secondary peaks between 60 and 80 kHz often present. These signals have considerably higher peak frequencies and amplitudes than previously measured peak frequencies between 30 and 60 kHz and amplitudes of 180–190 dB for dolphins in tanks (Evans, 1973). Au et al. (1974) attributed the dolphins' use of high frequencies and amplitudes to the high ambient noise environment of Kaneohe Bay. The Bay has an extremely high ambient noise level caused by snapping shrimp (Albers, 1965) and the noise energy extends beyond 100 kHz (Au et al., 1974).

Peak frequency and 3-dB bandwidth histograms of signals used by four bottlenose dolphins performing different echolocation tasks are presented in Fig. 4 (Au, 1980). Ten different click trains were used for the dolphins Ehiku and Heptuna, and 40 click trains for Ekahi and Sven; each click train consisted of 25 to 125 clicks. The frequency histograms indicate that the majority of clicks had peak frequencies between 110 and 130 kHz. The bandwidth histograms indicate that most of the signals had bandwidths greater than 25 kHz; bandwidths between 30 and 40 kHz were the most

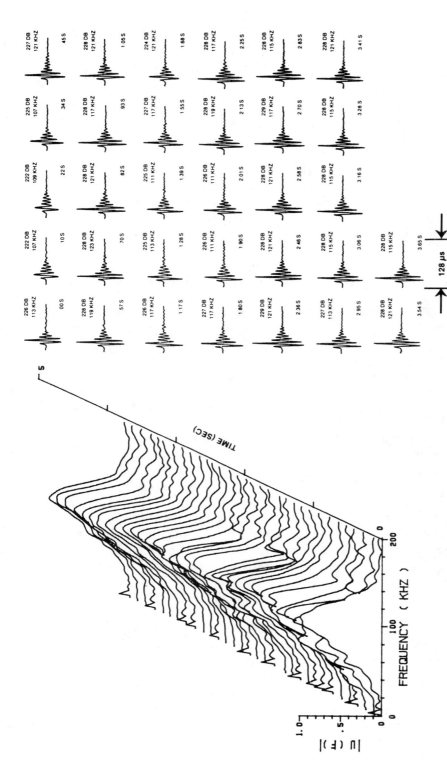

**FIGURE 3.** Example of bottlenose dolphin echolocation signals used in Kaneohe Bay. On the left are the frequency spectra (relative amplitude versus frequency) as a function of time, and on the right are the signal waveforms. From Au (1980). The peak-to-peak amplitude in dB re 1 $\mu$Pa and the peak frequency (frequency of peak energy) in kHz are given above each signal waveform, and the time of occurrence of the signals relative to the start of the click train is given below each click.

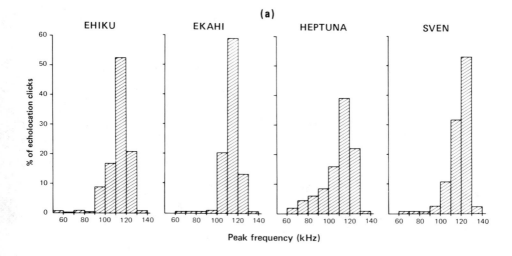

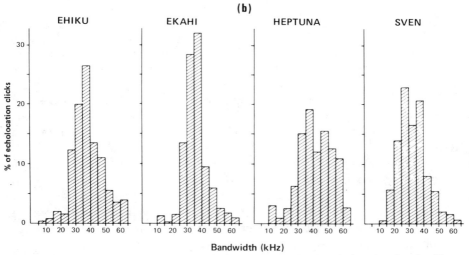

**FIGURE 4.** Peak frequency and 3-dB-bandwidth histogram for the echolocation signals of four bottlenose dolphins. From Au (1980).

common. The information capacity of a signal is generally proportional to its bandwidth.

Echolocation signals are projected from the dolphin's head in a directional beam. Norris et al. (1961) observed that a blindfolded dolphin could not detect targets below its jaws or at elevation angles greater than 90° above the rostrum. The transmission beam patterns in the vertical and horizontal planes for *Tursiops* were measured by Au (1980) and Au et al. (1978, 1986). The results of the three measurements on three different *Tursiops* are shown in Fig. 5, with the results of Au (1980) (dotted lines) and

322   ECHOLOCATION IN DOLPHINS

(a)                                    (b)

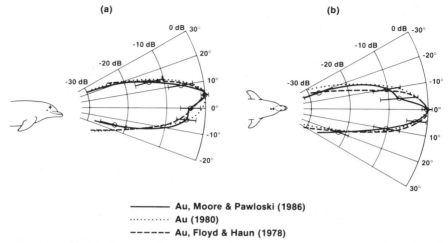

——— Au, Moore & Pawloski (1986)
·········· Au (1980)
–––––– Au, Floyd & Haun (1978)

**FIGURE 5.** Composite broadband transmission beam patterns of three bottlenose dolphins, in the (*a*) vertical plane and (*b*) horizontal plane.

Au et al. (1978) (dashed lines) overlaying the results of Au et al. (1986) (solid lines). The beams represented by the dashed and dotted lines were aligned with the beam represented by the solid line. The results of the three measurements were similar. The measurements of Au et al. (1986) indicated that echolocation signals are projected at an elevation angle of 5° above the animal's head. These measurements indicate that the acoustic energy radiated from a dolphin while echolocating is concentrated in a narrow beam that is aimed directly forward of the animal.

A narrow transmission beam allows a dolphin to scan across targets by moving its head from side to side as it echolocates. Such scanning can provide important information on the spatial characteristics of targets. A narrow beam also allows the dolphin to scan only a region of interest without receiving interfering echoes from objects in a different region. Finally, a narrow beam can be important in localizing and separating objects of interest.

Bottlenose dolphins typically echolocate in a pulse mode, sending out a signal and receiving the echo from the target before sending out another signal. Figure 6 shows the click interval as a function of target range for four different experiments with bottlenose dolphins. Also included in the figure is the two-way transit time, which is the time required for an acoustic signal to travel from the dolphin to the target and back. The lag time, which is defined as the time difference between the click interval and the two-way transit time, varied from 19 to 45 msec. This lag time indicates that the dolphin projects a signal, receives the target echo, and waits from 19 to 45 msec before projecting the next pulse. The lag time may be indicative of the time required by the dolphin to process each echo. It is

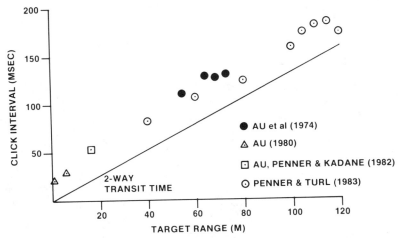

**FIGURE 6.** Click interval as a function of target range for bottlenose dolphin. The value of the two-way transit time for any target range can be read off the click interval scale.

interesting to note that animals often "lock-in" on the target range and will emit signals with the same click intervals on target-present and target-absent trials.

The amplitudes of dolphin echolocation signals are usually expressed in terms of peak-to-peak sound-pressure levels (SPL) instead of root-mean-square (rms) pressure. The evaluation of a time integral would be required to calculate the rms SPL. The amplitudes can vary considerably within a click train, between click trains (Au et al., 1974), and for different tasks (Au, 1980). The peak-to-peak source level (SPL at a reference 1 m from the animal) should depend on the signal-to-noise ratio of the received echoes that the animals require or desire to have to perform a given task. The echo level (SPL of the echo) from a target can be expressed by the equation

$$EL = SL - \text{(transmission loss)} - \text{(reflection loss)} \qquad (1)$$

where EL is the echo level in dB and SL is the source level in dB. Equation 1 can be simplified to

$$EL = SL - \text{(total loss)} \qquad (2)$$

The peak-to-peak source level, as a function of the total loss for different bottlenose dolphins performing different tasks, is shown in Fig. 7. The total loss in Fig. 7 is equated to the sum of the propagation loss due to spherical spreading and absorption plus the reflection loss given by TS (target strength). Equation 2 indicates that the echo level is inversely

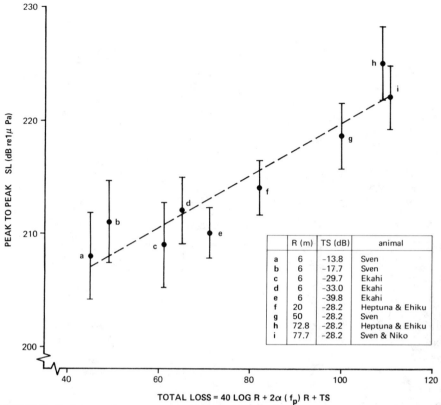

**FIGURE 7.** Peak-to-peak source levels of the echolocation signals used by several bottlenose dolphins in Kaneohe Bay as a function of the total signal losses. From Au (1980).

proportional to the total loss. The greater the loss, the less energy there is in the echo. The data in Fig. 7 indicate that the source levels increased as the total loss increased, suggesting that dolphins compensated for the amount of acoustic loss to the signal by changing its source level. The data also indicate that dolphins prefer to operate at a high signal-to-noise ratio, since the decrease in total losses from case i to case a exceeded 62 dB, yet the corresponding decrease in source level was only 12 dB.

The maximum average peak-to-peak source level in a trial was 227.5 dB for one of the trials in case h. The largest single click measured was 230 dB, emitted by Heptuna in case h. These source levels may seem inordinately high, considering the amount of power required to project high-intensity signals into the water. However, the levels are peak-to-peak and the durations of the signals are short. The rms pressures for these signals are typically 15 to 20 dB lower than the peak-to-peak amplitudes.

Bottlenose dolphins also emit a wide variety of other sounds not used for echolocation. Sound emissions can be classified into two broad categories of narrow-band, frequency-varying, continuous tonal sounds referred to as whistles and broadband echolocation clicks (Evans, 1967). Whistles appear to be used for intraspecific communications (Herman & Tavolga, 1980). These sounds are generally low-frequency emissions between 5 and 30 kHz. They are also referred to as squeaks, squawks, and squeals.

## B. The Reception System

A dolphin's echolocation capabilities are also dependent on the characteristics of its auditory system as well as its neurological processing skills. One of the more popular theories of sound reception by dolphins was proposed by Norris (1968), who postulated that the lower jaw acts acts as a receptor of sound. The sound enters through the thin oval (pan bone) area of the flared posterior end of the mandible (see Fig. 2). Electrophysiological measurements of evoked potentials in the inferior colliculus by Bullock et al. (1968) and cochlea potentials by McCormick, Weaver, Palin, and Ridgway (1970) indicated that maximum responses were obtained when a sound source was placed in the vicinity of the lower jaw and ears. According to McCormick et al., the upper jaw and most of the skull seem to be acoustically isolated from the ear, with the auditory meatus being vestigial.

Kellogg (1953) and Schevill and Lawrence (1953) were the first to obtain behavioral evidence that a bottlenose dolphin could hear sounds above 100 kHz. In a pioneering study, Johnson (1967) performed a carefully controlled psychophysical experiment to measure the auditory sensitivity of a bottlenose dolphin as a function of frequency. The dolphin was required to station in a sound stall and respond to the presence or absence of an acoustic signal. A go/no-go response paradigm was used and the signal intensity was controlled in a staircase manner. The dolphin's audiogram measured by Johnson is shown in Fig. 8 along with a human audiogram measured by Sivian and White (1933). Johnson's results indicate that a bottlenose dolphin can hear over a wide frequency range between 75 Hz and 150 kHz, with maximum sensitivity ($\pm 10$ dB) between 10 and 120 kHz. The dolphin and human audiograms are similar in shape, with the dolphin's shifted to higher frequencies by a factor of 10.

Johnson (1968) extended his auditory sensitivity measurement to include the presence of broadband masking noise. He measured the dolphin's capability to detect pure-tone signals masked by broadband noise. From his masked threshold data, he determined the critical ratio of the bottlenose dolphin as a function of frequency. The notion of critical ratio comes from the hypothesis that a pure-tone signal is masked only by a narrow band of noise that is centered about the signal frequency. When the signal power is equal to the noise power in this band, the subject will not be able to detect the signal. Since noise is usually expressed as a power

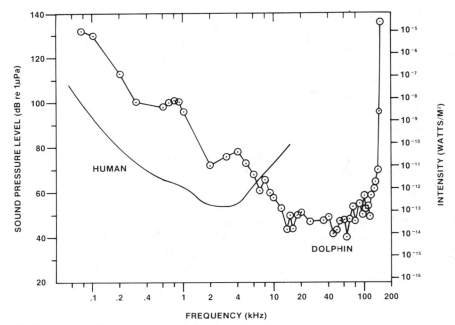

**FIGURE 8.** Auditory sensitivity of a bottlenose dolphin and human subjects. The left ordinate of sound-pressure level (SPL) in dB re 1 μPa is for the dolphin audiogram. The right ordinate of intensity in W/m² is for both the dolphin audiogram and the human audiogram of Sivian and White (1933). Adapted from Johnson (1967).

density (power per Hz), the width of this band can be estimated by taking the ratio of the signal power to noise density at the animal's hearing threshold in noise. The critical ratio as a function of frequency will then be related to the width of the internal auditory filter. Johnson's critical ratio results are shown in Fig. 9 along with monaural and binaural human data. The bandwidth of the dolphin's internal auditory filter increases almost proportionately with frequency, suggesting that the internal filter increases almost proportionately with frequency and that the dolphin's auditory system may be modeled as a bank of constant-$Q$ filters. It is this filter bank property of the dolphin's auditory system which allows the animal to perform frequency analysis of received acoustic signals. The dolphin's critical ratio seems to be an extension of the human critical ratio to higher frequencies. The results of Fig. 8 can also be used to indicate how much higher above the ambient noise a pure-tone signal needs to be for a dolphin to hear it.

A dolphin's ability to discriminate and recognize objects by means of echolocation is directly related to how well the animal discriminates differences in the frequency content of echoes. The frequency discrimina-

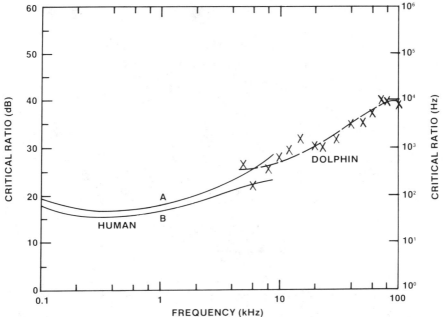

**FIGURE 9.** Critical ratios for a bottlenose dolphin (X) and humans (A for monaural and B for binaural). Adapted from Johnson (1968).

tion capability of bottlenose dolphins was studied by Jacobs (1972), Herman and Arbeit (1972), and Thompson and Herman (1975). In all three studies, the bottlenose dolphin was required to discriminate a constant-frequency (CF) pure-tone signal from a frequency-modulated (FM) signal having the same center frequency as the pure-tone signal. The amount of frequency modulation was continuously decreased until the animals could not discriminate the FM signal from the CF signal. The difference limen (DL) is the difference between the upper and lower frequency of the FM signal at the animals' threshold. The relative difference limen is the Weber fraction (DL/F), where $F$ is the center frequency of the FM signal. The results of these studies, plotted in terms of relative difference limens in percent, are shown in Fig. 10. The data of Herman and Arbeit (1972) and Thompson and Herman (1975) were obtained with the same animal but with slightly different behavior methodology. Their results, if averaged, would be a good representation of the dolphin's difference limens. The data plotted in Fig. 9 indicate that the dolphin's frequency discrimination sensitivity is similar to that in humans but shifted to higher frequencies by approximately a factor of 10.

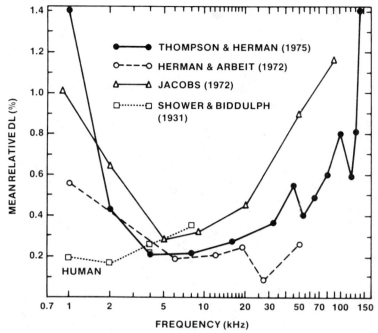

**FIGURE 10.** Frequency-discrimination capability of bottlenose dolphin and humans plotted in terms of the relative difference limen in percent. Adapted from Thompson and Herman (1975).

Another fundamental parameter of a sonar system is the spatial sensitivity pattern or the receiving beam pattern. It determines the amount of ambient noise and reverberation the sonar will receive, and may also affect the angular resolution of the sonar. The amount of ambient noise and reverberation a sonar receives is directly proportional to the width of the receiving beam.

Au and Moore (1984) measured the receiving beam pattern of a bottlenose dolphin in both the vertical and horizontal planes by determining the animal's masked threshold as the position of either the noise or signal sources varied in their angular position about the animal's head. The measurement in the vertical plane was made by having the dolphin turn on its side and station on a bite plate with a signal transducer located 3.5 m away directly in front of the animal. A masking transducer was placed at different positions about a 3.5 m radius arc centered at the bite plate. The amount of noise required to mask the signal transducer as a function of the position of the masking transducer was determined. The relative difference in the amount of noise required to mask the signal as a function of the angular position of the noise transducer is directly related to the received beam pattern. The measurement in the horizontal plane was made with the animal stationing on a horizontal bite plate. Two noise transducers spaced at fixed locations ±20° about the center line and a signal transducer transmitting a fixed level signal at a variable azimuthal position was used. The three transducers were located approximately 3.5 m from the bite plate. The amount of noise required to mask the signal transducer as a function of the position of the signal transducer is directly related to the received horizontal beam pattern. Two noise transducers were used in the horizontal plane to discourage the dolphin from performing a spatial

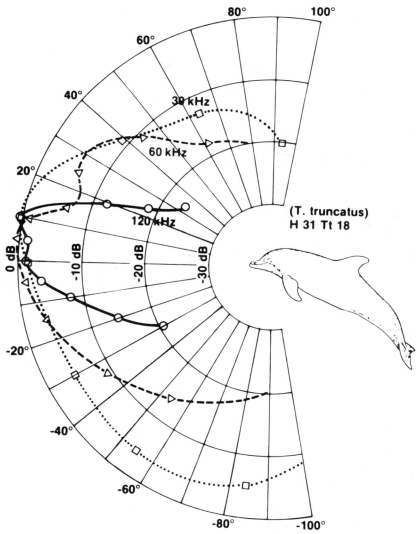

**FIGURE 11.** Bottlenose dolphin receiving vertical beam patterns for frequencies of 30, 60, and 120 kHz. From Au and Moore (1984).

filtering operation by internally steering the axis of its beam in order to maximize the signal-to-noise ratio. Their results in both planes for frequencies of 30, 60, and 120 kHz are shown in Figs. 11 and 12, respectively. The beams pointed directly forward and became narrower or more directional as the frequency increased. The vertical beams are asymmetrical, decreasing in intensity faster as a function of the angle above the animal's head than below. Such asymmetry is consistent with the theory that the animal

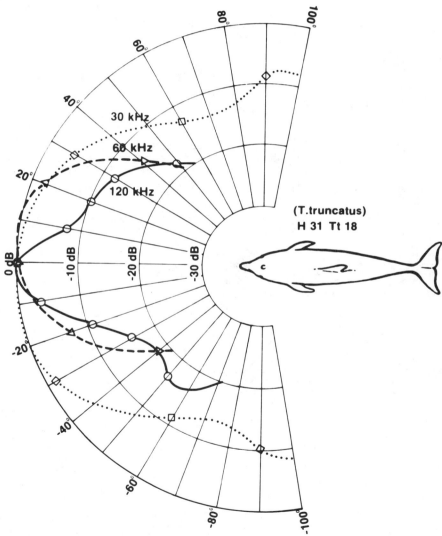

**FIGURE 12.** Bottlenose dolphin receiving horizontal beam patterns for frequencies of 30, 60, and 120 kHz. From Au and Moore (1984).

receives sound through its lower jaw. The horizontal beams were symmetrical about the longitudinal axis of the dolphin's body.

The ability to localize sound is important for an echolocator in order to resolve echoes from closely spaced targets or from different portions of an extended target, and to determine the relative position of targets within the acoustic beam. Renaud and Popper (1975) studied the sound-localization capability of a bottlenose dolphin in the horizontal and vertical planes by measuring the minimum audible angle (MAA). The animal was required to station on a bite plate facing two transducers located 18 m away at equal azimuth about a center line. On a given trial, pure-tone pulses were projected from one of the two transducers, and the animal was required to indicate which transducer emitted the sound. The angle between the two transducers was continuously reduced until the dolphin's threshold was reached. For the measurement in the vertical plane, the animal was required to turn on its side and station on a vertical bite bar. They measured MAA between 1 and 3° in both planes for pure-tone signals with frequencies between 20 and 90 kHz. At 6 and 100 kHz, the MAA increased to approximately 4°. Using simulated echo location clicks with a peak frequency of 64 kHz, MAAs of 0.7 to 0.8° were measured. The horizontal underwater pure-tone localization capability of dolphins is about as good as the in-air capabilities of several mammals, including humans (Popper, 1980). The dolphin's ability to localize the broadband click is considerably better than that of most animals.

## II. ECHOLOCATION CAPABILITIES

### A. Maximum Detection Range

The maximum detection range of two *Tursiops* was determined by Murchison (1980) in Kaneohe Bay using a 2.54-cm-diameter solid steel sphere and a 7.62-cm-diameter water-filled sphere. However, the results with the 7.62-cm sphere were affected by the presence of an underwater ridge along the bottom of the test range in the vicinity of the threshold range. Au and Snyder (1980) remeasured the maximum detection range of one of the dolphins (Sven) using a 7.62-cm-diameter water-filled sphere; a different part of Kaneohe Bay, where the bottom was relatively flat and the water depth between 5.8 and 6.1 m, was used. In both studies an overhead suspension system with a movable trolley and pulleys was used to vary target range between two poles spaced 200 m apart. The target was lowered into or raised out of the water by means of a nylon monofilament line that extended back to the experimenter's station.

A similar testing procedure was used by both Murchison (1980) and Au and Snyder (1980). A trial began with the dolphin at station. It was then cued to perform an echolocation search and determine the presence or absence of the target. The dolphin struck a specific paddle for a target-present response and another paddle for a target-absent response. A secondary bridge reinforcement tone and a fish reward were presented to the dolphin for each correct response. The dolphin was not reinforced for incorrect responses. The target presentation schedules were randomized according to a modified Gellerman series, with equal numbers of target-present and target-absent trials per session. Murchison used a fixed target range and 30 trials per animal in each session. Au and Snyder conducted 60 trials per session, with the trials divided into six 10-trial blocks. A different target range was used in each 10-trial block, with the ranges varying in increments of 5 or 2 m, depending on the specific ranges being tested.

The results of both experiments are displayed in Fig. 13 with correct detection and false alarm rates plotted as a function of the target range. The detection threshold ranges (range at 50% correct detection) for the 2.54-cm- and 7.62-cm-diameter spheres were 73 and 113 m, respectively. The animal's results are relatively consistent if target differences are considered in the sonar equation. For a noise-limited situation, the sonar equation can be expressed as

$$DT = (\text{echo intensity in dB}) - (\text{noise level in dB}) \qquad (3)$$

where DT is the detection threshold. The echo intensity is equal to the

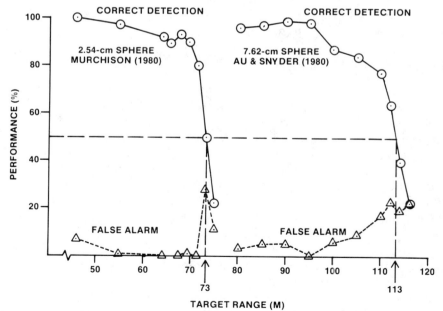

**FIGURE 13.** Dolphin target-detection performance as a function of range. The 2.54-cm sphere results are from Murchison (1980) and the 7.62-cm sphere results are from Au and Snyder (1980).

source level (SL) of the projected signal minus the two-way transmission loss (TL) plus the reflective strength of the target (TS). The noise level is equal to the ambient noise spectrum density (NL) minus the directivity index (DI) of the receiving beam. Therefore, the sonar equation can be expressed as

$$DT = (SL - 2TL + TS) - (NL - DI) \tag{4}$$

Since dolphins project short-duration broadband signals, the transient form of the sonar equation should be used (Urick, 1967). The transient form of the sonar equation is derived by substituting the following equation for the source level (Urick, 1967):

$$SL = 10 \log E - 10 \log T_e \tag{5}$$

where $E$ is the energy flux density of the projected signal and $T_e$ is the duration of the echo. If we assume that $E$, NL, DI, and DT were the same for both studies, then only differences in the transmission loss, target strength, and echo duration need be considered. If we let the subscript 1 refer to the 2.54-cm sphere results of Murchison and subscript 2 refer to the 7.62-cm results of Au and Snyder, the following equality can be derived by applying the sonar equation to both cases:

$$2TL_2 - TS_2 + 10 \log T_{e2} = 2TL_1 - TS_1 + 10 \log T_{e1} \tag{6}$$

For threshold ranges of 73 and 113 m, the two-way transmission losses are 81 and 92 dB, respectively. Results of target strength measurements on both targets (Au & Snyder, 1980), using a simulated dolphin echolocation signal, are shown in Fig. 14. Inserting the appropriate values for the transmission losses, target strength, and echo durations, we find that the left side of Eq. 3 is only 1.5 dB greater than the right side. This is good agreement considering the experiments were performed about two years apart.

## B. Target Detection in Noise

Au and Penner (1981) conducted a target-detection-in-noise experiment with two *Tursiops*. The animals were required to station in a hoop and echolocate a 7.62-cm stainless steel water-filled sphere at a range of 16.5 m. A noise source with a flat spectrum between 40 and 160 kHz was located between the animal and the target, 4 m from the hoop. Masking noise levels between 67 and 87 dB re 1 $\mu Pa^2/Hz$ in 5-dB increments were randomly used in blocks of 10 trials for a 100-trial session. The dolphins' performance as a function of the echo-energy-to-noise $[E_e/N_0]_{max}$ is shown in Fig. 15. The average value of the maximum source energy flux density

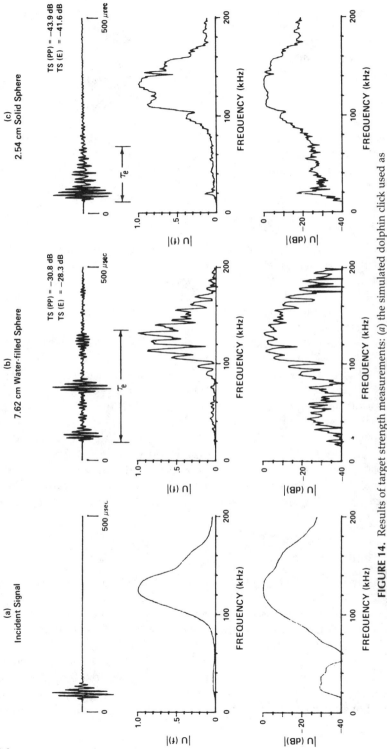

**FIGURE 14.** Results of target strength measurements: (*a*) the simulated dolphin click used as the incident signal, (*b*) and (*c*) the echoes from the 7.62-cm and 2.54-cm spheres, respectively. The echo waveforms are shown by the top traces, the frequency spectra on a linear scale by the middle traces, and on a log scale by the bottom traces. From Au and Snyder (1980).

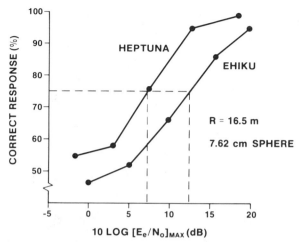

**FIGURE 15.** *Tursiops* target detection in noise performance as a function of the echo $S/N$. From Au and Penner (1981).

per trial was used in the calculations. The 75% correct response thresholds were at $[E_e/N_0]_{max}$ of approximately 7 dB for Heptuna and 12 dB for Ehiku. These signal-to-noise estimates represent the most conservative estimates possible since the maximum source energy flux density per trial was used. At the 82- and 87-dB noise levels, both dolphins began to guess; Ehiku did not emit any detectable signals during 20 and 41% of the trials, respectively. Heptuna did not emit any signals for 14% of the trials at the highest noise level. Therefore, the average of the maximum signal per trial between 67- and 77-dB noise levels were used to calculate $E_e/N_0$ at the 82- and 87-dB levels.

Moore, Hall, Friedl, and Nachtigall (1984) performed a backward masking experiment with an echolocating *Tursiops* in Kaneohe Bay. The dolphin was required to detect a water-filled aluminum cylinder 9 cm o.d., 21 cm long, located at a range of 46 m. Broadband masking noise was triggered by the dolphin's echolocation signal and could be temporally adjusted from coincidence with the target echo to delays of 700 $\mu$sec. Their results are shown in Fig. 16 with the percent correct versus the time delay between the target echo and the masking noise. The 70% correct response threshold was approximately 265 $\mu$sec.

## C.  Detection in Reverberation

The target detection capability of any echolocation system, man-made or biological, is limited by both interfering noise and reverberation. Reverberation differs from noise in several aspects. It is caused by the echolocator

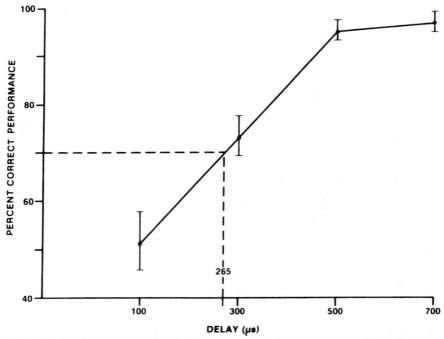

**FIGURE 16.** Psychometric function for backward masking as a function of masker delay. Threshold is shown for 70% correct response performance. From Moore, Hall, Friedl, & Nachtigall (1984).

itself, and is the total contribution of unwanted echoes scattered back from objects and inhomogeneities in the medium and on its boundaries. The spectral characteristics of reverberation are similar to those of the projected signal and its intensity is directly proportional to the intensity of the projected signal. Therefore, in a reverberation-limited situation, target detection cannot be improved by increasing the intensity of the projected signal. Target detection becomes dependent on the ability of the system to discriminate between the target of interest and false targets and clutter that contribute to the reverberation.

Murchison (1980) studied the effects of bottom reverberation on the target-detection capabilities of two *Tursiops* in Kaneohe Bay. A 6.35-cm-diameter solid steel sphere was used at depths varying from 1.2 to 6.3 m. At a depth of 6.3 m, the target was on the bottom. The animals' 50% correct detection threshold ranges for the different target depths are plotted in Fig. 17. As the target depth increased, the animals' detection ranges decreased, showing the effects of bottom reverberation. Murchison (1980) did not measure the bottom reverberation, so the echo-to-reverberation ratio (*E/R*) at the animals' performance threshold could not be determined.

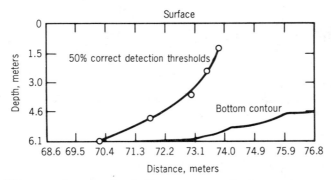

**FIGURE 17.** 50% correct detection threshold ranges for two *Tursiops* as a function of the depth of the 6.35-cm spherical target. From Murchison (1980).

Au and Turl (1983) investigated the capability of an echolocating *Tursiops* to detect targets placed near a clutter screen. The clutter screen consisted of forty-eight 5.1-cm-diameter cork balls with their centers spaced 15.2 cm apart in a 6 × 8 rectangular array. The cork balls were tied eight to a line and attached to a 1.9 × 1.9-m frame made of 3.2-cm-diameter water-filled PVC pipes. The targets were hollow aluminum cylinders (3.81-cm diameter and 0.32-cm wall thickness), 10.0, 14.0, and 17.8 cm long. By using cylinders of the same diameter and wall thickness, but of different lengths, the amount of energy in the echo returning to the animal could be varied without changing the echo structure, so that the echoes from each cylinder should "sound" the same to the animal.

Backscatter measurements of the targets and the clutter screen were made using a simulated dolphin echolocation signal. The measurement of the clutter screen was made with the transducer positioned so that its 3-dB beamwidth covered approximately the same area of the screen that the dolphin's beam would at a range of 6 m. Echoes from five consecutive pings with the smallest (10.0-cm) cylinder placed 10.2 cm in front of the clutter screen are shown in Fig. 18. The echoes from the cylinder are relatively similar from ping to ping. The backscatter from the clutter screen varied slightly from ping to ping. The echoes from the clutter screen were the results of a series of complex destructive and constructive interferences of scattered signals from individual balls and the frame.

Results of the dolphin's correct-detection performance as a function of the separation distance (ΔR) between the targets and the clutter screen are shown in Fig. 19. Only the target-present trials were used to generate the correct detection curves. The numbers in parentheses are the number of sessions at the various separation distances. The animal's accuracy decreased both as the separation distance decreased and as the targets got smaller. The false alarm rate was low (below 14.5%), being comparable to

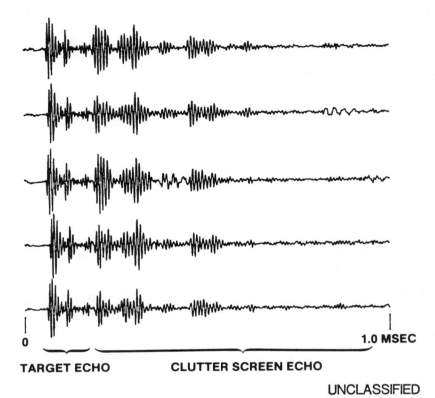

TARGET ECHO        CLUTTER SCREEN ECHO

**FIGURE 18.** Echoes from the 10-cm-long cylinder and the clutter screen, with a separation distance of 10.2 cm between the target and the clutter screen, for five consecutive pings. From Au and Turl (1983).

those obtained by Au and Snyder (1980) for a target-detection task as a function of range. This suggests that the dolphin was biased in favor of responding "target absent" and used a conservative criterion when reporting on the presence of the target.

The results for the case in which the targets were within the plane of the clutter screen ($\Delta R = 0$ cm) are shown in Fig. 20 as a function of the echo-to-reverberation ratio for the different targets. The linear least-squares lines fitted to the data indicate that the 50% detection threshold corresponded to $E/R_E$ values of 0.25 dB and $E/R_{pp}$ values of 2.5 dB. Therefore, the target echoes must be at least 2.5 dB avoe the reverberation echoes (on a peak-to-peak basis) before the animal can detect the targets. The straight lines indicate that the animal's target detection sensitivity is directly proportional to the $E/R$.

Au and Turl (1984) performed a follow-up experiment with the clutter screen in which the response bias of the dolphin was manipulated by

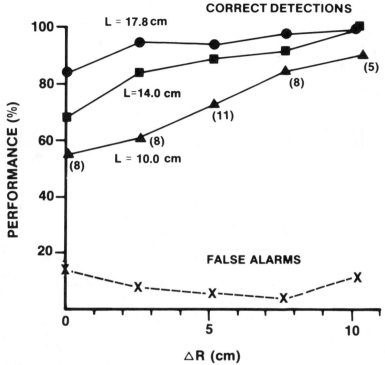

**FIGURE 19.** Dolphin's performance as a function of the clutter screen separation distance $\Delta R$, for the three targets. The numbers in parentheses are the number of sessions conducted at the different separation distances. From Au and Turl (1983).

varying the symmetry of the payoff matrix. The payoff matrix, a ratio of the number of pieces of fish for correct detections versus correct rejections, was varied in the following manner: 1:1, 1:4, 1:1, 4:1, 1:1, and 8:1. The targets were placed in the plane of the clutter screen ($\Delta R = 0$ cm). The results are plotted in an ROC (receiving–operating-characteristic) format in Fig. 21, with the ordinate representing the probability of detection and the abscissa the probability of false alarm. Also included are the ideal isosensitivity curves that best matched the dolphin's performance, excluding the results for the two smaller targets at the 8:1 payoff.

Changes in the animal's performance with changes in the payoff matrix were relatively systematic and predictable. As the payoff matrix increased from 1:1 to 4:1 and 8:1, the dolphin became progressively more liberal in reporting on the presence of a target, with a subsequent increase in the false-alarm rate. With the largest target, the animal became strongly biased towards the target-present response as the payoff matrix shifted to 4:1 and 8:1. However, its detection sensitivity remained relatively constant as its

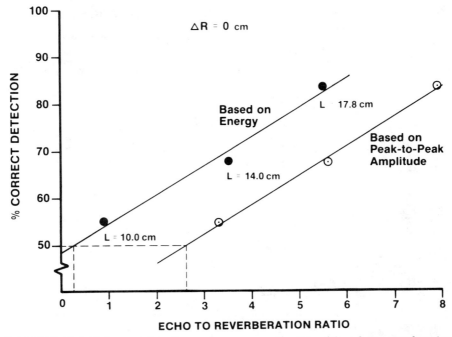

**FIGURE 20.** Dolphin's target-detection performance as a function of the echo-to-reverberation ratio for a zero separation distance (targets were within the plane of the clutter screen). The echo-to-reverberation ratios based on energy are indicated by closed circles and those based on peak-to-peak amplitudes are indicated by open circles. $L$ is the target length. From Au and Turl (1983).

response bias varied. With the two smaller targets, the dolphin shifted from being conservative to being relatively unbiased as the payoff matrix shifted from $1:1$ to $4:1$ and $8:1$. The animal's sensitivity also remained relatively constant except at the $8:1$ payoff. Very little difference in the animal's performance occurred when the payoff shifted from $1:1$ to $1:4$. The animal was already conservative at the $1:1$ payoff so that the shift to the $1:4$ payoff did not induce it to become more conservative.

## D. Target Recognition and Discrimination

Evans and Powell (1967) were first to demonstrate that an echolocating *Tursiops* could discriminate the thickness and material composition of metallic plates. Aluminum, copper, and brass circular disks of varying wall thickness and a diameter of 30 cm were used as targets. The blindfolded dolphin was required to discriminate the 0.22-cm-thick copper standard from a comparison target. Both targets were presented simultaneously in the same trial using a two-alternative, forced-choice technique and the

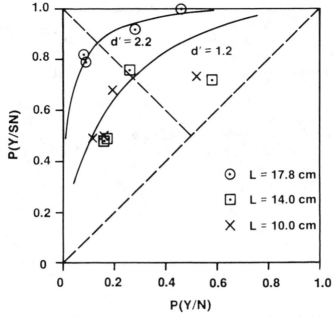

**FIGURE 21.** Dolphin variable payoff results. Points toward the left are for the 1:4 and 1:1 payoff matrices. Points on the right are for the 8:1 payoff condition. From Au and Turl (1984).

psychophysical method of constant stimuli. The dolphin was able to discriminate aluminum disks of 0.32, 0.64, and 0.79 cm thickness from the copper standard at a performance level greater than 95% correct. Brass disks of 0.64 and 0.98 cm thickness were discriminated from the copper standard with 100% correct performance. However, the animal was unable to discriminate the 0.32-cm-thick brass from the copper standard. The dolphin was also unable to discriminate the 0.16- and 0.27-cm-thick copper disks from the standard, but could discriminate the 0.32- and 0.64-cm copper disks at a 75 and 90% levels, respectively. According to Evans (1973) the experiment was replicated with another *Tursiops* and a Pacific white-sided dolphin (*Lagenorhychus obliquidens*) with comparable results.

Hammer and Au (1980) performed three experiments to investigate the target-recognition and discrimination capability of an echolocating *Tursiops*. Two hollow aluminum cylinders, 3.81 and 7.62 cm in diameter, and two coral rock cylinders of the same diameters, all 17.8 cm long, were used as standard targets. The coral rock targets were constructed of coral pebbles encapsulated in degassed epoxy. The targets were presented 6 and 16 m from the animal's pen. The dolphin was required to echolocate the target and respond to paddle A if it was one of the aluminum standards or paddle B if it was one of the coral rock standards. After baseline performance exceeded 95% correct with the standard targets, probe sessions were conducted to investigate the dolphin's ability to discriminate novel probe targets varying in structure and composition from the standards. All the probe targets were cylinders, 17.8 cm in length. Two probe targets were used in each probe session and only 8 of 64 trials of the session were used for probe trials, 4 for each probe target. The first experiment involved a general discrimination in which four aluminum (two solid and two hollow) and four nonmetal cylinders were used as probe targets. The dolphin reported all the probe targets as B (not A) even though several were aluminum and one of those was hollow with dimensions close to the larger aluminum standard. The results indicate that the dolphin could recognize the salient features of the standard aluminum target echoes and exclude all the other targets from this class.

The second experiment investigated the dolphin's ability to discriminate target wall thicknesses. Hollow aluminum probe targets with the same outer diameters but different wall thicknesses from the aluminum standards were used. The results showed that the dolphin could reliably discriminate wall thickness differences of 0.16 cm for the 3.81-cm o.d. cylinders and 0.32 cm for the 7.62-cm-o.d. cylinders.

In the third experiment, the dolphin's ability to discriminate material composition was tested using bronze, glass, and stainless steel probes that had the same dimensions as the aluminum standards. The results of the third experiment indicated that the dolphin could discriminate the bronze and steel cylinders from the aluminum, but classified the glass probe with the aluminum standard.

In a follow-up study, Schusterman, Kersting, and Au (1980) trained the same dolphin used by hammer and Au (1980) to discriminate between the aluminum and glass cylinders. Using a two-alternative, forced-choice response, the dolphin was required to strike paddle A when an aluminum cylinder was presented and paddle B when a glass cylinder was presented. After 30 sessions, the dolphin could perfectly discriminate the 3.61-cm aluminum and glass cylinders. However, the animal was never able to discriminate between the 7.62-cm aluminum and glass cylinders.

Au, Schusterman, and Kersting (1980) conducted an experiment to determine if an echolocating dolphin could discriminate between foam spheres and cylinders located 6 m from a hoop station. Three spheres and five cylinders of varying sizes but overlapping target strength (Table 1) were used so that target strength differences would not be a cue. Two spheres and two cylinders were used in each 64-trial session, with one target presented per trial. The dolphin was required to station in a hoop to echolocate.

Results of the sphere–cylinder discrimination experiment are shown in Fig. 22. The dolphin was able to discriminate between spheres and cylinders with an accuracy of at least 94% correct. Au et al. (1980)

**TABLE 1. Dimension-Measured Target Strength of Foam Targets Used in the Sphere (S) versus Cylinder (C) Discrimination Experiment**

| Target | Diameter (cm) | Length (cm) | Target Strength (dB) |
|---|---|---|---|
| $S_1$ | 10.2 | — | −32.1 |
| $S_2$ | 12.7 | — | −31.2 |
| $S_3$ | 15.2 | — | −28.7 |
| $C_1$ | 0.9 | 4.9 | −31.4 |
| $C_2$ | 0.5 | 3.8 | −32.3 |
| $C_3$ | 2.5 | 5.1 | −28.7 |
| $C_4$ | 3.8 | 3.8 | −30.1 |
| $C_5$ | 3.8 | 5.1 | −27.6 |

mistakenly postulated that the major cue was the larger surface-reflected component in the echoes from the spheres. However, when a horsehair mat was introduced in a session to absorb the surface-reflected component of the target echoes, the dolphin still performed the task perfectly. Therefore, the dolphin probably performed the target discrimination based on differences in the echoes. Target strength for a finite-length foam cylinder increases logarithmically with frequency and is constant with frequency for a sphere (Urick, 1967).

Nachtigall, Murchison, and Au (1980) conducted an experiment requiring a blindfolded echolocating *Tursiops* to discriminate between the shapes

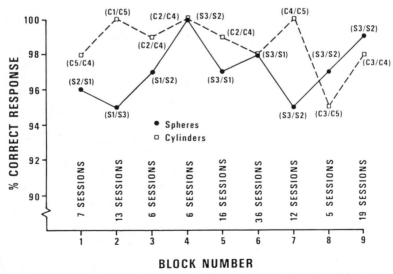

**FIGURE 22.** Results of sphere versus cylinder discrimination. From Au, Schusterman, and Kersting (1980).

of foam cylinders and cubes. The animal was trained to station at the entrance of a water-filled trough which extended from a fiberglass circular tank. The targets were presented 2 m from the animal, 40 cm below the surface, and 38 cm apart. Three different-sized cylinders were repeatedly paired with each of three different sized cubes. Once the ability of the animal to discriminate cylinders from cubes was well established, a probe technique was used to examine the effects of changing target aspect. Baseline performance trials were conducted on 56 of the 63 trials per session, but on the other seven trials one of the targets was presented either rotated or laid down horizontally. The results of the experiment, shown in Fig. 23, indicate that two of the probe orientations did not affect the animal's ability to discriminate the targets. However, the animal could not discriminate the targets when the probes were in the flat-face forward orientation. The dolphin most likely received echoes varying in amplitude when scanning across the flat surfaces of the cubes or the tops of the cylinders and received relatively uniform amplitude echoes when scanning across the curved portion of the cylinders.

## III DISCUSSION AND CONCLUSIONS

The echolocation capabilities of dolphins provide them with an invaluable tool for survival in an underwater environment. Their sonar can be used for the detection of both prey and predator, selection of desirable or

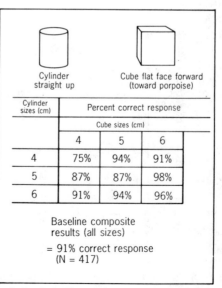

**FIGURE 23.** Results of cylinder versus cube discrimination. The performance results are in terms of percent correct choice of the cylinders. From Nachtigall, Murchison, and Au (1980).

specific prey, obstacle avoidance, and underwater navigation. For example, the detection range for fish can be calculated by considering the target strength of fish and the target detection data presented in Fig. 22. The maximum side-aspect target strength of a single fish is given by Love (1971) as

$$TS_s = 22.8 \log L - 2.8 \log \lambda - 22.1 \tag{7}$$

where $L$ is the length of the fish and $\lambda$ is the wavelength of the acoustic signal are in meters. Love also gives the dorsal-aspect target strength of an individual fish as

$$TS_d = 19.1 \log L + 0.9 \log \lambda - 23.9 \tag{8}$$

If we assume a length of 15.2 cm and a signal frequency of 120 kHz, we have from Eqs. 7 and 8, $TS_s = -35$ dB and $TS_d = -41$ dB. A dolphin in Kaneohe Bay should be able to detect this single fish at a range of approximately 73 m for the dorsal aspect and 87 m for the side aspect. In a quieter environment, the detection range will be longer. If we consider a school of fish, the echo amplitude will increase by a factor of approximately $\sqrt{N}$, where $N$ is the number of fish in the school, and the detection range will increase considerably. If we consider a predator shark 4 m in length, the side-aspect target strength will be approximately $-3$ dB, and the detection range will be approximately 250 m in Kaneohe Bay.

Considerable progress has been made in understanding dolphin echolocation, yet there are many inherent difficulties associated with this area of research. Limited communications between subjects and experimenters, a common problem in all animal research, make feedback from subjects difficult to obtain. Most experiments performed with the dolphin have been limited to binary responses. Therefore, experiments must be carefully designed so that the experimenter can infer from the subject's binary responses how the subject is being affected or is perceiving a stimulus. Of course, there is the possibility that the animal may be answering a different question than what the experimenter thinks is being asked. For instance, the Hammer and Au (1980) experiment was designed to have the dolphin indicate whether probe targets were more similar to either the A or B standards. Instead, the dolphin seemed to perceive the task as a "Not A" one, responding to paddle B for all targets that it could differentiate from the A standards.

Performing research in an aquatic environment presents many difficulties. Special considerations must be given to designing acoustically noninterfering animal stationing and stimulus mounting devices. Extreme precautions must be taken to ensure that probes used in electrophysiological measurements are not shorted out by conducting saltwater. The aquatic environment also makes monitoring the position and performing acoustic measurements on a free-swimming dolphin in its natural habitat difficult. Therefore, our knowledge of the natural history and echolocation behavior of dolphins in open water is limited, and studies are constrained by the environment.

Electrophysiological measurements are typically made using noninvasive techniques. Few invasive electrophysiological experiments have been conducted in the United States. The cost of procuring and maintaining dolphins and the politically sensitive issue concerning terminal experiments with dolphins have worked against invasive experiments. Most recent experiments have involved noninvasive cortical evoked potential measurements (Ridgway et al., 1980). Such is not the case for bat echolocation research in which considerable information has been obtained from electrophysiological measurements within the inner ear (Neuweiler, 1980) and cerebral cortex (Suga & O'Neill, 1980) of subjects. Although significant knowledge and understanding can be gained from electrophysiological experiments, they cannot be a substitute for controlled acoustic behavioral experiments.

The limited availability of subjects and the high expenses of maintaining and training marine mammals make broad-based dolphin research difficult to perform. Only a few researchers are fortunate enough to have both the proper facilities and the animals to conduct meaningful research.

Difficulties associated with dolphin echolocation and auditory research have been surmounted by careful experimental design and application of psychological testing procedures. Research clearly indicates that the dolphins possess keen echolocation capabilities and may be the "premier" echolocators, even if man-made sonar is considered. Dolphins can successfully echolocate in environments that are difficult—if not impossible—for man-made sonar (e.g., noisy and shallow water or under-ice environments that are full of clutter). They can discriminate and recognize target features such as structure, material composition, and shape.

The dolphin's unique echolocation capabilities can be attributed to the properties of the signals used, the dolphin's acoustic perception capabilities, and its capabilities to process acoustic information. The dolphin can transmit high intensity (up to 230 dB re 1 $\mu$Pa), broadband (30 to 40 kHz bandwidth) signals over a large range of peak frequencies (30 to 120 kHz). The signals are transmitted in a directional beam that allows the animal to localize and scan across objects of interest. Dolphins can hear over a wide frequency range (0.1 to 150 kHz) and detect low-intensity sounds (down to 40 dB re 1 $\mu$Pa). They can discriminate fine angular differences (1 to 3°) in the direction of received sounds. They can also discriminate fine differences in sound frequency (DL/$F \sim 0.2$). Sounds are received in a directional beam and processed by an auditory filter system which may be modeled as a bank of constant-$Q$ filters. A directional beam and a filter bank system are useful in limiting the amount of interfering noise and spurious signals received.

Dolphin echolocation and auditory research is in its infancy when compared to auditory research with humans and other animals. Yet considerable progress has been made in understanding the echolocation process during the past 20 years. Auditory research with humans and other animals has been invaluable in defining parameters to measure, suggesting experimental procedures, and providing various theories of audition to consider. Echolocation research will continue at NOSC with the goals enumerated in the beginning of this paper, for we have much to learn from and about these truly remarkable animals.

## REFERENCES

Albers, V. M. (1965). *Underwater acoustics handbook* (Vol. 2). University Park: Pennsylvania State University Press.

Au, W. W. L. (1980). Echolocation signals of the Atlantic bottlenose dolphin (*Tursiops truncatus*) in open waters. In R. G. Busnel & J. F. Fish (Eds.), *Animal sonar systems* (pp. 855–858). New York: Plenum Press.

Au, W. W. L., Floyd, R. W., & Haun, J. E. (1978). Propagation of Atlantic bottlenose dolphin (*Tursiops truncatus*). *Journal of the Acoustical Society of America, 64,* 411–422.

Au, W. W. L., Floyd, R. W., Penner, R. H., & Murchison, A. E. (1974). Measurement of echolocation signals of the Atlantic bottlenose dolphin, *Turisops truncatus* Montagu, in open waters. *Journal of the Acoustical Society of America, 56,* 1280–1290.

Au, W. W. L., & Moore, P. W. B. (1984). Receiving beam patterns and directivity indices of the Atlantic bottlenose dolphin *Tursiops truncatus. Journal of the Acoustical Society of America, 75,* 255–262.

Au, W. W. L., Moore, P. W. B., & Pawloski, D. (1986). Echolocating transmitting beam of the Atlantic bottlenose dolphin. *Journal of the Acoustical Society of America, 80,* 688–691.

Au, W. W. L., & Penner, R. H. (1981). Target detection in noise by echolocating Atlantic bottlenose dolphins. *Journal of the Acoustical Society of America, 70,* 687–693.

Au, W. W. L., Penner, R. H., & Kadane, J. (1982). Acoustic behavior of echolocating Atlantic bottlenose dolphin. *Journal of the Acoustical Society of America, 71,* 1269–1275.

Au, W. W. L., Schusterman, R. J., & Kersting, D. A. (1980). Sphere-cylinder discrimination via echolocation by *Tursiops truncatus.* In R. G. Busnel & J. F. Fish (Eds.), *Animal sonar systems* (pp. 859–862). New York: Plenum Press.

Au, W. W. L., & Snyder, K. J. (1980). Long-range target detection in open waters by an echolocating Atlantic bottlenose dolphin (*Tursiops truncatus*). *Journal of the Acoustical Society of America, 68,* 1077–1084.

Au, W. W. L., & Turl, C. W. (1983). Target detection in reverberation by an echolocating Atlantic bottlenose dolphin (*Tursiops truncatus*). *Journal of the Acoustical Society of America, 73,* 1676–1681.

Au, W. W. L., & Turl, C. W. (1984). Dolphin biosonar detection in clutter: Variation in the payoff matrix. *Journal of the Acoustical Society of America, 76,* 955–957.

Bullock, T. H., Grinnel, A. D., Ikezona, D., Kaneda, E., Katsuki, Y., Nomoto, M., Sato, O., Suga, N., & Yanagisawa, K. (1968). Electrophysiological study of central auditory mechanisms in cetaceans. *Zeitschrift fuer Vergleichende Physiologie, 59,* 117–156.

Diercks, K. J., Trochta, R. T., Greenlaw, R. C. F., & Evans, W. E. (1971). Recording and analysis of dolphin echolocation signals. *Journal of the Acoustical Society of America, 49,* 1729–1732.

Dormer, K. J. (1979). Mechanism of sound production and air recycling in delphinids: Cineradiographic evidence. *Journal of the Acoustical Society of America, 65,* 229–239.

Evans, W. E. (1967). Vocalization among marine animals. In W. N. Tavolga (Ed.), *Marine bioacoustics* (Vol. 2, pp. 159–186). New York: Pergamon Press.

Evans, W. E. (1973). Echolocation by marine delphinids and one species of freshwater dolphin. *Journal of the Acoustical Society of America, 59,* 191–199.

Evans, W. E., & Powell, B. A. (1967). Discrimination of different metallic plates by an echolocating delphinid. In R. G. Busnel (Ed.), *Animal sonar systems: Biology and bionics* (pp. 363–383). Jouy-en-Jouy: Laboratoire de Physiologie Acoustique.

Griffin, D. R. (1944). Echolocation in blind men, bats and radar. *Science, 100,* 589–590.

Hammer, C. E., Jr., & Au, W. W. L. (1980). Porpoise echo-recognition: An analysis controlling target characteristics. *Journal of the Acoustical Society of America, 68,* 1285–1293.

Herman, L. M., & Arbeit, W. R. (1972). Frequency difference limens in the bottlenose dolphin: 1–70 kHz. *Journal of Auditory Research, 12,* 109–120.

Herman, L. M., & Tavolga, W. N. (1980). The communication systems of cetaceans. In L. M. Herman (Ed.), *Cetacean behavior: Mechanisms and functions* (pp. 149–209). New York: Wiley (Interscience).

Hollien, H., Hollien, P. A., Caldwell, D. K., & Caldwell, M. C. (1976). Sound production by the Atlantic bottlenose dolphin *Tursiops truncatus. Cetology, 26,* 1–7.

Jacobs, D. W. (1972). Auditory frequency discrimination in the Atlantic bottlenose dolphin, *Tursiops truncatus* Montagu: A preliminary report. *Journal of the Acoustical Society of America, 53,* 696–698.

Johnson, C. S. (1967). Sound detection thresholds in marine mammals. In W. N. Tavolga (Ed.), *Marine bio-acoustics* (pp. 247–260). New York: Pergamon.

Johnson, C. S. (1968). Masked tonal thresholds in the bottlenose porpoise. *Journal of the Acoustical Society of America, 44,* 965–967.

Kellogg, W. N. (1953). Ultrasonic hearing in the porpoise *Tursiops truncatus. Journal of Comparative Physiological Psychology, 46,* 446–450.

Kellogg, W. N., & Kohler, R. (1952). Response of the porpoise to the ultrasonic frequencies. *Science, 116,* 250–252.

Love, R. H. (1971). Dorsal-aspect target strength of an individual fish. *Journal of the Acoustical Society of America, 49,* 816–823.

Mackay, R. S., & Liaw, H. M. (1981). Dolphin vocalization mechanism. *Science*, **212,** 676–677.

McBride, A. F. (1956). Evidence for echolocation by cetaceans. *Deep-Sea Research*, **3,** 153–154.

McCormick, J. G., Weaver, E. G., Palin, J., & Ridgway, S. H. (1970). Sound conduction in the dolphin ear. *Journal of the Acoustical Society of America*, **48,** 1418–1428.

Moore, P. W. B., Hall, R. W., Friedl, W. A., & Nachtigall, P. E. (1984). The critical interval in dolphin echolocation: What is it? *Journal of the Acoustical Society of America*, **76,** 314–317.

Morris, R. J. (1986). The acoustic faculty of dolphins. In M. M. Bryden & R. Harrison (Eds.), *Research on dolphins* (pp. 369–399). London & New York: Oxford University Press (Clarendon).

Murchison, A. E. (1980). Detection range and range resolution of echolocating bottlenose porpoise (*Tursiops truncatus*). In R. G. Busnel & J. F. Fish (Eds.), *Animal sonar systems* (pp. 945–947). New York: Plenum Press.

Nachtigall, P. E., Murchison, A. E., & Au, W. W. L. (1980). Cylinder and cube shape discrimination by an echolocating blindfolded bottlenose dolphin. In R. G. Busnel & J. F. Fish (Eds.), *Animal sonar systems* (pp. 43–70). New York: Plenum Press.

Neuweiler, G. (1980). Auditory processing of echoes: Peripheral processing. In R. G. Busnel & J. F. Fish (Eds.), *Animal sonar systems* (pp. 519–548). New York: Plenum Press.

Norris, K. S. (1968). The evolution of acoustic mechanisms in odontocete cetaceans. In E. Drake (Ed.), *Evolution and environment* (pp. 297–324). New Haven, CT: Yale University Press.

Norris, K. S. (1986). Sound production in dolphins. *Marine Mammal Science*, **2,** 233–235.

Norris, K. S., Dormer, K. J., Pegg, J., & Liese, G. (1971). The mechanism of sound production and air recylcing in porpoise: A preliminary report. In *Proceedings of the Eighth Conference on Biological Sonar and Diving Mammals* (pp. 113–129). Menlo Park, CA: Stanford Research Institute.

Norris, K. S., & Harvey, G. W. (1974). Sound transmission in the porpoise head. *Journal of the Acoustical Society of America*, **56,** 659–664.

Norris, K. S., Prescott, J. H., Asa-Dorian, P. V., & Perkins, P. (1961). Experimental demonstration of echolocation behavior in the porpoise *Tursiops truncatus* (Montagu). *Biological Bulletin (Woods Hale, Massachusetts)*, **120,** 163–176.

Penner, R. H., & Turl, C. W. (1983). Bottlenose dolphin (*Tursiops truncatus*): Difference in the pattern of interpulse intervals. *Journal of the Acoustical Society of America*, **74,** S74.

Popper, A. N. (1980). Sound emission and detection by delphinids. In L. M. Herman (Ed.), *Cetacean behavior: Mechanisms and functions* (pp. 1–52). New York: Wiley.

Purves, P. E. (1967) Anatomical and experimental observations on the cetacean sonar system. In R. G. Busnel (Ed.), *Animal sonar systems: Biology and bionics* (pp. 197–270). Jouy-en-Jouy: Laboratoire de Physiologie Acoustique.

Purves, P. E., & Pilleri, G. (1983). *Echolocation in whales and dolphins*. London: Academic Press.

Renaud, D. L., & Popper, A. N. (1975). Sound localization by the bottlenose porpoise *Tursiops truncatus*. *Journal of Experimental Biology, 63*, 569–585.

Ridgway, S. H., Carder, D. A., Green, R. F., Gaunt, A. S., Gaunt, S. L. L., & Evans, W. E. (1980). Electromyographic and pressure events in the nasolaryin system of dolphins during sound production. In R. G. Busnel & J. F. Fish (Eds.), *Animal sonar systems* (pp. 239–249). New York: Plenum Press.

Schevill, W. E. (1956). Evidence for echolocation by cetaceans. *Deep-Sea Research, 3,* 153–154.

Schevill, W. D., & Lawrence, B. (1953). High-frequency auditory response of a bottlenosed porpoise *Turisops truncatus* (Montagu). *Journal of the Acoustical Society of America, 124*, 147–165.

Schusterman, R. J. (1980). Behavioral methodology in echolocation. In R. G. Busnel & J. F. Fish (Eds.), *Animal sonar system* (pp. 11–41). New York: Plenum Press.

Schusterman, R. J., Kersting, D. A., & Au, W. W. L. (1980). Stimulus control of echolocation pulses in *Tursiops truncatus*. In R. G. Busnel & J. F. Fish (Eds.), *Animal sonar systems* (pp. 983–986). New York: Plenum Press.

Shower, E. G., & Biddulph, R. (1931). Differential pitch sensitivity of the ear. *Journal of the Acoustical Society of America, 3*, 275–287.

Suga, N., & O'Neill, W. E. (1980). Auditory processing of echoes: Representation of acoustic information from the environment in the bat cerebral cortex. In R. G. Busnel & J. F. Fish (Eds.), *Animal sonar systems* (pp. 589–611). New York: Plenum Press.

Thompson, R. K. R., & Herman, L. M. (1975). Underwater frequency discrimination in the bottlenose dolphin (1–140 kHz) and the human (1–8 kHz). *Journal of the Acoustical Society of America, 57*, 943–948.

Urick, R. J. (1967). *Principal of underwater sound*. New York: McGraw-Hill.

# 11

## LOCALIZATION OF AUDITORY AND VISUAL TARGETS FOR THE INITIATION OF SACCADIC EYE MOVEMENTS

*Martha F. Jay*

*Department of Ophthalmology, Northwestern University Medical School, Chicago, Illinois*

*David L. Sparks*

*Neurobiology Research Center and Department of Physiology and Biophysics, University of Alabama, Birmingham, Alabama*

Research in this laboratory is related to the broad question of how sensory signals are translated into commands for the control of movement. An intriguing aspect of this question is how information from more than one modality is used to initiate movements. We approached this problem by studying saccadic (quick, high-velocity) eye movements directed toward either auditory or visual targets and by investigating the role of neurons in the superior colliculus in the initiation of these movements. In this paper, we present the results of behavioral studies designed to compare the metrics (accuracy, latency, velocity, etc.) of saccades directed to auditory or visual stimuli. If the sensory signals function merely to trigger common pools of motor elements into patterns of activity determined by the intrinsic properties of the motor cells, then the parameters of the movements should not depend upon the modality (auditory or visual) of the saccade target. However, if the sensory signals do more than serve as a trigger signal, then the pattern of sensory-induced activity could influence the profile of activity observed in the motor elements and, thereby, indirectly affect movement parameters. In this case, behavioral studies comparing movements to auditory and visual targets could reveal similarities and differences in the auditory and visual signals used to initiate saccades.

## BACKGROUND

In this section we review previous studies in which subjects were asked to direct saccadic eye movements to the source of auditory targets. This earlier work suggests that saccades to auditory targets differ from saccades to visual targets in latency, velocity, accuracy, duration, and the number of saccades required to look to the target. Differences in the performance of human and monkey subjects have also been noted.

### Latency

For human subjects, the latency of saccades to auditory targets is reported to be longer than the latency to visual targets (Zahn, Abel, & Dell'Osso, 1978; Zambarbieri, Schmid, Prablanc, & Magenes, 1981), but the converse seems to be true for monkey subjects (Whittington, Hepp-Reymond, & Flood, 1981). In an attempt to determine if this apparent species difference is due to differences in apparatus or test conditions, one goal of the present study was to compare the latencies of both monkey and human subjects to auditory and visual targets using the same apparatus and sensory stimuli.

The eccentricity of auditory and visual targets has been found to affect the latency of saccades to both auditory and visual targets, but in opposite directions. The latency of a saccade to a visual target presented 5° from the fixation stimulus is about 200 msec and the latency increases to 250 msec for targets presented 40° in the periphery (White, Eason, & Bartlett, 1962). In contrast, the reaction time for saccades to auditory targets 5° from the fixation target is around 380 msec but decreases to 245 msec for targets at an eccentricity of 30° (Zahn et al., 1978; Zambarbieri et al., 1981). If the decreased latency of sound-induced saccades with increased target eccentricity is because sound localization is more efficient in the periphery, all saccades to peripheral auditory targets should have short latencies, regardless of the size of the movement (i.e., regardless of the original position of the eye). Alternatively, the latency difference could be related to the programming of an appropriate movement rather than the encoding of target position. This possibility was examined in our experiments by comparing the latencies of saccades of different amplitudes directed toward a single speaker location.

### Peak Velocity

For human subjects, saccades to auditory targets have a lower velocity than saccades to visual targets (Zahn et al., 1978; Zambarbieri et al., 1981). Also, spontaneous saccades in the dark or those produced in the light, but against a homogeneous visual field, are slower than saccades to visual targets (Becker & Fuchs, 1969; Jurgens, Becker, Rieger, & Widderich, 1981). The velocity of saccades to continuously present visual targets increases as a function of saccadic amplitude for movements up to approximately 20° (Fuchs, 1967), but for saccades to remembered targets, saccadic velocity continues to increase for saccades greater than 25° in amplitude (Becker & Fuchs, 1969). In contrast, Zambarbieri et al. (1981) report that the maximum velocity of saccades to auditory targets is significantly lower than the maximum velocity observed on visual trials. These findings suggest differences in the formation of motor commands for these various types of eye movements.

### Duration

The amplitude and duration of visual saccades are linearly related. Monkeys complete a 20° saccade to a visual target in approximately 35–40 msec (Fuchs, 1967). Human subjects take twice as long for a comparable saccade (Robinson, 1964). The duration/amplitude function for acoustically induced saccades, studied only in humans, is also a linear function but has a steeper slope than that observed for visual saccades (Zambarbieri et al., 1981).

### Saccadic Trajectory

Oblique saccades (having both horizontal and vertical components) are usually straight, rather than curved. Often, this requires the duration of the smaller component to be prolonged (Evinger, Kaneko, & Fuchs, 1981; Guitton & Mandl, 1980). Guitton and Mandl (1980) reported that this linkage is not strictly followed for spontaneous movements in the dark or saccades to auditory stimuli.

### Accuracy

The accuracy of saccades to sounds has not been studied in detail. Brown, Schessler, Moody, and Stebbins (1982) measured the minimum audible angle for localization in the vertical and horizontal planes. They found that the acuity of vertical localization depended upon the high-frequency content of the signal. With broad bandwidth signals, localization in the vertical plane was nearly as accurate as localization in the horizontal plane. Humans have been found to be superior to monkeys in detecting interaural time (Houben & Gourevitch, 1979; Zwislocki & Feldman, 1956) and intensity differences (Houben & Gourevitch, 1979; Klumpp & Eady, 1956; Mills, 1960). Since interaural time and intensity differences are cues for localization in the horizontal plane, monkeys should be less precise in their judgments of azimuth than humans. Monkeys have a higher upper frequency limit than humans, 40 versus 20 kHz, but this increased sensitivity to high frequencies yields no advantage in detecting interaural intensity differences, and is unlikely to confer an advantage in determining elevations (Stebbins, 1973, 1975; Stebbins, Green, & Miller, 1966).

## Multiple Saccades

Although on a small percentage of trials a small, short-latency corrective movement is required for foveal acquisition of the target, visual targets are usually acquired with a single saccade. In contrast, the acquisition of an auditory target usually requires multiple saccades (Zambarieri et al., 1981). The intervals between the onset of the multiple saccades are clustered in two groups: one with intersaccadic delays shorter than the initial saccade latency, and a second group with intervals approximately equal to the latency of the first saccade. Short intersaccadic intervals have been interpreted as evidence that the second movement is preprogrammed as a unit with the first (Becker, 1976; Zambarbieri et al., 1981). However, if increases in double saccades to auditory targets reflect less precise programming of auditory saccades instead of a strategy of preprogramming two or more successive saccades, then most intersaccadic delays should fall into the longer latency range.

## METHODS

### Subjects

Three rhesus monkeys and three human laboratory volunteers served as subjects.

### Head Restraint

During the experimental sessions, a crown attached to bolts inserted into the skull was used to stabilize the heads of the monkeys. Head stabilization of human subjects was attained via individually formed bite bars.

### Recording of Eye Position

For the monkeys, three loops of fine-gauge, insulated wire were sutured to the sclera of one eye (Judge, Richmond, & Chu, 1980). A modified search coil technique was used for measuring human eye position (Collewijn, van der Mark, & Jansen, 1975). One eye was irrigated with local anesthetic (0.5% Ophthaine) and an annulus contact lens (Skalar Medical) containing coils of copper wire was inserted. The lens was only comfortable for a short time so recording sessions were limited to 30 min. During training (monkeys) and data collection sessions (humans and monkeys), subjects were seated within two alternating magnetic fields in spatial and phase quadrature. Signals generated by movement of the eye coil within the magnetic fields were phase detected to produce voltages proportional to horizontal and vertical eye position with a sensitivity of at least 0.25° (Fuchs & Robinson, 1966).

## Auditory and Visual Targets

A green light-emitting diode (LED) served as the visual stimulus and a 20- to 20,000-Hz wide-band, white-noise burst was the auditory stimulus. The sound intensity was 80–90 dB SPL and was projected from a 7.62-cm speaker (Realistic, 40-1381) into a room lined with either heavy drapes or acoustic foam (Sonex, Illbruck).

As illustrated in Fig. 1, subjects were placed in a darkened room, with their heads fixed in the center of a 72-inch-diameter semicircular track that pivoted at both ends (Cal Tech Central Engineering Services). A speaker was attached to the track and a LED was placed in the center of the speaker. Computer-controlled stepping motors moved the speaker along the track to produce changes in azimuth and rotated the hoop to alter the elevation of the stimulus. Three additional LEDs were mounted 24° apart

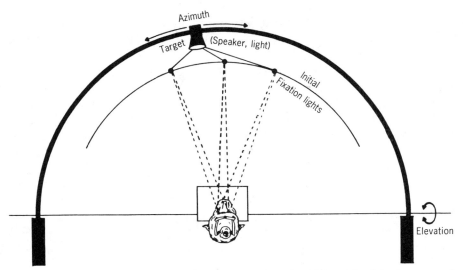

**FIGURE 1.** Experimental apparatus. The subject was seated, with head fixed, in the center of a semicircular track that pivoted on both ends. A speaker with a light-emitting diode attached at the center was mounted on the track and could be moved by computer-controlled stepping motors to produce changes in the elevation or azimuth of the auditory or visual targets. Three initial fixation lights were placed 24° apart in the horizontal plane. From Jay and Sparks (1987a).

in front of the subject and served as fixation points at the beginning of each trial.

A LSI-11/03 laboratory computer (Digital Equipment) was used to move the targets to specified locations, initiate the trials, reward the monkeys for correct eye movements, and store eye movement data in digital form on magnetic tape (3-msec sampling rate). All subsequent data analysis was performed using a second laboratory computer (LSI-11/73).

## Behavioral Tasks

Each trial began with the activation of one of the three fixation lights. If, after acquiring the fixation target, the subject maintained fixation for a variable interval (0.5–2.0 sec), the fixation light was extinguished and simultaneously a peripheral auditory or visual target was presented. The monkey received a liquid reward (0.1 mL orange juice) if it looked toward the target within 700 msec of its onset. A square 16° window of acceptable eye positions was allowed around the auditory targets and a 10° window was used for visual targets and fixation points. Using these criteria, animals were rewarded for most saccades directed toward the targets but were not rewarded for random movements (usually in the wrong quadrant) away from the fixation target. Human subjects were instructed to make fast, accurate eye movements to the targets; no rewards were provided.

Monkeys were trained to make saccades to visual targets. After this task was mastered, auditory targets were gradually introduced. Initially, if the subject did not acquire the auditory target within 700 msec, a LED was illuminated near the center of the speaker, permitting visually induced corrective saccades. This second visual signal was gradually eliminated as the subjects became proficient in making saccades to sounds.

Three different sets of target locations were used on separate days. In series A, auditory and visual targets were randomly presented along the horizontal plane (±30° in 10° increments). Both human and monkey subjects were tested with this paradigm, designed to provide target arrays that were similar to those used in previously published studies.

In series B, both the azimuth (±30° from a point directly in front of the subject) and the elevation (±20°) of targets were varied in 10° increments. Due to the large number of target positions, only monkey subjects were tested with this task. The purpose was to contrast auditory and visual saccades to targets off the horizontal plane, where auditory sound localization ability could be diminished.

With series C, the horizontal target position was varied between 24° left, center, and 24° right; the elevation was either ±20° or ±10°. In addition, the initial fixation point was varied between 24° left, center, and 24° right. Target series C, used with both human and monkey subjects, was designed to separate the effects of the position of a target relative to the head and the eye movement required to look to the target upon response latency.

## Data Analysis

To compensate for the nonlinearity of the eye coil system, a correction was made of every horizontal and vertical eye position sample (Sullivan & Kertesz, 1979). The data were then analyzed using a program that automatically measured saccades. Most saccades could be reliably framed by defining saccade onset as the sample in which eye velocity exceeded 35°/sec and saccade offset as the first sample when the velocity fell below 25°/sec. All saccade definitions were visually scanned and, if necessary, the velocity criteria were altered to bracket unusually slow eye movements.

Figure 2 illustrates the pattern of stimulus and response events occurring on a typical trial and describes the operational definition of the various dependent variables. An analysis of variance (ANOVA) for unequal groups was used in experiments A, B, and C to investigate the effects of target position, modality, and fixation direction upon saccade latency.

## RESULTS

### Latency

When targets were confined to 6 locations along the azimuth (target series A), the latencies of eye movements to auditory and visual stimuli differed for most of the human and monkey subjects. Little difference was observed on auditory and visual trials for one monkey (Blanche) and one human subject (DLS). For the other human subjects, the latencies of saccades to auditory targets (filled symbols) were generally shorter than saccade latencies to visual targets (open symbols). The opposite was true for two of the three monkeys (Fig. 3). For all subjects, the saccades with the longest latencies were those to auditory targets placed ±10° from the primary position. Response latency decreased for auditory trials and increased for visual trials with increasing target eccentricity. For human subjects, 10° saccades to sounds had longer latencies than those to lights, but auditory latencies were generally shorter than visual ones for larger eye movements. For two of the three monkeys, auditory latencies were longer than the visual ones for all target positions.

**Latency: Elevation Effects.** When 34 targets varying both in elevation and azimuth (target series B) were presented, the difference between latencies of saccades to auditory and visual targets decreased as target elevation increased (Fig. 4). In one monkey (Bubba), the intermodality effect was

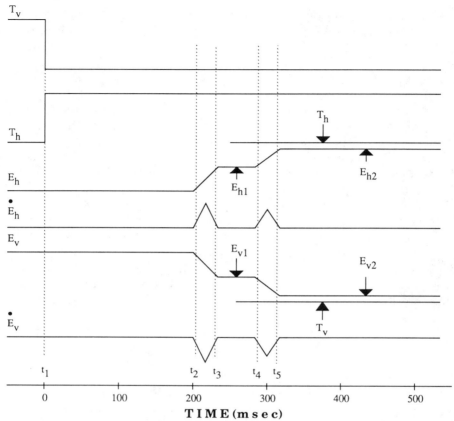

**FIGURE 2.** Schematic diagram illustrating stimulus and response events on a single trial and the definition of saccade parameters. Primary and secondary saccades are shown. $T_v$, Vertical target position; $T_h$, horizontal target position; $E_h$, horizontal eye position; $E_{h1}$, Horizontal eye position at the end of the first saccade; $E_{h2}$, horizontal eye position at the end of the second saccade; $\dot{E}_h$, horizontal eye velocity; $E_v$, vertical eye position; $E_{v1}$, vertical eye position at the end of the first saccade; $E_{v2}$, vertical eye position at the end of the second saccade; $\dot{E}_v$, vertical eye velocity; $t_1$, time at onset of eccentric target; $t_2$, the primary saccade exceeds the velocity threshold used to define saccade onset; $t_3$, offset of the primary saccade; $t_4$, time at onset of the secondary saccade; $t_5$, time at offset of the secondary saccade. Saccade latency is defined as $(t_2 - t_1)$. Intersaccadic interval is defined as $(t_4 - t_3)$. Error is defined as $\sqrt{(T_h - E_{h2})^2 + (T_v - E_{v2})^2}$.

statistically significant ($p < 0.05$) for all target positions except for those presented at $\pm20°$ and $\pm30°$ along the azimuth with an elevation of $\pm20°$. With Stella, all target combinations produced a significant effect, yet this latency difference between the auditory and visual targets decreased as elevation was increased. For all monkeys, the longest latencies were for auditory targets placed 10° above or below the primary position. Auditory targets placed 10° left or right in the horizontal plane resulted in the next

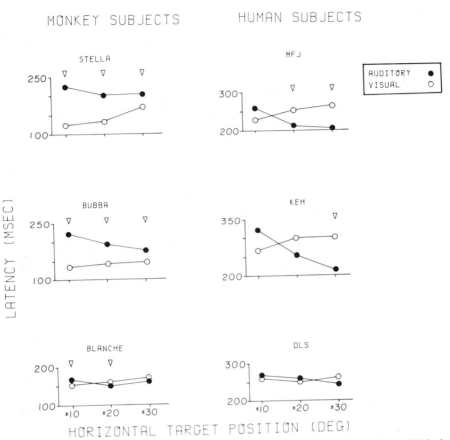

EXPERIMENT A

MONKEY SUBJECTS          HUMAN SUBJECTS

**FIGURE 3.** Auditory and visual saccadic latencies for human and monkey subjects. With the monkeys (left column), the latency of eye movements to sounds was longer than that to visual targets. The opposite relation was found for human subjects (right column). All data shown are from experiment A (exclusively horizontal targets). ●, Auditory trials; ○, visual trials. Statistically significant intermodality differences are indicated by open triangles at the top of the panel (ANOVA, $p < 0.05$).

highest latencies. The eye movement required to look toward these positions were the smallest of the target array, a topic that was explored further with target series C. When the latencies to horizontal targets were compared with those obtained with experiment A, the values were almost identical (Fig. 4, dashed lines).

**Latency: Effects of Saccade Size.** By varying the initial fixation point between straight ahead, 24° left of center, and 24° right of center, saccades of different sizes and directions could be generated to the same auditory

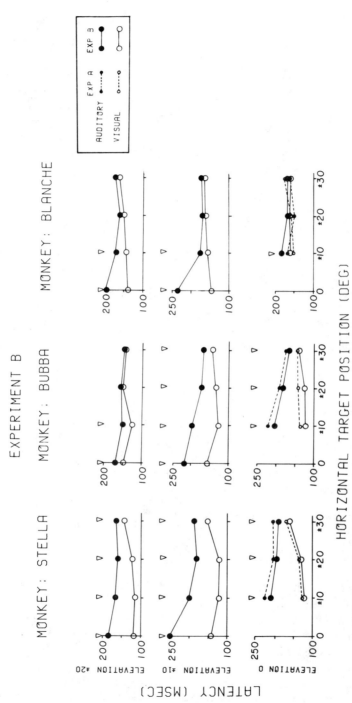

**FIGURE 4.** Saccadic latencies when targets varied in both azimuth and elevation. Data from monkeys during target series B. Azimuth is shown along the abscissa and the three target elevations are illustrated in separate rows. Latency values reproduced from experiment A in the bottom row (---). Statistically significant differences between auditory and visual trials indicated by open triangles at the top of each panel ($p < 0.05$).

target. For the two monkeys and three humans tested with series C, the point of initial fixation significantly affected auditory latencies (Fig. 5, monkey subjects). The longest latencies were for straight vertical movements to auditory targets, regardless of the initial fixation direction. This effect was most pronounced for targets 10° above or below the initial fixation position but was still evident for targets with an elevation of ±20°.

For example, with the monkey Stella, when the initial fixation direction was 24° left of center (Fig. 5, open circles), the longest auditory latencies were to speaker locations 10° above or below that point (horizontal target position −24°, target elevation ±10° and ±20°). When saccades to that

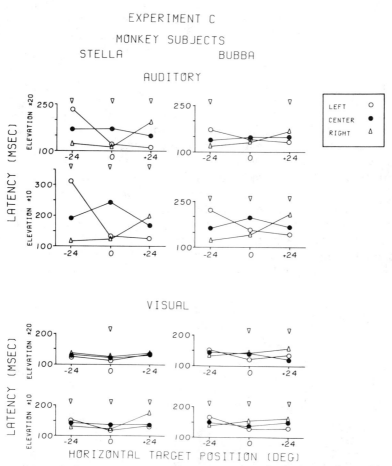

**FIGURE 5.** Saccadic latencies varying with movement amplitude for monkey subjects. For auditory trials (top), the largest response latencies were for target positions closest to the fixation point. This effect was not seen on visual trials (bottom). ○, Left fixation trials; ●, central fixation trials; and △, right fixation trials. Statistically significant intermodality differences are indicated by open triangles at the top of each panel ($p < 0.05$).

same speaker location were initiated from the right fixation point (triangles), the mean latency was reduced by over 200 msec. With the speaker placed 10° above or below the primary position (horizontal target position 0°, elevation ±10°), the longest latencies were for saccades initiated from straight ahead (filled circles). The auditory latency to this same speaker location was considerably reduced when initial gaze was directed toward either the left (open circles) or right (open triangles) fixation light. When the speaker was 10° above or below the right fixation point, the longest latencies were for eye movements initiated from the fixation point that was closest to the target (open triangles). The latencies were reduced for central fixation and were even shorter when the left fixation light was in use and the target remained 24° to the right of center. While there was some variability between subjects, this fixation effect was seen in all cases. The larger the eye movement, the shorter the latency of the auditory response regardless of the position of the auditory target relative to the head.

## Multiple Saccades

In about 20% of all auditory trials, the subjects used multiple saccades to acquire the target (Fig. 6). The monkeys employed more multiple saccades when the targets varied in both azimuth and elevation (series B) than when just horizontal positions were employed (series A). In contrast to the high percentage of multiple saccades on auditory trials, the frequency was less than 7% on visual trials for the monkeys. The human subjects did not use multiple saccades to acquire visual targets.

For multiple saccade trials to auditory targets, a consistent finding among all subjects was that the amplitude of the first saccade was approximately equal to the second (Fig. 7). On visual trials, where only the monkeys made multiple saccades, the second eye movements were

PERCENT MULTIPLE SACCADES

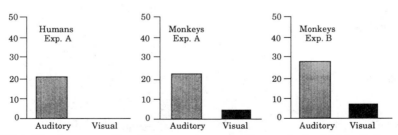

**FIGURE 6.** The frequency of multiple saccades on auditory and visual trials. Left column, human subjects during target series A; center column, monkey subjects; right column, monkey subjects during target series B. Auditory trials are shown with open bars and visual trials are those with filled bars.

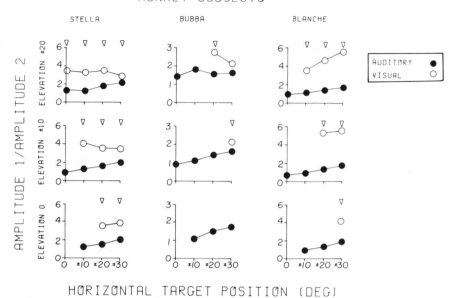

FIGURE 7. The relative amplitudes of first and second saccades for monkeys during experiment B. ●, Auditory trials; ○, visual trials. Horizontal target position is shown along the abscissa and elevation is represented separately in the rows. Open triangles indicate statistically significant intermodality differences ($p < 0.05$).

considerably smaller than the initial ones. The corrective, or second saccades, on visual trials were only one-third to one-fourth the size of the initial saccades. Of the target positions where multiple saccades on both auditory and visual trials were observed, the ratio between the amplitude of the first and second saccades was significantly greater for visual than auditory trials in most cases (Fig. 7).

While the corrective saccades on auditory trials were large compared to those on visual trials, the intersaccadic interval was shorter when auditory targets were presented (Fig. 8). The interval on auditory trials averaged approximately 200 msec for monkeys and was only 75 msec for the human subjects. For target positions where multiple saccades were initiated to both auditory and visual stimuli (seen only with monkeys), the intersaccadic interval was significantly shorter for auditory than visual trials for 3 out of 6 target positions in experiment A and for 14 out of 18 positions in experiment B ($p < 0.05$). No consistent trends were noted for the interval between successive saccades as a function of target eccentricity or elevation.

EXPERIMENT B

MONKEY SUBJECTS

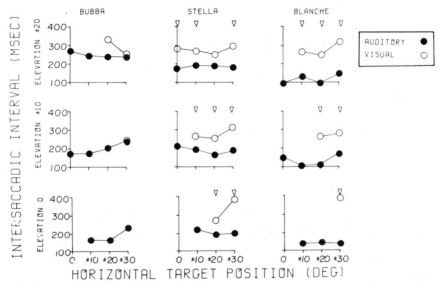

**FIGURE 8.** Intersaccadic intervals for experiment B. All data from monkey subjects. ●, Auditory trials; ○, visual trials. Statistically significant intermodality differences are indicated by open triangles at the top of each panel ($p < 0.05$).

### Velocity and Duration

For both monkeys and humans, the peak velocities of saccades to sounds were lower than those to lights. This difference increased with saccade size (Fig. 9). Velocity saturation was observed for both auditory and visual saccades. For the human subjects, the difference in saturation velocity between auditory and visually initiated saccades was roughly 30–70°/sec. This difference was much greater for the monkeys (130–350°/sec). The peak velocity for visual saccades was 300–400°/sec for humans and 700–1000°/sec for the monkeys. As expected from the velocity/amplitude relation, the duration of saccades to auditory targets was greater than the duration of amplitude-matched saccades to visual targets for all subjects.

### Accuracy

For all target positions, the final error of saccades to auditory targets was greater than that to visual ones. Accuracy was tabulated as the total (horizontal and vertical) difference between the final eye position on a trial and the actual target position. For trials where multiple saccades occurred,

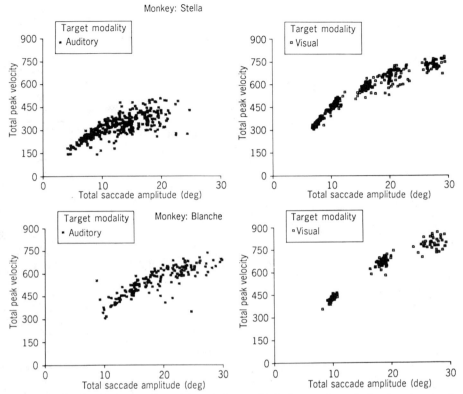

**FIGURE 9.** Peak velocity varying with saccade amplitude for monkeys during target series A trials. Left, auditory trials; right, visual trials.

accuracy determinations were made after the second saccade. With horizontal targets (series A), auditory trials were 3.2° less accurate than visual ones for the monkeys (Fig. 10) and this difference was 2.4° for the human subjects (Fig. 11). While the total error for auditory targets was slightly higher for monkeys than humans, 5.8° versus 4.6°, most of this increase was due to vertical inaccuracies on the part of the monkeys. The horizontal auditory error was 3.7° for monkeys and 3.6° for humans, while the vertical auditory error was 3.8° and 2.2°, respectively. Horizontal error generally increased with target eccentricity for both modalities. The same was true for vertical error when visual targets were presented. In contrast, vertical error for auditory trials appeared to decrease or remain about the same as the speaker was moved further out along the azimuth.

When the error values were compared for the monkeys looking to targets varying only in azimuth (series A, 6 positions) and on trials in which both elevation and azimuth were varied (series B, 34 positions), no

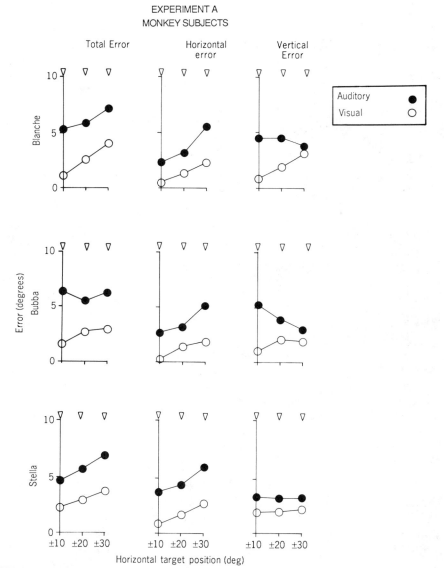

**FIGURE 10.** Saccadic error for monkey subjects during experiment A. Left column, the total distance between the end position of each saccade and the actual target position; center column, the horizontal error for each target position; right column, vertical error. ●, Auditory trials; ○, visual trials. Statistically significant intermodality differences are indicated by open triangles at the top of each panel.

change in accuracy was noted. The total auditory error for experiment B was 5.7°, compared to 5.8° in the horizontal experiment. Similarly, horizontal error was 3.8° in B and 3.7° in A for auditory trials. Vertical auditory error was also little affected by increasing the number of target positions (3.5° in B and 4° in A).

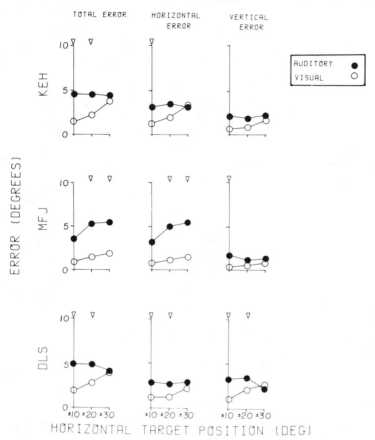

**FIGURE 11.** Saccadic error for human subjects during target series A. Columns, rows, and symbols are as described in Fig. 8.

## DISCUSSION

### Comparisons with Previous Studies

We examined the characteristics of saccades to targets varying in both azimuth and elevation and studied the effect of varying initial eye position, a condition that requires saccades of different amplitudes and directions to targets at fixed positions in space. Previous studies measured saccades to auditory targets that varied only in azimuth. Our data, from monkey and human subjects, confirmed most of the findings previously reported for human subjects (Zahn et al., 1978; Zambarbieri et al., 1981): (1) For a given amplitude movement, saccades to auditory targets have lower peak velocities and longer durations than saccades to visual targets; (2) auditory saccades are slightly less accurate than visual saccades; and (3) the latency of auditory saccades decreases as the eccentricity of the target increases, while the latency of saccades to visual targets increases as a function of eccentricity.

We found that saccades to auditory targets varying in elevation were almost as accurate as saccades to targets varying in azimuth, despite differences in the sensory signals used to localize targets in azimuth and elevation. The behavioral task used in the present experiments (generating a saccade toward the location of acoustic stimuli) is quite different from tasks used in psychophysical studies of sound localization. Nonetheless, our results are consistent with findings of Brown and colleagues (1982) who measured the minimum audible angle for localization in the vertical and horizontal planes for macaque monkeys. With broad bandwidth stimuli, the acuity of vertical localization was quite good: The minimum audible angle was less than 4°, a value that approximates the maximum sensitivity found in the horizontal plane (Brown, Beecher, Moody, & Stebbins, 1980). Our data are consistent with their suggestion that localization of broad bandwidth signals in the vertical plane is nearly as accurate as localization in the horizontal plane.

Contrary to previous reports, we found that for human subjects the latency of saccades to eccentric auditory targets was less than the latency of large saccades to visual targets. As discussed below, this was not true for saccades to targets near the fixation point. Procedural differences could account for this discrepancy. In one previous study (Zambarbieri et al., 1981) subjects were told to fixate as accurately as possible, a condition likely to yield longer latencies than when instructions are to look as quickly as possible toward the saccade target. In the other experiment (Zahn et al., 1978), subjects were instructed to look as quickly and accurately as possible—instructions similar to those given human subjects in our experiments. However, in the Zahn et al. study, the fixation light was continuously present. In our experiments, the fixation light was extinguished when an eccentric target appeared. Since recent experiments (Fischer & Boch, 1983) indicate that saccadic reaction times are significantly reduced if the fixation target is removed before the onset of a saccade target, the continued presence of the fixation stimulus could produce longer latencies.

The only previous study of saccades to auditory targets using monkeys as subjects (Whittington et al., 1981) found that eye movements to auditory targets had shorter latencies than eye movements to visual targets. We obtained the opposite result. A major procedural difference in the two studies is that eye position at the start of each trial was not constrained in the Whittington study. Since we found that the latency of saccades was more dependent upon the amplitude of the movement required to look to the target than upon the location of the target in space, the lack of constraints upon initial eye position could account for the latency differences. Also, it should be noted that, perhaps due to extensive training, the latencies of visually guided saccades of our monkey subjects were unusually short.

Studies using both humans and monkeys as subjects consistently report that the latencies of auditory saccades decrease as target eccentricity increases, while the latencies of visual saccades increase as a function of eccentricity. Based upon previous studies, the decrease in response latency for more eccentric auditory targets is not due to differences in either detection time or delays associated with attentional shifts. Detection of auditory stimuli, as indicated by a manual key release response, is independent of target location (Zambarbieri et al., 1981). In addition, the time taken to shift auditory attention increases, rather than decreases, as a linear function of the angular target position for distances up to 90° (Rhodes, 1987).

Two possible explanations for the decrease in auditory saccade latency as a function of target eccentricity are considered. Since these explanations are not incompatible, both could contribute to the observed latency/eccentricity relation. The first is based upon observations of Brown and colleagues (1982). In a study of the minimum audible angle, they noted that localization acuity was reduced at lateral test positions and that as the test position was displaced from the midline, monkeys were more likely to guess. We observed that the percentage of trials with more than one saccade increased with target eccentricity. Thus, it is possible that the initial short latency response to eccentric targets is a guess, based upon inaccurate or incomplete localization information, and must be corrected subsequently by one or more saccades.

A second possible explanation for the latency data is that the time required to generate motor command signals for saccadic eye movements depends upon the amplitude of the movement. In experiment C, eye movements of different amplitudes were made to targets at the same spatial location. If the decrease in latency for movements to eccentric targets were due to the shorter processing time required to encode the position of eccentric stimuli, then changes in initial fixation position should have little effect upon response latency. But we found that initial eye position did influence saccadic latency. Small saccades to auditory targets took longer to be initiated than larger ones, regardless of the spatial position of the target. Similar results were obtained by Perrot, Ambarsoom, and Tucker (1987) in studies measuring the latency and accuracy of head movements used to localize auditory targets. The mean latency for sources located 60° from the initial head position was 371 msec but increased to 453 msec if the auditory source was within 10° of the starting position.

## Comparison of Monkey and Human Performance

Except for slight differences in elevational errors, saccadic accuracy for the human and monkey subjects did not differ significantly. From dichotic experiments (Houben & Gourevitch, 1979; Klumpp & Eady, 1956; Mills, 1960), it was expected that humans would be superior on this task. However, based upon more recent psychophysical data (Brown et al., 1982), the similarity in human and monkey localization behavior is not unexpected. Brown and colleagues (1982) note that while the reduced dimensions of the macaque pinna (compared to the human's) would require frequencies in the range of 8–32 kHz for the expression of elevation-dependent pinna transformations, monkeys do detect stimuli at frequencies as high as 40 kHz. They discuss, too, the likelihood that monkeys are able to use elevation-dependent reflections off the torso as cues for vertical localization of auditory stimuli.

For both human and monkey subjects, multiple saccades occurred on approximately 20% of the auditory trials. The monkeys also made multiple saccades on a small percentage of visual trials. With the monkeys, the interval between the primary and corrective saccades was approximately 200 msec, an interval similar to the latency of the primary saccade. For human subjects, the intersaccadic intervals averaged about 75 msec, well below the 250 msec expected for an initial saccade. This suggests that, for monkeys, the corrective saccade is programmed separately after completion of the initial saccade. Human subjects may preprogram two consecutive saccades.

The peak velocity of saccadic eye movements (to both auditory and visual saccades) is higher for monkey subjects than it is for human subjects. The difference in peak velocity for auditory and visual saccades of comparable size was greater for monkeys than for humans. The neural bases of these differences are considered in the next section.

## Relation to Neurophysiological Studies

Movements that orient the eyes and head toward the location of auditory or visual stimuli require complex transformations of sensory signals into motor commands. A representation of the location of the target in space as well as commands specifying the metrics of the movement needed to direct gaze to the target must be formed. Moreover, signals of target location are computed, at least initially, in a different manner for each sensory system. Information about the location of a visual target is based on the site of retinal activation, the positions of the eyes in the orbits, along with the current orientation of the head and body. Auditory targets are localized in head coordinates and are based on interaural differences in the intensity and timing of acoustic stimuli. Auditory and visual signals indicating the presence of a saccade target could remain in separate sensory coordinates and not be transformed into motor coordinates until the various sensory pathways converge onto a final common pathway. Conversely, all sensory signals could first be translated into common coordinates and then converge onto a single premotor pathway for the generation of saccades.

We performed experiments (Jay & Sparks, 1987a) designed to determine whether auditory and visual saccades share a common premotor pathway by recording the activity of neurons in the superior colliculus while monkeys were generating saccades to auditory or visual targets. The intermediate and deeper layers of the superior colliculus are sites where auditory, visual, and somatosensory signals converge upon a brain region containing neurons that generate commands for the initiation of saccades (Sparks, 1986). Thus, auditory, visual, and somatosensory signals, originally encoded in different frames of reference, could converge in the superior colliculus to become translated into motor commands in a common coordinate frame. Saccade-related burst (SRB) cells in the superior colliculus generate a discrete high-frequency burst of activity beginning approximately 20 msec before visually triggered saccades (Sparks, Holland, & Guthrie, 1976, Sparks & Mays, 1980). We found that SRB neurons discharging before saccades to visual targets also discharged before saccades to auditory targets. This finding indicates that auditory and visual signals, originally encoded in different coordinates, have already been converted into the same coordinates and are sharing a motor circuit at the level of the superior colliculus. Even though premotor elements in the superior colliculus are being shared, results of the present study demonstrated that saccades to auditory and visual targets differ in latency, velocity, accuracy, and the number of saccades required to acquire the target. These findings indicate that the sensory signals do more than trigger a common pool of motor elements into patterns of activity determined by the intrinsic properties of individual motor cells or by predetermined characteristics of a network of motor elements. If the sensory signals served as mere triggers, the characteristics of the movement would be independent of the type of sensory stimulus that triggered the movement. Thus, we infer that inputs to the neural circuits involved in the generation of saccadic command signals influence the form (e.g., frequency and/or duration of the saccade-related burst) of the motor command. Moreover, sensory influences upon motor signals have been observed at the level of the superior colliculus. The discharge of SRB neurons is altered under conditions that modify the quality of the sensory signal. Saccade-related bursts observed in collicular neurons have a lower frequency before saccades to the location of a visual target that is no longer present (Rohrer, White, & Sparks, 1987) and before saccades to targets viewed by an amblyopic eye (Sparks, Gurski, Mays, & Hickey, 1986). In both cases, the reduction in burst frequency is correlated with a reduction in saccadic velocity.

In other neurophysiological experiments we found that the responses of auditory cells in the primate superior colliculus depend upon the position of the auditory stimulus in space and upon the positions of the eyes in the orbits (Jay & Sparks, 1984, 1987b). These findings were interpreted as support for the hypothesis that the activity of acoustically responsive neurons in the superior colliculus is encoded in motor coordinates and specifies the trajectory of the movements required to look to an auditory stimulus rather than the location of the stimulus in head coordinates. In this context, the finding that saccadic latency depends upon the amplitude of the movement required to look to a target rather than the spatial location of the target is of interest. Neurophysiological data are consistent with the hypothesis that the observed differences in saccadic latency are not due to differences in the time required to localize spatially the auditory target but arise because of variations in the time needed to construct different motor commands.

In summary, these studies of sensory-guided eye movements generated new information about the accuracy of localization of auditory targets that vary in both elevation and azimuth and provided unexpected clues about the contribution of sensory signals to the actual format of motor commands. However, much remains to be learned about the translation of sensory signals into commands for the control of movements.

## ACKNOWLEDGMENTS

This work was supported by National Institutes of Health Grants R01-EY01189, R01-EY05486 and P30-EY03039.

## REFERENCES

Becker, W. (1976). Parallel information processing and continuous information uptake in the saccadic eye movement system. *Pfluegers Archiv, Abstract Supplement*, **362**, R48.

Becker, W., & Fuchs, A. F. (1969). Further properties of the human saccadic system: Eye movements and correction saccades with and without visual fixation points. *Vision Research*, **9**, 1247–1258.

Brown, C. H., Beecher, M. D., Moody, D. B., & Stebbins, W. C. (1980). Localization of noise bands by Old World monkeys. *Journal of the Acoustical Society of America*, **68**, 127–132.

Brown, C. H., Schessler, T., Moody, D., & Stebbins, W. (1982). Vertical and horizontal sound localization in primates. *Journal of the Acoustical Society of America*, **72**, 1804–1811.

Collewijn, H., van der Mark, F., & Jansen, T. C. (1975). Precise recordings of human eye movements. *Vision Research*, **15**, 447–450.

Evinger, C., Kaneko, C. R. S., & Fuchs, A. F. (1981). Oblique saccadic eye movements of the cat. *Experimental Brain Research,* **41,** 370–379.

Fischer, B., & Boch, R. (1983). Saccadic eye movements after extremely short reaction times in the monkey. *Brain Research,* **260,** 21–26.

Fuchs, A. F. (1967). Saccadic and smooth pursuit eye movements in the monkey. *Journal of Physiology (London),* **191,** 609–631.

Fuchs, A. F., & Robinson, D. A. (1966). A method for measuring horizontal and vertical eye movement chronically in the monkey. *Journal of Applied Physiology,* **21,** 1068–1070.

Guitton, D., & Mandl, G. (1980). Oblique saccades in the cat: A comparison between the durations of horizontal and vertical components. *Vision Research,* **20,** 875–881.

Houben, D., & Gourevitch, G. (1979). Auditory lateralization in monkeys: An examination of two cues serving directional hearing. *Journal of the Acoustical Society of America,* **66,** 1057–1063.

Jay, M. F., & Sparks, D. L. (1984). Auditory receptive fields in primate superior colliculus shift with changes in eye position. *Nature (London),* **309,** 345–347.

Jay, M. F., & Sparks, D. L. (1987a). Sensorimotor integration in the primate superior colliculus: I. Motor convergence. *Journal of Neurophysiology,* **57,** 22–34.

Jay, M. F., & Sparks, D. L. (1987b). Sensorimotor integration in the primate superior colliculus: II. Coordinates of auditory signals. *Journal of Neurophysiology,* **57,** 35–55.

Judge, S. J., Richmond, B. J., & Chu, F. C. (1980). Implantation of magnetic search coils for measurement of eye position: An improved method. *Vision Research,* **20,** 535–538.

Jurgens, R., Becker, W., Rieger, P., & Widderich, A. (1981). Interaction between goal-directed saccades and the vestibular-ocular reflex (VOR) is different from interaction between quick phase and the VOR. In A Fuchs & W. Becker (Eds.), *Progress in oculomotor research* (pp. 11–18). Amsterdam: Elsevier/North-Holland.

Klumpp, R. G., & Eady, H. R. (1956). Some measurements of interaural time difference thresholds. *Journal of the Acoustical Society of America,* **28,** 859–860.

Mills, A. W. (1960). Lateralization of high frequency tones. *Journal of the Acoustical Society of America,* **32,** 132–134.

Perrott, D. R., Ambarsoom, H., & Tucker, J. (1987). Changes in head position as a measure of auditory localization performance: Auditory psychomotor coordination under monoaural and binaural conditions. *Journal of the Acoustical Society of America,* **82,** 1637–1645.

Rhodes, G. (1987). Auditory attention and the representation of spatial information. *Perception & Psychophysics,* **42,** 1–14.

Robinson, D. A. (1964). The mechanics of human saccadic eye movement. *Journal of Physiology (London),* **174,** 245–264.

Rohrer, W. H., White, J. M., & Sparks, D. L. (1987). Saccade-related burst cells in the superior colliculus: Relationship of activity with saccadic velocity. *Society of Neuroscience Abstracts,* **13,** 1092.

Sparks, D. L. (1986). The neural translation of sensory signals into commands for the control of saccadic eye movements: The role of the primate superior colliculus. *Physiological Reviews, 66,* 118–171.

Sparks, D. L., Gurski, M. R., Mays, L. E., & Hickey, T. L. (1986). Effects of long-term and short-term monocular deprivation upon oculomotor function in the rhesus monkey. *Advances in the Biosciences, 57,* 191–197.

Sparks, D. L., Holland, R., & Guthrie, B. L. (1976). Size and distribution of movement fields in the monkey superior colliculus. *Brain Research, 113,* 21–34.

Sparks, D. L., & Mays, L. E. (1980). Movement fields of saccade-related burst neurons in the monkey superior colliculus. *Brain Research, 190,* 39–50.

Stebbins, W. C. (1973). Hearing of old world monkeys. *Journal of Physical Anthropology, 38,* 357–364.

Stebbins, W. C. (1975). Hearing of the anthropoid primates: A behavioral analysis. In B. Tower (Ed.), *The nervous system: Vol. 3. Human communication and its disorders* (pp 113–124). New York: Raven Press.

Stebbins, W. C., Green, S., & Miller, F. L. (1966). Auditory sensitivity of the monkey. *Science, 153,* 1646–1647.

Sullivan, M. J., & Kertesz, A. (1979). Signal detection via phase-locked sampling in a magnetic search coil eye movement monitor. *IEEE Transactions on Biomedical Engineering,* **BME-26,** 50–52.

White, C. T., Eason, R. G., & Bartlett, N. R. (1962). Latency and duration of eye movements in the horizontal plane. *Journal of the Optical Society of America, 52,* 210–213.

Whittington, D. A., Hepp-Reymond, M.-C, & Flood, W. (1981). Eye and head movements to auditory targets. *Experimental Brain Research, 41,* 358–363.

Zahn, J. R., Abel, L. A., & Dell'Osso, L. R. (1978). Audio-ocular response characteristics. *Sensory Processes, 2,* 32–37.

Zambarbieri, D., Schmid, R., Prablanc, C., & Magenes, G. (1981). Characteristics of eye movements evoked by the presentation of acoustic targets. In A. Fuchs & W. Becker (Eds.), *Progress in oculomotor research* (pp. 559–566). Amsterdam: Elsevier/North Holland.

Zwislocki, J., & Feldman, R. S. (1956). Just noticeable differences in dichotic phase. *Journal of the Acoustical Society of America, 28,* 860–864.

# 12

# EYE–HAND COORDINATION AND VISUAL CONTROL OF MOVEMENT: STUDIES IN BEHAVING ANIMALS

*Apostolos P. Georgopoulos*

*The Philip Bard Laboratories of Neurophysiology, Department of Neuroscience, The Johns Hopkins University School of Medicine, Baltimore, Maryland*

The function of eye–hand coordination and visual control of movement has been studied in both human subjects and experimental animals. Psychophysical experiments in healthy subjects and observations in brain-damaged patients (Jeannerod, 1986) have provided important information concerning these functions and the possible role of neural structures involved in its control. On the other hand, studies in experimental animals have provided valuable insights into the neural mechanisms that may underlie the specification and control of parameters of visuomotor coordination.

A powerful technique for studying the neurophysiological basis of behavior is the recording of the activity of brain cells in animals performing particular tasks (Lemon, 1984). First, animals are trained to perform in certain tasks using reward-based operant conditioning techniques. Then, the activity of single cells in selected brain areas is recorded during task performance using microelectrodes inserted into the brain through the dura. This is the most direct technique and one with a fine grain for studying the neural mechanisms of behavior.

The first stage in these studies is the training of naive animals to perform appropriate behavioral tasks. This training serves three goals: First, it prepares the animal for the subsequent recording sessions; second, it provides an insight into how the behavior is being shaped from the naive condition to the "learned" condition; and third, it permits a comparison between the behavioral capacities of animals and human subjects, so that findings from animals can be extended to human subjects. I illustrate these interwoven aspects with a series of studies that were aimed at elucidating some of the cerebrocortical mechanisms involved in arm movements directed to visual targets in immediate extrapersonal space.

## I. VISUALLY GUIDED REACHING MOVEMENTS

Reaching movements involve well-coordinated and tightly coupled motions about the shoulder and elbow joints (Soechting & Lacquaniti, 1981). The behavioral goal of reaching movements is to move and position the hand at a desired point in extrapersonal space. Therefore, they can be regarded as vectors with origin at their starting point, direction defined from their starting and end points, and amplitude equal to the distance between the start and end points. We focused on the direction of these movements; in particular, we wished to find out (a) how animals learn to make movements in a desired direction, (b) how the brain specifies the direction of a movement in space, (c) how a movement direction changes when the target of the movement changes while the movement is planned or while it is evolving, and (d) what brain mechanisms underlly the eye–hand coordination in (b) and (c). In designing these experiments we had to choose the experimental animal, the brain area to study, the behavioral apparatus, and the appropriate experimental design. We chose rhesus monkeys as experimental animals because of their proximity to human subjects with respect to reaching movements. These animals, like other primates, possess keen vision and proficient visuomotor coordination; in fact, reaching movements toward visual objects of interest are very common within the monkey's behavioral repertoire. We chose the motor cortex as the area to study because it is intimately connected to the motor system in its output side (i.e., the brain-stem and spinal motor structures), but it also receives a wealth of input from other cortical areas and from the thalamus. Finally, the behavioral devices we used were of two kinds: One allowed movements of a handle on a two-dimensional (2-D) planar working surface (Georgopoulos, Kalaska, & Massey, 1981) whereas the other permitted free movements in 3-D space; the former movements resembled drawing movements on a board, whereas the latter resembled natural pointing movements (Schwartz, Kettner, & Georgopoulos, 1988).

## II. STUDIES OF VISUALLY GUIDED MOVEMENTS IN 2-D SPACE

### A. Motor Learning

We wished to train monkeys to move a handle in a planar working surface toward lighted targets (Georgopoulos et al., 1981). The apparatus used for that purpose is shown in Fig. 1. The animal sits on a primate chair positioned at A and faces a 25 × 25-cm plane (B) tilted 15° from the horizontal toward the animal. The working surface consists of frosted plexiglass onto which a red laser beam is back-projected through a system of mirrors controlled by a pair of galvanometers. The beam appears on the plane as a small red dot and can be positioned anywhere within the plane;

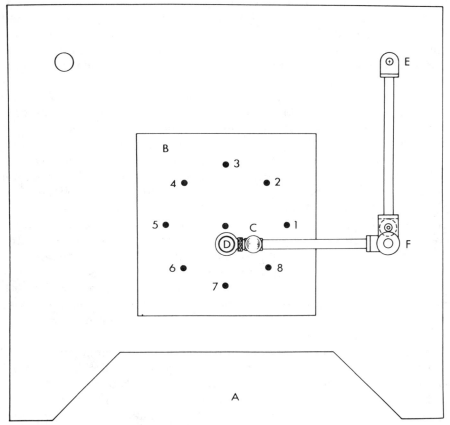

**FIGURE 1.** Schematic drawing of the 2-D apparatus used to study "drawing" movements of monkeys. See text for explanation.

eight positions are shown in Fig. 1. A handle is attached to the apparatus so that it can move freely and with very low friction on the working surface. A transparent plexiglass circle (D) is attached to the distal end of the handle; the X-Y position of the center of that circle on the plane is computed with 0.125 mm resolution from the angles of the two joints (E, F) of the handle.

The objective is to train the animal to grasp the manipulandum at C and capture within the circle D a lighted target. Then, if the light is turned on first in the center and then at various peripheral locations in different trials, the resulting movements will be of the same origin and amplitude but of different directions. In fact, this was our initial experimental goal. The problem is that these "drawing" movements are not natural for monkeys. Therefore, it takes several weeks for the animal to be trained. The first step is to familiarize the animal with the apparatus so that it grasps the handle, moves it around on the plane, and uses it as a tool to capture lighted targets within the circle D (see Fig. 1). In the beginning the animal is rewarded with drops of apple juice for merely grasping the handle and moving it around. Fortunately, rhesus monkeys are very inquisitive animals and indeed they grasp and move the manipulandum around during almost all the training time (3–4 hr/day). It is upon that continuous and apparently haphazard motion of the handle that the shaping of behavior is then applied by rewarding the animal (through a computer program) only when it moves the handle within a circular area centered around the light that is on. This area is indicated to the animal by the diameter of the circle D in Fig. 1. In the beginning of training the diameter of this circle is 50 mm; as the training progresses, progressively smaller circles are used, down to the 10-mm-diameter circle used in the neural recording experiments. At all times during the training period the light appears first in the center and then in one of several peripheral locations. When the animal has learned to move the handle to the light that is on, the conditions of reward change so that there is no reward for holding at the center light but only for moving from the center to the peripheral light and holding there for a period of time. Initially, the animal has to hold at the peripheral light for 20–40 msec but ultimately holding periods of 500–1000 msec are used. Similarly, holding at the center before the peripheral light comes on is short, but at the final stages of training center hold times from 1.5 to 5 sec are required.

Recording the movement trajectories during the training period has provided an objective way to visualize and quantify the motor learning. Figure 2 shows a lateral view of a monkey working in the apparatus. The insert on the lower right shows a plot of a movement trajectory together with the accuracy windows required at the center and the peripheral target. The working space occupies a large portion the extrapersonal space in front of the animal. Figure 3 shows families of trajectories made by one

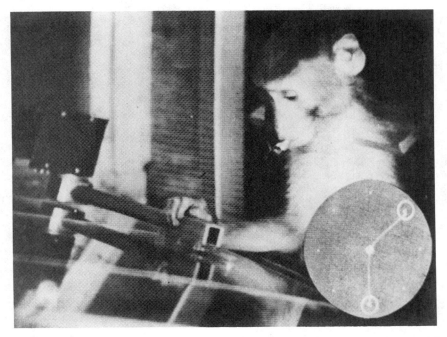

**FIGURE 2.** A monkey working in the 2-D apparatus and traces of trajectories of some of the movements, lateral view. Adapted from Georgopoulos, Kalaska, and Massey (1981) and reproduced with permission by the publisher.

animal during three days along the training period. It can be seen that the spatial variability of the trajectories decreased substantially with training time. In fact this decrease is exponential, as is the case with other learning curves. A training curve from one animal is shown in Fig. 4. In general, it takes 30–40 days for a monkey to become proficient in this task although there is variability between different animals.

**Transfer of the Skill to the Other Hand.** The animals are trained to operate the handle using only one hand; when they become proficient, they are trained in the task using the other hand. It is interesting that when they begin using the nontrained hand the spatial variability of the trajectories is high, similar to that exhibited during the beginning of the original training. This is in spite of the fact that now the animal knows the meaning of the lights as targets of the movement. However, the motor learning progresses very rapidly in this case and low levels of trajectory variability are achieved with the new hand within a few days, compared to several weeks of training in the case of the original hand. A curve

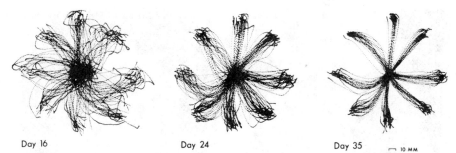

Day 16                          Day 24                          Day 35    ⌐ 10 MM

**FIGURE 3.** Families of movement trajectories in 2-D space during three days of training of one monkey. Each point in a trajectory denotes the X-Y position of the handle sampled every 10 msec with a precision of 0.125 mm. Notice the reduction of spatial variability of the trajectories from day 16 to day 35 of training. Adapted from Georgopoulos, Kalaska, and Massey (1981) and reproduced with permission by the publisher.

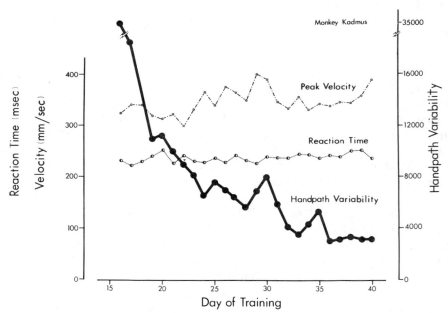

**FIGURE 4.** Training curve of one animal. The spatial variability of the movement trajectories is reduced with training but the reaction times and peak velocities of the movements remain approximately the same. The units of spatial variability are arbitrary units of area; the method of measurement of the spatial variability is described in Georgopoulos, Kalaska, and Massey (1981). From Georgopoulos et al. (1981) reproduced with permission by the publisher.

illustrating the accelerated learning of the second hand is shown in Fig. 5. This finding indicates that there are substantial savings in the motor learning by the new hand. The neural substrate of this phenomenon is unknown.

## B. Neuronal Recordings

When the animal is well trained a recording chamber is attached onto the skull overlying the arm area of the motor cortex and the activity of single cells in that area is recorded while the animal performs in the task. Microelectrode penetrations are made daily for several days so that the arm area is being sampled many times. The experimental variable of interest in these studies was the direction of the movement trajectory in space. How does the activity of single cells in the motor cortex vary with respect to the direction of movement in space? A clear answer to this question was obtained (Georgopoulos, Kalaska, Caminiti, & Massey, 1982): Cell activity is highest with movements in a particular direction (i.e., the cell's preferred

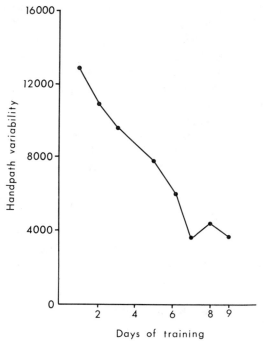

FIGURE 5. Accelerated motor learning with the nontrained hand. Data from one animal. A. P. Georgopoulos, J. F. Kalaska, and J. T. Massey (unublished observations).

direction) and decreases progressively with movements made in directions farther and farther away from the preferred one. This is illustrated in Fig. 6A which shows in raster form the impulse activity of one cell recorded in the arm area of the motor cortex during movements in eight directions (center diagram). It can be seen that the activity of the cell varies in an orderly fashion with the direction of the movement. In fact, this variation in cell activity is a linear function of the cosine of the angle formed between the direction of a particular movement and the cell's preferred direction (Fig. 6B: This is a broad and symmetric directional tuning function centered on the cell's preferred direction. Preferred directions differ among different cells and range throughout the directional continuum.

### C. Neuronal Population Coding

It can be seen in the example of Fig. 6 that a cell is engaged during movements in any direction. This indicates that, conversely, a movement in a particular direction will engage all the directionally tuned cells. Therefore, the initiation of that particular movement is a function of the neuronal ensemble. How is information in a directionally heterogeneous neuronal ensemble combined to produce a specific signal for a particular movement direction? We proposed a simple coding scheme for this population operation (Georgopoulos, Caminiti, Kalaska, & Massey, 1983; Georgopoulos, Kalaska, Crutcher, Caminiti, & Massey, 1984). Assume that a cell "votes" in the direction of its preferred direction, and that the strength of that "vote" is proportional to the change in cell activity associated with the particular movement direction under consideration. Thus the motor command for movement direction can be regarded as a population of cell vectors with direction and length defined above. Then the outcome of the population operation (the "population vector") is taken to be the vector sum of these cell vectors. Indeed, the population vector was found to point in or near the direction of the movement. This is illustrated in Fig. 7 for eight movement directions. All eight clusters represent the same population consisting of 241 directionally tuned cells whose vectorial contributions are shown as thin continuous lines. It can be seen that the population vector (dashed line with arrow) points closely in the direction of the movement indicated in the drawing at the center of the figure. A more important finding was that the population vector calculated every 20 msec during the reaction time predicted well the direction of the upcoming movement (Georgopoulos et al., 1984). This is illustrated in Fig. 8. Moreover, the population vector predicts well the direction of the upcoming movement even when the movement is delayed for a period of time (Georgopoulos, Crutcher, & Schwartz, 1989).

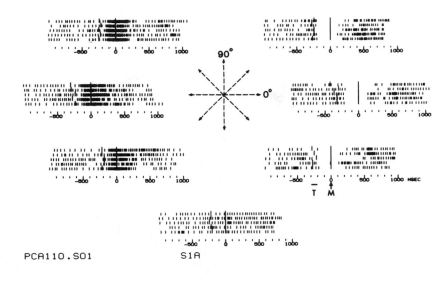

PCA110.SO1     S1A

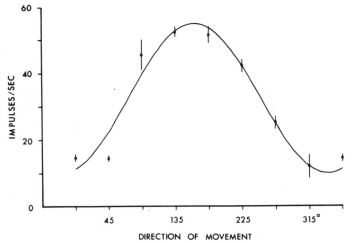

DIRECTION OF MOVEMENT

**FIGURE 6.** Broad directional tuning in 2-D space of a cell recorded in the arm area of the motor cortex. (A) Impulse activity during five trials with movements in the directions indicated in the drawing at the center. Short vertical bars indicate the occurrence of an action potential. Rasters are aligned to the onset of movement (M). Longer vertical bars preceding the onset of movement indicate the onset of the target (T); those following the movement indicate the entrance to the target window (see Fig. 2) and the delivery of reward. (B) Average frequency of discharge (±SEM) from the onset of the stimulus until the entry to the target window is plotted against the direction of movement. Continuous curve is a cosine function fitted to the data using multiple regression analysis; the function is of the form: $D = b + k \cos (\theta - \theta_0)$, where $D$ is the discharge rate, $b$ and $k$ are constants, $\theta$ is the direction of the movement, and $\theta_0$ is the cell's preferred direction. From Georgopoulos, Kalaska, Caminiti, and Massey (1982); reproduced with permission by the publisher.

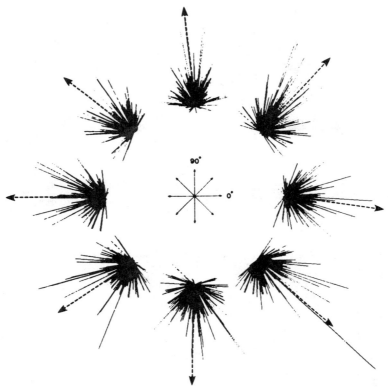

**FIGURE 7.** Neuronal population vector (heavy dashed lines with arrow) calculated for 8 movement directions. All clusters represent the same neuronal population composed of individual cell vectors (thin lines, $N = 241$ cells). The dotted lines in the center indicate the direction of movement. From Georgopoulos, Caminiti, Kalaska, and Massey (1983); reproduced with permission by the publisher.

## III. STUDIES OF VISUALLY GUIDED MOVEMENTS IN 3-D SPACE

The results of the studies described above indicated that the coding of the direction of movement in the arm area of the motor cortex is a distributed process. However, these findings were based on recordings in animals that were trained for relatively long periods of time in the 2-D apparatus; would the same results hold for free movements in 3-D space?

### A. Motor Learning

Monkeys were trained to reach freely toward visual targets in 3-D space. An apparatus was used that allowed unobstructed reaching to targets in 3-D space. A schematic drawing of the apparatus is shown in Fig. 9. It

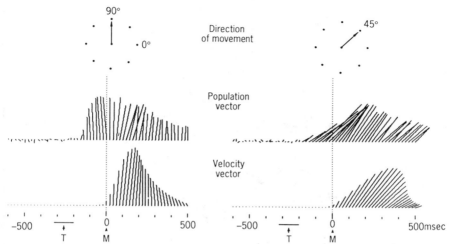

**FIGURE 8.** Population vectors computed every 20 msec for movements in two different directions. Instantaneous velocity vectors are also shown. Notice that the population vector lengthens well before the movement begins and points in the approximate direction of the upcoming movement. From Georgopoulos, Kalaska, Crutcher, Caminiti, and Massey (1984); reproduced with permission by the publisher.

consisted of a metal plate and metal rods threaded through holes cut in the plate. The front end of each rod housed a red button 16 mm in diameter which could be lighted and, if depressed, closed a switch. The buttons were arranged so that one was in the center and eight were in the corners of an imaginary cube. The central light was at shoulder level in the midsagittal plane with its front end 15 cm away from the animal. The remaining buttons were placed at 12.5-cm distances from the center at the corners of an imagined parallelepiped rotated in the horizontal plane 18° toward the performing arm. Thus, movements could be made in eight directions (from the center to peripheral targets) at approximately equal angular intervals. Similarly to the 2-D case, the light in the center was turned on first and the monkey was required to push it for a period of time. Then one of the peripheral lights came on and the animal had to move toward it and push it to receive a liquid reward. Unlike the 2-D case, animals learned this task quickly, within 10–15 days. In fact, there was no truly motor learning involved because monkeys are proficient in reaching freely in 3-D space; the training time was devoted mostly to learning to hold still for a period of time in a given light. Thus, this experimental arrangement provided an almost natural situation in which to study the relations of motor cortical activity to the direction of reaching arm movements.

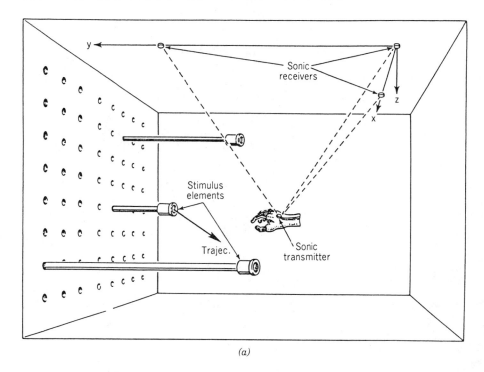

(a)

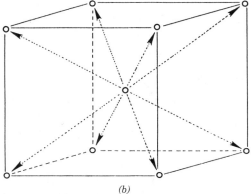

(b)

**FIGURE 9.** Schematic drawing of the apparatus used to study free reaching movements in 3-D space. (A) Monkeys reached toward and pushed lighted buttons mounted at the end of metal rods threaded through a heavy metal plate. The movement trajectory was monitored using an ultrasonic system. (B) Schematic diagram of the location of the 9 buttons used. Dotted lines indicate directions of movements. From Schwartz, Kettner, and Georgopoulos (1988); reproduced with permission by the publisher. Copyright by the Society for Neuroscience.

## B. Neuronal Recordings

The results obtained from this experiment (Georgopoulos, Schwartz, & Kettner, 1986; Schwartz et al., 1988) were very similar to those obtained previously in the 2-D case. The activity of single cells studied during performance in this task was broadly tuned to the direction of the movement. This is illustrated in Fig. 10*A*. Moreover, the directional tuning function was identical for both the 2-D and the 3-D cases: The activity of a cell was highest with movements in the cell's preferred direction and decreased progressively with movements made in directions farther and farther away from the preferred one. This decrease followed the cosine of the angle formed between the direction of a particular movement and the cell's preferred direction. This principle is illustrated in Fig. 10*B*.

## C. Neuronal Population Coding

The same distributed code for movement direction that was discovered for the 2-D case (Georgopoulos, Caminiti, et al., 1983; Georgopoulos et al., 1984) also held well in the 3-D case (Georgopoulos, Kettner, & Schwartz, 1988; Georgopoulos et al., 1986).

## IV. EYE–HAND COORDINATION: EFFECTS OF CHANGE IN TARGET LOCATION

In the experiments described above the movements were directed to a stationary target. We wished to learn how the movement would be affected if its target changed unpredictably during the reaction or movement time. Is the process that generates the aimed movement all or none or is it modifiable? This is an important question, for many times the object of interest toward which the hand is directed changes location, and therefore it is interesting to know whether the motor system can follow the target and change the movement in midflight, or whether an appreciable lag exists between switching motor patterns.

## A. Behavioral Observations

The 2-D apparatus described above (Fig. 1) was used in these experiments. Two peripheral lights in opposite locations were used (e.g., at 12 and 6 o'clock). The monkey held the handle in the center for a variable period of time after which one of the two peripheral targets was turned on; it remained on for 50, 100, 150, 200, 250, 300 or 400 msec and then was turned off and the other target was turned on. The animal was rewarded after holding at the second light for 0.5 sec. Trials in which the first light stayed on without changing location were interspersed among those in which the

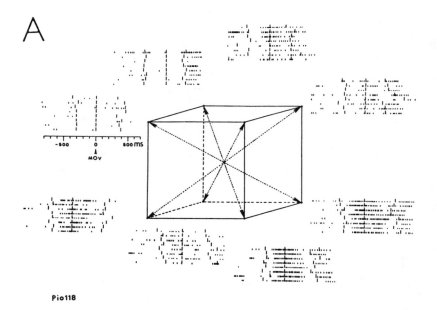

Pio118

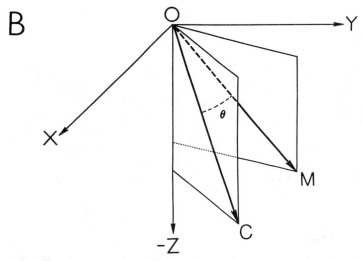

**FIGURE 10.** Broad directional tuning in 3-D space of a motor cortical cell. (*A*) Impulse activity is shown in raster form for eight trials in eight movement directions indicated in the drawing at center. MOV, onset of movement. (*B*) Principle of directional tuning. **C** is the preferred direction of the cell whose rasters are shown in (*A*); **M** is the direction of a movement; $\theta$ is the angle between **C** and **M**. The cell activity is varies in a linear fashion with cos $\theta$: It is maximum when cos $\theta$ = 1, that is, when **C** and **M** coincide; minimum when cos $\theta$ = −1, that is, when **C** and **M** are in opposite directions; and in between when $1 > \cos \theta > -1$, that is, when **M** is between the direction of **C** and its opposite. Adapted from Georgopoulos, Schwartz, and Kettner, 1986, and reproduced with permission by the publisher. Copyright 1986 by the AAAS.

target location changed so that the animal could not predict whether the target will change location in a certain trial; moreover, the animal could not predict for how long the first target would stay on in the trials in which it changed location. The effect of the change in target location on the movement trajectory is shown in Fig. 11. The first target was at 12 o'clock and the second target at 6 o'clock. Single trajectories of arm movements are shown in the lower trace. It can be seen that the arm moved initially toward the first target, then changed direction and moved toward the second target. The eye movements were also monitored in these experiments using implanted electrooculographic (EOG) electrodes. In the upper trace of Fig. 11 trajectories of eye movements are shown. In contrast to arm movements, the eyes moved all the way to the first (and then to the second target) or directly to the second target.

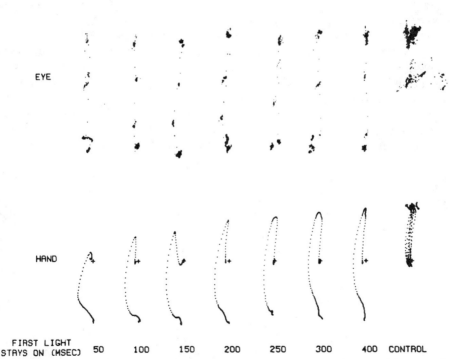

FIRST LIGHT
STAYS ON (MSEC)   50    100   150   200   250   300   400  CONTROL

**FIGURE 11.** Modification of hand movement trajectories (bottom) in the target-change task. The first target light was at 12 o'clock; it stayed on for the time indicated and then changed location to 6 o'clock. Single trajectories are shown (X-Y data points were recorded every 10 msec). On the far right (control), seven trajectories to the target at 12 o'clock are superimposed. Eye movements (upper) were recorded simultaneously with hand movements using EOG. From Georgopoulos, Kalaska, and Massey (1981); reproduced with permission.

There are several aspects of these experiments that are noteworthy. First of all, how consistently is a movement made toward the first target? This question was examined by varying systematically the duration of the first target and observing the arm movement toward it. It was found that the occurrence of a movement toward the first target depends on how long the first light stays on. This is illustrated in Fig. 12 which plots the probability of occurrence of a movement toward the first target versus the duration of presentation of the first target. It can be seen that this probability increases steeply and becomes one when the first light stays on for at least 100 msec.

The second question concerns the temporal relations between the direction of the target and the duration of the movement. Figure 13 illustrates the finding that the duration of the arm movement toward the first target was a linear function of the time for which the first target stayed on. This is was observed in all three animals studied, as shown in Fig. 13. It is noteworthy that this relation was preserved even when the animals adopted a different strategy for a particular target light and did not move immediately toward it when it appeared, but waited for a period of time. In that case, there were no movements toward that target for the shortest interstimulus interval of 50 msec, but the duration of the movement toward the first target, when it was made at a later time, was still a linear function of the time for which that light stayed on. Figure 14 illustrates the results obtained from one monkey. Filled triangles indicate the duration of movement toward the first target when the monkey moved promptly after the target's appearance, whereas open triangles indicate the duration of movement toward another target light which the monkey chose to ignore for a period of time. It can be seen that in the latter case the duration of the movement was still a function of the time for which the first target light stayed on but the line was transposed to the right. These results indicate

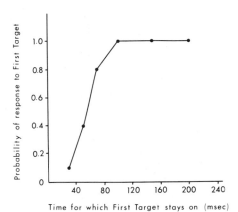

**FIGURE 12.** Probability of response to first of two targets presented in quick succession. From A. P. Georgopoulos, J. F. Kalaska, and J. T. Massey (unpublished observations).

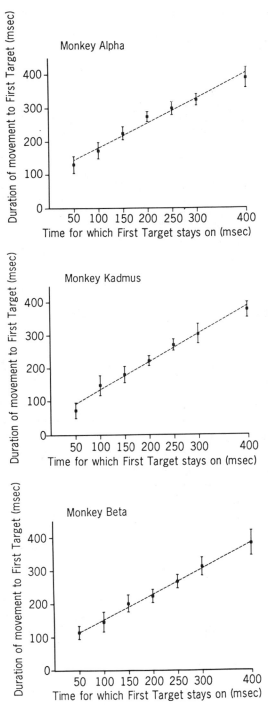

**FIGURE 13.** Duration of movement toward the first of two targets (ordinate) presented in succession at specified interstimulus intervals (abscissa). Data from three monkeys. From A. P. Georgopoulos, J. F. Kalaska, and J. T. Massey (unpublished observations).

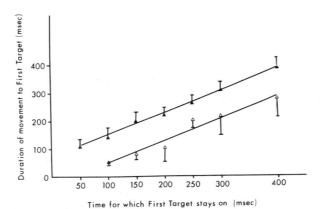

**FIGURE 14.** Duration of movement (mean ± SD) toward the first of two targets (ordinate) presented in succession at specified interstimulus intervals (abscissa); ▲, movements in which the monkey moved promptly toward the first target; △, movements in which the monkey waited for a period of time. From A. P. Georgopoulos, J. F. Kalaska, and J. T. Massey (unpublished observations).

that the temporal dependence of the duration of the movement upon the duration of the first light is strong and independent of the strategy adopted.

The third question concerns the spatial characteristics of the trajectory of the first movement. In most cases tested, the two target lights were located along the same line, so that the first and second movements were in opposite directions. However, when the two target lights were arranged in such a way that the corresponding movements were not in opposite directions, a modification of the direction of the trajectory of the first movement was observed in some trials. An example is shown in Fig. 15. The first light (A) was at approximately 2 o'clock, and the second light (B) was at 90° angle from it. The two continuous lines in Fig. 15 show examples of trajectories made directly to each of the lights in control trials. The dotted lines plot trajectories obtained at the interstimulus interval indicated in the plot (50, 100, and 300 msec). It can be seen that the trajectory deviated toward the direction of the second target, the more so the shorter the interstimulus interval was. These results indicate that the motor command for the direction of movement generated after the appearance of the first target can be modified by the second target when the targets shift early in time.

The fourth question concerns the velocity characteristics of the trajectories of the movements toward the second target, after their reversal. It can be seen in Fig. 11 (lower part) that these movements were faster than those made toward the first target. This is shown in Fig. 16: The peak velocity of the movements toward the second target was approximately 3

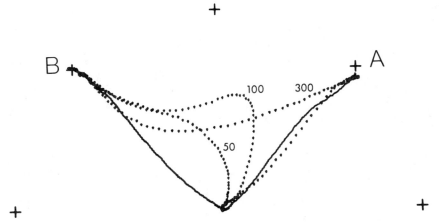

**FIGURE 15.** Modification of the direction of movement elicited by the first target (A) when the second target (B) was at an angle of 90°. Solid lines indicate control trajectories of movements made directly to targets A and B. Dotted lines are trajectories of movements made at the stated interstimulus interval (X-Y data points were sampled every 10 msec). From Georgopoulos, Kalaska, and Massey (1981); reproduced with permission.

times that of the movement toward the first target. This was true even for the interstimulus interval of 50 msec, in which case the movement toward the first light was small, and, therefore, the amplitude of the second movement was very similar to that of the movement made directly to the second target in control trials. These results can be explained by the strategy that monkeys and human subjects used to solve the problems posed by this target-shift task. This strategy was revealed by an analysis of the forces exerted on the manipulandum during performance of the task (Massey, Schwartz, & Georgopoulos 1986).

The problem is that if the trajectory is to be interrupted at different points along its extent, the dynamics of the hand and the manipulandum will differ in these points, reflecting different velocities and accelerations; therefore, the braking forces will have to be precalculated in accordance with the anticipated kinematic and dynamic conditions. However, the subjects ignored the differences between the mechanical conditions associated with different interruption points, and, instead, produced a large braking force in the opposite direction in response to the second target that stopped the limb and moved it to the second target. These braking forces were excessive for the short movements toward the first target but were effective for all cases and did not require individual adjustments for particular trials. Therefore, the computational load was reduced at the expense of increased spending of mechanical energy; in other words, a simple and mechanically effective but energy-inefficient strategy underlay the efficient eye–hand coordination in this task.

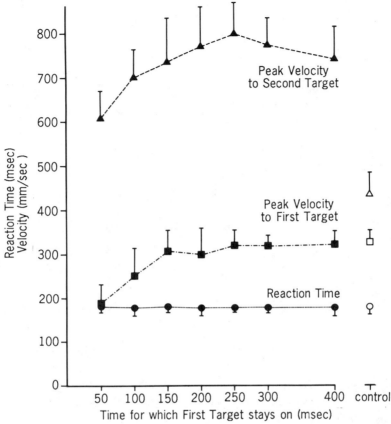

**FIGURE 16.** Reaction times to first target (●); peak velocities of first movement (■); peak velocities of second movement (▲). Data points are means (±SD) from one experiment (first target at 12 o'clock, second target at 6 o'clock). From A. P. Georgopoulos, J. F. Kalaska, and J. T. Massey (unpublished observations).

## B. Neuronal Recordings

When the location of visual stimuli is changed in quick succession, the patterns of cell activity in motor cortex follow these changes with remarkable temporal fidelity (Georgopoulos, Kalaska, Caminiti, & Massey, 1983) and attest to the efficient engagement of that structure during eye–hand coordination. An example is shown in Fig. 17. The location of the first and second target, and the required movement direction, are shown in the top of the figure. The pattern of cell activity associated with movements directly to these targets consisted of an increase in cell activity when moving to target A and of a decrease in activity when moving to target B. When the first target was target A, the pattern of cell activity was that associated with movement to that target; this pattern lasted for a time

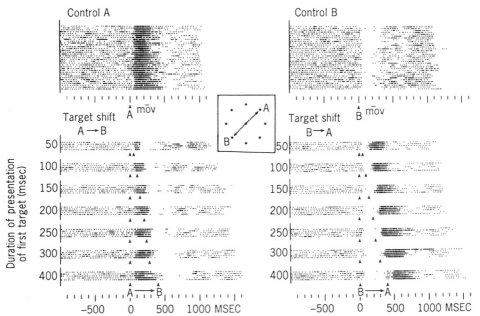

**FIGURE 17.** Changes in discharge of a cell in the arm area of the motor cortex during the target-shift task (see text). Beneath each control raster are seven rasters of trials during which the target changed location after 50–400 msec, as indicated. In the control trials, there was an increase in cell activity following presentation of target A, and a decrease following presentation of target B. These changes in activity were preserved when targets A or B were presented as the first targets in target shift trials; after an initial period that depended on the duration of presentation of the first target, the pattern of cell activity changed to that appropriate for the movement elicited by the second target. *mov*, range of movement onset. From Georgopoulos, Kalaska, Caminiti, and Massey (1983); reproduced with permission.

proportional to the time for which the first target stayed on, and then became like the pattern of activity associated with movement to target B. Similar changes were observed when the first and second targets were targets B and A, respectively. These temporal relations between the duration of presentation of the first target, the duration of change in cell activity associated with the presentation of the first target, and the duration of the hand movement toward that target are illustrated in Fig. 18. It can be seen that the temporal associations are linear. These results indicate that when a sequence of hand movements is generated toward targets that change location, the motor cortical activity follows well the changes in visual information and generates the appropriate motor commands without temporal smearing. This apparently underlies, at least in part, the efficient eye–hand coordination observed at the behavioral level. It is noteworthy that similar findings were obtained in area 5 of the parietal

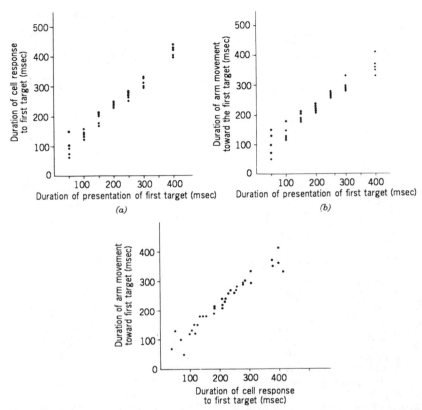

**FIGURE 18.**  Linear relations between the time for which the first target stayed on, and neural and behavioral events associated with it. Each data point is the value from a single trial. Data are derived from the target shift trials displayed on the left side of Fig. 17. Neural and behavioral events were collected simultaneously. Duration of the movement toward the first target was from the onset of movement until the reversal of the trajectory toward the second target. From Georgopoulos, Kalaska, Caminiti, and Massey (1983); reproduced with permission.

cortex (Kalaska, Caminiti, & Georgopoulos 1981), an area that is involved mainly in the proprioceptive monitoring of the motion of the limb, as discussed above.

## V.  NEURAL MECHANISMS OF EYE–HAND COORDINATION

### A.  Effects of Lesions

A crucial role of the posterior parietal areas 5 and 7 in the control of visually guided reaching is suggested by the effects of their lesions in human subjects and subhuman primates. Apparent disturbance of visually guided movements has been observed in many cases of brain damage but in several of these cases the defect does not seem to be with eye–hand coordination per se but rather with impaired visual function or impaired hand movement. Defects in eye–hand coordination itself have been described in patients with lesions of the posterior parietal cortex. Experiments in the monkey have shown that lesions of the posterior parietal cortex result in defective reaching to visual targets in space (Ettlinger & Kalsbeck, 1962; LaMotte & Acuna, 1975). This is illustrated in Fig. 19. It has also been shown that interruption of the pathways that link the parietal–occipital cortex with the frontal cortex results in defective visual control of manipulatory movements of the hand (Haaxma & Kuypers, 1975).

However, the most interesting cases are furnished by human patients, especially those suffering from the syndrome of optic ataxia (Hécaen & de Ajuriaguerra, 1954; Rondot, de Recondo, & Dumas, 1977), first described by Balint (1909). Patients with optic ataxia usually do not have impaired

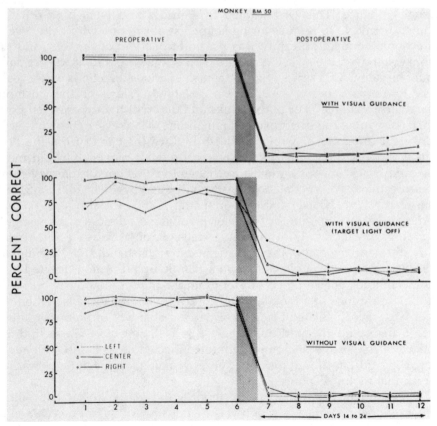

**FIGURE 19.** Reaching defects after posterior parietal lesion. Performance of a monkey on consecutive tests of each reaching task before and after the lesion is shown. From LaMotte and Acuna (1975); reproduced with permission.

vision or impaired hand or arm movements; yet, they cannot reach accurately to visual targets. The syndrome is complex, and several variants of it exist (Hécaen & de Ajuriaguerra, 1954; Perenin & Vighetto, 1983; Rondot et al., 1977). The brain damage in the cases of optical ataxia has been localized in the parietal cortex (angular and/or supramarginal gyrus), its underlying white matter, and/or the posterior part of the corpus callosum. Although some of the defects in eye–hand coordination in that syndrome can be attributed to the destruction of the parietal cortex itself, others are best understood as disconnection syndromes produced by the interruption of pathways linking the posterior parietal and prestriate cortex to the precentral premotor and motor areas on the same (ipsilateral) and the contralateral side. These pathways have been summarized succinctly by Ferro, Bravo-Marques, Castro-Caldas, and Antunes (1983).

Eye–hand coordination in general need not be constrained to pointing or manipulatory movements alone. In fact, a most interesting case is that in which a spatial pattern of arm movements is planned and executed under visual guidance. This was studied by Petrides and Iversen (1979) who trained monkeys to remove a ring-shaped sweet from a wire with several directional bends. The candy was placed at one end of the wire and the monkeys moved the sweet along the path of the wire until it was free and could be eaten. This series of movements in various directions was an easy task to normal monkeys. However, bilateral lesions of the posterior parietal cortex (area 7) or of the banks and depth of the superior temporal sulcus disrupted the performance dramatically. Monkeys with such lesions had great difficulty or were unable to remove the candy from the bent wires: When a directional turn was encountered, monkeys would move the candy back and forth unable to change the direction of the movement and to generate the spatial motor pattern dictated by the spatial configuration of the wire. These studies demonstrated the crucial role of parietal cortex in visuospatial motor behavior involving sequences of movements in various directions. Unfortunately, lesions of frontal areas were not tried.

A different question was investigated by Halsband and Passingham (1982) who trained monkeys to turn a handle in a direction dictated by a visual stimulus of a certain color; for example, to pull a handle in response to the appearance of a blue light or to turn it counterclockwise when the light was yellow. Bilateral lesions of parietal area 5 *or* of premotor lateral area 6 disrupted the performance in this task severely. These findings suggest a role of the anatomically interconnected areas 5 and 6 (Jones & Powell, 1970) in the translation of a visual stimulus into a learned directional motor response.

A role of the cortical area just posterior to the genu of the arcuate sulcus in conditional response selection was demonstrated by Petrides (1987). In that study monkeys were trained to perform two different movements, each in response to a different stimulus. This ability to select a motor response according to the appropriate stimulus was severely affected following discrete lesions of the postarcuate area mentioned above. This effect on the selection of motor response was dissociated from a different defect in associative visuomotor performance in which the response was based on the property of another visual stimulus. In each trial the monkey viewed two boxes one of which was lit in a random fashion; the animal was trained to move to the box that was lit when one of two objects was presented and to move to the box that was unlit when another object was presented. Since the location of the lit (or unlit) box varied from trial to trial, the movement itself was different in response to the presentation of the same object but the association of a given object to the quality of the target of the movement (lit or unlit) was invariant. Performance in this task was seriously affected by discrete lesions of the cortex just anterior, but not posterior, to the arcuate sulcus. These results underscore the importance of the periarcuate cortex in stimulus–response processing and provide clear evidence for different roles in this process of the pre- and postarcuate parts of that area.

## B.  Neuronal Recordings

There are several levels at which the neural mechanisms of eye–hand coordination can be studied. Commonly, animals are trained to move their hands toward visual targets in reaction time tasks, and then the activity of single cells in particular areas of the brain is recorded during the performance of these movements, as described in the preceding sections. The salient finding of such studies from several laboratories is that cells in various brain structures are engaged in these tasks. From this it follows that performance in eye–hand coordination tasks is the result of cooperation of various brain regions.

Mountcastle, Lynch, Georgopoulos, Sakata, and Acuna (1975) described a class of cells in the posterior parietal cortex (areas 5 and 7) of the behaving monkey that were activated only when the monkey reached toward visual targets. Another class of neurons recorded in area 7 is of particular interest, for they seemed to be specifically related to eye–hand coordination. These neurons were only moderately active during tracking by the hand of a moving object in the dark, or during tracking of a moving object by eyes alone, but were intensely active during combined eye–hand tracking (Mountcastle et al., 1975). Unfortunately, these types of neurons have not been studied further. A different class of area 7 neurons possess functional properties that make them good candidates for the function of monitoring the motion of the limb through the peripheral visual field. These neurons are sensitive to moving visual stimuli and their receptive field typically spares the foveal region (Motter & Mountcastle, 1981). By contrast, neurons in area 5, which lies just anterior to area 7, are sensitive to passive manipulations of the limb (Mountcastle et al., 1975; Sakata, Takaoka, Kawarasaki, & Shibutani, 1973), although certain cells are engaged only during active reaching (Mountcastle et al., 1975). It is thus possible that parietal areas 7 and 5 are concerned with processing visual and somesthetic feedback from the moving limb, respectively, and may be involved in the "cross-modal comparison" postulated by Prablanc, Pélisson, and Goodale (1986).

Changes in the neural activity of several motor structures are observed during the reaction time of visually triggered aimed hand movements. These structures include the motor cortex, the arcuate premotor area, the supplementary motor area, the area 6 on the lateral convexity of the hemisphere medial to the medial limb of the arcuate sulcus, the basal ganglia (putamen, globus pallidus, subthalamic nucleus), the cerebellum, and possibly other areas not yet studied. These early changes in neural activity are commonly interpreted to mean that these structures are involved in the generation of the movement, but the neuronal changes may not be specifically involved in eye–hand coordination. This is exemplified by the results of an experiment in which the meaning of a visual stimulus for the upcoming reaching movement was dissociated from the location of that stimulus as the target of the movement (Godschalk, Lemon, Kuypers, & van der Steen, 1985). Monkeys were trained to reach toward a food object placed at different locations. Recording of single cell activity in the postarcuate premotor area revealed changes in neuronal activity following the presentation of the target. These changes lasted throughout the period preceding the triggering of the movement; therefore, they could relate to the visual stimulus, the movement, or both. Now, the actual movement trajectory was dissociated from the location of the visual stimulus by training the monkeys to move to each of three different locations depending on whether one, two, or three lights were lit at the same place in front of them. It was found that the changes in cell activity were related to the direction of the actual movement that followed and not to the visual stimulus itself. In general, it seems that the patterns of activity in motor structures relate to the movement and not to the sensory modality of the stimulus. This was shown in studies in which movements of the hand were triggered by any of three different sensory stimuli, namely a somesthetic, a visual, or an auditory stimulus (Lamarre, Busby, & Spidalieri, 1983). It was found that cell discharge in the motor cortex was related to the movement; no differences were found in the cell activity among the different stimuli used. However, responses of cells to stimuli of different modalities have been described in premotor frontal areas (Tanji & Kurata, 1985; Vaadia, Benson, Hienz, & Goldstein, 1986; Weinrich & Wise, 1982).

## ACKNOWLEDGMENTS

This work was supported by USPHS Grant NS17413 and ONR Contract N00014-88-K-0751.

# REFERENCES

Balint, R. (1909). Seelenlahmung des "Schauens," optische Ataxie, rämliche Störung der Aufmarksamkeit. *Monatschrift für Psychiatrie und Neurologie,* **25,** 51–81.

Ettlinger, G., & Kalsbeck, J. E. (1962). Changes in tactile discrimination and in visual reaching after successive and simultaneous bilateral posterior ablations in the monkey. *Journal of Neurology, Neurosurgery and Psychiatry,* **25,** 256–268.

Ferro, J. M., Bravo-Marques, J. M., Castro-Caldas, A., & Antunes, L. (1983). Crossed optic ataxia: Possible role of the dorsal splenium. *Journal of Neurology, Neurosurgery and Psychiatry,* **46,** 533–539.

Georgopoulos, A. P., Caminiti, R., Kalaska, J. F., & Massey, J. T. (1983). Spatial coding of movement: A hypothesis concerning the coding of movement direction by motor cortical populations. *Experimental Brain Research, Supplement,* **7,** 327–336.

Georgopoulos, A. P., Crutcher, M. D., & Schwartz, A. B. (1989). Cognitive spatial motor processes. 3. Motor cortical prediction of movement direction during an instructed delay period. *Experimental Brain Research,* **75,** 183–194.

Georgopoulos, A. P., Kalaska, J. F., Caminiti, R., & Massey, J. T. (1982). On the relations between the direction of two-dimensional arm movements and cell discharge in primate motor cortex. *Journal of Neuroscience,* **2,** 1527–1537.

Georgopoulos, A. P., Kalaska, J. F., Caminiti, R., & Massey, J. T. (1983). Interruption of motor cortical discharge subserving aimed arm movements. *Experimental Brain Research,* **49,** 327–340.

Georgopoulos, A. P., Kalaska, J. F., Crutcher, M. D., Caminiti, R., & Massey, J. T. (1984). The representation of movement direction in the motor cortex: Single cell and population studies. In G. M. Edelman, W. E. Gall, W. M. Cowan (Eds.), *Dynamic aspects of neocortical function* (pp. 501–524). New York: Wiley.

Georgopoulos, A. P., Kalaska, J. F., & Massey, J. T. (1981). Spatial trajectories and reaction times of aimed movements: Effects of practice, uncertainty, and change in target location. *Journal of Neurophysiology,* **46,** 725–743.

Georgopoulos, A. P., Kettner, R. E., and Schwartz, A. B. (1988). Primate motor cortex and free arm movements to visual targets in three-dimensional space. II. Coding of the direction of movement by a neuronal population. *Journal of Neuroscience,* **8,** 2928–2947.

Georgopoulos, A. P., Schwartz, A. B., & Kettner, R. E. (1986). Neuronal population coding of movement direction. *Science,* **233,** 1416–1419.

Godschalk, M., Lemon, R. N., Kuypers, H. G. J. M., & van der Steen, J. (1985). The involvement of monkey premotor cortex neurones in preparation of visually cued arm movements. *Behavioral Brain Research,* **18,** 143–157.

Haaxma, R., & Kuypers, H. G. J. M. (1975). Intrahemispheric cortical connexions and visual guidance of hand and finger movements in the rhesus monkey. *Brain,* **98,** 239–260.

Halsband, V., & Passingham, R. (1982). The role of premotor and parietal cortex in the direction of action. *Brain Research,* **240,** 368–372.

Hécaen, H., & de Ajuriaguerra, J. (1954). Balint's syndrome (psychic paralysis of visual fixation) and its minor forms. *Brain,* **77,** 373–400.

Jeannerod, M. (1986). Mechanisms of visuomotor coordination: A study in normal and brain-damaged subjects. *Neuropsychologia, 24,* 41–78.

Jones, E. G., & Powell, T. P. S. (1970). An anatomical study of converging sensory pathways within the cerebral cortex of the monkey. *Brain, 93,* 793–820.

Kalaska, J. F., Caminiti, R., & Georgopoulos, A. P. (1981). Cortical mechanisms of two-dimensional aimed arm movements. III. Relations of parietal (areas 5 and 2) neuronal activity to direction of movement and change in target location. *Society for Neuroscience Abstracts, 7,* 563.

Lamarre, Y., Busby, L., & Spidalieri, G. (1983). Fast ballistic arm movements triggered by visual, auditory, and somesthetic stimuli in the monkey. I. Activity of precentral cortical neurons. *Journal of Neurophysiology, 50,* 1343–1358.

LaMotte, R. H., & Acuna, C. (1975). Defects in accuracy of reaching after removal of posterior parietal cortex in monkeys. *Brain Research, 139,* 309–326.

Lemon, R. N. (1984). *Methods for neuronal recording in conscious animals.* Chichester: Wiley.

Massey, J. T., Schwartz, A. B., & Georgopoulos, A. P. (1986). On information processing and performing a movement sequence. *Experimental Brain Research, Supplement, 15,* 242–251.

Motter, B. C., & Mountcastle, V. B. (1981). The functional properties of the light-sensitive neurons of the posterior parietal cortex studied in waking monkeys: Foveal sparing and opponent vector organization. *Journal of Neuroscience, 1,* 1–26.

Mountcastle, V. B., Lynch, J. C., Georgopoulos, A. P., Sakata, H., & Acuna, C. (1975). Posterior parietal association cortex of the monkey: Command functions for operations within extrapersonal space. *Journal of Neurophysiology, 38,* 871–908.

Perenin, M. T., & Vighetto, A. (1983). Optic ataxia: A specific disorder in visuomotor coordination. In A. Hein & M. Jeannerod (Eds.), *Spatially oriented behavior* (pp. 305–326). New York: Springer-Verlag.

Petrides, M. (1987). Conditional learning and the primate frontal cortex. In E. Perecman (Ed.), *The frontal lobes revisited* (pp. 91–108). New York: IRBN Press.

Petrides M., & Iversen, S. D. (1979). Restricted posterior parietal lesions in the rhesus monkey and performance on visuospatial tasks. *Brain Research, 161,* 63–77.

Prablanc, C., Pélisson, D., & Goodale, M. A. (1986). Visual control of reaching movements without vision of the limb. I. Role of retinal feedback of target position in guiding the hand. *Experimental Brain Research, 62,* 293–302.

Rondot, P., de Recondo, J., & Dumas, J. L. R. (1977). Visuomotor ataxia. *Brain, 100,* 355–376.

Sakata, H., Takaoka, Y., Kawarasaki, A., & Shibutani, H. (1973). Somatosensory properties of neurons in the superior parietal cortex (area 5) of the rhesus monkey. *Brain Research, 64,* 85–102.

Schwartz, A. B., Kettner, R. E., & Georgopoulos, A. P. (1988). Primate motor cortex and free arm movements to visual targets in three-dimensional space. I. Relations between single cell discharge and direction of movement. *Journal of Neuroscience, 8,* 2913–2927.

Soechting, J. F., & Lacquaniti, F. (1981). Invariant characteristics of a pointing movement in man. *Journal of Neuroscience, 1,* 710–720.

Tanji, J., & Kurata, K. (1985). Contrasting neuronal activity in supplementary and precentral motor cortex of monkeys. I. Responses to instructions determining motor responses to forthcoming signals of different modalities. *Journal of Neurophysiology, 53,* 129–141.

Vaadia, E., Benson, D. A., Hienz, R. D., & Goldstein, M. H., Jr. (1986). Unit study of monkey frontal cortex: Active localization of auditory and of visual stimuli. *Journal of Neurophysiology, 56,* 934–952.

Weinrich, M., & Wise, S. P. (1982). The premotor cortex of the monkey. *Journal of Neuroscience, 2,* 1329–1345.

# DEVELOPMENT

# 13

# VISION FOLLOWING LOSS OF CORTICAL DIRECTIONAL SELECTIVITY

*Tatiana Pasternak*

*Center for Visual Science and Department of Neurobiology and Anatomy, University of Rochester, Rochester, New York*

The analysis of motion perception in terms of neural mechanisms, anatomical or physiological, is hampered by the difficulties of cross organism comparisons. Psychophysical data are most often gathered from human subjects, while cats and nonhuman primates are used for physiological studies of neural mechanisms. The work described here examined the role of directionally selective neurons in motion perception by obtaining psychophysical and physiological results in the same subjects. We applied psychophysical techniques commonly used with human observers to the study of motion perception in normal cats and in cats with greatly reduced directional selectivity in their visual cortex. With this approach we were able to demonstrate the importance of cortical directional selectivity for the discrimination stimulus direction. We also found that directionally selective neurons appear to be critically important for other previously unsuspected visual functions, namely, discrimination of differences in speed and differences in flicker rate.

The goal of this chapter goes beyond familiarizing the reader with properties of motion mechanisms of cats and demonstrating a relationship between cortical directional mechanism and visual behavior. I also hope to show that cats provide an excellent animal model for the study of mammalian motion mechanisms, and offer the possibility of combining sophisticated psychophysical and physiological techniques in studying the same subject. The chapter is organized as follows. A brief review of the existing psychophysical and physiological data on the response to motion will be followed by a detailed comparison of motion mechanisms of normal cats and primates. Finally, the motion perception of cats deficient in directionally selective mechanisms wil be discused.

Recent psychophysical and physiological studies have provided a detailed description of the visual system's response to image motion (for recent review, see Nakayama, 1985). In both types of studies the direction of target motion and its speed were among the most important stimulus manipulations affecting response, and many striking parallels between psychophysical and physiological response to stimulus motion have emerged (see Sekuler, Pantle, & Levinson, 1978). Psychophysical studies have demonstrated the existence of mechanisms sensitive to opposite directions of motion (Pantle & Sekuler, 1969; Sekuler & Ganz, 1963). These directionally selective mechanisms are most sensitive at low spatial frequencies (Thompson, 1984; Watson, Thompson, Murphy, & Nachmias, 1980) and show relatively narrow tuning for stimulus speed (Pantle & Sekuler, 1968; Tolhurst, Sharpe, & Hart, 1973). Physiological studies showed that the response of many cortical neurons to motion is also affected by the direction of motion. Such neurons increase their firing rate in response to a particular direction of motion, but respond very little if at all in the opposite direction (e.g., Hubel & Wiesel, 1959). This type of selectivity is often encountered in the striate cortex (e.g., Orban, Kennedy, & Maes, 1981) and is even more common in such extrastriate visual areas as the mediotemporal (MT) area in the monkey (Maunsell & Van Essen, 1983; Mikami, Newsome, & Wurtz, 1986; Zeki, 1974) and the lateral suprasylvian area of the cat (e.g., Hubel & Wiesel, 1969; Rauschecker, von Grunau, & Poulin, 1987; Spear & Baumann, 1975). Directionally selective neurons in these extrastriate areas appear to respond best to relatively low spatial frequencies moving at high speeds (Maunsell & Van Essen, 1983; Morrone, DiStefano, & Burr, 1986) and are tuned to a narrow range of stimulus speeds (Morrone et al., 1986; Movshon, 1975).

Because of the similarity between the physiological and psychophysical response to stimulus direction, it has been commonly assumed that the perception of motion is somehow mediated by directionally selective neurons. However, such parallels are only suggestive; they do not demonstrate that the behavioral response depends on this physiological mechanism. Several years ago my laboratory began a series of studies aimed at examining more directly the relation between cortical directional selectivity and motion perception. We used cats reared for prolonged periods of time in an environment illuminated every 125 msec by a brief stroboscopic flash (i.e., at 8 Hz). Such rearing conditions irreversibly eliminate directional selectivity in most neurons in the striate cortex while leaving other receptive field properties relatively intact (Cynader & Chernenko, 1976). Similar effects on directional selectivity have been found in cats reared at lower strobe frequencies (2 Hz or lower) (Cremieux, Orban, Duysens, & Amblard, 1987; Cynader, Berman, & Hein, 1973; Kennedy & Orban, 1983;

Pasternak, Movshon, & Merigan, 1979). However, at these lower frequencies the deficit was not limited to directional selectivity, but involved profound effects on other receptive field properties including responsivity, orientation selectivity, and spatial resolution. The finding of selective directional deficits following rearing in 8-Hz stroboscopic illumination has been replicated more recently (Pasternak, Schumer, Gizzi, & Movshon, 1985) and extended to other cortical areas (Spear, Tong, McCall, & Pasternak, 1985) as well as the superior colliculus (M. Cynader, unpublished).

Modern psychophysical studies in humans have introduced many novel approaches and procedures to the study of motion perception. While some of these procedures (e.g., direction-specific adaptation used by Sekuler and colleagues; e.g., Sekuler & Ganz, 1963) are difficult to use with animals; other procedures, such as forced-choice detection and discrimination paradigms (e.g., Watson et al., 1980), are relatively easy to adapt to use with animals. In our studies of motion mechanisms in cats we used such forced-choice procedures. This not only allows a detailed analysis of spatiotemporal properties of motion mechanisms of cats, but also permits a comparison of the motion response of cats and human observers. We manipulated a variety of stimulus parameters including contrast, spatial and temporal frequency, and speed and direction of motion. We have found that properties of motion-detecting mechanisms in normal cats parallel those of primates. The following section briefly compares the spatiotemporal sensitivity and motion perception of cat and primate. The similarities in response to motion in two organisms as diverse as cat and human suggest some general principles of encoding of motion in the mammalian visual system.

**Procedural Details.** In all behavioral experiments described in this chapter I used a variation of the technique introduced by Berkley (1970). The cats were tested in a two-alternative forced-choice procedure, in which, on each trial they were confronted with two stimuli (e.g., two identical gratings moving in opposite directions). They were rewarded for making a nose-pressing response toward the positive stimulus (e.g., rightward motion) (for details, see Pasternak, 1986, 1987a; Pasternak & Leinen, 1986). The stimuli were sinusoidal gratings that either were stationary, drifted, or flickered sinusoidally in counterphase. Figure 1 shows an example of stimuli used in experiments described below. In addition to stimulus direction we manipulated four major stimulus parameters; spatial frequency, speed, temporal frequency, and contrast. The spatial frequency (c/deg) of the grating is defined as a number of cycles of the grating (a single dark and bright bar) per 1° of visual angle, while temporal frequency (c/sec or Hz) of the grating is defined as a number of cycles of the grating passing by a single point in space per second. The speed of the drifting grating is expressed in degrees per second or in hertz and depends on the spatial and temporal frequency of the grating: speed

## DETECTION

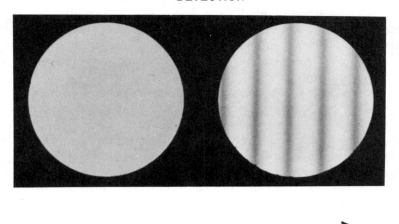

## DISCRIMINATION

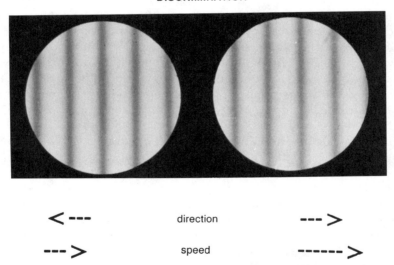

< ---          direction          --- >

--- >          speed          ------ >

**FIGURE 1.** Visual stimuli used in the experiments. *Top:* Stimulus configuration in a detection paradigm. The cats viewed a grating and a blank of the same mean luminance were rewarded for responding toward the grating. The location of the grating was randomized from trial to trial. *Bottom:* Stimulus configuration for the direction and speed discrimination tasks. In the direction discrimination experiment the cats viewed two identical targets that moved at the same speed but opposite directions of motion (indicated by the arrows) and were rewarded for the response toward the rightward moving target. During the speed-discrimination experiment the cats viewed two identical targets moving at different speeds (indicated by arrow length) and were rewarded for responding toward the faster moving grating.

(deg/sec) = temporal frequency (c/sec)/spatial frequency (c/deg). Contrast of the grating is defined as $L_{max} - L_{min}/L_{max} + L_{min}$, where $L_{max}$ is the luminance of the bright bar of the grating and $L_{min}$ is the luminance of the dark bar.

## I. SPATIOTEMPORAL VISION OF CATS AND PRIMATES

The spatial vision of cats, tested with gratings modulated sinusoidally in space and time, is remarkably similar to that of other mammals including human observers (see Uhlrich, Essock, & Lehmkuhle, 1981). Figure 2 shows the contrast sensitivity of cats and the effects on sensitivity of counterphase temporal modulation. The acuity (or the high-frequency cut off frequency) of cats reaches about 4 c/deg and is reduced at high temporal modulations, while low-frequency sensitivity peaks at moderate temporal frequencies. Similar spatiotemporal interactions have been reported for human observers (e.g., Robson, 1966). Figure 3 compares the contrast sensitivity of cats and macaque monkeys measured under similar testing conditions. The temporal contrast sensitivity of cats is nearly identical to that of primates when both are tested at optimal spatial frequencies (shown on the left of Fig. 3). The main difference in performance between the two species becomes apparent only when one compares their spatial resolution (right side of Fig. 3). The peak sensitivity and acuity of cats are at spatial frequencies that are about 10 times lower than those of primates. Anatomical and physiological factors that account for this difference have been discussed elsewhere (e.g., see Pasternak & Merigan, 1981). These similarities and differences in spatial and temporal vision of cats and primates should be kept in mind in the analysis of properties of motion mechanisms of the two species.

## II. MOTION MECHANISMS IN CATS AND HUMANS

### A. Detection of Motion and Flicker

Studies of directional mechanisms in humans, pioneered by Levinson and Sekuler (1975), show that sensitivity to moving gratings is higher than sensitivity to counterphase gratings (Kelly, 1979; Stromeyer, Madsen, Klein, & Zeevi, 1978; Watson et al., 1980). This difference approaches a factor of 2 at low spatial and high temporal frequencies, and decreases at high spatial and low temporal frequencies (Watson et al., 1980). Although there are several interpretations of this difference in sensitivity, most authors agree that the superior sensitivity to moving gratings is due to the activity of directionally selective mechanisms, which are less sensitive to slowly moving high spatial frequency targets (Levinson & Sekuler, 1975; Pasternak, 1986; Watson et al., 1980; Wilson, 1985; also see below). A comparison of the sensitivity of cats to moving and counterphase gratings

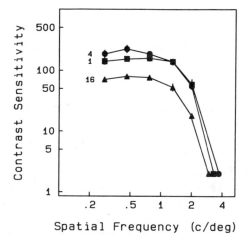

**FIGURE 2.** Spatial contrast sensitivity (1/threshold contrast) functions of a normal cat measured at 1, 4, and 16 Hz. Threshold contrast is defined as the contrast at which the animal detected the target with probability of 0.75. Stimulus: sinusoidal grating flickered in counterphase. Contrast: $L_{max} - L_{min}/L_{max} + L_{min}$, where $L_{max}$ is the luminance of the bright bar and $L_{min}$ is the luminance of the dark bar of the grating. Luminance: 75 cd/m²; target size: 10° deg; equal luminance surround; viewing distance, 35 cm; stimulus onset: modulated by a half cycle of a raised cosine of 1.25 Hz. Numbers next to each curve refer to the temporal frequency. Error bars: standard error of the mean.

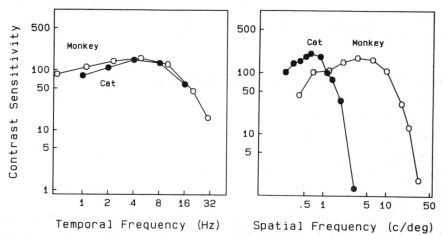

**FIGURE 3.** The comparison of cat (●), and monkey (○) temporal (left) and spatial (right) contrast sensitivity functions. The data for the spatial functions were measured with stationary gratings, under similar conditions. Temporal contrast sensitivities were measured at 0.28 c/deg for the cat and 0.7 c/deg for the monkey. Macaque monkey data were provided by William Merigan.

revealed a pattern of results nearly identical to that of human observers (Pasternak, 1986).Figure 4 shows that cats are more sensitive by a factor of 2 to moving than counterphase gratings at low spatial frequencies as well as at high temporal and spatial frequencies. No difference in sensitivity was found at high spatial and low temporal frequencies. Figure 5 summarizes the relative sensitivity to drifting and counterphase gratings in cats (Pasternak, 1986) and humans (Watson et al., 1980). In both species, sensitivity to flicker is comparable to that of motion only at low temporal and high spatial frequencies; otherwise, motion sensitivity is greater. These data suggest that the sensitivity of directional mechanisms in cats, as in humans, is maximal at low spatial and high temporal frequencies. A similar conclusion can be drawn from the direction-discrimination experiments described below.

## B. Direction Discrimination in Cats and Humans

A dependence of the sensitivity of directional mechanisms in humans on spatiotemporal stimulus parameters has been shown in direction-discrimination experiments by Watson et al. (1980). They found that while at low spatial frequencies subjects were able to discriminate grating direction at threshold contrasts, at high spatial and low temporal frequencies sensitivity for direction was lower than that for simple detection. They proposed that directional mechanisms become insensitive under the latter stimulus conditions. Similar studies performed in cats revealed a parallel pattern of results (Pasternak & Leinen, 1986). The results of the above studies, shown in Fig. 6, demonstrate nearly identical contrast sensitivities

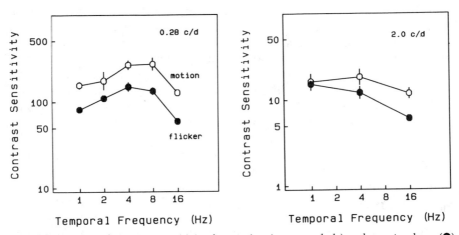

**FIGURE 4.** Temporal contrast sensitivity for moving (open symbols) and counterphase (●) grating at low (left) and high (right) spatial frequencies. Stimulus conditions identical to those described in Fig. 2.

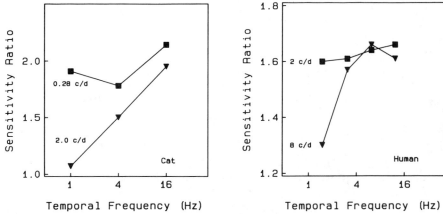

**FIGURE 5.** The comparison of cat and human contrast sensitivity to moving and counterphase flickering gratings. The difference in sensitivity expressed as sensitivity ratios (moving/counterphase flicker) shown for the cat (left) and the human observer (right). The ratios for the cat were calculated from the data shown in Fig. 4, while the ratios for the human were calculated from data by Watson, Thompson, Murphy, and Nachmias (1980). The ratio of 1 indicates identical sensitivity for moving and counterphase gratings, while the ration of 2 indicates that sensitivity for motion is higher than that for flicker by a factor of 2. Numbers next to each curve refer to the spatial frequency of the grating in cycles/degree. Note that in cats and in humans the superior sensitivity to motion is nearly absent at high temporal and low spatial frequency.

for detection and discrimination of direction at low spatial frequencies, with poorer direction sensitivity at higher spatial and lower temporal frequencies. Figure 7 shows these results as ratios (direction sensitivity/ detection sensitivity) to facilitate comparison with the human data of Watson et al. (1980) and Thompson (1984). The pattern of results was similar in the two organisms, although absolute differences in sensitivity are much larger for the cat than for the human observer. The maximal difference in humans does not exceed 0.3 log units, while in the cat the difference between the sensitivity of directional and detection mechanisms may exceed a factor of 10. As we will see below, this difference makes the cat an excellent model for the study of motion mechanisms.

Speed is another fundamental dimension of the moving target that has been examined in both species. The measurement of difference thresholds for speed in human observers has revealed a relative independence of speed discrimination of suprathreshold grating contrasts (McKee, Nakayoma, & Silverman, 1986). Moreover, observers are more accurate in discriminating differences in speed than differences in flicker rate for counterphase gratings of the same temporal frequency. Figure 8 shows the results of experiments, carried out in my laboratory, that examined the accuracy of both speed and temporal frequency discrim-

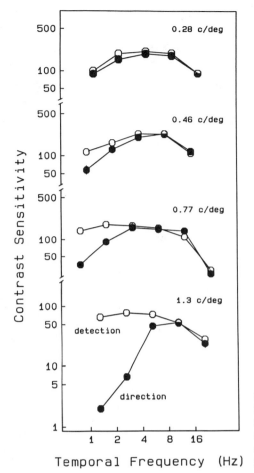

**FIGURE 6.** Contrast sensitivity for detecting the moving grating (○) and discriminating its direction of motion (●) measured over a range of temporal and spatial frequencies. See Fig. 2 for stimulus conditions.

ination over a range of contrasts in human observers and in cats (Pasternak, 1987a). The subjects were presented with two targets moving at different speeds and were to choose the faster moving or faster flickering stimulus. Gratings in the speed and flicker experiments were matched for temporal and spatial frequencies. To eliminate possible cues due to differences in apparent contrast between the standard and the comparison grating, contrast of the two stimuli on each trial was varied ±15% around the mean. The results for cats and humans were very similar. In both species the accuracy of discrimination, expressed as Weber fractions (speed increment/base speed) was relatively independent of grating contrast. Moreover, like humans, cats are less accurate in discriminating flicker rate than speed. Thus, the mechanisms underlying speed and temporal frequency discrimination appear to have similar properties in cats and humans.

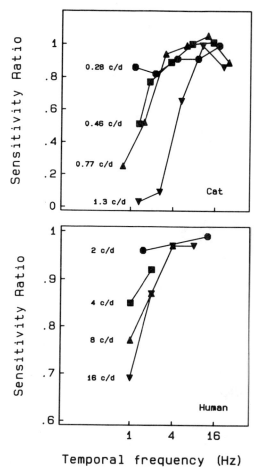

**FIGURE 7.** The comparison of differences in sensitivities for detecting the grating and discriminating its direction in the cat (top) and the human (bottom). Differences in sensitivity are expressed as sensitivity ratios (direction discrimination/detection). The ratios for the cat were calculated from the data shown in Fig. 6. The ratios for humans were calculated from Watson, Thompson, Murphy, and Nachmias (1980) for 2 c/deg and from Thompson (1984) for other spatial frequencies. Numbers next to individual curves refer to the spatial frequency of the target in cycles/degree. The ratio of 1 indicates identical sensitivity for detection and direction discrimination, while the ratio less than 1 indicates that sensitivity for detection is higher than that for direction. Note that the ratios are lowest at high spatial and low temporal frequencies for both cat and human observers, indicating the reduced sensitivity of directional mechanisms.

## C. Differences between Cats and Primates

The comparisons shown above emphasize similarities in the properties of motion mechanisms in cat and primate in response to variation in such fundamental dimensions as contrast and spatial and temporal frequencies.

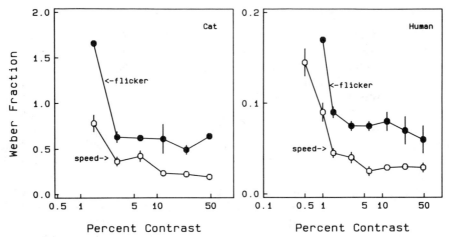

**FIGURE 8.** Weber fractions for discriminating differences in speed of moving gratings (threshold increment in speed/base speed, ○) and for discriminating differences in temporal frequencies of counter-phase gratings (●) plotted as a function of grating contrast. Moving and counterphase gratings were identical temporal and spatial frequency. Cat: 0.28 c/deg, 2.24 Hz; human: 2/cdeg, 4 Hz.

Obviously, there are also important differences between the two species. The most obvious is the substantially lower spatial resolution of the cat than in the primate. A second difference is the significantly lower accuracy of speed and temporal frequency judgments in cats than in humans. A third difference is the great disparity in sensitivity between detecting and directional mechanisms in the cat at higher spatial, lower temporal frequency. While the first two differences constitute a certain limitation, the third offers an important advantage in using the cat in the study of neural mechanisms of motion perception (see below).

## III. VISION OF CATS WITH LOSS OF CORTICAL DIRECTIONAL SELECTIVITY

Psychophysical measurements described above provided a detailed description of motion mechanisms in normal cats. The next series of studies was aimed at establishing the relation between motion mechanisms studied psychophysically and cortical directional selectivity. We measured motion thresholds in cats with greatly reduced directional selectivity in cortical neurons. The following section provides a detailed analysis of visual function in such animals. It will be demonstrated that directional selectivity is required, not only for the ability to discriminate stimulus direction, but also for the ability to discriminate speeds and temporal frequencies.

## A. Physiological Properties of Single Cortical Neurons

Single-unit recordings were obtained from the striate cortex (in J. A. Movshon's laboratory) and lateral suprasylvian area (in P. Spear's laboratory) in cats reared in 8-Hz stroboscopic illumination and tested behaviorally in my laboratory (Pasternak et al., 1985; Spear et al., 1985). Directional selectivity in area 17 was almost absent (over 90% loss); other receptive field properties, including orientation selectivity and temporal frequency response, were indistinguishable from those of normal cats. The only other property affected by strobe rearing was spatial frequency response. Both the optimal spatial frequency and the spatial resolution of area 17 neurons were lower than those in normal cats by about an octave. Recordings from the lateral suprasylvian cortex, an area that in normal cats contains a high proportion of directionally selective cells, showed a 90% loss of directional selectivity, but without abnormalities in other receptive field properties (Spear et al., 1985). M. Cynader (personal communication, 1987) recorded from area 18 and the superior colliculus of such animals and found similarly reduced directional selectivity, while other receptive field properties remained nearly intact.

## B. Spatial and Temporal Vision

Measurements of spatial contrast sensitivity revealed a sensitivity loss in the strobe-reared cats at higher spatial frequencies and approximately an octave loss in visual acuity (left side of Fig. 9). This result was consistent with the octave decrease in the spatial frequency response of area 17 neurons described above. Figure 9 shows the temporal contrast sensitivity of these animals measured at a low spatial frequency. There is a slight sensitivity loss relative to normals at higher temporal frequencies ($>4$ Hz), but little, if any, decrease at lower temporal frequencies. The intact sensitivity at low spatial and temporal frequencies parallels our physiological finding of normal sensitivity of area 17 neurons. Thus, our initial physiological and psychophysical measurements showed that the sensitivity of detecting mechanisms in strobe-reared cats was within the normal range at low spatial and low to moderate temporal frequencies. Moreover, psychophysical and physiological measurements failed to reveal abnormalities in response to other stimulus dimensions (see Pasternak & Leinen, 1986; Pasternak et al., 1985). Thus, any deficit in motion discriminations at these frequencies is likely to be due to the severe reduction in cortical directionality.

## C. Detection of Motion and Flicker by Nondirectional Mechanisms

The next set of experiments explored the difference in sensitivity to moving and counterphase gratings normally found in humans and cats (see above). This difference has commonly been attributed to the detection of

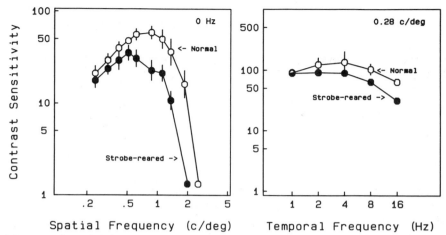

**FIGURE 9.** Spatial (left) and temporal (right) contrast sensitivity functions for normal and strobe-reared cats. Each data point in the curve on the left is a geometric mean of sensitivities for five normal and eight strobe-reared cats. Each data point in the curve on the right is a geometric mean of sensitivities for four normal and four strobe-reared cats.

the oppositely moving components of the counterphase gratings by independent directionally selective mechanisms (Levinson & Sekuler, 1975; Watson et al., 1980). Wilson (1985), on the other hand, has proposed that counterphase flicker is detected by a separate flicker mechanism. He explained the difference in sensitivity by the superior sensitivity of directionally selective mechanism over flicker-detecting mechanism. Although fundamentally different, both models predict that motion and flicker sensitivity should converge in the absence of directionally selective mechanisms. The results were consistent with this prediction. The five strobe-reared cats showed nearly identical sensitivities to motion and flicker (Pasternak, 1986). Figure 10 illustrates this result for one of the animals (compare with the data for the normal cat shown in Fig. 4). Although the sensitivity of these cats to counterphase gratings was within the range of normal values, their sensitivity to moving gratings was reduced by approximately 0.3 log unit. This resulted in nearly equal sensitivity for motion and flicker. These data support the idea that directionally selective mechanisms are responsible for the difference in sensitivity to the two types of temporal modulation by elevating sensitivity to motion. However, they are inconsistent with with the idea, originally proposed by Levinson and Sekuler (1975), that this difference is due to detection of counterphase gratings by directionally selective mechanisms. Since the main effect of reduced directional selectivity was a two-fold threshold elevation for detecting moving gratings with no change in sensitivity for counterphase gratings, it appears that rather than detecting

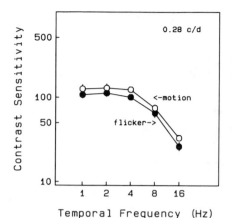

**FIGURE 10.** Contrast sensitivity for moving (O) and counterphase (●) grating plotted as a function of temporal frequency for one of the strobe-reared cats (cat 811).

counterphase flicker, directionally selective mechanisms provide the superior sensitivity for moving targets.

## D. Direction Discrimination

The data presented above showed that the effects of reduced directional selectivity on detection of moving gratings were rather subtle (<0.3 log unit). At this point we knew that animals with loss of directional selectivity could detect moving patterns at very low contrasts (<1%), but we did not know whether these cats were able to identify the direction of this motion. In the next study we measured contrast sensitivity for discriminating the direction of motion (Pasternak & Leinen, 1986; Pasternak et al., 1985). Contrast thresholds for discriminating between two identical gratings moving in opposite directions were compared to contrast thresholds for detecting the grating. The two types of thresholds were obtained in two separate experiments but usually on the same day. The performance of one of the strobe-reared cats is shown in Fig. 11.

While normal cats (see figure 4) had nearly identical contrast sensitivity for detecting the 0.28-c/deg drifting grating and discriminating its direction of motion, animals with severe loss of directional selectivity required 10 times higher contrast to discriminate a grating's direction than to detect it. This difference between detection and discrimination increased with spatial frequency and at 0.77 c/deg the animal was unable to identify stimulus direction above chance levels, although it was still able to detect the grating at relatively low contrasts. This type of results was found in all four cats tested in this experiment. Thus, strobe-reared cats showed severe deficits in identifying stimulus direction, although there was no effect on

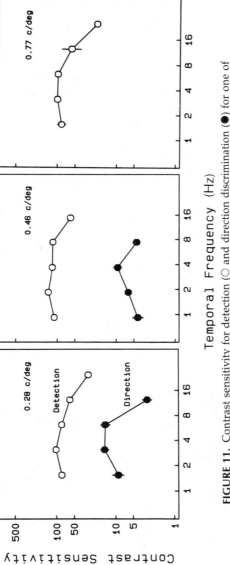

**FIGURE 11.** Contrast sensitivity for detection (○) and direction discrimination (●) for one of the strobe-reared cats (cat 812) as a function of temporal frequency. Compare with the normal data shown in Fig. 6.

421

detection of flicker and only very subtle effects on detection of moving gratings. Since both neural and behavioral deficits were limited to responses to the direction of stimulus motion, it is reasonable to assume that the deficit in direction discrimination resulted from a loss of cortical directional selectivity. Further, it is likely that the residual sensitivity of strobe-reared cats for direction was due to the limited and/or reduced contrast sensitivity of the remaining neurons that encode stimulus direction.

The residual contrast sensitivity of strobe-reared cats for discriminating direction was sufficient that these animals should be able to discriminate directions at very high contrasts. Indeed, when tested with high-contrast isotropic random dot patterns or gratings, these animal performed as well as normal cats. They were not only able discriminate opposite directions over a wide range of speeds (2 to 64 deg/sec), but were able to discriminate stimulus directions at the slowest discriminable speed (Pasternak & Leinen, 1986). Moreover, their difference thresholds for the direction of movement, measured with drifting random dot patterns, were identical to those of normal cats (see Pasternak & Merigan, 1984, for the performance of normal cats). Thus, even a small number of directionally selective neurons can support normal direction discriminations, provided that the targets are of high contrast. Indeed, casual observation of these cats revealed no obvious abnormalities. Their visuomotor behavior was indistinguishable from normal.

The finding that the normal cats were unable to discriminate stimulus direction, for fine gratings moving slowly, suggests a possible location of neurons signaling stimulus direction to the animal. Physiological studies of areas 17 and 18 have shown that the two areas have different spatiotemporal properties, with area 17 neurons responding to higher spatial and lower temporal frequencies. Area 18 neurons prefer low spatial frequencies modulated at higher temporal rates (Movshon, Thompson, & Tolhurst, 1978). Moreover, the percentage of directional cells in area 17 appears to be lower than in area 18 (Orban et al., 1981). Since optimal stimulus parameters for discriminating direction in psychophysical studies are consistent with the properties of area 18 neurons, it is reasonable to propose that these neurons are much better suited for signaling stimulus direction than neurons in area 17. This possibility is indirectly supported by the complete loss of directional sensitivity of strobe-reared cats at spatial frequencies above 0.7 c/deg, spatial frequencies at which area 18 neurons are largely unresponsive. Since the directional selectivity loss in these animals is not limited to area 17 but extends to area 18 (M. Cynader, personal communication; also see Kennedy & Orban, 1983), it is likely that the residual sensitivity of strobe-reared cats for direction is mediated by the few remaining directionally selective neurons in area 18.

In summary, these experiments have provided the first demonstration of the involvement of cortical directional selectivity in discrimination of direction. Conversely, it appears that even a small number of directionally selective neurons can support normal direction discriminations if targets are of sufficiently high contrasts. It is likely that these remaining neurons are located in area 18 and beyond.

## E. Discrimination of Speeds and Temporal Frequencies

The experiments described above showed that strobe-reared cats provide a useful animal model for the study of the role of directional selectivity in visual behavior. The next set of experiments examined the role of directionally selective neurons in an animal's ability to discriminate stimulus speeds. We took advantage of the large difference between the sensitivity of directional and nondirectional mechanisms shown by strobe-reared cats. Since these animals could discriminate directions only at high contrasts, we examined the contrast dependence of their ability to discriminate differences in speed. We reasoned that if these cats were unable to discriminate either direction or speed at low contrasts, then it is likely that information about speed is encoded by directionally selective mechanisms.

The study of speed discrimination is complicated by the fact that for a given spatial frequency target any change in speed must be accompanied by a change in temporal frequency [speed (deg/sec) = temporal frequency (c/sec)/spatial frequency (c/deg)]. To determine whether the animals make genuine speed discriminations, it is necessary to rule out possible use of temporal frequency cues. McKee et al. (1986) have demonstrated that observers are able to make pure speed discriminations by showing that relatively large changes in temporal frequencies (produced by changes in spatial frequencies) had no effect on the accuracy of speed discrimination. In the present study we measured the discrimination of the speed of moving gratings and of the temporal frequency of gratings flickered in counterphase (Pasternak, 1987b). The performances for one of the normal and one of the strobe-reared cats are shown in Fig. 12; those of the other three cats were nearly identical. Weber fractions are plotted as a function of stimulus contrast. The broken vertical line indicates the contrast required to detect the presence of the grating and the inverted triangle indicates the contrast needed to discriminate its direction of motion. Both normal and strobe-reared cats were less accurate in discriminating temporal frequencies than speeds. The lower accuracy of temporal frequency discriminations makes it unlikely that the animals relied on temporal frequency differences to make speed discriminations. One interesting aspect of these results was that the two functions parallel each other and that cats with reduced directionality were unable to discriminate either speeds or temporal frequency at contrasts too low to discriminate stimulus direction. Note that at these low contrasts the cats were still able to detect the presence of

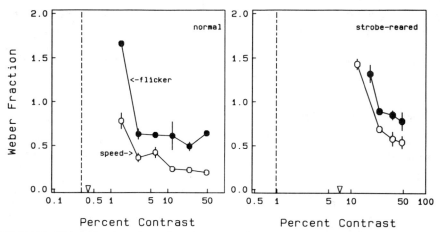

**FIGURE 12.** Discrimination of speeds and temporal frequencies for the normal and one of the strobe-reared cats (cat 812). The data for the normal cat are replotted from Fig. 8. Contrast threshold for detecting the grating is indicated by a vertical broken line, while contrast threshold for discriminating its direction is shown by an inverted triangle. Note that the two thresholds in the normal cat are very similar, while those in the strobe-reared cat are separated by nearly a log unit.

the grating. It thus appears that the ability to discriminate speed and flicker differences is determined by the sensitivity of directional mechanisms rather than the sensitivity of detecting mechanisms. When the sensitivity of directional mechanisms becomes too low, discrimination of speed is not possible.

Flicker discrimination thresholds also appear to depend upon the sensitivity of directional mechanisms; their contrast dependence parallels that for speed discrimination. The inability of strobe-reared cats to discriminate flicker differences is surprising given that neither behavioral nor physiological measurements revealed abnormalities in temporal frequency response in this temporal frequency range. It is unlikely that this result was due to some other neural abnormality that we failed to detect during physiological recordings, since the normal cat tested under conditions of reduced sensitivity for direction (high spatial, low temporal frequency; see above) showed impaired speed and flicker discrimination nearly identical to that of strobe-reared cats (see Fig. 6 in Pasternak, 1987b). The dependence of speed and temporal frequency discrimination upon directional mechanisms, and the high correlation ($r = 0.95$; see Pasternak, 1987b) between speed and flicker thresholds suggest that speed and temporal frequency signals are processed by a common directionally selective mechanism. The lower accuracy of flicker discrimination and recent finding of longer temporal summation of flicker signals (Pasternak, 1987a) suggest that flicker discrimination may depend upon directionally selective velocity signals.

## IV. SUMMARY AND CONCLUSIONS

1. Motion mechanisms in cats and primates show a similar dependency on such major stimulus parameters as contrast, direction, speed, and temporal frequency. The similarity in response to motion in two such diverse organisms suggests some general principles of encoding of motion signals in the mammalian visual system.

2. Cats reared under 8-Hz stroboscopic illumination have neural deficits, which at low spatial frequencies are largely limited to directional selectivity. Such cats provide a useful model for examining the role of directional selectivity in perception.

3. These animals are as capable as normal cats at detecting the presence of flickering patterns, but they show a small sensitivity loss for detecting moving patterns of the same temporal frequency. The sensitivity loss for detecting motion results in a convergence of sensitivity for motion and flicker unlike that of normal cats and humans. This result provides direct support for the hypothesis that the superior sensitivity for motion over flicker is due to the activity of directional mechanisms.

4. Although such animals detect moving targets at contrasts only slightly below normal, they require 10 times higher contrasts to identify the direction of motion of these targets. It is argued that the residual sensitivity for direction is likely to be determined by the activity of a small number of remaining directionally selective neurons. This result provides the first demonstration of the involvement of directionally selective neurons in direction discrimination. Since these cats had no deficits in discriminating direction at high stimulus contrast, it appears that even a small number of directionally selective cells can support normal direction discrimination with highly visible targets. The spatiotemporal properties of residual directional mechanisms in strobe-reared cats suggest that the neurons signalling stimulus direction to the animal are located in area 18 and beyond.

5. Strobe-reared cats are unable to discriminate differences in speed at contrasts too low to discriminate directions of motion. Their discrimination of temporal frequencies paralleled that of speed and was also limited by the sensitivity of directional mechanisms. It appears then that speed and flicker discriminations depend on directionally selective velocity mechanisms.

## ACKNOWLEDGMENTS

This research was supported by grants from the National Eye Institute, EY 04118, EY06175 (to TP) and EY 01319 (Core Grant to the Center for Visual Science). I thank Bill Merigan and Walt Makous for comments on the manuscript.

## REFERENCES

Berkley, M. A. (1970). Visual discrimination in the cat. In W. C. Stebbins (Ed.), *Animal psychophysics: The design and conduct of sensory experiments* (pp. 231–247). New York: Appleton-Century-Crofts.

Blake, R., & Camisa, J. M. (1977). Temporal aspects of spatial vision in the cat. *Experimental Brain Research, 28*, 325–333.

Blake, R., Halopigian, K., & Wilson, H. R. (1987). Spatial frequency discrimination in cats. *Journal of the Optical Society of America,* 1443–1449.

Cremieux, J., Orban, G., Duysens, J., & Amblard, B. (1987). Response properties of area 17 neurons in cats reared in stroboscopic illumination. *Journal of Neurophysiology, 57*, 1511–1535.

Cynader, M., Berman, N., & Hein, A. (1973). Cats reared in stroboscopic illumination: Effects on receptive fields in visual cortex. *Proceedings of the National Academy of Sciences of the U.S.A. 70*, 1353–1354.

Cynader, M., & Chernenko, G. (1976). Abolition of directional selectivity in the visual cortex of the cat. *Science, 193*, 504–505.

Hubel, D. H., & Wiesel, T. N. (1959). The cat's striate cortex. *Journal of Physiology (London), 148*, 574–591.

Hubel, D. H., & Wiesel, T. N. (1969). Visual area of the lateral suprasylvian gyrus (Clare-Bishop area) of the cat. *Journal of Physiology (London), 202*, 251–260.

Kelly, D. H. (1979). Motion and vision. II. Stabilized spatio-temporal threshold surface. *Journal of the Optical Society of America, 69*, 1340–1349.

Kennedy, H., & Orban, G. A. (1983). Response properties of visual cortical neurons in cats reared in stroboscopic illumination. *Journal of Neurophysiology, 49*, 686–704.

Levinson, E., & Sekuler, R. (1975). The independence of channels in human vision selective for direction of movement. *Journal of Physiology (London), 250*, 347–366.

Maunsell, J. H. R., & Van Essen, D. C. (1983). Response properties of single units in middle temporal visual area of the macaque. *Journal of Neurophysiology, 49*, 1127–1147.

McKee, S., Nakayama, K., & Silverman, G. H. (1986). Presize velocity discrimination despite variations in temporal frequency and contrast. *Vision Research, 26*, 609–619.

Mikami, A., Newsome, W. T., & Wurtz, R. H. (1986). Motion selectivity in macaque visual cortex. I. Mechanisms of direction and speed selectivity in extrastriate area MT. *Journal of Neurophysiology, 55*, 1308–1327.

Morrone, M. C., DiStefano, M., & Burr, D. C. (1986). Spatial and temporal properties of neurons of the lateral suprasylvian cortex of the cat. *Journal of Neurophysiology,* **56,** 969–986.

Movshon, J. A. (1975). Velocity tuning of single units in cat striate cortex. *Journal of Physiology(London),* **249,** 445–468.

Movshon, J. A., Thompson, I. D., & Tolhurst, D. J. (1978). Spatial and temporal contrast sensitivity of neurons in areas 17 and 18 of cat's cortex. *Journal of Physiology (London),* **283,** 101–120.

Nakayama, K. (1985). Biological image motion processing: a review. *Vision Research,* **25,** 625–661.

Orban, G. A., Kennedy, H., & Maes, H. (1981). Response to movement of neurons in areas 17 and 18 of the cat: Direction selectivity. *Journal of Neurophysiology,* **45,** 1059–1073.

Pantle, A., & Sekuler, R. (1968). Velocity sensitive elements in human vision: Initial psychophysical evidence. *Vision Research,* **8,** 445–450.

Pantle, A., & Sekuler, R. (1969). Contrast response of human visual mechanisms sensitive to orientation and direction of motion. *Vision Research,* **9,** 397–406.

Pasternak, T. (1986). The role of cortical directional selectivity in detection of motion and flicker. *Vision Research,* **26,** 1187–1194.

Pasternak, T. (1987a). Temporal summation is longer for temporal frequency than for speed discrimination. *Investigative Ophthalmology & Vision Science,* **28**(Suppl.), 297.

Pasternak, T. (1987b). Discrimination of differences in speed and flicker rate depends on directionally selective mechanisms. *Vision Research,* **27,** 1881–1890.

Pasternak, T., & Leinen, L. (1986). Pattern and motion vision in cats with selective loss of cortical directional selectivity. *Journal of Neuroscience,* **6,** 938–945.

Pasternak, T., & Merigan, W. H. (1981). Luminance dependence of spatial vision of the cat. *Vision Research,* **21,** 1333–1339.

Pasternak, T., & Merigan, W. H. (1984). Effects of stimulus speed on direction discriminations. *Vision Research,* **24,** 1349–1355.

Pasternak, T., Movshon, J. A., & Merigan, W. H. (1979). Creation of directional selectivity in adult strobe-reared cats. *Nature (London),* **292,** 834–836.

Pasternak, T., Schumer, R. A., Gizzi, M. S., & Movshon, J. A. (1985). Abolition of cortical directional selectivity affects visual behavior in cats. *Experimental Brain Research,* **61,** 214–217.

Rauschecker, J. P., von Grunau, M. W., & Poulin, C. (1987). Centrifugal organization of direction preferences in the cat's lateral suprasylvian visual cortex and its relation to flow field processing. *Journal of Neuroscience,* **7,** 943–958.

Robson, J. G. (1966). Spatial and temporal contrast sensitivity functions of the visual system. *Journal of the Optical Society of America,* **56,** 1141–1142.

Sekuler, R., & Ganz, L. (1963). Aftereffect of seen motion with the stabilized retinal image. *Science,* **139,** 419–420.

Sekuler, R., Pantle, A. J., & Levinson, E. (1978). Physiological basis of motion perception. In R. Held, N. W. Leibowitz, & H. L. Teuber (Eds.), *Handbook of Sensory Physiology* (Vol. 8, pp. 67–96). Berlin: Springer.

Spear, P. D., & Baumann, T. P. (1975). Receptive field characteristics of single neurons in lateral suprasylvian visual area of the cat. *Journal of Neurophysiology,* **38,** 1403–1420.

Spear, P. D., Tong, L., McCall, M. A., & Pasternak, T. (1985). Developmentally induced loss of direction selective neurons in cat' lateral suprasylvian visual cortex. *Developmental Brain Research,* **20,** 281–285.

Stromeyer, C. F., Madsen, J. C., Klein, S., & Zeevi, Y. Y. (1978). Movement-selective mechanisms in human vision sensitive to high spatial frequencies. *Journal of the Optical Society of America,* **68,** 1002–1005.

Thompson, P. (1984). The coding of velocity of movement in the human visual system. *Vision Research,* **24,** 41–45.

Tolhurst, D. J., Sharpe, C. R., & Hart, G. (1973). The analysis of the drift rate of moving sinusoidal gratings. *Vision Research,* **13,** 2545–2555.

Uhlrich, D. J., Essock, E. A., & Lehmkuhle, S. (1981). Cross-species correspondence of spatial contrast sensitivity functions. *Behavioral Brain Research,* **2,** 291–299.

Watson, A. B., Thompson, P. G., Murphy, B. J., & Nachmias, J. (1980). Summation and discrimination of gratings moving in opposite directions. *Vision Research,* **20,** 341–348.

Wilson, H. R. (1985). A model for direction selectivity in threshold motion perception. *Biological Cybernetics,* **51,** 213–222.

Zeki, S. (1974). Functional organization of a visual area in the posterior bank of the superior temporal sulcus of the rhesus monkey. *Journal of Physiology (London),* **236,** 549–573.

# 14

## THE PERCEPTION OF MUSICAL PATTERNS BY HUMAN INFANTS: THE PROVISION OF SIMILAR PATTERNS BY THEIR PARENTS

*Sandra E. Trehub*

*University of Toronto, Mississauga, Ontario*

The study of musical pattern perception in infancy might be viewed as a rather esoteric enterprise, of interest perhaps to scholars of aesthetics, music, or other seemingly optional aspects of human endeavor. For reasons as yet unknown, however, music is no less universal than language, appearing in every known culture. And while the creation or production of music may engage a minority of individuals, all of us *listen* to music, whether actively or passively, by choice or circumstance. There have been suggestions that musical and linguistic capacities use common organizing principles to impose structure on their disparate input and that these principles may be innately specified (Lerdahl & Jackendoff, 1983).

There are further reasons why the study of music perception in infancy is, or should be, of general interest to the developmentalist. As will become apparent in the pages that follow, preverbal infants are remarkably precocious in this domain, exhibiting a disposition for global, relational (i.e., adult-like) processing of musical patterns (Trehub, 1985, 1987). What is even more remarkable is that parents, in their interactions with young infants, behave as if they have intuitive knowledge of these skills, giv-

The preparation of this chapter was assisted by grants from the Natural Sciences and Engineering Research Council of Canada and the University of Toronto.

**429**

ing unusual emphasis to musical elements in the temporal and pitch patterning of their vocalizations (M. Papoušek & Papoušek, 1981). The nature and possible functions of this parental attunement to infant perceptual abilities will be discussed in the final portion of this chapter.

Finally, the study of auditory pattern perception in human infancy is relevant to those concerned with comparative perception. Just as the development of specific conditioning techniques has been necessary to "bridge the language gap between animal subject and human experimenter" (Stebbins, Coombs, & Prosen, 1984, p. 1), so are similar methodological concerns central to the study of preverbal human listeners with limited cognitive and motoric skills. Moreover, the perception of temporally patterned signals is as important to nonhuman species as it is to humans. In this regard, it is interesting to note that musical tunes have been used to investigate the processing of complex auditory relations in songbirds, rats, and monkeys (D'Amato & Salmon, 1982, 1984; Hulse, Cynx, & Humpal, 1984; Hulse, Page, & Braaten, 1989). One tentative hypothesis to emerge from this realm is that, in contrast to humans, auditory pattern processing in nonhuman species may be restricted to species-specific signals (D'Amato & Salmon, 1984).

## THE UNDERLYING QUESTIONS

For a number of years, my colleagues and I have been attempting to specify the skills that infants bring to the task of listening to complex auditory sequences. Musical patterns are simply complex auditory sequences that are multidimensional and highly structured, and thus are particularly well suited for the study of auditory information processing. Despite ever-increasing documentation of music perception abilities in adults (see Deutsch, 1982; Dowling & Harwood, 1986; Sloboda, 1985), the origin of these abilities remains unclear. Do human listeners have particular propensities to structure complex auditory input that are independent of listening experience? Or are such abilities attributable to exposure, formal or informal, to the music of our culture?

In our study of music perception in infancy, we have considered sequences of sounds as well as sets (i.e., categories) of sounds, and have focused on the perception of structure or perceptual organization. Recent attempts to identify principles of perceptual organization (e.g., grouping, selective attention) in adult vision and audition (see Kubovy & Pomerantz, 1981) have provided conceptual impetus for the study of auditory pattern perception and its development. Pomerantz (1981) notes that "one of the earliest steps in perception must be grouping, whereby as yet unidentified parts, or blobs of sensory input, are perceptually linked into potential objects to be recognized by later processing mechanisms. . . . In this way, higher-order units are perceived without prior recognition of their components" (p. 179). It is important to ascertain whether such principles of perceptual organization are operative in infancy and how their operation is transformed and enhanced by age-related changes in mental structure and experience.

## THE PROCEDURE

Our general methodological approach to the study of auditory pattern perception in infancy is to present auditory sequences that exceed the immediate memory span of the infant and to ascertain which deviations from the original pattern are detectable (i.e., what information is retained). In this way, we can gain access to infants' strategies for organizing perceptual information, such as the segmentation and chunking of input, "disortions" of input to simplify coding, and the synthesis of global representations from local details. Musical patterns are particularly useful in this regard because the psychological dimensions that characterize musical patterns (pitch, duration, loudness, and timbre) are equally characteristic of nonmusical patterns.

Specifically, we train infants (6 months or older) to turn toward a loudspeaker (the sound source) when they hear a *change* in sound. This is neither as difficult nor as arbitrary as it may seem. The infant's limited response repertoire and even more limited ability to learn arbitrary associations between stimuli and responses leave us with no alternative but to devise tasks that capitalize on naturally occurring behaviors. One of these is the infant's propensity to turn toward a sound source. We maximize the likelihood of this response by reinforcing its occurrence to "appropriate" (i.e., altered) auditory signals. The challenge lies in getting the infant to turn only in the case of sound change and to refrain from turning to the standard stimulus.

The infant sits on the parent's lap in one corner of a sound-attenuating booth, listening to a repeating melody presented from a loudspeaker 45° to one side (see Fig. 1). Initially, the infant turns frequently toward the loudspeaker, with its repeating melody, but soon redirects interest from this repeating melody to the tester, who manipulates interesting puppets directly in front of the infant. The tester records, by means of a hand-held button box, when the infant is looking directly ahead (in principle, ready for a test trial) and when the infant turns 45° to the sound source (either correct or false-positive responses). At randomly determined intervals, when the infant is facing directly ahead (i.e., watching the tester), we present (i.e., substitute) a melody that embodies some change, substantial or subtle, to the original melody. On such occasions, infants tend to turn toward the source of the novel or altered pattern. Head turns in conjunction with sound changes are reinforced by the illumination and activation of animated toys near the loudspeaker; turns in the absence of such changes are unreinforced.

**FIGURE 1.** Left side: Infant looks directly ahead, ignoring the familiar repeating pattern from the loudspeaker. Right side: Infant turns when the pattern changes, receiving visual reinforcement for a correct response.

During an initial training phase, we require infants to meet a criterion of four successive correct responses to sound change. In the subsequent test phase, we present approximately 30 test trials, half of which involve a change in sound, the other half being control trials with no change (delivered automatically and in random order). To preclude the possibility of inadvertent cuing by the tester or parent, both wear headphones that deliver masking sounds. The procedure lends itself reasonably well to the signal-detection model (Green & Swets, 1966), with the proportion of head turns on change trials providing an estimate of the probability of a hit and the proportion of turns on no-change trials providing an estimate of the probability of a false alarm. If the mean $d'$ is significantly greater than zero, this indicates that the infants can detect the change in question. (For further details on data analysis, see Thorpe, Trehub, Morrongiello, & Bull, 1988.)

## THE ROLE OF CONTOUR IN MELODY PERCEPTION

Adults' recognition of familiar melodies does not depend upon specific notes or pitches but rather on *relations* between component pitches (Attneave & Olson, 1971; Bartlett & Dowling, 1980), specifically, the pattern of *intervals*. Intervals refer to the distance between tones in a melody, which can be designated in semitones. In Western European music, the *semitone* is the smallest interval between two notes, representing the frequency ratio of $2^{1/12}$ or 1.06 : 1.00 and dividing the octave into 12 equal intervals. Sequences of tones with the same intervals but different notes or pitches are considered to be *transpositions* and are, under many circumstances, indistinguishable from one another. For example, despite hearing the NBC chimes thousands of times over many years, listeners are nevertheless unable to recall or recognize the exact pitches (Attneave & Olson, 1971). The implication is that our representation of familiar melodies is abstract, involving exact interval information about the relations between adjacent pitches but little information about the pitches themselves (see Fig. 2).

Our representation of unfamiliar melodies is even more abstract. Not only do we discard precise pitches, as is the case with familiar melodies, we also dispense with precise interval information (see Fig. 2). What remains intact is configural information about successive directional changes in pitch (ups and downs), or the *melodic contour* (Dowling, 1978). The contour of a melody refers to the pattern of successive pitch changes (rising, falling, staying the same) within a melody, with the *direction* of such changes being relevant rather than their extent (see Dowling & Harwood, 1986).

What about infants? Some years ago, we established, with a quite different procedure, that 5-month-old infants could discriminate between different melodic patterns and that they did so on the basis of relational as opposed to absolute information (Chang & Trehub, 1977a). After infants' cardiac decelerative response had habituated to a six-tone pattern, we presented them with two types of changes: transpositions, which had all new tones but preserved the interval relations of the original pattern; and

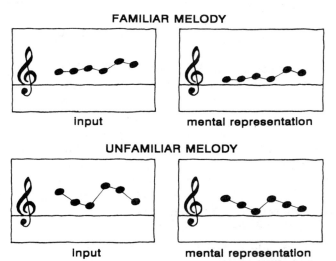

**FIGURE 2.** Adults' representation of familiar melodies includes information about intervals, not about exact pitch levels (upper panel). Their representation of unfamiliar melodies specifies the direction of pitch changes (lower panel).

contour changes, in which the same tones as the transpositions were presented in permuted order. Thus, both comparison patterns had identical component tones but one preserved the relations inherent in the standard or original pattern whereas the other did not. The outcome was that response recovery (i.e., dishabituation) did not occur for the transposed pattern but was clearly evident for the pattern with relational changes, implying that infants had encoded relational aspects of the original melody. It was impossible to determine, however, whether contour or interval changes were critical to infants' representation of the original melody because these changes co-occurred in the permuted comparison.

In a subsequent study with the conditioned headturn procedure (Trehub, Bull, & Thorpe, 1984), we tested 8- to 10-month-old infants for their discrimination of various changes in a six-tone melody, including transpositions (same intervals and contour, different component tones), contour-preserving transformations (same contour, different intervals), and contour-violating transformations (same component tones reordered) (see Fig. 3). On the basis of our previous findings (Chang & Trehub, 1977a), we expected that infants would fail to discriminate transpositions from the original melody. Such failure, if coupled with success on the other changes, would imply that infants encoded interval relations between notes, as is the case for adults and children with familiar or overlearned melodies (Bartlett & Dowling, 1980; Trehub, Morrongiello, & Thorpe, 1985). On the other hand, success would imply that infants' representation included exact pitches, in line with the performance of those celebrated listeners with "absolute" or "perfect" pitch (Ward & Burns, 1982). Yet another possibility was that infants would fail to detect transpositions *and* contour-preserving transformations, succeeding only on contour-violating transformations. This pattern of findings would imply that infants encoded the contour of auditory sequences, as is the case for adults and children with unfamiliar melodies (Dowling, 1978; Morrongiello, Trehub, Thorpe, & Capodilupo, 1985). When the retention interval was very brief (800 msec), infants detected all of the changes, performing best, however, on the contour changes. The minimal memory demands of this situation enabled infants to use exact pitch cues. A somewhat longer retention interval (2.6 sec) led to a different pattern of findings. Infants did not respond to transpositions or to contour-preserving transformations but did detect contour-violating transformations (see Fig. 4). Under the more difficult conditions, then, it became evident that infants encoded and retained contour information at the expense of exact pitch or interval information. In this study, the change in contour had been accomplished by reordering the four internal tones of the six-tone melody. We established, in further research, that contour changes were detectable even if any single tone of a six-tone melody was altered (Trehub, Thorpe, & Morrongiello, 1985).

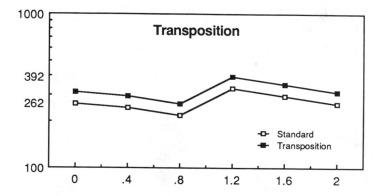

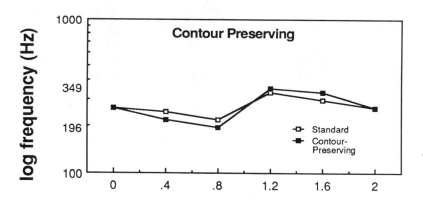

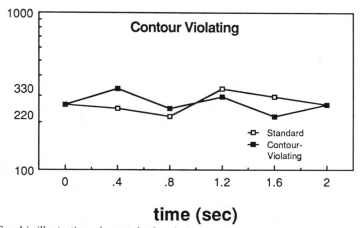

**FIGURE 3.** Graphic illustration of a standard melody and its transformations. Melodies from Trehub, Bull, and Thorpe (1984).

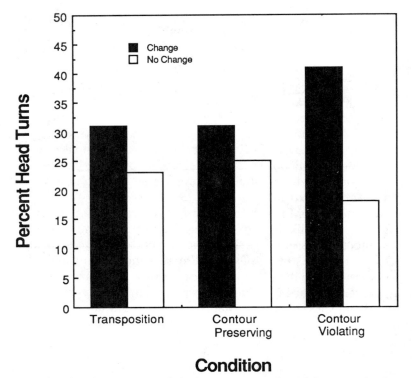

**FIGURE 4.** Infants' performance on melody discrimination tasks with long retention intervals. Data from Trehub, Bull, and Thorpe (1984, Experiment 2). Responses on change and no-change trials differ significantly only for contour-violating changes.

In the aforementioned studies, infants had detected contour changes in the context of a fixed repeating pattern (i.e., the original melody). Their failure to respond to transpositions and to contour-preserving changes could have been due to an inability to detect such changes or to infants' perception of some *similarity* between the original and comparison patterns. Even if infants were unable to detect the changes under these conditions, this would not preclude the possibility of perceiving the changes as well as the similarities under other conditions. Our concern was whether infants could behave in an adult-like fashion, establishing perceptual equivalence classes or categories for discriminable patterns that shared a common contour. To accomplish this goal, we required infants to discriminate between *sets* of melodies that contrasted in contour (Trehub, Thorpe, & Morrongiello, 1987). In one condition, the standard set included five-tone melodies that were transpositions of one another (same intervals, different pitches); exemplars of the comparison set differed in contour from the standard set but were also presented in various transpositions. In a second condition, within-category variations among members of the standard or contrasting set preserved contour but altered interval size (and pitches). The task for infants was to respond to changes in contour (for which they were reinforced) but to ignore changes in pitch in one condition and changes in interval size in the other. We found that infants detected contour changes in both variable contexts, indicating that they could go beyond simply discriminating between tone sequences to *categorize* them on the basis of relational properties such as common intervals or contour. Moreover, the comparable levels of performance in both variable contexts increased the likelihood that infants were operating on the basis of common contour in both conditions rather than common intervals in one condition and contour in the other.

These studies confirmed that infants' representation of melodies is abstract and adult-like, with contour playing a critical role. Although infants had successfully extracted the pattern of successive directional changes in pitch (i.e., the contour) in the five- and six-tone melodies of the previous studies, what remained unclear was how much information infants required to perceive the pitch of adjacent tones as rising or falling. Thorpe (1986) tackled the problem by requiring infants to discriminate between sets of transposed two-tone sequences, one set rising in pitch, the other falling. Infants succeeded in differentiating between rising and falling two-tone sequences, regardless of whether the two tones were separated by six semitones or by as little as one semitone. (It is interesting to note that some of the parents, tested under the same conditions as their infants, were unable to discern whether the patterns separated by one or two semitones were rising or falling.) These findings are in accord with our previous evidence in highlighting infants' orientation to relational aspects of auditory sequences. Moreover, the results indicate that infants can extract pitch directional information from the smallest interval relevant to Western European music (i.e., the semitone).

## GOOD STRUCTURE OR FORM

Adults' detection of changes to brief melodies is superior for melodies that conform to Western musical structure than for those that do not (Cohen, 1982; Cuddy, Cohen, & Mewhort, 1981). This is also the case for school-age children (Zenatti, 1969). The facilitation for "good" (i.e., appropriate in the Western sense) melodies likely reflects a stable representation of such melodies (Dowling & Harwood, 1986; Krumhansl, Bharucha, & Kessler, 1982), which, in turn, may be attributable to formal or informal exposure to the music of our culture. Thus, infants should be unlikely to exhibit differential performance for melodies that are "good" or "poor" in this sense.

We tested infants 9–11 months of age and children 4–6 years of age for their detection of a semitone change in one position of a five-tone melody (Trehub, Cohen, Thorpe, & Morrongiello, 1986). In one condition, the standard or repeating background melody (C-E-G-E-C) conformed to Western musical scales. In another, the standard was a similar five-tone melody (same contour, four tones identical to the conforming melody) that embodied one wrong note in terms of Western scale structure (C-E-G$^\#$-E-C). The outcome was that preschool children more readily detected deviations from the well-structured melody (i.e., superior detection of semitone changes) than from the poorly structured melody but infants performed equivalently on both patterns. This finding was consistent with the notion that incidental exposure to Western music rather than innate predispositions (Demany & Armand, 1984) promoted the internalization of Western scale structure and enhanced performance for the "good" melody.

In a subsequent extension of the study with infants (Cohen, Thorpe, & Trehub, 1987), we increased the difficulty of the task by changing from the fixed background sequences of Trehub et al. (1986) to sets of transposed melodies as in Trehub, Thorpe, and Morrongiello (1987). One group of infants was presented with the well-structured standard (or repeating background) sequence and the less well-structured comparison, which deviated from the standard sequence by one semitone. In another condition, the same poorly structured melody served as the standard and the same well-structured melody as comparison (i.e., the reverse situation). Surprisingly, infants detected the semitone change only in the context of the "good" melody (as background), suggesting that their representation of the latter melody may be more stable or complete than that of the poorly structured melody. Why might this be? Our "good" melody was based on the tones of the major triad, which is regarded, in music theory, as the prototype of tonal structure (Schenker, 1906/1954). The major triad tones embody simple ratio relations (4 : 5 : 6), compared to the complex ratio relations (16 : 20 : 25) of the "poor" melody. Performance enhancement for the major triad tones could derive from infants' exposure to the small ratios in harmonic complex sounds (e.g., speech) in the natural environment (Terhardt, 1974), or from pitch processing mechanisms that are sensitive to periodicity (Creel, Boomsliter, & Powers, 1970).

Such asymmetric performance has parallels in the adult literature. For example, it is easier for adults to detect deviations from well-structured melodic (Bharucha, 1984), rhythmic (Bharucha & Pryor, 1986), or linguistic (Bharucha, Olney, & Schnurr, 1985) patterns than from less well-structured patterns, even when deformations from the poorly structured patterns yield "good" patterns. Presumably, a superior standard pattern generates a coherent representation against which to evaluate comparison stimuli.

Having established the salience of contour (Thorpe, 1986; Trehub et al., 1984; Trehub, Thorpe, & Morrongiello, 1987) and of intervals when "good form" prevails (Cohen et al., 1987), we attempted to establish whether infants could use exact pitch information when contour cues were unavailable. After familiarizing infants with a set of three pitches, with order randomized on each repetition (obliterating contour cues), we tested for their detection of a comparison set with three different pitches (Cohen, 1986). To compensate for the withdrawal of contour information, we selected, as the standard set, three tones with favored status, namely the component tones of the major triad. The result was that infants differentiated between sets of pitches in the context of variable contours, whether the comparison set consisted of tones with similar musical relations (i.e., different sets of major triad tones) or very different relations. In further research that is currently under way, there is some indication that infants may have greater difficulty detecting the comparison pitch set (even major triad tones) when the standard set embodies tones with relations that are less significant musically (e.g., complex ratio relations).

## TEMPORAL PATTERNING

In listening to auditory sequences, we *group* elements within the overall patterns, whether or not the patterns incorporate variations in pitch, duration, or stress (Fraisse, 1982). One consequence of such grouping processes is that listeners hear the temporal intervals between groups of elements as longer than identical intervals within any group (Bolton, 1894). Temporal grouping promotes the retention of auditory sequences (Huttenlocher & Burke, 1976) and the perception of speech (Bailey, 1983; Martin, 1972). Moreover, temporal grouping is considered to be the most basic component of music understanding (Lerdahl & Jackendoff, 1983).

We and other researchers documented that infants can discriminate between contrasting temporal or rhythmic sequences (Chang & Trehub, 1977b; Demany, McKenzie, & Vurpillot, 1977; Jusczyk, Pisoni, Walley, & Murray, 1980; Morrongiello, 1984; Morrongiello & Trehub, 1987). We then attempted to establish whether infants would exhibit grouping processes similar to those observed in adults. In a series of studies, we tested infants, preschool children, and adults for their detection of temporal intervals between elements of auditory patterns (Thorpe, 1985; Thorpe & Trehub, 1989; Thorpe et al., 1988). The standard pattern had three tones of one pitch, waveform, or intensity followed by three tones of contrasting pitch, waveform, or intensity (XXX000), with all tones and intertone intervals 200 msec in duration. The contrasting pattern had an incremented (lengthened) intertone interval following the third tone (XXX 000, *between* tone groups) or following the fourth tone (XXX0 00, *within* a tone group). Since equally spaced tones are grouped by their similarity in pitch or other distinctive characteristics (Wertheimer, 1923/1938), this should yield a structure of two groups of three tones for the standard pattern. Thus, an extended intertone interval *between* groups would conserve the global temporal structure (i.e., two groups of three tones) and, therefore, should be relatively difficult to detect. On the other hand, an equivalently extended interval *within* a tone group would alter the structure (i.e., the Gestalt principle of proximity conflicting with that of similarity) and, consequently, should be more noticeable. Indeed, infants, preschool children, and adults detected the extended within-group intervals more readily than the between-group intervals, implying that they had grouped the elements of the standard pattern on the basis of similarity. Thus, immature as well as mature listeners impose rhythmic patterning on melodic sequences that embody equal durations of tones and intertone intervals (see Fig. 5).

The rhythmic patterning of an auditory sequence confers a temporal identity on that sequence just as the pattern of pitches gives it a melodic identity. And just as the identity of a melody is independent of specific pitches, it is also independent of specific rates or tempos. Having established that infants perceive the similarity of melodies across changes in pitch and interval size (Trehub et al., 1984; Trehub, Thorpe, & Morrongiello, 1987), we attempted to establish whether they could conserve the temporal or rhythmic patterning of a tone sequence across changes in tempo or rate of presentation (Trehub & Thorpe, 1989). We required them to discriminate between sets of three-tone patterns of 2, 1 (XX X) or 1, 2 (X XX) form, and also between sets of four-tone patterns embodying 2, 2 (XX XX) or 3, 1 (XXX X) form, each sequence being presented at five different rates or tempos. Infants discriminated between the contrasting sets of three- and four-tone patterns, indicating their ability to categorize sequences on the basis of temporal structure and to generalize across variations in tempo.

input          mental representation

**FIGURE 5.** Infants' representation of a pattern with equally spaced tones. Grouping leads to enlargement of the silent interval between tone groups.

## PITCH AND TIMBRE

The melodies in most of the aforementioned studies with infants were composed of pure tones or sine waves, so that my use of the psychological dimension, *pitch*, as opposed to the acoustic dimension, *frequency*, is inconsequential. In contrast to pure tones with a single frequency component, periodic complex sounds embody several simultaneous frequencies. The pitch of a complex sound is signaled by the fundamental or lowest tone (in the case of adults), with overtones or harmonics at integer multiples of the fundamental (see Moore, 1982). Before one can generalize the previous findings with pure tones to infants' perception of complex sounds, it is necessary to establish that infants perceive the pitch of complex sounds as do adults. Recently, Clarkson and Clifton (1985) demonstrated that infants respond equivalently to spectrally distinct tonal complexes that signal a common pitch for adults, and differentially to tonal complexes that signal contrasting pitches for adults. These findings imply the presence of comparable pitch processing mechanisms in infants and adults.

Beyond the consideration of pitch is the more amorphous psychological dimension of *timbre*, which refers to the overall quality that differentiates one complex sound from another of identical pitch, loudness, and duration. The pattern of energy distribution across the overtones of a complex sound gives rise to qualities that distinguish various musical instruments and various spoken vowels or consonants. These qualities of "flute-ness," "violin-ness," [a]-ness, or [b]-ness are conserved over a considerable range of pitch and loudness (see Dowling & Harwood, 1986).

Timbre perception is critical to auditory pattern processing and has been the subject of considerable research with adults (e.g., Plomp, 1976; Slawson, 1968). Trehub, Endman, and Thorpe (in press) explored infants' ability to use timbre as a basis for categorizing complex tones. They constructed one set of complex tones with energy emphasis corresponding roughly to the overtone peaks (formants) in the vowel [a], and a contrasting set with energy peaks similar to those in the vowel [i]. Members of both sets embodied variations in fundamental frequency in one condition, intensity in a second, and duration in a third. Infants discriminated between the contrasting harmonic structures in the context of all three variations, indicating their ability to conserve the timbre or distinctive quality of complex, variable sounds. In a further experiment, infants exhibited the ability to conserve fundamental frequency, intensity, or duration across variations in harmonic structure.

On the basis of these various studies of melodic, temporal, and timbral aspects of patterns, it is clear that infants go well beyond detecting subtle changes in musical patterns, exhibiting global, relational strategies or organizing principles that are characteristic of adults. This is not to suggest, however, that infant abilities in this domain are equal to those of adults, for infants are necessarily constrained by limited processing resources and even more limited experience.

## MUSIC DESIGNED FOR INFANTS

If one were to create a musical genre with an infant audience in mind, how could the foregoing research be used as a guide in this endeavor? First, in establishing an appropriate pitch range, the previous research is of little assistance because pitch range was not varied systematically. Nevertheless, one can play it safe by selecting parameters that simply worked. Consider the prototypic major triadic melody, which was used successfully to demonstrate contour discrimination in the context of pitch and interval changes (Trehub, Thorpe, & Morrongiello, 1987) and semitone discrimination in the context of fixed or transposed sequences (Cohen et al., 1987; Trehub et al., 1986). The basic version of this pattern, C-E-G-E-C, began with $C_4$ or middle C, yielding component frequencies of 262, 330, and 392 (for frequencies of the tempered scale, see Backus, 1969, p. 134). The studies with variable sets of standard and comparison sequences had component frequencies ranging from 233 Hz ($B^b_3$) to 587 Hz ($D_5$). It is interesting to note, parenthetically, that although the major triad structure occurs very frequently in Western music, it occurs considerably more frequently in nursery songs (Cohen et al., 1987).

Second, the selection of contours would be critical because of their demonstrated role in infants' perception and retention of melodies. Again, contour was not examined parametrically but the patterns that infants discriminated in complex (i.e., variable) contexts had simple rise–fall contours (Cohen et al., 1987; Trehub, Thorpe, & Morrongiello, 1987) or even simpler unidirectional (rising, falling, level) contours (Thorpe, 1986; Trehub & Thorpe, 1989). Moreover, the finding that infants could conserve contour over pitch and interval changes would make it possible to add moderate interest or novelty to the rise–fall and unidirectional contours by repeating them with new starting pitches (higher or lower, with identical intervals), new intervals, or both.

Third, consider the temporal structure that would be appropriate for infant listeners. Most of the melodies in the foregoing research embodied temporally equidistant tones with onset-to-onset times of 400 msec between adjacent tones (200 msec duration, 200 msec silence), resulting in overall melody durations of 0.6 sec for two-tone sequences (Thorpe, 1986), 1.8 sec for five-tone sequences (Cohen et al., 1987; Trehub et al., 1986; Trehub, Thorpe, & Morrongiello, 1987), and 2.2 sec for six-tone sequences (Trehub et al., 1984; 1985). This translates to a rate of 2.5 tones per second during periods of melody presentation. Intermelody intervals or unfilled pauses ranged from 0.8 sec (e.g., Trehub et al., 1984, experiment 1) to 1.4 sec (e.g., Trehub, Thorpe, & Morrongiello, 1987). Variations in tempo could be introduced with the assurance that infants would be able to conserve the identity of the underlying pattern across such variations (Thorpe et al., 1986). Also, moderate rhythmic variations might be desirable to capitalize on infants' ability to differentiate between contrastive rhythms (Chang & Trehub, 1977b). Even with uniformly spaced notes, infants might be expected to impose rhythmic structures by grouping tones of similar pitch or timbre (Demany, 1982; Thorpe, 1985).

Finally, let us consider the instrument of choice. Here I depart from the strategy of using research experience as a guide, declining to advance the cause of the tone synthesizer. A practical, general-purpose alternative is the human voice, whose universality and flexibility in the production of pitch, amplitude, and timing variations are commendable. The pitch or fundamental frequency of the human voice varies with the speaker's age, sex, and other individual characteristics. Modal values are approximately 130 Hz for young adult males, an octave above that (260 Hz) for young adult females, and 300–400 Hz for children (Ohala, 1981). Thus, the pitch range of female and child voices is consistent with that recommended on the basis of the melody perception research.

To summarize, then, the design features of infant music should embody pitch levels in the vicinity of the octave beginning with middle C (262 Hz), simple contours that are unidirectional or have few changes in pitch direction (e.g., rise–fall), slow tempos (approximately 2.5 notes/sec), and simple rhythms. Moreover, moderate variations or embellishments on these basic design features would be acceptable, as is the human voice as the mode of transmission.

This collection of features does not characterize formal or informal musical styles in our own culture or in others (for cross-cultural perspectives, see Blacking, 1973; Titon, 1984). Perhaps this should not surprise us because "the audience is the ultimate composer of any musical tradition" (Pantaleoni, 1985, p. 403), and the audiences for most of the musical traditions studied to date are adults at work, play, or worship. Among the exceptions are some lullabies and nursery songs (Bernstein, 1976; M. Papousek & Papousek, 1981), which do incorporate a number of these design criteria. Nevertheless, it is likely that musical notation systems and commercialization have resulted in significant alterations to these forms so that the examination of improvised materials might be preferable. Since melody is often considered to have originated in the prosody of speech (Treitler, 1982), it might be instructive to examine parental speech to infants for the presence of some of the proposed features.

## PARENTAL SPEECH TO PRELINGUISTIC INFANTS

There is evidence that parents, blissfully unaware of infant perceptual skills, behave as if they have such knowledge, delivering stimulation that is particularly well matched to the perceptual abilities outlined earlier. For reasons that are less than clear, the mere presence of an awake newborn or young infant elicits speech directed toward that infant (Fernald & Simon, 1984; Rheingold & Adams, 1980; Rosenthal, 1982; Stern, Spieker, Barnett, & MacKain, 1983). Infant-directed speech, *babytalk* (i.e., talk to babies), or *motherese*, embodies considerable simplification in structure and form compared to speech addressed to adults (Ferguson, 1964; Fernald, 1984; Sherrod, Friedman, Crawley, Drake, & Devieux, 1977; Snow, 1977), but the hallmark of such speech is its patterning of stress and intonation, or its prosody (Fernald, 1984, 1985; Fernald & Kuhl, 1987; M. Papoušek & Papoušek, 1981; Sachs, 1977; Stern et al., 1983; Stern, Spieker, & MacKain, 1982).

Detailed analyses of the intonation and stress patterns (i.e., prosody) of maternal speech to newborns and young infants (Fernald & Simon, 1984; M. Papousek & Papousek, 1981; M. Papousek, Papousek, & Bornstein, 1985; Stern et al., 1983) have revealed the presence of several distinctive features: heightened pitch, increased pitch range, rhythmicity, slower tempo, shorter utterances, smooth pitch transitions, simple pitch contours, and prosodic repetition. The brevity of infant-directed utterances confers a sense of unity or coherence to the pitch contour of each utterance, setting these contours apart from the complex, nondistinctive contours of utterances to adults (Stern et al., 1982). Indeed, Fernald and Simon (1984) found that the patterns characterizing maternal speech to newborns rarely occurred when the same mothers spoke to adults.

So different are the features of infant-directed from adult-directed speech that M. Papoušek and Papoušek (1981) characterize them as *musical* elements as opposed to suprasegmental or paralinguistic elements. Parents tend to use simple contours, generally unidirectional (rising, falling, one-level, or two-level) but sometimes bell-shaped (rise–fall or fall–rise), repeating the contours with altered lexical or segmental content (Fernald, 1984; Fernald & Simon, 1984; M. Papoušek, Papoušek, & Harris, 1986). H. Papousek and Papousek (1984) provide an illustrative case of a mother who repeated one contour 37 times within 3 min, varying her utterance length (1–7 syllables) and semantic content (11 different meanings). Thus, melodic contours provide a relatively stable frame for shifting content.

There are indications that babytalk utterances are often poorly articulated (M. Papoušek, 1987), with the majority having no discernible syntactic structure (M. Papoušek & Papoušek, 1981). Many of these utterances are monosyllabic with elongated vowels (Fernald & Simon, 1984; M. Papoušek & Papoušek, 1981); others are "contentless" pitch contours or melodies (M. Papoušek, 1987). This suggests that, in the case of babytalk, the pitch contour *is* the utterance as opposed to its accompaniment, and segmental or lexical items represent *optional* content for the prosodic envelope. Such babytalk contours are contrasted with adult-directed contours, which embody a narrow range of fundamental frequency and multiple changes in the direction of pitch movement (Bolinger, 1970; Fernald & Simon, 1984).

The average pitch or fundamental frequency of a mother's voice moves upward by three or four semitones when she talks to her infant (Fernald & Simon, 1984; Jacobson, Boersma, Fields, & Olsen, 1983; M. Papousek et al., 1985). The maximum maternal pitch is approximately 437 Hz for speech to 4-month-old infants (Stern et al., 1983) and the minimum remains unchanged (Garnica, 1977), so that the resulting babytalk range may extend well over an octave, coinciding with the range recommended for the hypothetical musical genre.

The rhythmicity and slow tempo of babytalk (Beebe, Feldstein, Jaffe, Mays, & Alson, 1985; Fernald, 1985; Fernald & Simon, 1984; Koester, 1987; Stern & Gibbon, 1979) also set it apart from speech observed in most other contexts but not from the proposed music for infants. Indeed, such rhythmicity in adult speech is ordinarily confined to ritualized or pathological contexts (Jaffe, Anderson, & Stern, 1979) and is evident, as well, in children's chants (Hargreaves, 1986; Moorhead & Pond, 1978).

The duration of infant-directed utterances is reported to be about 1.1 sec compared to 2.2 sec for talk to adults (Fernald & Simon, 1984; M. Papoušek et al., 1985) and 1.8 sec for the five-tone sequences in Trehub et al. (1986; Trehub, Thorpe, & Morrongiello, 1987) and Cohen et al. (1987). The average tempo is approximately 3.5 syllables per second of vocalization (compared to 2.5 tones per second for the musical stimuli), although approximately a third of the utterances to 3-month-olds are monosyllabic, with an articulation rate of about 1.8 syllables per second (M. Papoušek et al., 1985). Pauses between utterances are about 1–1.6 sec (comparable to the intermelody intervals), depending upon the age of the listener (Fernald & Simon, 1984; Stern et al., 1983). Thus, the tempo and rhythmic nature of parental speech fall within values used inadvertently in music perception research with infants.

There is some suggestion that parents use specific contour or melody types in distinctive caretaking contexts: rising and bell-shaped contours to capture or maintain their infant's attention (M. Papoušek & Papoušek, 1981; Stern et al., 1982), falling contours for soothing or promoting sleep (M. Papoušek & Papoušek, 1981), and more variable contours to heighten the level of positive affect (Stern et al., 1982). The tempo and rhythm of parental vocal activity are also context- or state-dependent: slow, rhythmic bursts of stimulation for attentive infants; increases in tempo for inattentive infants; progressively decreasing tempo to induce sleep; and more varied rhythms and tempos for fussy infants (Koester, 1987; M. Papoušek & Papoušek, 1981).

These adjustments of pitch contour, rhythm, and tempo are dependent not only on context but also on the listener's age and presumed processing abilities. It is notable that pitch patterning variables (higher pitch, greater pitch range, distinctive contours) and repetition may reach a peak of usage at approximately 4 months of age, gradually declining thereafter (Stern et al., 1983).

## THE LISTENER'S PERSPECTIVE ON BABYTALK

There is every indication that babies are highly responsive to babytalk. They vocalize to it (B. J. Anderson, Vietze, & Dokecki, 1977; Mayer & Tronick, 1985; Stevenson, VerHoeve, Roach, & Leavitt, 1986; Wolff, 1963), using higher pitch with mothers than with fathers (Lieberman, 1967). They imitate the intonation contours of maternal utterances as early as 6 weeks of age, even though they are unable to produce utterances of comparable duration (Lieberman, Ryalls, & Robson, 1982, in Lieberman, 1984). Infants smile in response to babytalk, especially when it is high in pitch (Mayer & Tronick, 1985; Wolff, 1963). Indeed, by the second week of life, the human voice elicits infant smiling more reliably than any stimulus in any modality, but such smiles are confined to irregular (REM) sleep (Wolff, 1963). Human vocalizing also represents the first and most effective stimulus configuration to elicit smiling in waking infants; this occurs at approximately 3 weeks of age (Wolff, 1963).

Infants differentiate their mother's voice from that of a stranger very early in life, although estimates for this achievement range from birth to 4 months of age (Brown, 1979; DeCasper & Fifer, 1980; Mehler, Bertoncini, Barrière, & Jassik-Gerschenfeld, 1978; Mills & Melhuish, 1974). It is notable that Mehler et al. (1978) found differentiation of mother's from stranger's voice when both were presented in the babytalk register but not with flat intonation (monotone). Infants of 2 months also differentiate between the rising and falling intonation contours of female speakers, displaying greater attention to the rising contours (Sullivan & Horowitz, 1983).

Fernald (1985) offered 4-month-olds a choice of infant- or adult-directed speech, each of which was presented contingent upon a head turn to a particular side (left for one type of speech, right for the other). In this context, 4-month-olds registered their "preference" for the recorded babytalk of unfamiliar adults. In a further study, Fernald and Kuhl (1987) attempted to isolate the features of infant-directed speech that underlie its salience. In a design similar to that of Fernald (1985), they presented 4-month-old infants with pairs of synthesized pure-tone patterns that preserved various prosodic aspects of infant- and adult-directed speech: the pitch and temporal patterning (amplitude held constant), the amplitude and temporal patterning (frequency held constant), or the temporal patterning alone (frequency and amplitude held constant). Infants turned more frequently toward the loudspeaker that emitted synthetic babytalk contours only when such contours preserved the pitch patterning of natural babytalk. This finding lends support to the view that the essence of babytalk is its melody.

## UNIVERSALITY AND CONTINUITY OF PARENTAL SPEECH ADJUSTMENTS

Are such speech adjustments related to culture, infant caretaking experience, sex, and age? The answer appears to be negative on all counts. Babytalk modifications have been documented for many languages including English, Arabic, Spanish, German, Kwara'ae, and Mandarin (Ferguson, 1964; Fernald & Simon, 1984; Grieser & Kuhl, 1988; M. Papoušek, 1987; Sachs, Brown & Salerno, 1976; Watson-Gegeo & Gegeo, 1986). Tone languages such as Mandarin are of particular interest since tone differences (e.g., level, rising, falling–rising, falling) signal distinctions in lexical meaning, raising the possibility that the use of babytalk contours could generate violations of lexical tone rules (M. Papoušek, 1987). Indeed, Chinese mothers use the same unidirectional and bidirectional contours as English-speaking mothers, deftly circumventing the potential conflict by favoring nonverbal, lexically ambiguous utterances and avoiding typical Mandarin tones (Grieser & Kuhl, 1988; M. Papousek, 1987). Another provocative example is found in the integration of speech and song in central Thailand, where the Thai dialect includes five tones: low, middle, high, ascending, and descending (List, 1961). For the chants and songs of everyday life, speech tones play the most prominent role, with song melody subserviant. It is not uncommon, however, for Thai lullabies, like the lullabies of many cultures, to have refrains sung to nonsense syllables (List, 1961), which is a clever solution to the potential conflict between tone rules and favored melodic forms.

Motherhood and experience are also irrelevant to the production of babytalk, which has been observed in the infant-directed speech of primiparous as well as multiparous mothers (Fernald & Simon, 1984), fathers (H. Papoušek & Papoušek, 1984; Parke, Grossman, & Tinsley, 1981), and strangers, both male and female (Rheingold & Adams, 1980). Even preschool children modify their speech in similar ways when talking to infants or to "baby" dolls, whether or not they have younger siblings (E. S. Anderson, 1986; Sachs & Devin, 1976; Shatz & Gelman, 1973; Watson-Gegeo & Gegeo, 1986; Weeks, 1971).

Although there is some attenuation of prosodic babytalk parameters beyond the first year of life (Remick, 1976; Stern et al., 1983), certain characteristics remain in evidence over the next few years (Ferguson, 1964; Garnica, 1977; Snow, 1977). For example, mothers continue to use high pitch, an expanded pitch range, and rising contours in conversations with their 2-year-olds but less so with their 5-year-olds (Garnica, 1977). Increasingly, however, modifications geared to enhancing the content of verbal messages become more prominent in speech to older infants and young children. For example, mothers increase the duration of critical content words for 2- and 5-year-olds and use many diminutives and child-created words (Jocíc, 1978). Their speech exhibits increased emphasis on the articulatory distinctions between similar sounds, but only for listeners who are beginning to produce their first words, not for younger infants or for children who are already producing multiword utterances (Malsheen, 1980). Mothers also produce an increasing proportion of grammatical utterances for 4-month-olds, 1-year-olds, and 2-year-olds (Cross & Morris, 1980). Finally, they reduce the length and syntactic complexity of their utterances, speak slowly and fluently, and increase their verbal redundancy (Broen, 1972; Newport, 1977; Phillips, 1973; Remick, 1976; Snow, 1972). Thus, although such adjustments still function to gain and hold children's attention, their role is progressively subordinated to and coordinated with the verbal content of parental messages.

## FUNCTIONS OF PARENTAL SPEECH ADJUSTMENTS

The apparent universality of the babytalk mode or register raises questions about its function. One possibility is that babytalk is used simply to regulate infant attention and arousal (B. J. Anderson et al., 1977). H. Papoušek and Papoušek (1979) note that, initially, parents are the principal source of external stimulation for infants, mediating most interactions with the environment. In this role, they evaluate the infant's state, and, striving to maintain it at an optimal level, continue, modify, or discontinue their stimulation. Parental vocalization during the neonatal period, with its steady rhythms and long pauses, can be viewed as preventing overstimulation in contrast to the goal, at 4 months, of promoting alertness, interest, and positive affect with the heightened use of pitch patterning (Stern et al., 1983).

Another possibility is that such behavior, in addition to its arousal regulatory function, expresses aspects of the emotional state of the caregiver (Stern et al., 1982; Trevarthen & Marwick, 1986). Voice and speech patterns are known to be affected by emotional arousal, with happiness and activation associated with substantial increases in pitch and pitch variability (Scherer, 1982). It is interesting that musical sequences with large pitch variations and rising contour are rated as pleasant or happy (Scherer & Oshinsky, 1977); those with slow tempos as serene, delicate, and sentimental (Gundlach, 1932; Weden, 1972); and those with high pitch as animated or playful (Gundlach, 1932; Hevner, 1936; Weden, 1972). Even preschool children agree that melodies with high pitch or with rising contours connote happiness (Trehub, Cohen, & Guerriero, 1987). Clear links of babytalk to maternal and infant affect can be seen in some cultures where mothers use an adult style (including shouting) for reprimanding their infant, and babytalk for the infant's amusement or pacification (Demuth, 1986).

Caregivers may go beyond the expression of their own feelings in attempts to share the feelings of their charges. Trevarthen and Marwick

(1986) characterize the mother as entering "into the ebb and flow of infant emotions" (p. 281), shadowing or echoing by imitating the infant's vocalizations and facial expressions (p. 287). Lester, Hoffman, and Brazelton (1985) refer to "coordinated cycles of affective displays" (p. 24) or rhythmic synchrony between mothers and their 3- to 5-month-old infants. The vocalizations of infants often trigger maternal vocalizations (B. J. Anderson et al., 1977; Keller & Scholmerich, 1987), leading to simultaneous vocalization of the interactants (but see Mayer & Tronick, 1985; Stevenson et al., 1986, for evidence of turn-taking). There are indications, however, that simultaneous vocalization is more likely under conditions of higher arousal, and turn-taking when more neutral affect prevails (Stern, Jaffe, Beebe, & Bennett, 1975). A substantial proportion of maternal vocalizations are imitations of infant vocalizations, with the pitch and melodic contours being the most frequently imitated features (H. Papoušek & Papoušek, 1987). The other side of the coin is that infants produce frequent vocal matches of maternal utterances, with contour being the most common target for matching (H. Papoušek & Papoušek, 1987).

Yet another view of babytalk is that it is part of a repertoire of intuitive parenting or tutorial behaviors designed to foster the cognitive as well as social growth of infants (H. Papoušek & Papoušek, 1984, 1987; M. Papoušek & Papoušek, 1981; M. Papoušek et al., 1985, 1986). For the most part, parents have little conscious awareness and control of these behaviors that are emitted so effortlessly and universally in the course of caretaking activities (H. Papoušek & Papoušek, 1987; Stern, Hofer, Haft, & Dore, 1985). Indeed, it is difficult to evoke the full set of babytalk adjustments in simulated or role-play contexts, implying that cues or feedback from the infant may be critical (Fernald & Simon, 1984; Jacobson et al., 1983; Murray & Trevarthen, 1986). On the other hand, young children seem to be capable of producing credible babytalk for inanimate "babies" such as dolls or even rocks (Demuth, 1986; Sachs & Devin, 1976), suggesting that an appropriate affective state may be the important mediating variable. Thus, although the infant's presence may be necessary for triggering the requisite emotional state and resultant babytalk melodies in adults, children may experience the appropriate feelings in simulated as well as "live" contexts.

There are various means by which babytalk contours can support tutorial functions. Specific contour types seem to promote visual contact between parent and infant and modify infant state in ways that influence the possibilities for learning (H. Papoušek & Papoušek, 1984, 1987). For example, rising contours typically accompany maternal encouragement for goal-directed activity. The influence of vocal encouragement can be seen in the mother's ability to elicit infant exploratory activity in novel or potentially fear-provoking situations (Campos & Barrett, 1984). Moreover, modest infant successes are reinforced with specific contours of praise (H. Papoušek & Papoušek, 1984; Stern et al., 1982, 1983).

Finally, babytalk provides an ideal medium for linguistic tutorials. The propensity of mothers to imitate their infants and to engage in repetitive vocal play promotes interest in speech and in the exercise of developing vocal skills. The attention-getting properties of babytalk provide a continuing framework for the delivery of language lessons geared to the infant's developmental status. The dialogic character of much mother–infant vocal play also provides a model for mature conversation, and the multimodal nature of stimulation delivered in conjunction with babytalk (e.g., mouth movements, facial expressions, rhythmic movements) guides infants' production and interpretation of such behavior. This is not to say that parents simply engage in teaching, and infants learn as a direct consequence of such teaching. On the contrary, the Papoušeks argue that biological preadaptations guide infants' interest in skill acquisition as well as parents' propensity to share their own knowledge in didactically appropriate ways (e.g., H. Papoušek & Papoušek, 1987).

It is hardly surprising, then, that infants engage in extensive imitative activity early in life (Meltzoff, 1986), that rise–fall contours predominate in infant vocalizations between 1 month and 1 year of age (Delack & Fowlow, 1978), that infants exhibit precocious association of specific speech sounds with relevant mouth movements (Kuhl & Meltzoff, 1982; MacKain, Studdert-Kennedy, Spieker, & Stern, 1983), that they link happy and sad sounding voices with the corresponding facial expressions (Walker, 1982), and male and female voices with appropriate faces (Spelke & Owsley, 1979). It is also the case that the pitch and temporal patterning of language emerges well before the first words (Crystal, 1973; deBoysson-Barries, Sagart, & Durand, 1984; Lewis, 1951; Nakazima, 1962).

In an insightful paper on linguistic input to prelinguistic infants, Sachs (1977) advanced the speculation that babytalk "is not simply a culturally-transmitted, functionless ritual" but rather "a species-typical pattern whose evolution has been determined by the sensory capacities of infants and by the infant's requirements for the development of normal communication with its social world" (p. 51). She suggested that the claim of species typicality could be substantiated by three lines of evidence: (1) cross-cultural universality, (2) the tuning of adult productions to infant sensitivity, and (3) an accelerated ontogeny of behavior in this domain. Evidence in support of her position, although relatively sparse in 1977, is currently overwhelming.

## SOME UNANSWERED QUESTIONS

Further details regarding the links between infant perceptual skills, parental input, and the acquisition of productive competence remain to be uncovered. There are many intriguing questions that have not been addressed in the research to date. For example, do the melodic and rhythmic aspects of babytalk have properties in common with the graded vocal signals of other species? Could these emotionally charged vocalizations of human caretakers originate in limbic structures, as may be the case for vocal communication in nonhuman species (Robinson, 1976)? Is appropriate babytalk evident in parents with affective disorders? If not, what are the implications of aberrant input for the child's acquisition of species-typical babytalk? When and how do parents effect the transition from species-specific babytalk contours to culturally specified intonation and stress patterns? Do some of the emotion-inducing aspects of music have their roots in early vocal play between infants and their caretakers? Is the evolution and preservation of musical ability (perception and production) simply an accident of its harmlessness (Granit, 1977), an inadvertent consequence of shared auditory mechanisms with speech (Lerdahl & Jackendoff, 1983), or a result of its support for critical caretaking and tutorial functions in early life? These questions are likely to occupy researchers and theorists for many years to come.

## REFERENCES

Anderson, B. J., Vietze, P., & Dokecki, P. R. (1977). Reciprocity in vocal interactions of mothers and infants. *Child Development, 48,* 1676–1681.

Anderson, E. S. (1986). The acquisition of register variation by Anglo-American children. In B. B. Schieffelin & E. Ochs (Eds.), *Language socialization across cultures* (pp. 153–166). London & New York: Cambridge University Press.

Attneave, F., & Olson, R. K. (1971). Pitch as a medium: A new approach to psychophysical scaling. *American Journal of Psychology, 84,* 147–166.

Backus, J. (1969). *The acoustical foundations of music.* New York: Norton.

Bailey, P. J. (1983). Hearing for speech: The information transmitted in normal and impaired speech. In M. E. Lutman & M. P. Haggard (Eds.), *Hearing science and hearing disorders* (pp. 1–34). London: Academic Press.

Bartlett, J. C., & Dowling, W. J. (1980). Recognition of transposed melodies: A key-distance effect in developmental perspective. *Journal of Experimental Psychology: Human Perception and Performance, 6,* 501–515.

Beebe, B., Feldstein, S., Jaffe, J., Mays, K., & Alson, D. (1985). Interpersonal timing: The application of an adult dialogue model to mother-infant vocal and kinesic interactions. In T. M. Field & N. A. Fox (Eds.), *Social perception in infants* (pp. 217–247). Norwood, NJ: Ablex.

Bernstein, L. (1976). *The unanswered question.* Cambridge, MA: Harvard University Press.

Bharucha, J. J. (1984). Anchoring effects in music: The resolution of dissonance. *Cognitive Psychology, 16,* 485–518.

Bharucha, J. J., Olney, K. L., & Schnurr, P. P. (1985). Detection of coherence-disrupting and coherence-conferring alterations in text. *Memory and Cognition, 13,* 573–578.

Bharucha, J. J., & Pryor, J. H. (1986). Disrupting the isochrony underlying rhythm: An asymmetry in discrimination. *Perception & Psychophysics, 40,* 137–141.

Blacking, J. (1973). *How musical is man?* Seattle: University of Washington Press.

Bolinger, D. (1970). Relative height. In D. Bolinger (Ed.), *Intonation: Selected readings* (pp. 137–157). Hammondsworth, England: Penguin Books.

Bolton, T. L. (1894). Rhythm. *American Journal of Psychology, 6,* 145–238.

Broen, P. A. (1972). The verbal environment of the language-learning child. *ASHA Monographs,* No. 17.

Brown, C. J. (1979). Reactions of infants to their parents' voices. *Infant Behavior and Development, 2,* 295–300.

Campos, J. J., & Barrett, K. C. (1984). Toward a new understanding of emotions and their development. In C. E. Izard, J. Kagan, & R. B. Zajonc (Eds.), *Emotions, cognition, and behaviour* (pp. 229–263). London & New York: Cambridge University Press.

Chang, H. W., & Trehub, S. E. (1977a). Auditory processing of relational information by young infants. *Journal of Experimental Child Psychology, 24,* 324–331.

Chang, H. W., & Trehub, S. E. (1977b). Infants' perception of temporal grouping in auditory patterns. *Child Development, 48,* 1666–1670.

Clarkson, M. G., & Clifton, R. K. (1985). Infant pitch perception: Evidence for responding to pitch categories and the missing fundamental. *Journal of the Acoustical Society of America, 77,* 1521–1528.

Cohen, A. J. (1982). Exploring the sensitivity to structure in music. *Canadian University Music Review, 3,* 15–30.

Cohen, A. J. (1986, April). *Infants' memory for sets of frequencies.* Paper presented at meetings of the International Conference on Infant Studies, Los Angeles, CA.

Cohen, A. J., Thorpe, L. A., & Trehub, S. E. (1987). Infants' perception of musical relations in short transposed tone sequences.. *Canadian Journal of Psychology, 41,* 33–47.

Creel, W., Boomsliter, P. C., & Powers, S. R. (1970). Sensations of tone as perceptual forms. *Psychological Review, 77,* 534–545.

Cross, T. G., & Morris, J. E. (1980). Linguistic feedback and maternal speech: Comparisons of mothers addressing infants, one-year-olds and two-year-olds. *First Language, 1,* 98–121.

Crystal, D. (1973). Non-segmental phonology in language acquisition: A review of the issues. *Lingua, 32,* 1–45.

Cuddy, L. L., Cohen, A. J., & Mewhort, D. J. K. (1981). Perception of structure in short melodic sequences. *Journal of Experimental Psychology: Human Perception and Performance, 7,* 869–883.

D'Amato, M. R., & Salmon, D. P. (1982). Tune discrimination in monkeys (*Cebus apella*) and in rats. *Animal Learning & Behavior, 10,* 126–134.

D'Amato, M. R., & Salmon, D. P. (1984). Processing of complex auditory stimuli (tunes) by rats and monkeys (*Cebus apella*). *Animal Learning & Behavior, 12,* 184–194.

deBoysson-Bardies, B., Sagart, L., & Durand, C. (1984). Discernible differences in the babbling of infants according to target language. *Journal of Child Language, 11,* 1–15.

DeCasper, A. J., & Fifer, W. P. (1980). Of human bonding: Newborns prefer their mothers' voices. *Science, 208,* 1174–1176.

Delack, J. B., & Fowlow, P. J. (1978). The ontogenesis of differential vocalization: Development of prosodic contrastivity during the first year of life. In N. Waterson & C. Snow (Eds.), *The development of communication* (pp. 93–110). Chichester: Wiley.

Demany, L. (1982). Auditory stream segregation in infancy. *Infant Behavior and Development, 5,* 261–276.

Demany, L., & Armand, F. (1984). The perceptual reality of tone chroma in early infancy. *Journal of the Acoustical Society of America, 76,* 57–66.

Demany, L., McKenzie, B., & Vurpillot, E. (1977). Rhythm perception in early infancy. *Nature (London), 266,* 718–719.

Demuth, K. (1986). Prompting routines in the language socialization of Basotho children. In B. B. Shieffelin & E. Ochs (Eds.), *Language socialization across cultures* (pp. 51–79). London & New York: Cambridge University Press.

Deutsch, D. (1982). *The psychology of music.* New York: Academic Press.

Dowling, W. J. (1978). Scale and contour: Two components of a theory of memory for melodies. *Psychological Review, 85,* 341–354.

Dowling, W. J., & Harwood, D. L. (1986). *Music cognition.* Orlando, FL: Academic Press.

Ferguson, C. (1964). Baby talk in six languages. *American Anthropologist, 66,* 103–114.

Fernald, A. (1984). The perceptual and affective salience of mothers' speech to infants. In L. Feagans, C. Garvey, & R. Golinkoff (Eds.), *The origins and growth of communication* (pp. 5–29). Norwood, NJ: Ablex.

Fernald, A. (1985). Four-month-old infants prefer to listen to motherese. *Infant Behavior and Development, 8,* 181–195.

Fernald, A., & Kuhl, P. K. (1987). Acoustic determinants of infant preference for motherese. *Infant Behavior and Development, 10,* 279–293.

Fernald, A., & Simon, T. (1984). Expanded intonation contours in mothers' speech to newborns. *Developmental Psychology, 20,* 104–113.

Fraisse, P. (1982). Rhythm and tempo. In D. Deutsch (Ed.), *The psychology of music* (pp. 149–180). New York: Academic Press.

Garnica, O. K. (1977). Some prosodic and paralinguistic features of speech to young children. In C. E. Snow & C. A. Ferguson (Eds.), *Talking to children: Language input and acquisition* (pp. 63–88). London & New York: Cambridge University Press.

Granit, R. (1977). *The purposive brain.* Cambridge, MA: MIT Press.

Green, D. M., & Swets, J. A. (1966). *Signal detection theory and psychophysics.* New York: Wiley.

Grieser, D. L., & Kuhl, P. K. (1988). Maternal speech to infants in a tonal language: Support for universal prosodic features in motherese. *Developmental Psychology, 24,* 14–20.

Gundlach, R. H. (1932). A quantitative analysis of Indian music. *American Journal of Psychology, 44,* 133–145.

Hargreaves, D. J. (1986). *The developmental psychology of music.* London & New York: Cambridge University Press.

Hevner, K. (1936). Experimental studies of the elements of expression in music. *American Journal of Psychology, 48,* 248–268.

Hulse, S. H., Cynx, J., & Humpal, J. (1984). Cognitive processing of pitch and rhythm structures by birds. In H. L. Roitblat, T. C. Bever, & H. S. Terrace (Eds.), *Animal cognition* (pp. 183–198). Hillsdale, NJ: Erlbaum.

Hulse, S. H., Page, S. C., & Braaten, R. F. (1989). An integrative approach to serial pattern learning: Music perception and comparative acoustic perception. In W. C. Stebbins & M. Berkley (Eds.), *Comparative perception: Vol. 2. Communication.* New York: Wiley (Interscience).

Huttenlocher, J., & Burke, D. (1976). Why does memory span increase with age? *Cognitive Psychology, 8,* 1–31.

Jacobson, J. L., Boersma, D. C., Fields, R. B., & Olson, K. L. (1983). Paralinguistic features of speech to infants and small children. *Child Development, 54,* 436–442.

Jaffe, J., Anderson, S., & Stern, D. (1979). Conversational rhythms. In D. Aronson & R. Rieber (Eds.), *Psycholinguistic research.* Hillsdale, NJ: Erlbaum.

Jocíc, M. (1978). Adaptation in adult speech during communication with children. In N. Waterson & C. Snow (Eds.), *The development of communication* (pp. 159–171). Chichester: Wiley.

Jusczyk, P. W., Pisoni, D. B., Walley, A., & Murray, J. (1980). Discrimination of relative onset time of two-component tones by infants. *Journal of the Acoustical Society of America, 67,* 262–270.

Keller, H., & Scholmerich, A. (1987). Infant vocalizations and parental reactions during the first four months of life. *Developmental Psychology, 23,* 62–67.

Koester, L. S. (1987, April). *Multimodal repetitive stimulation in parent-infant interactions: A look at micro-rhythms.* Paper presented at meetings of the Society for Research in Child Development, Baltimore, MD.

Krumhansl, K., Bharucha, J., & Kessler, E. (1982). Perceived harmonic structure of chords in three related musical keys. *Journal of Experimental Psychology: Human Perception and Performance, 8,* 24–36.

Kubovy, M., & Pomerantz, J. R. (Eds.). (1981). *Perceptual organization.* Hillsdale, NJ: Erlbaum.

Kuhl, P. K. (1987). Perception of speech and sound in early infancy. In P. Salapatek & L. Cohen (Eds.), *Handbook of infant perception: Vol. 2. From perception to cognition* (pp. 275–382). San Diego, CA: Academic Press.

Kuhl, P. K., & Meltzoff, A. N. (1982). The bimodal perception of speech in infancy. *Science, 218,* 1138–1144.

Lerdahl, F., & Jackendoff, R. (1983). *A generative theory of tonal music.* Cambridge, MA: MIT Press.

Lester, B. M., Hoffman, J., & Brazelton, T. B. (1985). The rhythmic structure of mother-infant interaction in term and preterm infants. *Child Development, 56,* 15–27.

Lewis, M. M. (1951). *Infant speech.* London: Routledge & Kegan Paul.

Lieberman, P. (1967). *Intonation, perception, and language.* Cambridge, MA: MIT Press.

Lieberman, P. (1984). *The biology and evolution of language*. Cambridge, MA: Harvard University Press.

List, G. (1961). Speech melody and song melody in central Thailand. *Ethnomusicology*, **5**, 16–32.

MacKain, K., Studdert-Kennedy, M., Spieker, S., & Stern, D. (1983). Infant intermodal speech perception is a left-hemisphere function. *Science*, **219**, 1347–1349.

Malsheen, B. J. (1980). Two hypotheses for phonetic clarification in the speech of mothers to children. In G. H. Yeni-Komshian, J. F. Kavanaugh, & C. A. Ferguson (Eds.), *Child phonology: Vol. 2. Perception* (pp. 173–184). New York: Academic Press.

Martin, J. G. (1972). Rhythmic (hierarchical) versus serial structure in speech and other behavior. *Psychological Review*, **79**, 487–509.

Mayer, N. K., & Tronick, E. Z. (1985). Mothers' turn-giving signals and infant turn-taking in mother-infant interaction. In T. M. Field & N. A. Fox (Eds.), *Social perception in infants* (pp. 199–216). Norwood, NJ: Ablex.

Mehler, J., Bertoncini, J., Barrière, M., & Jassik-Gerschenfeld, D. (1978). Infant recognition of mother's voice. *Perception*, **7**, 491–497.

Meltzoff, A. N. (1986). Imitation, intermodal representation, and the origins of mind. In B. Lindstrom & R. Zetterstrom Eds.), *Precursors of early speech* (pp. 245–265). New York: Stockton Press.

Mills, M., & Melhuish, E. (1974). Recognition of mother's voice in early infancy. *Nature (London)*, **252**, 123–124.

Moore, B. C. J. (1982). *An introduction to the psychology of hearing* (2nd ed.). London: Academic Press.

Moorhead, G. E., & Pond, D. (1978). *Music of young children*. Santa Barbara, CA: Pillsbury Foundation.

Morrongiello, B. A. (1984). Auditory temporal pattern perception in 6- and 12-month-old infants. *Developmental Psychology*, **20**, 441–448.

Morrongiello, B. A., & Trehub, S. E. (1987). Age-related changes in auditory temporal perception. *Journal of Experimental Child Psychology*, **44**, 413–426.

Morrongiello, B. A., Trehub, S. E., Thorpe, L. A., & Capodilupo, S. (1985). Children's perception of melodies: The role of contour, frequency, and rate of presentation. *Journal of Experimental Child Psychology*, **40**, 279–292.

Murray, L., & Trevarthen, C. (1986). The infant's role in mother-infant communications. *Journal of Child Language*, **13**, 15–29.

Nakazima, S. (1962). A comparative study of the speech developments of Japanese and American English in childhood. *Studia Phonologica*, **2**, 27–46.

Newport, E. L. (1977). Motherese: The speech of mothers to young children. In N. J. Castellan, Jr., D. B. Pisoni, & G. R. Potts (Eds.), *Cognitive theory* (Vol. 2, pp. 177–217). Hillsdale, NJ: Erlbaum.

Ohala, J. (1981). The nonlinguistic components of speech. In J. K. Darby, Jr. (Ed.), *Speech evaluation in psychiatry* (pp. 39–49). New York: Grune & Stratton.

Pantaleoni, H. (1985). *On the nature of music*. Oneonta, NY: Welkin Books.

Papoušek, H., & Papoušek, M. (1979). The infant's fundamental adaptive response system in social interaction. In E. B. Thoman (Ed.), *Origins of the infant's social responsiveness* (pp. 175–208). Hillsdale, NJ: Erlbaum.

Papoušek, H., & Papoušek, M. (1984). Learning and cognition in the everyday life of human infants. In J. S. Rosenblatt, C. Beer, M.-C. Busnel, & P. J. B. Slater (Eds.), *Advances in the study of behavior*, **14**, 127–163. (Vol. 14, pp. 127–163) Orlando, FL: Academic Press.

Papoušek, H., & Papoušek, M. (1987). Intuitive parenting: A dialectic counterpart to the infant's integrative competence. In J. D. Osofsky (Ed.), *Handbook of infant development* (2nd ed., pp. 669–720). New York: Wiley.

Papoušek, M. (1987, April). *Models and messages in the melodies of maternal speech in tonal and nontonal languages.* Paper presented at meetings of the Society for Research in Child Development, Baltimore, MD.

Papoušek, M., & Papoušek, H. (1981). Musical elements in the infant's vocalization: Their significance for communication, cognition, and creativity. In L. P. Lipsitt (Ed.), *Advances in infancy research* (Vol. 1, pp. 163–224). Norwood, NJ: Ablex.

Papoušek, M., Papoušek, H., & Bornstein, M. H. (1985). The naturalistic vocal environment of young infants: On the significance of homogeneity and variability in parental speech. In T. M. Field & N. A. Fox (Eds.), *Social perception in infants* (pp. 269–297). Norwood, NJ: Ablex.

Papoušek, M., Papoušek, H., & Harris, B. J. (1986). The emergence of play in parent-infant interactions. In D. Gorlitz & J. F. Wohlwill (Eds.), *Curiosity, imagination, and play: On the development of spontaneous cognitive and motivational processes* (pp. 214–246). Hillsdale, NJ: Erlbaum.

Parke, R. D., Grossman, K., & Tinsley, B. R. (1981). Father-mother-infant interaction in the newborn period: A German-American comparison. In T. M. Field (Ed.), *Culture and early interactions.* Hillsdale, NJ: Erlbaum.

Phillips, J. (1973). Syntax and vocabulary of mothers' speech to young children: Age and sex comparisons. *Child Development*, **44**, 182–185.

Plomp, R. (1976). *Aspects of tone sensation.* New York: Academic Press.

Pomerantz, J. R. (1981). Perceptual organization in information processing. In M. Kubovy & J. R. Pomerantz (Eds.), *Perceptual organization* (pp. 141–180). Hillsdale, NJ: Erlbaum.

Remick, H. (1976). Maternal speech to children during language acquisition. In W. von Raffler-Engel & Y. Lebrun (Eds.), *Baby talk and infant speech.* Lisse, Netherlands: Swets & Zeitlinger.

Rheingold, H., & Adams, J. L. (1980). The significance of speech to newborns. *Developmental Psychology*, **16**, 397–403.

Robinson, B. W. (1976). Limbic influences on human speech. *Annals of the New York Academy of Sciences*, **280**, 761–771.

Rosenthal, M. K. (1982). Vocal dialogues in the neonatal period. *Developmental Psychology*, **18**, 17–21.

Sachs, J. (1977). The adaptive significance of linguistic input to prelinguistic infants. In C. E. Snow & C. A. Ferguson (Eds.), *Talking to children: Language input and acquisition* (pp. 51–61). London & New York: Cambridge University Press.

Sachs, J., Brown, R., & Salerno, R. A. (1976). Adults' speech to children. In W. von Raffler-Engel & Y. Lebrun (Eds.), *Baby talk and infant speech.* Lisse, Netherlands: Swets & Zeitlinger.

Sachs, J., & Devin, J. (1976). Young children's use of age-appropriate speech styles in social interaction and role-playing. *Journal of Child Language,* **3,** 81–98.

Schenker, H. (1954). *Harmony* (O. Jones, Ed., & E. M. Borgese, Trans.). Cambridge, MA: MIT Press. (Original work published 1906).

Scherer, K. R. (1982). The assessment of vocal expression in infants and children. In C. E. Izard (Ed.), *Measuring emotions in infants and children* (pp. 127–163). London & New York: Cambridge University Press.

Scherer, K. R., & Oshinsky, J. S. (1977). Cue utilization in emotion attribution from auditory stimuli. *Motivation and Emotion,* **1,** 331–346.

Shatz, M., & Gelman, R. (1973). The development of communication skills: Modifications in the speech of young children as a function of listener. *Monographs of the Society for Research in Child Development,* **38** (5, Serial No. 152).

Sherrod, K. B., Friedman, S., Crawley, S., Drake, D., & Devieux, J. (1977). Maternal language to prelinguistic infants: Syntactic aspects. *Child Development,* **48,** 1662–1665.

Sloboda, J. A. (1985). *The musical mind: The cognitive psychology of music.* London & New York: Oxford University Press (Clarendon).

Slawson, A. W. (1968). Vowel timbre and musical timbre as functions of spectrum envelope and fundamental frequency. *Journal of the Acoustical Society of America,* **43,** 87–101.

Snow, C. E. (1972). Mothers' speech to children learning language. *Child Development,* **43,** 549–565.

Snow, C. E. (1977). The development of conversation between mothers and babies. *Journal of Child Language,* **4,** 1–22.

Spelke, E. S., & Owsley, C. J. (1979). Intermodal exploration and knowledge in infancy. *Infant Behavior and Development,* **2,** 13–28.

Stebbins, W. G., Coombs, S., & Prosen, C. (1984). Comparative psychoacoustics: New directions. In C. I. Berlin (Ed.), *Hearing science: Recent advances* (pp. 1–46). San Diego, CA: College-Hill Press.

Stern, D. N., & Gibbon, J. (1979). Temporal expectancies of social behaviors in mother-infant play. In E. B. Thoman (Ed.), *Origins of the infant's social responsiveness* (pp. 409–429). Hillsdale, NJ: Erlbaum.

Stern, D. N., Hofer, L., Haft, W., & Dore, J. (1985). Affect attunement: The sharing of feeling states between mother and infant by means of intermodal fluency. In T. M. Field & N. A. Fox (Eds.), *Social perception in infants* (pp. 249–268). Norwood, NJ: Ablex.

Stern, D. N., Jaffe, J., Beebe, B., & Bennett, S. L. (1975). Vocalizing in unison and in alternation: Two modes of communication within the mother-infant dyad. *Annals of the New York Academy of Sciences,* **263,** 89–100.

Stern, D. N., Spieker, S., Barnett, R. K., & MacKain, K. (1983). The prosody of maternal speech: Infant age and context related changes. *Journal of Child Language,* **10,** 1–15.

Stern, D. N., Spieker, S., & MacKain, K. (1982). Intonation contours as signals in maternal speech to prelinguistic infants. *Developmental Psychology,* **18,** 727–735.

Stevenson, M. B., VerHoeve, J. N., Roach, M. A., & Leavitt, L. A. (1986). The beginning of conversation: Early patterns of mother-infant vocal responsiveness. *Infant Behavior and Development, 9,* 423–440.

Sullivan, J. W., & Horowitz, F. D. (1983). The effects of intonation on infant attention: The role of rising intonation contour. *Journal of Child Language, 10,* 521–534.

Terhardt, E. (1974). Pitch, consonance, and harmony. *Journal of the Acoustical Society of America, 55,* 1061–1069.

Thorpe, L. A. (1985). *Auditory-temporal organization: Developmental perspectives.* Unpublished doctoral dissertation, University of Toronto.

Thorpe, L. A. (1986, April). *Infants categorize rising and falling pitch.* Paper presented at meetings of the International Conference on Infant Studies, Los Angeles, CA.

Thorpe, L. A., & Trehub, S. E. (1989). The duration illusion and auditory grouping in infancy. *Developmental Psychology. 25,* 122–127.

Thorpe, L. A., Trehub, S. E., Morrongiello, B. A., & Bull, D. (1988). Perceptual grouping by infants and preschool children. *Developmental Psychology, 24,* 484–491.

Titon, J. T. (Ed.). (1984). *Worlds of music: An introduction to the music of the world's peoples.* New York: Schirmer Books.

Trehub, S. E. (1985). Auditory pattern perception in infancy. In S. E. Trehub & B. A. Schneider (Eds.), *Auditory development in infancy* (pp. 183–195). New York: Plenum Press.

Trehub, S. E. (1987). Infants' perception of musical patterns. *Perception & Psychophysics, 41,* 635–641.

Trehub, S. E., Bull, D., & Thorpe, L. A. (1984). Infants' perception of melodies: The role of melodic contour. *Child Development, 55,* 821–830.

Trehub, S. E., Cohen, A. J., & Guerriero, L. (1987, April). *Children's understanding of the emotional meaning of music.* Paper presented at meetings of the Society for Research in Child Development, Baltimore, MD.

Trehub, S. E., Cohen, A. J., Thorpe, L. A., & Morrongiello, B. A. (1986). Development of the perception of musical relations: Semitone and diatonic structure. *Journal of Experimental Psychology: Human Perception and Performance, 12,* 295–301.

Trehub, S. E., Endman, M. W., & Thorpe, L. A. (in press). Infants' perception of timbre: Classification of complex tones by spectral structure. *Journal of Experimental Child Psychology.*

Trehub, S. E., Morrongiello, B. A., & Thorpe, L. A. (1985). Children's perception of familiar melodies: The role of intervals, contour, and key. *Psychomusicology, 5,* 39–48.

Trehub, S. E., & Thorpe, L. A. (1989). Infants' perception of rhythm: Categorization of auditory sequences by temporal structure. *Canadian Journal of Psychology, 43,* 217–229.

Trehub, S. E., Thorpe, L. A., & Morrongiello, B. A. (1985). Infants' perception of melodies: Changes in a single tone. *Infant Behavior and Development, 8,* 213–223.

Trehub, S. E., Thorpe, L. A., & Morrongiello, B. A. (1987). Organizational processes in infants' perception of auditory patterns. *Child Development, 58,* 741–749.

Treitler, L. (1982). The early history of music writing in the West. *Journal of the American Musicological Society*, **35**, 237–279.

Trevarthen, C. & Marwick, H. (1986). Signs of motivation for speech in infants, and the nature of a mother's support for development of language. In B. Lindblom & R. Zetterstrom (Eds.), *Precursors of early speech* (pp. 279–308). New York: Stockton Press.

Walker, A. S. (1982). Intermodal perception of expressive behaviors by human infants. *Journal of Experimental Child Psychology*, **33**, 514–535.

Ward, W. D., & Burns, E. M. (1982). In D. Deutsch (Ed.), *The psychology of music* (pp. 431–451). New York: Academic Press.

Watson-Gegeo, K. A., & Gegeo, D. W. (1986). Calling out and repeating routines in Kwara'ae children's language socialization. In B. B. Schieffelin & E. Ochs (Eds.), *Language socialization across cultures* (pp. 17–50). London & New York: Cambridge University Press.

Weden, L. (1972). A multidimensional study of perceptual-emotional qualities in music. *Scandinavian Journal of Audiology*, **13**, 241–257.

Weeks, T. (1971). Speech registers in young children. *Child Development*, **42**, 1119–1131.

Wertheimer, M. (1938). Laws of organization in perceptual forms. In W. D. Ellis (Ed.), *A source book of Gestalt psychology*, London: Routledge & Kegan Paul. (Original work published in German, 1923)

Wolff, P. H. (1963). Observations on the early development of smiling. In B. M. Foss (Ed.), *Determinants of infant behavior*, Vol. 2. London: Methuen.

Zenatti, A. (1969). *Le développement génétique de la perception musicale* [The development of musical perception]. Paris: Centre National de la Recherche Scientifique.

# 15

## EXPERIMENTALLY INDUCED AND NATURALLY OCCURRING MONKEY MODELS OF HUMAN AMBLYOPIA

*Ronald G. Boothe*

Yerkes Regional Primate Research Center, Departments of Psychology and Ophthalmology, Emory University, Atlanta, Georgia

One of the most prominent features of primate visual perception is the presence of single binocular vision (Polyak, 1957). The separate information picked up by the two eyes is fused by the brain into a single percept that includes information about the locations of objects in three-dimensional space. Binocular vision is a highly complex process that involves both motor and sensory components. The brain must control the positions of the two eyes so that the object of regard remains imaged simultaneously onto the fovea of each retina. The brain must also make a detailed comparison of the disparities between the positions of the images in the two eyes to achieve sensory fusion. Both of these processes require highly organized functional connections within those parts of the brain that deal with binocular information.

These neural connections are not mature at the time of birth. Instead they are formed during an extended period of postnatal development. The specific connections that get formed during this postnatal period are influenced to a surprisingly large degree by the visual stimulation produced in the two eyes. I will use the term *developmental neural plasticity* to refer to the fact that the brain's functional connectivity can be influenced by early visual experience.

Developmental neural plasticity has been demonstrated in a number of studies of visual deprivation in animals. For example, if one eye of a

**461**

monkey is deprived of patterned visual input during early development, obvious anatomical and physiological changes can be detected within various parts of the brain's visual system (for recent reviews of this extensive literature, see Boothe, Dobson, & Teller, 1985; Boothe, Vassdal, & Schneck, 1986; von Noorden, 1985a; Wiesel, 1982). It may be useful, in terms of putting these visual deprivation effects into a broader perspective, to keep in mind that they are probably just an extreme case of a more general phenomenon. Primates have been able to adapt to a wide range of visual environments (Walls, 1942/1963). They have been able to achieve a high degree of compatibility with the environment in which they find themselves, in part, through maintaining a prolonged period of developmental neural plasticity. This allows the brain the opportunity to adjust itself to the environmental conditions in which it finds itself developing. This is probably a particularly effective method for adjusting the neural mechanisms that underlie binocular functions. For example, the distance between the two eyes in the adult, an important factor that needs to be taken into account in binocular visual functions, cannot be prespecified because it depends on factors such as nutrition during childhood. An extended period of plasticity allows the brain's binocular neurons to adjust their inputs appropriately for these kinds of factors. It is only in those rare cases where the visual environment present during development is deficient or abnormal that this plasticity becomes nonadaptive.

It is not only monkeys, but also humans who are susceptible to the effects of visual deprivation during development. For example, it is estimated that 2–5% of the human population suffer from amblyopia, a developmental clinical disorder in which the brain fails to establish appropriate functional connections with one or the other of the two eyes (National Eye Institute (NEI), 1983–1987). Amblyopia provides a striking example of developmental neural plasticity. It is exposure to an abnormal or impoverished visual input to one eye during development that causes the brain's connections to form improperly and leads to the perceptual deficits that become labeled as amblyopia.

In the remainder of this chapter, I summarize some of the primary research findings obtained from monkeys that relate to human amblyopia. I include major findings from other laboratories where pertinent, but emphasize results from my own laboratory. A more comprehensive review of this general topic can be found in Boothe, Dobson, and Teller (1985). However, before discussing monkey results, I will first describe in somewhat more detail what is meant by the clinical term amblyopia.

## I. HUMAN AMBLYOPIA

Amblyopia is a clinical ophthalmology term that refers to a group of disorders of visual function (NEI, 1983–1987). There are several types of amblyopia, but they are all acquired during childhood and thought to be environmentally determined. A tentative diagnosis of amblyopia is made when a patient has poor acuity in one eye which cannot be corrected with glasses; an ophthalmological examination fails to reveal anything wrong with the inside of the eye; and there is no evidence of neurological problems, disease, or trauma that might account for the malfunction. The diagnosis is strengthened if a family history reveals evidence that there was a period during infancy or childhood when the affected eye received abnormal or impoverished visual stimulation. This abnormal visual input during development is taken to be the causative factor of the disorder and the subtypes of amblyopia are diagnosed based on this presumptive cause. For example, if an amblyopic patient was known to have strabismus (crossed eyes) during childhood, the resulting disorder would be labeled strabismic amblyopia; if the patient was known to have an anisometropia (refractive error in one eye), it would be labeled an anisometropic amblyopia.

The first step in treating amblyopia is to eliminate the presumptive causative factor(s). For example, a strabismus can be corrected by conducting surgery on the extraocular muscles, and an anisometropia can be eliminated by prescribing glasses (von Noorden, 1985b). This must be done during childhood because if the disorder is left untreated the amblyopia eventually becomes permanent and untreatable in the adult. Even if an amblyopic eye is corrected of its initial problems during childhood, it will often not recover normal vision unless it receives additional treatment. The most common form of additional treatment is occusion therapy in which the child is compelled to wear an occluding patch on the other eye, thus enforcing use of the amblyopic or "lazy" eye (NEI, 1983–1987).

## II. ESTABLISHING A MODEL FOR FURTHER STUDIES

Many of the experiments that are needed to answer questions about the neural mechanisms that underlie amblyopia would not be feasible to conduct on human subjects. These questions can be addressed by conducting experiments with an appropriate animal model. However, it is not sufficient to just conduct studies with animals and then generalize the results to human amblyopia. An important and too often neglected first step is to establish that the animal studies do in fact model the essential aspects of human amblyopia that are being studied.

The experiments to be described in this chapter all used macaque monkeys as subjects. There are a number of factors that make macaques a good choice for these kinds of studies. First of all, adult macaque monkeys and humans exhibit similar visual functions. They respond very similarly on a large number of psychophysical tests of visual functions including acuity (Grether, 1941), contrast sensitivity (DeValois, Morgan, & Snodderly, 1974; Harwerth & Smith, 1985; Williams, Boothe, Kiorpes, & Teller, 1981), meridional sensitivity (Boltz, Harwerth, & Smith, 1979;

Williams et al., 1981), temporal sensitivity (Harwerth & Smith, 1985; Merigan, Pasternak, & Zehl, 1981; Symmes, 1962), oculomotor behavior (Boltz & Harwerth, 1979; Fuchs, 1967; Motter & Poggio, 1984; Skavenski, Robinson, Steinman, & Timberlake, 1975; Snodderly, 1987; Snodderly & Kurtz, 1985), and various aspects of binocular functions (Harwerth & Boltz, 1979; Julesz, Petrig, & Buttner, 1976; Sarmiento, 1975). There are other species of animals that respond similarly to humans on some of these functions, but no nonprimate species would be as similar to humans overall as monkeys, and among the kinds of monkeys that have been studied, the various species of macques appear to be the most similar to humans.

Macaque monkeys are particularly well suited to model the acuity loss that occurs in a human amblyopic eye. Table 1 lists some representative acuity values for various kinds of animals. Note that mammals other than primates have poorer acuity than humans. In fact, spatial resolution of normal eyes in nonprimate mammals is as poor as acuity in many human amblyopic eyes. Primates have been able to achieve their fine acuity through specialized anatomical and physiological mechanisms in the eye and brain (Polyak, 1957; Walls, 1942/1963). An example of this would be the fovea centralis of the primate retina. Moderate levels of acuity loss that occur in amblyopia are likely to be associated with malfunctions in precisely those mechanisms that have evolved to subserve high acuity levels. Lower mammals lack these specialized mechanisms and for this reason have only limited utility as animal models of the moderate acuity losses that most frequently accompany human amblyopia.

Some bird species, such as falcons and eagles, have spatial resolution that is as good or better than humans (see Table 1). However, the neural

**TABLE 1.**

| Animal | Acuity (cycles/deg)[a] | Reference |
|---|---|---|
| Mouse | 0.5 | Sinex, Burdette, & Pearlman (1979) |
| Hooded rat | 1 | Dean (1981) |
| Tree shrew | 2 | Petry, Fox, & Casagrande (1984) |
| Cat | 3.5–7 | Berkley (1976) |
| Galago | 6 | Langston, Casagrande, & Fox (1986) |
| Macaque monkey | 50 | See Fig. 3 |
| Humans | 50 | See Fig. 3 |
| Falcon | 73 | Reymond (1987) |
| Eagle | 143 | Reymond (1985) |

[a] A description of the cycles/degree units that are used to designate acuity is presented in Section III of the text.

pathways of their visual systems are somewhat different from those of mammals. Studies of comparative perception in these species can be useful from the perspective of learning about convergent evolution of similar functions, but the most appropriate animal models for studying the neural mechanisms underlying human amblyopia are those species that have the most similar visual systems. In this regard, the anatomical structure of the visual pathways of macaque monkeys is very similar to that of humans (Boycott & Dowling, 1969; Polyak, 1957).

Another important factor that needs to be taken into consideration in choosing an animal to study is the fact that amblyopia is a developmental disorder. Therefore, it is best studied in species that undergo prenatal and postnatal visual development that is similar to that in humans. Macaque monkeys provide a good model in this regard also. It is possible to directly relate major neuroembryological events that take place in formation of the monkey visual system to similar events in human neuroembryology (Booth, 1988). Many behavioral visual functions have also been tested and found to be similar in neonatal monkeys and humans, and it is possible to relate the time courses for development of a number of visual functions between monkeys and humans by using a 4 to 1 adjustment of the age scales (Boothe, Dobson, & Teller, 1985; Boothe, Kiorpes, Regal, & Lee, 1982; Gunderson & Swartz, 1985; Mendelson, 1982).

There are some methodological issues that arise in trying to study development of visual functions in monkeys. First, how does one ask a nonverbal subject, such as a monkey, what it can see? Second, how does one relate results obtained with a particular set of procedures from monkeys to the results obtained with a different set of procedures from humans? Two psychophysical procedures have been adapted and refined specifically for the purpose of solving these methodological issues.

Preferential looking is based upon the fact that infants direct their eyes toward contours in the visual environment (Fantz, Fagan, & Miranda, 1975; Teller, 1979; see also Dobson, this volume). An infant is positioned in front of a display that contains contours on one side and a blank field on the other side. If the infant can see the contours, it tends to direct its eyes preferentially toward the side of the display containing contours (see Fig. 1). This procedure can be used to measure a visual function such as acuity by putting a patterned stimulus on one side of the display and measuring the smallest pattern size that elicits preferential looking. This method has the advantage that perceptual functions can be measured in infant humans and monkeys by using virtually identical equipment and procedures (Teller, Morse, Borton, & Regal, 1974; Teller, Regal, Videen, & Pulos, 1978).

Preferential looking procedures become inefficient as infants get older. Neither 36-week-old monkeys nor 36-month-old humans will sit passively and generate preferential looking data. At these older ages, operant reinforcement procedures are more efficient. The subject is induced to

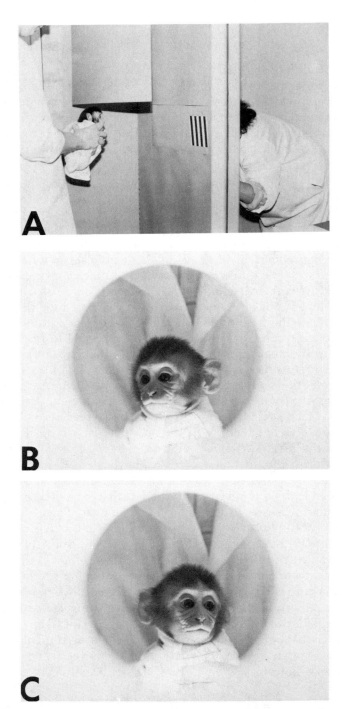

**FIGURE 1.** This figure shows the preferential looking procedure being conducted with an infant monkey. (*A*) One experimenter (the holder) holds the infant in front of the display, while a second experimenter (the observer) observes the infant through a peephole in the display. A screen blocks a holder's view of the display. The observer judges whether or not

make some response, such as pointing with its arm toward a visual stimulus. For example, to measure acuity the subject would be given a reward for responding to the side of a visual display that presents a patterned stimulus. Once training is complete, acuity can be estimated by determining the smallest pattern element size that consistently elicits correct responses.

The specific operant methods that are used to train and test monkeys in my laboratory are based on a face-mask cage design first described by Sackett, Tripp, Milbrath, Gluck, and Pick (1971). The monkeys are tested in a cage that has a mask, molded to the shape of an infant monkey's face, mounted on one wall as illustrated schematically in Fig. 2. There are eye holes in the face mask so that the monkey can look out of the cage at a visual display. An arm hole is present underneath the face mask and the monkey can reach out of the cage and pull one of the two grab bars that are positioned in front of a computer-controlled video display monitor. The monkey is taught to pull one of the two bars on each trial, according to what visual stimuli are presented on the display. Correct bar pulls result in liquid reinforcement or a food pellet being delivered to the mouth hole. Incorrect bar pulls result in a short time-out period of about 10 sec during which the display is turned off and further bar pulls have no effect. Various devices such as shutters, apertures, and lenses can be positioned directly in front of the eye holes in the face mask. They are used to control factors such as refractive error, pupil size, and monocular testing of either eye.

## III. NORMAL DEVELOPMENT OF ACUITY AND CONTRAST SENSITIVITY

In this section, I will elaborate on the processes that govern normal development of acuity and contrast sensitivity, two of the primary visual functions disrupted in an amblyopic eye. An understanding of some of the factors involved in normal development of these functions will provide us with a background against which we can evaluate the abnormal development that occurs in amblyopic eyes.

There are a number of measures of spatial resolution, or acuity. One of the simplest, and one that can be easily measured in animals as well as humans, is grating acuity. A grating is produced that is made up of equal-width, high-contrast, dark and bright stripes. The width of the stripes is decreased until the observer can no longer detect that stripes are present. The finest stripe width that can be seen is called grating acuity. We can specify grating acuity in terms of the size of the visual angle

---

the infant monkey can see the patterned stimulus (in this example, vertically oriented stripes) on the display based on the head and eye movements of the monkey. (*B*) The observer's view of the monkey during a trial when a pattern that the monkey can easily see is presented on the monkey's right. (*C*) Observer's view on a trial when the pattern is presented on monkey's left.

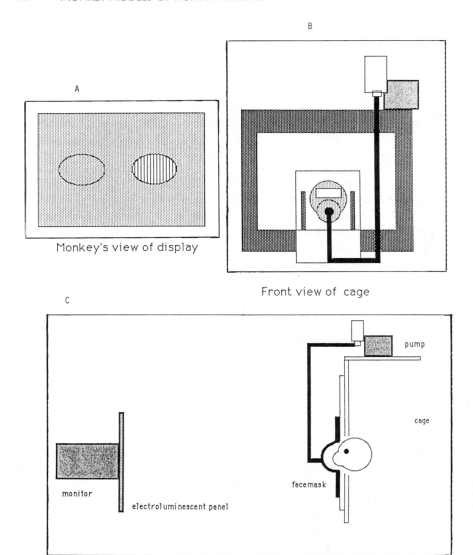

A

Monkey's view of display

B

Front view of cage

C

monitor

electroluminescent panel

pump

cage

facemask

Cross section testing room

**FIGURE 2.** This figure illustrates the face-mask cage procedure that is used in the operant studies of visual perception in monkeys. The monkey is allowed to free roam inside the cage and can initiate a trial at any time by placing its face into the mask. While its face is in the mask, the monkey can view a visual display that is generated on a video monitor. The display is surrounded by an electroluminescent panel that provides a uniform background matched to the display in luminence and color. Two grab bars are present on the cage wall underneath the mask. The monkey is trained to pull the bar corresponding to the side of the display that contains the correct visual stimulus (the left bar if stripes are present on the left, and the right bar if stripes are present on the right). When the monkey pulls the correct bar, a pump delivers apple juice or dispenses a banana pellet to a mouth hole in the face mask. Incorrect bar pulls result in a time-out period signaled by an audible tone during which the display goes blank and further bar pulls are ignored.

subtended by the stripes. If each stripe subtends 1 min of arc, then there will be 30 cycles of dark and bright stripes per degree of visual angle (30 cycles/deg). By convention, a grating acuity of 30 cycles/deg corresponds to a Snellen acuity of 20/20. For clinical purposes, an acuity poorer than 20/20 (fewer than 30 cycles/deg) is considered to be abnormal.

The magnitude of postnatal improvement in primate acuity is striking. I illustrate this in Fig. 3, taken from Boothe, Kiorpes, and Carlson (1985), which summarizes the results from a number of laboratories concerning the time courses of acuity development in humans and macaque monkeys. Acuity starts out at similar levels in neonates (about 1 cycle/deg) and asymptotes toward similar levels in adults (about 50 cycles/deg), an improvement by a factor of 50. This figure demonstrates that there is a protracted period of postnatal development before mature adult functioning is achieved. Note that this figure plots age in weeks for monkeys and in months for humans in accord with the 4 to 1 ratio discussed earlier.

What factors are responsible for the large postnatal improvement in spatial resolution in prmates? Obvious candidates include structural changes in the eye's optics to allow a higher quality retinal image, growth of the eyeball and subsequent magnification of the image on the regina, and changes in packing density of the photoreceptors in the eye. Each of these factors probably play a small role, but previous analyses have

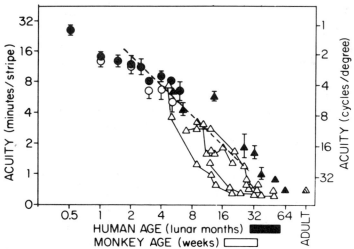

**FIGURE 3.** This figure summarizes the results from a number of studies conducted in my laboratory and in the laboratories of D. Teller and V. Dobson. Acuity is plotted as a function of age, which is specified in months for humans and in weeks in monkeys. The left ordinate specifies grating acuity in terms of minutes/stripe where a 1-min stripe corresponds to 20/20 Snellen acuity. The right ordinate specifies acuity in terms of cycles/degree. This figure is reprinted from Boothe, Kiorpes, and Carlson (1985).

demonstrated that none of them, either separately or in combination, can account for more than a fraction of the total postnatal improvement in acuity (Blakemore & Vital-Durand, 1986; Boothe, 1983).

Adult acuity levels probably depend primarily on refinements in electrophysiological processing by neural circuits in the eye and/or brain. Blakemore and Vital-Durand (1981, 1986) have shown that the acuity limits of individual neurons in the lateral geniculate nucleus, a structure that serves as the primary relay between the retina and the primary visual cortex in primates, is as poor in neonates as behavioral acuity. Furthermore, these neurons exhibit a time course for the postnatal development in spatial resolution that is very similar to postnatal development of behavioral acuity.

It is not only acuity that is immature at birth, but also sensitivity to contrast across a range of spatial frequencies. This can be demonstrated by making measurements of a contrast-sensitivity function, which is a more general measure than grating acuity of an organism's abilities to see spatial patterns. Grating acuity determines only the finest stripe width that can be seen, and provides no information about how well the organism can see other stripe sizes that are larger than the acuity limit. Contrast-sensitivity measurements present the observer with gratings that vary in both stripe width and in contrast. The grating luminance is modulated sinusoidally across space. (This is for technical reasons that are beyond the scope of this chapter. For further discussion, see Enroth-Cugell & Robson (1984).) A particular stripe width is shown and its contrast is varied until it can just be seen. Then the same measurement is repeated at a number of other stripe widths. In this way, an observer's relative sensitivity to spatial contrast over a range of stripe widths can be determined. The contrast cannot be increased beyond 100%, and therefore as the stripe width is made narrower a limit will be reached beyond which the observer cannot detect the stripes even with maximum contrast. This limit is roughly analogous to a simple measurement of grating acuity.

The development of contrast sensitivity is illustrated in Fig. 4, taken from Boothe, Kiorpes, Williams, and Teller (1988), which shows contrast-sensitivity functions obtained longitudinally from a single infant monkey, AB, during the first postnatal year. Note that the abscissa on the contrast-sensitivity function shown in Fig. 4 is specified in terms of cycles/degree of spatial frequency. Low spatial frequencies correspond to wide stripe widths and high spatial frequencies to narrow stripe widths. The ordinate is specified in units of contrast sensitivity which is the reciprocal of the contrast needed to see the grating. A value of 1 corresponds to a grating that cannot be seen unless it contains 100% contrast. Sensitivity values higher than one indicate that the grating can be seen with less than 100% contrast.

The smooth curves drawn through the data points in Fig. 4 are best fitting curves described in Boothe et al. (1988). The developmental changes

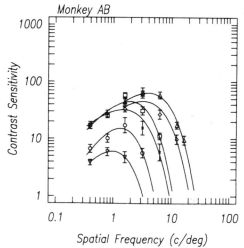

**FIGURE 4.** Development of the spatial contrast-sensitivity function for infant monkey AB. Contrast sensitivity is plotted as a function of spatial frequency at each of a number of ages: 10 weeks ($\nabla$), 11 weeks (o), 14 weeks (X), 15 weeks ($\square$), 26 weeks ($\diamond$), 38 weeks ($\triangle$. This figure is reprinted from Boothe, Kiorpes, William S, and Teller (1988).

that are depicted for monkey AB in this figure are typical of the developmental changes seen in normal monkeys (Boothe et al., 1988) and humans (Atkinson, Braddick, & Moar, 1977; Banks & Salapatek, 1978; Bradley & Freeman, 1982; Norcia, Tyler, & Hamer, 1988; Pirchio, Spinelli, Fiorentini, & Maffei, 1978). They can be characterized as shifts of the function upward (increased sensitivity to contrast), and rightward on the spatial frequency axis (increased spatial resolution).

It is interesting to speculate about what kinds of changes in neural processing might account for these developmental shifts of the contrast-sensitivity function. Current models of the neural basis of contrast sensitivity emphasize interactions between the center and surround regions within a single cell's receptive field. For example, Enroth-Cugell and Robson (1984) describe a model of retinal ganglion cell receptive fields in which the cell's contrast-sensitivity function shifts, upward in sensitivity and rightward toward higher spatial frequencies, as the size and strength of its surround are increased relative to its center. There is some empirical evidence that these kinds of changes in receptive field properties occur in the primate visual system during normal development. For example, Blakemore and Vital-Durand (1986) report evidence that surround strengths of neurons in the monkey lateral geniculate nucleus increase during postnatal development.

However, the behavioral contrast-sensitivity functions as depicted in Fig. 4 are too broad to be accounted for on the basis of receptive fields of single neurons in the early stages of the visual system. Instead, the behavioral function probably reflects the upper envelope of sensitivity of a number of different classes of cells that have different sized receptive fields (Wilson, 1988). For example, the behavioral sensitivity to low spatial frequencies may be dependent on a class of neurons with large receptive fields, sensitivity to high spatial frequencies dependent on a different class of neurons with small receptive fields. We will use the term *channel* to refer to a class of neurons that have similar sized receptive fields. In this terminology the behavioral contrast-sensitivity function reflects the upper envelope of multiple channels. The developmental results shown in Fig. 4 can be related to this kind of a multichannel model by assuming that the channels that respond to low spatial frequencies mature first and those responding to high frequencies last. In the following sections we will attempt to relate some of the effects of amblyopia to this multichannel model of contrast sensitivity.

## IV. AMBLYOPIA ASSOCIATED WITH STRABISMUS

Strabismus refers to an ocular misalignment in which the eyes are either crossed or walleyed. The most frequent type of amblyopia in humans is that associated with strabismus, that is, strabismic amblyopia. In this section I will discuss studies with monkeys that pertain to both strabismus (an oculomotor problem) and its associated amblyopia (a sensory problem).

Until recently there were no reports of naturally occurring strabismus in monkeys. This lack of widespread occurrence of strabismus in monkeys seems surprising given the general similarity of human and monkey visual systems that we discussed above in Section II. Jampolsky (1978) discussed this issue in an extensive review of the topic and speculated that there must be something fundamentally different about the monkey oculomotor system that makes it impervious to the conditions that lead to misalignment in humans. Jampolsky was not completely correct in his conclusion that strabismus does not occur naturally in monkeys. My laboratory has located several monkeys with naturally occurring strabismus during the past few years (Kiorpes & Boothe, 1981; Kiorpes, Boothe, Carlson, & Alfi, 1985). Nevertheless, large numbers of these monkeys have not been found to date, and it is interesting to speculate about whether there are any interspecies differences in oculomotor systems that could account for the widespread occurrence of strabismus only in humans.

There is one obvious difference between human oculomotor systems and those of monkeys that can be seen even at the gross structural level. Monkeys have an eye muscle, the accessory lateral rectus, that is not normally present in either great apes or humans (Swindler & Wood, 1973). The accessory lateral rectus in monkeys receives innervation from a ventral subgroup of abducens motoneurones which is similar to the innervation of the retractor bulbi in lower mammals, suggesting a homology for these two muscles (Spencer & Porter, 1981). The retractor system in lower mammals functions in reflex withdrawal of the eye in response to touching the front of the eye (Delgado-Garcia, Evinger, & Baker, 1980). In other characteristics the accessory lateral rectus in monkeys is similar to the human lateral rectus, and for these reasons, Spencer and Porter (1981) speculate that the monkey accessory lateral rectus may reflect a transition whereby the functions of the retractor bulbi in lower mammals are taken over by the lateral rectus in humans.

It is interesting to speculate that human strabismus might be a result of the fact that during the course of evolution humans have lost the nictitating membrane and retractor bulbi muscle that make up the retractor system in lower mammals. One of the important questions yet to be answered about the naturally strabismic monkeys that have been located is whether or not their accessory lateral rectus muscle systems are abnormal. One possibility is that their oculomotor systems have become more human-like and this has caused the monkeys to become susceptible to the same factors that cause misalignment in humans. Studies currently underway in my laboratory are attempting to address these kinds of issues.

Human strabismus tends to run in families and may have a genetic component. McKusick (1983) includes human strabismus in his listing of autosomal dominant phenotypes, although, given the confusing literature available on this topic (von Noorden, 1985b), it seems unlikely that this classification will be maintained. An analysis of the pedigrees of the naturally strabismic monkeys I have located strongly indicates that strabismus in monkeys also runs in genetic lines, but the exact mode of inheritance has not yet been determined. It might be possible to discover the modes of inheritance by conducting selective breeding studies with naturally strabismic monkeys.

In the case of strabismic amblyopia the strabismus (an oculomotor problem) and the amblyopia (a sensory processing problem) are both present and likely to be inextricably linked together. The brain cannot know where to point its eyes in order to have the two foveas pointing at the same object unless the brain is receiving appropriate sensory input from both eyes. On the other hand, if the eyes are not aligned properly, the brain may not be able to receive appropriate sensory input from the two eyes. A complete understanding of strabismic amblyopia will only come from studies of the interactions between improper ocular alignment and improper sensory processing. The naturally strabismic monkeys, in which both of these deficits are present, provide an ideal model for these studies. Naturally strabismic monkeys are currently being tested behaviorally in my lab to document their oculomotor deficits and determine whether their sensory deficits are similar to those present in humans with strabismic amblyopia (Boothe, Kiorpes, & Carlson, 1985; Eggers & Boothe, 1987;

Joosse, Wilson, & Boothe, 1988; Quick & Boothe, 1988; Quick, Joosse, & Boothe, 1988.

In addition to performing studies on monkeys that have a naturally occurring strabismus, it is possible to produce a misalignment of the eyes in monkeys experimentally by doing surgery on or by injecting a paralytic drug into the extraocular muscles. Several laboratories have conducted studies in which a strabismus was experimentally induced in infant monkeys by using these kinds of methods (Boothe, Kiorpes, & Carlson, 1985; Harwerth, Smith, Boltz, Crawford, & von Noorden, 1983; Kiorpes & Boothe, 1980; von Noorden, 1973; von Noorden & Dowling, 1970). Two critical factors appear to be the primary determinants of whether or not an amblyopia develops in these animals: age at surgery and type of fixation. If the strabismus is produced after 12 weeks of age, amblyopia does not usually occur, indicating that the period of developmental neural plasticity for strabismic amblyopia is about over by this age. If the strabismus is produced prior to 12 weeks of age, the outcome depends on whether the animal consistently uses the unmanipulated eye for fixation or alternates between the two eyes for fixation. In animals that consistently fixate with the unmanipulated eye, the opposite (deviated) eye usually develops an amblyopia. Animals that alternate approximately equally between the two eyes for fixation do not develop an amblyopia in either eye.

It is also possible to induce a strabismus in a monkey indirectly by disruption of the normal binocular visual input during development (Quick, Tigges, Gammon, & Boothe, 1989). This is illustrated by the photograph in Fig. 5. This monkey was raised with an occluder contact lens on one eye that disrupted the normal binocular visual input to the brain. The effect of this rearing condition was to produce an amblyopia in the occluded eye (see Section V), but also had the effect of causing an eye misalignment to develop in this monkey.

## V. AMBLYOPIA THAT IS ASSOCIATED WITH STIMULUS DEPRIVATION AND DEFOCUS

Stimulus-deprivation amblyopia results from conditions in which all light to one eye is blocked or in which one eye is deprived of all patterned input. The rearing procedure that has been used most frequently with animals to study the effects of pattern deprivation involves surgically suturing the eyelids closed. The lids act as diffusers, allowing light to pass to the retina but removing all contours and patterns from the retinal image, thereby effectively removing all spatial frequencies from the visual input. This lid-suture rearing condition most closely simulates human stimulus-deprivation amblyopia that is associated with conditions such as a ptosis (drooping eyelid) that covers the pupil during childhood. Several studies have reported that monocular lid suture during development leads to

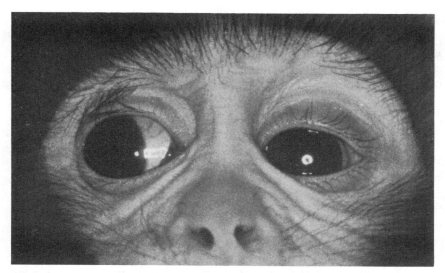

**FIGURE 5.** This monkey was raised with an occluder that blocked light input to its left eye. This photograph shows the monkey viewing a peripheral target presented to its right side. The right eye fixates the target but the left eye does not move from its central position, resulting in a walleyed strabismus.

severe acuity deficits (Hendrickson, Boles, & McLean, 1977; von Noorden, Dowling, & Ferguson, 1970), and contrast-sensitivity deficits (Harwerth, Crawford, Smith, & Boltz, 1981; Harwerth et al., 1983) in the lid-sutured eye later in life after it has been opened. Sensitivity to spatial contrast is essentially obliterated over almost the entire normal range of spatial frequencies. Residual pattern perception, if present at all, operates only over a range that is poorer than the resolution levels present at birth.

Stimulus-deprivation amblyopia is also associated with conditions in which all light input to an eye is blocked. Examples include an amblyopia that develops in children with an uncorrected dense unilateral cataract, or "occlusion amblyopia" that sometimes develops from excessive occlusion therapy of a child's eye during clinical treatment of ophthalmic disorders. These conditions have been simulated in monkeys by placing opaque occluders in front of one eye. The effects on visual development are severe: By six months of age acuity in the continuously occluded eye of a monkey has deteriorated to levels worse than present at birth (O'Dell, Gammon, Fernandes, Wilson & Boothe, 1989).

In anisometropic amblyopia, one eye receives varying amounts of optical defocus. An anisometropia can be produced in a monkey by defocusing one eye experimentally. Optical defocus operates differently from the pattern deprivation that is produced by lid suture or occlusion in that it produces a graded effect on the various spatial frequencies that are focused onto the retina. Small amounts of defocus remove only the high spatial frequencies from the retinal image (i.e., the fine spatial detail) while larger amounts of defocus attenuate progressively lower frequencies (more coarse features). A retina with moderate defocus will receive near normal stimulation at low spatial frequencies but will be subjected to a loss of input from higher spatial frequencies.

Two methods have been used to simulate anisometropic amblyopia in monkeys. Smith, Harwerth, & Crawford (1985) raised monkeys with helmets that allowed them to attach a spherical lens in front of one eye. Other studies administered drops of atropine, a drug that produces optical blur for targets at close viewing distances, to one eye of infant monkeys during rearing (Boothe, Kiorpes, & Hendrickson, 1982; Hendrickson et al., 1987; Kiorpes et al., 1987; Movshon et al., 1987). The defocused eye received impoverished stimulation that became progressively worse at higher spatial frequencies and developed a pattern of contrast sensitivity deficits with similar characteristics. The experimentally defocused eyes of monkeys from the studies that used both methods developed deficits in the high-frequency portions of their contrast-sensitivity functions that persisted after the defocusing conditions were terminated.

Recently my laboratory has been involved with studies of monkeys that have been reared with various combinations of defocus and occlusion to their two eyes (Gammon, Boothe, Chandler, Tigges, & Wilson, 1985). These monkeys are being raised under conditions that mimic clinical treatments given to children that have unilateral infantile cataracts. The monkeys receive a lensectomy, a surgical procedure that removes the natural lens from one eye, soon after birth, just as would be done for a child with a congenital or infantile cataract. The eye with its lens missing is called an aphakic eye. Following the lensectomy, some monkeys are raised with no further treatment to their aphakic eye. These monkeys receive severely defocused input to the aphakic eye and provide a model for studying aphakic amblyopia.

Other monkeys are reared under conditions that simulate additional treatments that would ordinarily be given to children with an aphakic eye. In some of these monkeys the aphakic eye is provided with a high-power, extended-wear contact lens that produces well-focused input at a particular viewing distance. Other monkeys wear a lower power lens that only partially corrects the defocus in the aphakic eye. Finally, in some cases the opposite eye is occluded, either continuously or part-time, to mimic treatment by occlusion therapy. A photograph of a monkey receiving one of these treatments is shown in Fig. 6. This monkey has had the natural lens removed surgically from its right eye, and this eye is shown wearing an extended-wear contact lens. The other eye is shown wearing an opaque occluder lens. Detailed information about procedures of contact lens rearing in monkeys is presented in Fernandes, Tigges, Tigges, Gammon, and Chandler (1988).

Results from animals wearing various combinations of these lenses reveal a gradation of deprivation effects that are correlated with the spatial frequency content present in their retinal images during rearing. Occluded eyes and uncorrected aphakic eyes, which receive little or no patterned input during rearing, have the poorest spatial resolution. Aphakic eyes that wear an optical correction such that they receive only mild defocus show mild deficits, confined primarily to the high spatial frequencies. Aphakic eyes that wear an optimal optical correction can achieve spatial resolution that is within the normal range if the opposite eye is continuously occluded as in the monkey shown in Fig. 6 (Gammon et al., 1985; 1988; O'Dell et al., 1989; Quick et al., 1987). However, in this latter case the occluded eye deteriorates to acuity levels poorer than present at birth.

**FIGURE 6.** Photograph of a monkey with an aphakic right eye. The aphakic eye is wearing an extended-wear contact lens that brings objects at close distances into focus. The opposite eye is shown wearing an occluder contact lens that simulates patching therapy.

Experiments are in progress with various combinations of optical correction and part-time occlusion. The goal of these studies is to try to find rearing conditions in which both eyes develop normal spatial resolution simultaneously.

## VI. APPLICATION OF MULTICHANNEL COMPETITION MODELS TO AMBLYOPIA

The perceptual deficits associated with amblyopia that we have been discussing in this chapter are correlated with specific anatomical and physiological changes in the brain. The most widely studied brain regions are the lateral geniculate nucleus (LGN) of the thalamus which receives direct retinal input, and the primary visual cortex to which the LGN projects. The LGN is a layered structure with each layer receiving input from only one of the two eyes. Neurons in layers of the LGN that receive retinal inputs from a deprived eye have somas that are shrunken in size compared to cells in layers receiving input from a nondeprived eye.

Axons from the LGN terminate in layer IV of primary visual cortex and the segregation of left eye and right eye input is maintained by ocular dominance columns. In normal animals the ocular dominance columns associated with the two eyes are equal in width. However, following monocular deprivation, columns formed by terminals from the deprived LGN layers are contracted in width while the columns from the other eye are expanded. The changes in soma size in the LGN are considered to be secondary to those changes in cortical column widths. Throughout the layers of primary visual cortex, electrophysiological recordings from neurons reveal a shift in the ocular dominance distribution such that most cells show a preference for being activated by the nondeprived eye.

These anatomical and physiological results are often discussed in terms of models that involve binocular competition (Wiesel, 1982). According to these models, axon terminals carrying signals from the two eyes compete with one another during development at some central level, probably layer IV in the primary visual cortex, for functional connections to postsynaptic sites. The terminals carrying signals from the deprived eye are unable to compete successfully. Therefore, by the end of the period of developmental neural plasticity, postsynaptic cortical cells end up being able to be driven only by signals from the nondeprived eye.

In Section III we discussed a multichannel model of normal development in which the behavioral contrast-sensitivity function is assumed to be the upper envelope of a number of different neural channels. It is possible to explain many of the deprivation effects that we have discussed in this chapter by combining a competition model with a multichannel model.

The advantages of a multichannel competition model are most apparent in trying to explain the effects of defocus rearing. Several lines of evidence demonstrate that defocus affects only a subpopulation of neurons. For example, in studies of monkeys raised under conditions of atropinization, a subpopulation of cortical neurons that responded to high spatial frequencies had a large amount of ocular dominance shift away from the formerly defocused eye. On the other hand, a subpopulation of cortical neurons that responded to only low spatial frequencies included large numbers of cells that responded to each eye (Movshon et al., 1987). Neuroanatomical studies of these same animals also produced evidence that defocus affects only a subpopulation of visual neurons (Hendrickson et al., 1987). Neurons in the parvocellular layers of the LGN and the cortical layers in which these neurons terminate show effects that are similar to those seen following monocular lid-suture rearing. However, neurons in the magnocellular layers of the LGN and their cortical terminations are less affected by defocus rearing.

A multichannel competition model assumes that there is a population of neural elements in the brain that, in normally reared monkeys, include cells with the capability of responding to both eyes. In cases of stimulus-deprivation amblyopia, none of these neurons are stimulated effectively by the deprived eye and few, if any, of them develop appropriate functional connections with the deprived eye. In cases of defocus to one eye, it is only a subpopulation of these neurons (those channels that respond to high spatial frequencies) that are deprived of their normal input, and it is only this subpopulation that fails to form functional connections with the defocused eye.

Multichannel competition models along these lines have not been widely applied to the spatial resolution losses that accompany strabismic amblyopia. However, a basis for applying these kinds of models comes from studies of visual scanning patterns in human infants (Banks & Salapatek, 1983; Haith, 1980). Measurements of scanning patterns reveal that infants devote most of their eye movements to scanning near and across contours in the visual environment. It is contours that are responsible for the high spatial frequencies that are present in the visual environment. One possible developmental hypothesis would be that during normal development the infant's brain is guiding movements of its eyes in patterns that are designed to maximize input of high spatial frequencies to the foveas of the fixating eyes. In an infant with normal binocular alignment both eyes would be fixating the same contours in the visual environment, and would therefore receive identical (maximized) input of high spatial frequencies. However, in an infant with strabismus the high-frequency input would be maximized only for the fixating eye. Models along these lines are consistent with the finding that subjects with alternating fixation develop good acuity in both eyes, whereas subjects who consistently fixate with only one eye during the developmental period end up with poor resolution in the nonfixating eye. Ikeda (1979) proposed a model of experimentally induced strabismic amblyopia that was based on a similar line of reasoning from studies with kittens. Quantitative predictions have never been derived from models of this type for humans or monkeys.

We do not yet have a good theoretical basis for understanding many of the results from animal models of aphakic amblyopia. Most of these animals receive various amounts of defocus in one eye and occlusion in the opposite eye. Neither eye is normal, and furthermore the subpopulations of cells that are affected by the deprivation are likely to be different for the two eyes. These conditions lead to complicated competitive interactions between various subpopulations of neurons that are not yet well understood.

## VI. SUMMARY AND CONCLUSIONS

With the exception of a few species that live underground, in caves, or deep in the ocean, all vertebrate animals depend on vision for their survival (Polyak, 1957; Walls, 1942/1963). Information acquired from the environment through their organs of sight is used to satisfy basic needs such as locating food and water, identifying mating partners for reproduction, and avoidance of predators. Primates, in particular, depend upon vision more than any of their other senses, and devote a large part of their brains to the processing of visual information.

The adult primate visual system forms through an extended period of prenatal and postnatal development. We have discussed several examples in which the postnatal visual environment, that is, what the developing organism sees, influences the formation of neural circuits in the brain and thereby adult sensory processing. The examples on which we have focused our attention in this chapter involve abnormal visual environments that lead to deficits in adult visual perception. When these deficits occur in humans they are called amblyopia. The conditions that lead to amblyopia in humans can be simulated in monkeys and produce similar symptoms. Abnormal modifications to the visual environment that we have discussed include producing a misalignment of the optical axes of the two eyes, blocking out all light to an eye, removal of some or all spatial frequencies from an eye's image, and various combinations of these treatments.

Each of these rearing conditions alters the visual input in specific ways during development, and leads to corresponding perceptual deficits in visual processing in adults. Many of these results are consistent with a multichannel competition model of neural processing in which a subpopulation of neural elements in one or both eyes are affected by the visual deprivation.

The visual deprivation effects that have been described in this chapter are specific examples of a more general phenomenon called developmental neural plasticity. Even though neural plasticity can be maladaptive in specific conditions such as those that lead to amblyopia, its operation is probably generally adaptive. Neural plasticity allows primates to use information from the environment to guide brain development, and by this mechanism they are able to mold their brains to function well in a wide range of environments. Prominent theories of perception (e.g., Gibson, 1979) have taken as their starting point the high degree of compatibility between perceiving visual organisms and their visual environments. The studies described in this chapter reveal that primates achieve this high degree of compatibility, in part, through developmental neural plasticity.

## ACKNOWLEDGMENTS

I thank C. O'Dell, M. Quick, E. Bergman, and all of the students in the psychobiology graduate seminar at Emory University for providing me with comments about previous versions of this chapter. I thank J. Torbit for preparing the manuscript for publication, and F. Kiernan for photographic assistance. My own research that is described in this chapter was supported by NIH Grants EY05975, EY06436, and by RR00165 to the Yerkes Regional Primate Research Center which is fully accredited by the American Association for Accreditation of Laboratory Animal Care.

## REFERENCES

Atkinson, J., Braddick, O., & Moar, K. (1977). Development of contrast sensitivity over the first 3 months of life in the human infant. *Vision Research, 17,* 1037–1044.

Banks, M. S., & Salapatek, P. (1978). Acuity and contrast sensitivity in 1-, 2-, and 3-month-old human infants. *Investigative Ophthalmology & Visual Science, 17,* 361–365.

Banks, M. S., & Salapatek, P. (1983). Infant visual perception. In P. H. Mussen, M. Haith & J. J. Campos (Eds.), *Handbook of child psychology: Biology and infancy* (4th ed., vol. 2, pp. 435–571). New York: Wiley.

Berkley, M. (1976). Cat visual psychophysics: Neural correlates and comparisons with man. *Progress in Psychobiology and Physiological Psychology, 6,* 63–119.

Blakemore, C., & Vital-Durand, F. (1981). Postnatal development of the monkey's visual system. *CIBA Foundation Symposium, 86,* 152–171.

Blakemore, C., & Vital-Durand, F. (1986). Organization and postnatal development of the monkey's lateral geniculate nucleus. *Journal of Physiology* (London), *380,* 453–492.

Boltz, R. L., & Harwerth, R. S. (1979). Fusional vergence ranges of the monkey: A behavioral study. *Experimental Brain Research, 37,* 87–91.

Boltz, R. L., Harwerth, R. S., & Smith, E. L. (1979). Orientation anisotrophy of visual stimuli in rhesus monkey: A behavioral study. *Science, 205,* 511–513.

Boothe, R. (1983). Optical and neural factors limiting acuity development: Evidence obtained from a monkey model. *Current Eye Research, 2,* 211–215.

Boothe, R. (1989). Visual development: Central neural aspects. In E. Meisami & P. Timeras (Eds.), *Handbook of human growth and developmental biology.* (pp. 179–191). Boca Raton, FL: CRC Press.

Boothe, R., Dobson, V., & Teller, D. (1985). Postnatal development of vision in human and nonhuman primates. *Annual Review of Neuroscience, 8,* 495–545.

Boothe, R., Kiorpes, L., & Carlson, M. (1985). Studies of strabismus and amblyopia in infant monkeys. *Journal of Pediatric Ophthalmology and Strabismus, 22,* 206–212.

Boothe, R., Kiorpes, L., & Hendrickson, A. (1982). Anisometropic amblyopia in (*Macaca nemestrina*) monkeys produced by atropinization of one eye during development. *Investigative Ophthalmology & Visual Science, 22,* 228–233.

Boothe, R., Kiorpes, L., Regal, D., & Lee, C. (1982). Development of visual responsiveness in *Macaca nemestrina* monkeys. *Developmental Psychology, 18,* 665–670.

Boothe, R., Kiorpes, L., Williams, R., & Teller, D. (1988). Operant measurements of contrast sensitivity in infant macaque monkeys during normal development. *Visual Research, 28,* 387–396.

Boothe, R., Vassdal, E., & Schneck, M. (1986). Experience and development in the visual system: Anatomical studies. In W. T. Greenough & J. M. Juraska (Eds.), *Developmental neuropsychobiology* (pp. 295–315). New York: Academic Press.

Boycott, B., & Dowling, J. E. (1969). Organization of the primate retina: light microscopy. *Philosophical Transactions of the Royal Society of London, Series B, 255,* 109–184.

Bradley, A., & Freeman, R. D. (1982). Contrast sensitivity in children. *Vision Research, 22,* 953–959.

Dean, P. (1981). Visual pathways and acuity in hooded rats. *Behavioral Brain Research, 3,* 239–271.

Delgado-Garcia, J., Evinger, C., & Baker, R. (1980). Activity of identified motoneurons in the abducens and accessory abducens of the alert cat during eye movement and retraction. *Society for Neuroscience Abstracts, 6,* 16.

DeValois, R. L., Morgan, H. C., & Snodderly, D. M. (1974). Psychophysical studies of monkey vision. III. Spatial luminance contrast sensitivity tests of macaque and human observers. *Vision Research, 14,* 75–82.

Eggers, H. & Boothe, R. (1987). Naturally occurring accommodative esotropia in macaques. *Investigative Ophthalmology & Visual Science, 28,*(ARVO Suppl.), 103.

Enroth-Cugell, C., & Robson, J. R. (1984). Functional characteristics and diversity of cat retinal ganglion cells: Basic characteristics and quantitative description. *Investigative Ophthalmology & Visual Science, 25,* 250–267.

Fantz, R. L., Fagan, J. R., III, & Miranda, S. B. (1975). Early visual selectivity. In L. B. Cohen & P. Salapatek (Eds.), *Infant perception: From sensation to cognition* (pp. 249–345). New York: Academic Press.

Fernandes, A., Tigges, M., Tigges, J., Gammon, J., & Chandler, C. (1988). Management of extended-wear contact lenses in infant rhesus monkeys. *Behavior Research Methods, Instruments, & Computers, 20,* 11–17.

Fuchs, A. F. (1967). Saccadic and smooth pursuit eye movements in the monkey. *Journal of Physiology (London)*, **191,** 609–631.

Gammon, J. A., Boothe, R., Chandler, C., Tigges, M., & Wilson, J. (1985). Extended-wear contact lenses for vision studies in monkeys. *Investigative Ophthalmology & Visual Science*, **26,** 1636–1639.

Gammon, J. A., O'Dell, C., Quick, M., Fernandes, A., Wilson, J., Tigges, M., Tigges, J., & Boothe, R. (1988). Visual functions in monkeys modelling infantile aphakia-amblyopia. *Investigative Ophthalmology & Visual Science*, **29,**(ARVO Suppl.), 75.

Gibson, J. J. (1979). *The ecological approach to visual perception.* Boston, MA: Houghton Mifflin.

Grether, W. F. (1941). A comparison of visual acuity in the rhesus monkey and man.. *Journal of Comparative Physiology and Psychology*, **31,** 23–33.

Gunderson, V. M., & Swartz, K. B. (1985). Visual recognition in infant pigtailed macaques after a 24-hour delay. *American Journal of Primatology*, **8,** 259–264.

Haith, M. M. (1980). *Rules that babies look by.* Hillsdale, NJ: Erlbaum.

Harwerth, R. S., & Boltz, R. L. (1979). Behavioral measures of stereopsis in monkeys using random dot stereograms. *Physiology and Behavior*, **22,** 229–234.

Harwerth, R. S., Crawford, M. L. J., Smith, E. C., & Boltz, R. L. (1981). Behavioral studies of stimulus deprivation amblyopia in monkeys. *Vision Research*, **21,** 779–789.

Harwerth, R. S., & Smith, E. L., III. (1985). Rhesus monkey as a model for normal vision of humans. *American Journal of Optometry & Physiological Optics*, **62,** 633–641.

Harwerth, R. S., Smith, E. L., Boltz, R. L., Crawford, M. L. J., & von Noorden, G. K. (1983). Behavioral studies on the effect of abnormal early visual experience in monkeys: Spatial modulation sensitivity. *Vision Research*, **23,** 1501–1510.

Hendrickson, A. E., Boles, J., & McLean, E. (1977). Visual acuity and behavior of monocularly deprived monkeys after retinal lesions. *Investigative Ophthalmology & Visual Science*, **16,** 469–473.

Hendrickson, A. E., Movshon, J. A., Boothe, R. G., Eggers, H. M., Gizzi, M. S., and Kiorpes, L. (1987). Effects of early unilateral blur on the macaque's visual system. II. Anatomical observations. *Journal of Neuroscience*, **7,** 1327–1339.

Ikeda, H. (1979). Is amblyopia a peripheral defect? *Transactions of the Ophthalmological Societies of the United Kingdom*, **99,** 347–352.

Jampolsky, A. (1978). Unequal visual inputs and strabismus management: A comparison of human and animal strabismus. In *Symposium on Strabismus: Transactions of the New Orleans Academy of Ophthalmology*, pp. 358–492. St. Louis, MO: Mosby.

Joosse, M., Wilson, J., & Boothe, R. (1988). Monocular visual fields of macaque monkeys with naturally occurring strabismus. *Investigative Ophthalmology & Visual Science*, **29,**(ARVO Suppl.), 345.

Julesz, B., Petrig, B., & Buttner, U. (1976). Fast determination of stereopsis in rhesus monkey using random-dot stereograms. *Journal of the Optical Society of America*, **66,** 1090–

Kiorpes, L., & Boothe, R. (1980). The time course for the development of strabismic amblyopia in infant monkeys (*Macaca nemestrina*). *Investigative Ophthalmology & Visual Science, 19,* 841–845.

Kiorpes, L., & Boothe, R. (1981). Naturally occurring strabismus in monkeys (*Macaca nemestrina*). *Investigative Ophthalmology & Visual Science, 20,* 257–263.

Kiorpes, L., Boothe, R., Carlson, M., & Alfi, D. (1985). Frequency of naturally occurring strabismus in monkeys. *Journal of Pediatric Ophthalmology and Strabismus, 22,* 60–64.

Kiorpes, L., Boothe, R. G., Hendrickson, A. E., Movshon, J. A., Eggers, H. M., & Gizzi, M. S. (1987). Effects of early unilateral blur on the macaque's visual system. I. Behavioral observations. *Journal of Neuroscience, 7,* 1318–1326.

Langston, A., Casagrande, V. A., & Fox, R. (1986). Spatial resolution of the galago. *Vision Research, 26,* 791–796.

McKusick, V., (1983). *Mendelian inheritance in man.* Baltimore, MD: Johns Hopkins University Press.

Mendelson, M. J. (1982). Clinical examination of visual and social responses in infant rhesus monkeys. *Developmental Psychology, 18,* 658–664.

Merigan, W. H., Pasternak, T., & Zehl, D. (1981). Spatial and temporal vision of macaques after central retinal lesions. *Investigative Ophthalmology & Visual Science, 21,* 17–26.

Motter, B. C., & Poggio, G. F. (1984). Binocular fixation in rhesus monkey: Spatial and temporal characteristics. *Experimental Brain Research, 54,* 304–314.

Movshon, J. A., Eggers, H. M., Gizzi, M. S., Hendrickson, A. E., Kiorpes, L., & Boothe, R. G. (1987). Effects of early unilateral blur on the macaque's visual system. III. Physiological observations. *Journal of Neuroscience, 7,* 1340–1351.

National Eye Institute (NEI). (1983–1987). Report of the strabismus, amblyopia, and visual processing panel. In *Reports of the program panels* (Vol. 2, Part 5). Washington, DC: National Advisory Eye Council.

Norcia, A. M., Tyler, C. W., & Hamer, R. D. (1988). High visual contrast sensitivity in the young human infant. *Investigative Ophthalmology & Visual Science, 29,* 44–49.

O'Dell, C. D., Gammon, J. A., Fernandes, A., Wilson, J. R., & Boothe, R. G. (1989). Development of acuity in a primate model of human infantile unilateral aphakia. *Investigative Ophthalmology & Visual Science, 30,* 160–166.

Petry, H. M., Fox, R., & Casagrande, V. (1984). Spatial contrast sensitivity of the tree shrew. *Vision Research, 24,* 1037–1042.

Pirchio, M., Spinelli, D., Fiorentini, A., & Maffei, L. (1978). Infant contrast sensitivity evaluated by evoked potentials. *Brain Research, 141,* 179–184.

Polyak, S. (1957). *The vertebrate visual system.* Chicago, IL: University of Chicago Press.

Quick, M., & Boothe, R. (1988). *Measurement of binocular alignment in normal monkeys and in monkeys with strabismus.* Manuscript submitted for publication.

Quick, M., Joosse, M., & Boothe, R. (1988). Assessment of spatial vision and visual fields in naturally strabismic monkeys. *Society for Neuroscience Abstracts, 14,* 1244.

Quick, M., O'Dell, C., Gammon, J., Wilson, J., Tigges, M., Fernandes, A., & Boothe, R. (1987). Assessment of spatial vision in monkeys with experimentally induced aphakia. *Society for Neuroscience Abstracts, 13,* 1243.

Quick, M., Tigges, M., Gammon, J., & Boothe, R. (1989). Early abnormal experience induces strabismus in infant monkeys. *Investigative Ophthalmology & Visual Science, 30,* 1012–1017.

Reymond, L. (1985). Spatial visual acuity of the eagle *Aquila audax:* A behavioral, optical and anatomical investigation. *Vision Research, 25,* 1477–1491.

Reymond, L. (1987). Spatial visual acuity of the falcon, *Falco berigora:* A behavioural, optical and anatomical investigation. *Vision Research, 27,* 1859–1874.

Sackett, G. P., Tripp, R., Milbrath, C., Gluck, J., & Pick, H. (1971). A method for studying visually guided perception and learning in newborn macaque. *Behaviour Research Methods and Instrumentation, 3,* 233–236.

Sarmiento, R. F. (1975). The stereoacuity of the macaque monkey. *Vision Research, 15,* 493–498.

Sinex, D. G., Burdette, L. J., & Pearlman, A. L. (1979). A psychophysical investigation of spatial vision in the normal and reeler mutant mouse. *Vision Research, 19,* 853–858.

Skavenski, A. A., Robinson, D. A., Steinman, R. M., & Timberlake, G. T. (1975). Miniature eye movements of fixation in rhesus monkey. *Vision Research, 15,* 1269–1273.

Smith, E. L., III, Harwerth, R. S., & Crawford, M. L. J. (1985). Spatial contrast sensitivity deficits in monkeys produced by optically induced anisometropa. *Investigative Ophthalmology & Visual Science, 26,* 330–342.

Snodderly, D. M. (1987). Effects of light and dark environments on macaque and human fixational eye movements. *Vision Research, 27,* 401–416.

Snodderly, D. M., & Kurtz, D. (1985). Eye position during fixation task: Comparison of macaque and human. *Vision Research, 25,* 83–98.

Spencer, R. F., & Porter, J. D. (1981). Innervation and structure of extraocular muscles in the monkey in comparison to those of the cat. *Journal of Comparative Neurology, 198,* 649–666.

Swindler, D., & Wood, C. (1973). *An atlas of primate gross anatomy: Baboon, chimpanzee, and man.* Seattle: University of Washington Press.

Symmes, D. (1962). Self-determination of critical flicker frequencies in monkeys. *Science, 136,* 714–715.

Teller, D. Y. (1979). The forced-choice preferential looking procedure: A psychophysical technique for use with human infants. *Infant Behavior and Development, 2,* 135–153.

Teller, D. Y., Morse, R., Borton, R., & Regal, D. (1974). Visual acuity for vertical and diagonal gratings in human infants. *Vision Research, 14,* 1433–1439.

Teller, D. Y., Regal, D. M., Videen, T. O., & Pulos, E. (1978). Development of visual acuity in infant monkeys (*Macaca nemestrina*) during the early postnatal weeks. *Vision Research, 18,* 561–566.

von Noorden, G. K. (1973). Experimental amblyopia in monkeys. Further behavioral observations and clinical correlations. *Investigative Ophthalmology, 12,* 721–726.

von Noorden, G. K. (1985a). Amblyopia: A multidisciplinary approach: Proctor lecture. *Investigative Ophthalmology & Visual Science, 26,* 1704–1716.

von Noorden, G. K. (1985b). *Binocular vision and ocular motility: Theory and management of strabismus* (3rd ed.). St. Louis, MO: Mosby.

von Noorden, G. K., & Dowling, J. E. (1970). Experimental amblyopia in monkeys. II. Behavioral studies of strabismic amblyopia. *Archives of Ophthalmology (Chicago)*, **84,** 215–220.

von Noorden, G. K., Dowling, J. E., & Ferguson, D. C. (1970). Experimental amblyopia in monkeys. I. Behavioral studies of stimulus deprivation amblyopia. *Archives of Ophthalmology (Chicago)*, **84,** 206–214.

Walls, G. L. (1963). *The vertebrate eye and its adaptive radiation.* New York: Hafner. (Reprinted from the Cranbrook Institute of Science, Bloomfield Hills, MI, 1942)

Wiesel, T. N. (1982). Postnatal developmental of the visual cortex and the influence of environment. *Nature (London)*, **299,** 583–591.

Williams, R., Boothe, R., Kiorpes, L., & Teller, D. (1981). Oblique effects in normally reared monkeys (*Macaca nemestrina*): Meridional variations in contrast sensitivity measured with operant techniques. *Vision Research,* **8,** 1253–1266.

Wilson, H. (1988). Development of spatiotemporal mechanisms in infant vision. *Vision Research,* **28,** 611–628.

# 16

# BEHAVIORAL ASSESSMENT OF VISUAL ACUITY IN HUMAN INFANTS

*Velma Dobson*

*Department of Psychology, University of Pittsburgh, Pittsburgh, Pennsylvania*

## I. THE PROBLEM

Visual acuity is a measure of the finest detail that an individual can discriminate in the visual environment. Based on the type of stimulus used, the visual acuity measurement obtained can be classified as one of four types: recognition, resolution, localization, or detection acuity (Riggs, 1965). For human adults tested in clinical settings, the stimulus recommended for acuity assessment by the National Research Council–National Academy of Sciences Committee on Vision (1980) is the Landolt ring, which, like the familiar Snellen letter chart, provides a measure of recognition acuity. In laboratory settings, stimuli include individual point or line stimuli, which provide a measure of detection acuity; grating or checkerboard stimuli, which provide a measure of resolution acuity; and the so-called vernier or hyperacuity stimuli (Westheimer, 1979), which provide a measure of localization acuity.

In the typical behavioral acuity task, human adults are given verbal instructions. They then respond verbally or by pressing a button to indicate their ability to see the various stimuli presented. For example, in a recognition acuity task, the adult reads aloud the letters on the chart or indicates verbally the direction of the gap in each Landolt ring. In detection, resolution, and localization acuity tasks, stimuli are typically presented in a forced-choice paradigm, and the adult is required to indicate which of two temporal intervals or spatial locations contained the relevant stimulus, for example, the black dot, the grating, or the line with a gap in it. Frequently, hundreds of stimulus presentations are required to obtain a precise estimate of visual acuity with the forced-choice paradigm.

Although measurement of visual acuity in adult humans often involves verbal instructions and/or a verbal response, verbal communication is not required for acuity assessment. Many behavioral studies of visual acuity in animals have been conducted in which animals have been trained through operant procedures to respond to visual stimuli in a forced-choice paradigm. Bloom and Berkley (1977), for example, trained cats to indicate discrimination of a grating from a homogeneous field by pressing with their noses on a panel covering the grating. Mitchell, Griffen, Wilkinson, Anderson, and Smith (1976) also trained cats to respond to the grating target in a two-choice paradigm. Instead of pressing a panel, however, the cats were required to jump from a stand to the platform covered with the grating, in a procedure similar to that used by Lashley (1930) for studies of pattern vision in the rat. An example of recognition acuity assessment in animals can be found in the work of von Noorden, Dowling, and Ferguson (1970) who trained monkeys to press on a panel to indicate the "odd" Landolt ring of three rings presented simultaneously. Many other examples of visual acuity assessment in animals could be cited (see, for example, chapter by Boothe in this book), each of which varies slightly in the method of stimulus presentation and the response measure used. Despite their specific methodological differences, these studies have one thing in common—the fact that, as in psychophysical studies of visual acuity in adult humans, each acuity estimate is based on hundreds of stimulus presentations.

Like animals, human infants cannot read letters or respond verbally to visual acuity stimuli. Therefore, it would seem appropriate to explore the possibility of adapting for use with human infants the techniques that have allowed successful acuity assessment in animals. Unfortunately, however, very young infants do not have the motor capabilities to press on a panel to indicate their choice in a forced-choice task. Furthermore, even when they become old enough to be able to perform a panel-press task, it is impossible to maintain their cooperation and attention for the hundreds of trials typically used in forced-choice procedures with adult humans or animals. Thus, behavioral measurement of visual acuity in human infants has required the development and use of methods not typically employed in studies of adult humans or animals.

The goal of this chapter is to describe the field of behavioral* visual

---

* It is also possible to measure visual acuity using electrophysiological techniques, such as the electroretinogram (ERG) or the visually evoked potential (VEP). Discussion of these measures is beyond the scope of this chapter. For the interested reader, reviews of VEP acuity results in human infants can be found in Dobson and Teller (1978) and Norcia and Tyler (1985). Fiorentini, Pirchio, and Sandini (1984) and Odom, Maida, Dawson, and Romano (1983) have studied ERG acuity in infants.

acuity assessment in human infants, to illustrate its progression from its beginnings thirty years ago as a primarily descriptive science to its recent emergence into the realm of theoretical vision science. Three issues will be covered. The first and most basic issue is the methodological one: What methods have been developed that allow fast, yet accurate, behavioral estimation of visual acuity across a wide age range, from infancy to early childhood? The second issue is a parametric issue: How are the acuity results obtained from the methods that have been developed for use with infants affected by the parameters used during testing? The final issue is a theoretical one: What can the results of visual acuity testing in infants tell us about the nature of the mechanisms that limit acuity in the developing visual system?

## II. HISTORICAL PERSPECTIVE

Historically, the modern age of visual acuity testing in infants began in 1957, with a study of newborn infants by Gorman, Cogan, and Gellis (1957). These researchers noticed that infants, like adults and animals, show a characteristic pattern of eye movements, termed optokinetic nystagmus (OKN), when a pattern is moved across a large portion of their visual field. To measure visual acuity, Gorman et al. placed infants under a canopy of moving black-and-white stripes and watched each infant's eye movements to determine the finest stripes to which the infant showed an OKN response. Their results showed that newborn infants could resolve gratings of 1.5 cycles/deg,* which is substantially poorer acuity than the value of 30 cycles/deg shown by the typical adult human. Subsequently, Fantz, Ordy, and Udelf (1962) used the procedure of Gorman et al. to document the improvement in acuity that occurs in infants between 1 and 6 months of age, and Dayton, Jones, Aiu, Rawson, Steele, and Rose (1964) repeated the Gorman et al. study of newborn infants, but used the electrooculogram rather than visual observation to record the infants' eye movements.

Despite the success of these three research groups, the use of OKN as a measure of visual acuity in infants did not become popular either in the laboratory or in the clinic until recent years (see Maurer, Lewis, & Brent, in press, for recent OKN acuity results from infants). At least three factors contributed to the neglect of OKN for acuity assessment in infants. First, it is difficult to construct the large-size, high-quality grating stimuli needed for OKN acuity assessment. Artifacts in the stimuli can elicit an OKN response that is indistinguishable from that elicited by a grating target, with the result that acuity may be overestimated. Banks and Salapatek

---

* The term "cycles/degree" refers to the spatial frequency or the fineness of spacing of the black and white stripes used in assessment of resolution acuity. Specifically, the term refers to the number of cycles (one cycle equals one white and one black stripe) that are present in one degree of visual angle.

(1981) have suggested that poor stimulus quality may have accounted for the unusually good acuity values reported by Dayton et al. (1964) in newborn infants. A second factor involved in the lack of popularity of OKN for acuity assessment in infants is that the response measure (judgment of the presence versus absence of OKN) can be ambiguous, due to the irregularity and inconsistency of eye movements in young infants (Fantz et al., 1962). Finally, the third reason for the paucity of studies using OKN is that in 1962 Robert Fantz and his colleagues introduced preferential looking (PL), a technique that uses smaller-size grating stimuli and a different type of eye movement response for acuity assessment of infants.

The PL procedure is based on the fact that infants will fixate a pattern in preference to a homogeneous field (Berlyne, 1958; Fantz, 1958). Therefore, if an infant shows a consistent preference for a grating over a homogeneous field of equal space-average luminance, one can conclude that the infant can resolve the grating. Acuity can be estimated as the highest spatial frequency grating (finest stripes) that an infant, or a group of infants of the same age, fixate longer than or more frequently than they fixate a matched gray field. In a study in which they used both preferential looking and OKN to obtain the first normative behavioral visual acuity data on infants between 1 and 6 months of age, Fantz et al. (1962) pointed out that "in contrast (to the OKN procedure), the fixation test of acuity involved a simple, unambiguous criterion of fixation at each moment, and a quantitative criterion of differential response for each test" (p. 913).

The PL procedure was modified in 1974 by Teller and her colleagues into what has become known as the forced-choice preferential looking (FPL) procedure. In the FPL procedure, an observer, who is kept unaware of the left–right position of the grating and gray targets, uses the infant's fixation behavior to make a judgment concerning the left–right position of the grating. This means that FPL is an *objective* procedure, which yields forced-choice data similar in form to those obtained in adult psychophysical experiments (see Teller, 1979). In contrast, the observer in the PL procedure of Fantz et al. knew the location of the grating and made a *subjective* judgment concerning the length of time the infant fixated the grating, the number of times the infant fixated the grating, or whether the infant's first fixation after the targets were presented was toward the grating or toward the gray target.

With the development of the FPL procedure by Teller, Morse, Borton, and Regal (1974) the field of infant vision testing progressed rapidly. By simply varying the type of stimulus used, it was now possible to use psychophysical procedures similar to those used with adults to study not only visual acuity, but also color vision (see Teller & Bornstein, 1987, for review), rod vision (e.g., Brown, 1986; Hansen & Fulton, 1981; Powers, Schneck, & Teller, 1981), the sensitivity of the infant's visual system to temporally modulated stimuli (e.g., Hartmann & Banks, 1984; Regal, 1981), as well as many other aspects of vision in infants.

The most proliferative area of infant vision research has been the study of visual acuity and, specifically, the study of grating (resolution) acuity.* Initially, researchers used the FPL procedure to obtain normative data on visual acuity development between birth and 6 months, which yielded a number of replications and extensions of the original data of Fantz et al. (1962) (for reviews, see Atkinson & Braddick, 1981; Banks & Salapatek, 1983; Dobson & Teller, 1978; Held, 1979). As will be described in the remainder of this chapter, subsequent research has been aimed at (1) modifying the FPL procedure for use with infants greater than 6 months of age and toddlers, (2) investigating parameters that affect acuity in infants, and (3) using the results of acuity studies to draw conclusions about factors that underly the relatively poor acuity shown by infants.

## III. METHODOLOGICAL CONCERNS

### A. Early FPL Testing

The forced-choice, preferential looking (FPL) procedure (Teller, 1979; Teller et al., 1974) has been used in the majority of studies of visual acuity in infants and young children. Initially, the procedure was both labor-intensive and time-intensive, and was useful only with infants less than 6 to 9 months of age. Much of the research conducted since 1974 has been concerned with reducing the number of people required for testing, improving the efficiency of the procedure, and expanding the age range over which FPL is useful.

An early version of FPL acuity testing is shown in Fig. 1. The infant (a subject in a study by Allen, 1979) was held in front of a gray screen by an adult whose view of the screen was blocked. The screen contained two targets (not visible)—a black and white square-wave grating and a gray area equal to the grating in space-average luminance. An observer behind the screen, who was kept unaware of target position, used the infant's looking behavior to make a forced-choice judgment concerning the location of the grating on each trial. A third adult set up target position, recorded the observer's responses, and provided feedback to the observer concerning whether each judgment was correct. In a typical procedure, 80 to 100 trials (20 trials at each of 4 or 5 grating spatial frequencies) were used to estimate an infant's acuity threshold.

### B. Improvements in Methodology: Reduction in Personnel

The need for three adults for every acuity test made early FPL testing labor-intensive. The solution to this problem was to add video equipment

---

* There have been no modern studies of detection or recognition acuity in infants. Recently, however, two groups of researchers have measured the development of vernier (localization) acuity in infants (Manny & Klein, 1984; Shimojo, Birch, Gwiazda, & Held, 1984).

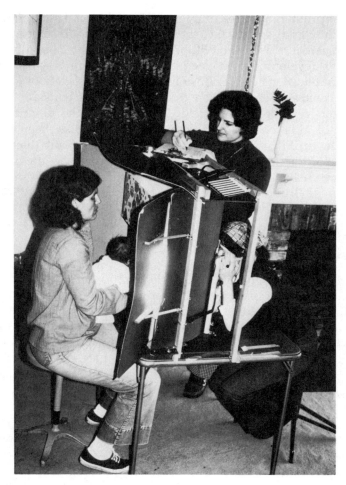

**FIGURE 1.** Early FPL testing. The infant is held in front of a gray screen by an adult whose view of the screen is blocked. A second adult, the observer, watches the infant's eye and head movements in response to black and white grating stimuli located on the left or right side of the screen (not visible in the figure). The third adult sets up the stimuli, records the observer's judgment of the left–right position of the grating stimulus on each trial, and provides the observer with feedback concerning the correctness of her judgment on each trial.

and a microcomputer to the FPL apparatus, and thereby reduce the number of adults needed to one (e.g., Brown, Dobson, & Maier, 1987; Packer, Hartmann, & Teller, 1984; Schneck, Hamer, Packer & Teller, 1984). Specifically, the microcomputer was programmed to set up the targets (through a mechanical interface with a disk containing various gratings), record footswitch responses made by the experimenter, and provide the experimenter with feedback concerning the correctness of the judgment on each trial. A video camera positioned behind the peephole in the screen placed an image of the infant's face on a television monitor that was visible to the person holding the infant. This meant that the person holding the infant could watch the infant's face, make a forced-choice judgment concerning target location on each trial, and indicate that judgment on a footswitch connected to the computer. The computer recorded the judgment, provided feedback to the experimenter, and set up the next trial.

## C. Improvements in Methodology: Reduction in Time

In addition to being labor-intensive, early FPL procedures were also time-consuming. With infants, time is an important consideration, since most infants do not remain cooperative for long periods, especially when viewing of repetitive stimuli is required. The time-consuming nature of FPL testing was due to two factors: (1) the large number of trials required by the two-alternative, forced-choice nature of the procedure and (2) the long intertrial interval needed to allow the experimenter to change the stimuli by hand. Use of microcomputers (e.g., Brown et al., 1987), slide projectors (e.g., Gwiazda, Brill, Mohindra, & Held, 1978), and CRT systems (e.g., Atkinson, Braddick, & Moar, 1977a) to present stimuli helped to reduce the intertrial test time. However, these tools did not reduce the number of trials required.

One way to reduce the number of trials required in a two-alternative, forced-choice procedure is to concentrate trials near threshold, to allow accurate bracketing of threshold without "wasting time" testing stimuli far above or far below threshold. Such accurate placement of trials is the goal of staircase procedures, which, in fact, have been used by a number of research groups to measure visual acuity in infants (e.g., Atkinson, Braddick, & Pimm-Smith, 1982; Gwiazda, Brill, Mohindra, & Held, 1980; D. L. Mayer, Fulton, & Hansen, 1982). Unfortunately, even when staircase procedures are used, there is a limit below which the number of trials cannot be reduced without significant losses in the accuracy of the resulting acuity estimate. Computer simulations (McKee, Klein, & Teller, 1985; Teller, 1983; 1985) have indicated that at least 60 trials are needed to estimate an infant's acuity to within an octave* (a degree of accuracy that is considered to be acceptable in studies of acuity in infants). Therefore, whether a staircase procedure or the method of constant stimuli is used, the lower limit on the number of trials required to obtain an acceptably accurate estimate of acuity in infants using the FPL procedure is still large, given the infant's limited attention span and limited availability for testing.

Although it is usually not difficult to obtain 60 trials from an infant tested in the laboratory, there are situations in which it would be beneficial to have a quicker method for acuity estimation in infants, for example, in a study requiring multiple acuity estimates under a variety of stimulus conditions, or for screening to assure that infants have normal acuity prior

---

* An octave is a halving or doubling of spatial frequency, for example, from 15 to 30 cycles/deg.

to their participation in studies of perceptual or cognitive abilities. Two procedures have been developed to increase the efficiency of acuity testing under conditions where time is limited. The first is a screening tool that is called the diagnostic grating procedure (Dobson, 1983; Dobson, Mayer, & Lee, 1980; Dobson, Salem, Mayer, Moss, & Sebris, 1985; Dobson, Teller, Lee, & Wade, 1978). The procedure provides not an estimate of acuity threshold but, rather, documentation of whether an infant's acuity is within the normal range. Each infant is tested with the FPL procedure, but with only one stimulus grating, a grating that has been determined empirically to be the finest grating detectable by 90 to 95% of infants of the same age. A small number of trials ($\geq 6$ and $\leq 20$) with this grating allows the experimenter to determine whether an infant's vision is within the normal range.

A second procedure for rapid acuity testing in infants is the acuity card procedure (McDonald, McDonald, Dobson, Sebris, Baitch, Varner, & Teller, 1985; see Teller, McDonald, Preston, Sebris & Dobson, 1986, for review). In contrast to the FPL procedure, where the observer's task is to use the infant's behavior to judge whether a grating stimulus is on the left or right, the observer's task in the acuity card procedure is to use the infant's overall response to a series of FPL-like grating stimuli to make a subjective judgment concerning the finest grating the infant can resolve. Results to date suggest that acuity estimates obtained in normal infants and young children (McDonald, Ankrum, Preston, Sebris, & Dobson, 1986; McDonald, Sebris, Mohn, Teller, & Dobson, 1986; Brown & Yamamoto, 1986; Dobson, Schwartz, Sandstrom, & Michel, 1987; Kohl, Rolen, Bedford, Samek, & Stern, 1986; McDonald et al., 1985; Mohn & van Hof-van Duin, 1986), infants and children born prior to term (Brown & Yamamoto, 1986; Mohn & van Hof-van Duin, 1986), and infants with visual or neurological disorders (Hertz, 1987; D. L. Mayer, Morrell, & Rodier, 1986; Mohn & van Hof-van Duin, 1986; Preston, McDonald, Sebris, Dobson, & Teller, 1987) are similar to those obtained with FPL testing. However, in contrast to FPL, the time required for acuity card testing averages only about 5 min. These results, coupled with reports indicating that a large percentage of infants can be tested successfully in clinical settings (Droste, Archer, & Helveston, 1987; D. L. Mayer et al., 1986; Sebris, Dobson, McDonald, & Teller, 1987) suggest that, for those willing to accept the results of a subjective test, the acuity card procedure can provide a useful means of estimating acuity in infants in settings in which time and/or the cooperation of the child is limited.

## D. Improvements in Methodology: Operant Procedures for Older Infants and Young Children

Another difficulty of early FPL testing was that it was generally useful only with infants less than 6 months of age (see Dobson & Teller, 1978, for review), although Gwiazda et al. (1978) reported that the use of projected stimuli in a darkened room allowed successful testing of infants up to 12 months of age.

To solve the problem of how to test older infants and young children, researchers turned to operant procedures that had been used successfully to test other sensory and cognitive functions of children in this age range. Moore and Wilson (1978) had shown that auditory function could be tested successfully by visual reinforcement audiometry (VRA). In VRA, older infants and toddlers are trained to make a head-turn response in response to auditory stimuli in order to activate an animated toy. For visual acuity testing, Mayer modified the procedure so that the child was taught to make a head-turn response to a grating stimulus. The resulting procedure was termed operant preferential looking (OPL) (D. L. Mayer & Dobson, 1980, 1982), because the method is actually a modification of the FPL procedure: The child is seated in front of the FPL apparatus and is required to indicate the left–right position of the grating by looking or pointing. The observer, watching the child through a peephole, uses the child's response to make a judgment concerning the left–right position of the stimulus. If the observer makes the correct judgment, the child is rewarded by activation of the animated toy. With the OPL technique, it was possible to obtain acuity data from infants and children between 6 months and 5 years of age. However, it was difficult to get 18- and 24-month-olds to learn the task and to maintain attention for long enough to complete the trials required for an acuity estimate.

An operant procedure that has been used in learning studies of toddlers is the panel-press procedure (e.g., Lipsitt, 1967), in which the child is reinforced for pressing the correct of several visually different panels. Birch, Gwiazda, Bauer, Naegele, and Held (1983) developed a panel-press procedure for acuity testing, in which the child receives a piece of cereal as a reward for pressing the panel containing a grating stimulus. (This procedure is similar to that used in acuity studies with cats (e.g., Bloom & Berkley, 1977), except that the toddlers press the panel with their hands while the cats press the panel with their noses.) With the panel-press procedure, Birch et al. (1983) were able to obtain acuity data from children between 7 months and 5 years of age. Although extensive training in the procedure was often required, the percentage of children in the difficult-to-test 1- to 2-year-old age range who completed acuity testing was considerably higher than that in the D. L. Mayer and Dobson (1982) study.

A third operant procedure was developed by Atkinson, French, and Braddick (1981). In this procedure, which is similar to that used by Volkmann and Prizer (1973) to test orientation discrimination in 2- and 3-year-olds and similar to many alley-running procedures used with animals, the child stands at the end of a divided alleyway and must choose which side has the correct grating stimulus at the far end. The child runs down the selected side of the alleyway and, if the correct side was chosen, the child is reinforced with a candy that was hidden under the target. This procedure requires that the child have reasonably good running skills, as well as the ability to learn the task, which may be why the youngest age for which Atkinson et al. (1981) report data is 36 months.

In summary, with the addition of operant reinforcement to the FPL procedure, it became possible to obtain acuity data from children from infancy up to the age at which they could be tested with procedures similar to those used with adults. As a result, normative data for the development of binocular grating acuity over the age range from birth to 5 years are now available (see Fig. 2).

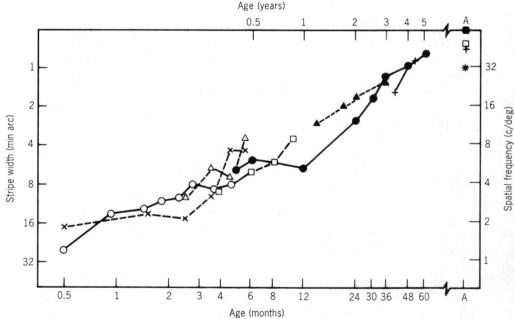

**FIGURE 2.** Estimates of acuity in children 2 weeks to 5 years of age obtained using behavioral measures of acuity. For children, the data shown are from X, Fantz, Ordy, and Udelf (1962), optokinetic nystagmus; △, Fantz, Ordy, and Udelf (1962), preferential looking; ○, Allen (1979), FPL; □, Gwiazda, Brill, Mohindra, and Held (1978), FPL; ●, D. L. Mayer and Dobson (1982), operant preferential looking; ▲, Birch, Gwiazda, Bauer, Naegele, and Held (1983), operant push panel; +, Atkinson, French, and Braddick (1981), preferential running. For adults the data are from +, Atkinson, French, and Braddick (1981); ■, D. L. Mayer and Dobson (1982), forced-choice procedure; □, D. L. Mayer and Dobson (1982), yes–no procedure. The * indicates the nominal adult Snellen acuity of 20/20 (6/6). The units of acuity (cycles/deg) on the ordinate are defined in the text. When acuity and age are plotted on log axes, there appears to be a linear improvement in acuity between 2 weeks and 5 years of age. Reprinted with permission from Dobson (1983).

## IV.  FINDINGS: WHAT FACTORS INFLUENCE ACUITY RESULTS IN INFANTS?

It is well established that visual acuity in adults is influenced both by stimulus parameters and by characteristics of the adult being tested (for reviews, see Lit, 1968; Riggs, 1965; Westheimer, 1965). Therefore, with the development of methodology to measure visual acuity in infants, it is not surprising that researchers began studying the effects of variations in stimulus parameters and observer characteristics on infants' acuity. The following sections are intended to give the reader a sampling of the types of studies that have been conducted to explore factors that influence acuity results in infants and young children. In reading these sections, it should be remembered that the field of visual acuity testing in infants is a relatively young one; thus, our knowledge of the effects of stimulus parameters and observer characteristics on acuity is far from exhaustive. This is further complicated by the fact that visual acuity is highly correlated with age in infants and young children, so that the effect of stimulus parameters and observer characteristics may vary as a function of the age of the child tested.

### A. Luminance

One of the striking features of acuity results from normal infants is the good agreement in the results reported by researchers in different laboratories. This agreement occurs despite variations in stimulus luminance of more than one log unit across labs (see Atkinson, Pimm-Smith, Evans, & Braddick, 1983, and Dobson & Teller, 1978, for summaries of stimulus luminance conditions in FPL acuity studies), and suggests that infants' acuity, like that of adults, may be independent of luminance across a range of daylight luminances.

We conducted two studies to examine the effects of variation in stimulus luminance on acuity in 2-month-old infants. In the first study (Dobson, Salem, & Carson, 1983), we used the FPL procedure to measure the infants' acuities at luminances between 0.16 and 500 cd/m$^2$. As predicted by the consistency of acuity scores across previous studies, the results indicated that infants showed little or no variation in acuity for variations in luminance above 10 cd/m$^2$. We were unable to test luminances below 0.16 cd/m$^2$, because at lower luminances, the acuity of the adult observer in FPL testing became too poor to detect the infant's responses to the grating stimuli.

In a second study (Brown et al., 1987), we added infrared illumination to the infant's face and placed an infrared-sensitive video camera behind the peephole. By watching the image of the infant's face transmitted to a television monitor by the camera, the observer was able to make FPL judgments at stimulus luminances down to 0.0025 cd/m$^2$. The results indicated that infants, like adults, show a significant reduction in acuity as luminance is reduced below about 10 log cd/m². However, the relative reduction in acuity is somewhat less in infants than in adults. Given the visual evoked potential (VEP) results of Fiorentini, Pirchio, and Spinelli (1980), which indicate that contrast sensitivity functions (CSFs) measured under low luminance conditions become adult-like by about 6 months whereas CSFs measured under higher luminance conditions do not become adult-like until after 6 months, it would be of interest to repeat our FPL studies with infants older than 2 months, to see if behavioral results also show differential rates of maturation of mechanisms controlling acuity to low versus high luminances.

## B. Stimulus Size

The size of the grating target used for acuity testing has varied from 7.5° (Atkinson, Braddick, & Braddick, 1974) to 48° of visual angle (Banks & Salapatek, 1978; see Atkinson et al., 1983, and Dobson & Teller, 1978, for summaries of stimulus size in FPL studies). The good agreement that has been found across studies in the acuity results of infants 2 months of age and older suggests that, for the target sizes used and the ages tested, stimulus size has little effect on acuity results.

The effect of stimulus size on acuity in young infants was investigated empirically by Atkinson et al. (1983). These researchers measured the acuities of 1-, 2-, and 3-month-old infants for grating stimuli presented in a circular field that measured either 19 or 10° in diameter. Three-month-olds showed no difference in acuity results for the two target sizes. However, 1- and 2-month-olds showed better acuity for the larger of the two target sizes. Two hypotheses have been advanced to account for the results of Atkinson et al. First, Atkinson (1984) has suggested that using larger targets helps to overcome the tendency of 1- and 2-month-olds to look at target edges rather than the target itself (Milewski, 1976). This hypothesis is attractive, because it can account for the finding that better acuity scores have been reported for 1-month-olds tested with large stimuli (e.g., Banks & Salapatek, 1978) or smaller stimuli in a matched surround (e.g., Allen, 1979) than have been reported for 1-month-olds tested with small stimuli embedded in a contrasting surround (e.g., Atkinson et al., 1977a; Gwiazda et al., 1978, 1980). However, this hypothesis cannot explain why 2-month-olds tested in previous studies show similar acuity scores, regardless of the stimulus configuration, whereas the 2-month-olds in the Atkinson et al. (1983) study showed significantly better acuity for the 19° than for the 10° stimulus. A second hypothesis to account for the improvement in acuity shown by the 1- and 2-month-olds for the 19° target is related to the finding that contrast threshold for grating stimuli in adults is related to the number of cycles (black and white stripes) in the target grating (e.g., Howell & Hess, 1978; Robson & Graham, 1981). With an increase in the number of cycles, spatial probability summation of the underlying neural elements may increase, thereby resulting in an improved acuity score (see D. L. Mayer, 1986, for further discussion). This hypothesis can account for the finding that the 1- and 2-month-olds showed better acuity for the 19° than for the 10° target in the Atkinson et al. (1983) study, but it does not explain why 1-month-olds show better acuity for small targets embedded in a matched surround (Allen, 1979) than for small targets embedded in a contrasting surround (Atkinson et al., 1977a; Gwiazda et al., 1978, 1980).

## C. Retinal Location

In adults, visual acuity is highest in the fovea and falls off progressively with increasing retinal eccentricity. Two groups of researchers have examined the effect of retinal location on grating acuity in infants. Sireteanu, Kellerer, and Boergen (1984) presented 1- to 12-month-old infants with a 12° grating centered 10° from midline and compared "peripheral" acuity with "best" acuity. Peripheral acuity was based on the direction of the infant's first fixation away from midline after the presentation of the grating 10° to the right or left of midline. Best acuity was calculated from the FPL observer's judgments of grating location, based on all cues provided by the infant during free viewing of the target, with no time limit set on the observer's observation period. The results of the two measures showed parallel acuity development from birth to 10 months, with acuity under free viewing being slightly better than that obtained by first fixation. At the oldest test age, acuity under free viewing showed an improvement over earlier values, whereas acuity for first fixation did not. Sireteanu et al. suggest that the difference found at the oldest test age indicates that peripheral acuity may have reached a plateau in development, whereas central acuity continues to improve with age. It is not clear, however, whether the difference in acuity values found under the two test conditions was due to differences in retinal location between targets or to the improvement in FPL results that occurs when an observer is allowed to use the infant's total response rather than just first fixation (Atkinson et al., 1977a; Volkmann & Dobson, 1976).

In the second study of grating acuity versus retinal location, Maurer, Jobson, and Lewis (1986) presented 2-, 5-, and 12-month-old infants with grating stimuli located at 10 or 30° from midline. For both stimulus locations, the observer's task was to report the direction of the first eye movement away from midline. The results indicated an improvement in acuity between 2 and 5 months, but no change between 5 and 12 months. Unlike adults, infants showed no difference in acuity at the two retinal locations.

In terms of defining the function that describes acuity as a function of retinal location in infants, these studies by Sireteanu et al., and Maurer et al. are but initial attempts. In both studies, the central fixation target and the peripheral grating target were quite large, so that it was impossible to determine precisely which part of the retina the infant was using to respond to the target on any trial. For example, the Sireteanu et al. stimuli were 12° in diameter, centered 10° from midline. This means that, assuming that infants were fixating with their foveas, their eye movement response to the target could have been based on stimulation of any part of the retina between 4 and 16° from the fovea.

The problem of specifying the retinal location infants were using in the Sireteanu et al. and Maurer et al. studies is made even more difficult by the fact that it is not clear what part of the retina infants use for fixation of targets. Infants' foveal photoreceptors are known to be immature (Abramov, Gordon, Hendrickson, Hainline, Dobson & La Bossiere, 1982; Hendrickson & Yuodelis, 1984; Yuodelis & Hendrickson, 1986), so that infants may use a nonfoveal part of the retina for fixation. The hypothesis that infants use nonfoveal retina for fixation in an FPL task is supported by D. L. Mayer, Fulton, and Hansen (1985) finding that infants who have ocular conditions that produce an absence of a fovea may show FPL grating acuity scores equivalent to those of normal infants.

The way to determine the effect of retinal position on infants' acuity would be to use small stimuli. However, to accomplish this, researchers must first figure out how to overcome infants' lack of interest in small-field stimuli.

### D. Target Distance and Optical Defocus

Visual acuity in adults is typically measured with targets at a distance of 4–6 m. With infants, however, the target distance used in most studies has been less than 1 m, in part because it is easier to get infants to pay attention to near stimuli. Comparison across studies suggests that, within the relatively small range of distances used, target distance does not significantly affect the acuity results of infants of a given age. This conclusion is supported by the data of Salapatek, Bechtold, and Bushnell (1976), which showed that 1- and 2-month-old infants have equal acuities at 30, 60, 90, and 150 cm. Recently, Cornell and McDonnell (1986) reported that the acuities of 6- to 36-week-old infants for stimuli presented at 6 m were similar to acuities reported in studies of infants of the same age tested at distances of less than 1 m. Thus, target distance is a parameter that does not appear to have a major effect on visual acuity in infants.

The finding of Salapatek et al., that acuity is not affected by target distance in young infants, was an unexpected result. Eleven years earlier, Haynes, White, and Held (1965) had shown that very young infants do not accomodate accurately to stimuli at different distances, and that newborn infants' focal distance appeared to be fixed at a mean distance of 19 cm. This led to the prediction that targets at different distances would not be focused equally well on the infant's retina, and the expectation that acuity would vary as a function of target distance in young infants. The Salapatek et al. results led to a reexamination of the effect that defocus would be expected to have on the acuity of the young infant (see Green, Powers, & Banks, 1980). Basically, Green et al. pointed out that, because the acuity of young infants is poor, the acuity versus distance results of Salapatek et al. are not in conflict with the poor accommodative ability shown by young infants in the Haynes et al. study. In adults, spatial frequencies above 3 cycles/deg serve as the most effective stimuli for accommodation (e.g., Owens, 1980). The acuity of young infants is poorer than 3 cycles/deg, so it is not surprising that they do not accommodate to changes in target distance. Similarly, it is not surprising that they do not show changes in acuity with distance, since their poor acuity probably prevents them from detecting the defocus of the higher spatial frequencies that occurs as target distance is varied.

The insensitivity to defocus predicted by Green et al. for young infants was confirmed in a study we conducted with 6-week-old infants (Powers & Dobson, 1982). In this study, the acuity of infants was assessed with the FPL procedure while they wore defocussing lenses of the following powers: $-14$, $-3$, $0$, $+6$, and $+14$ diopters (D). As shown in Fig. 3, the

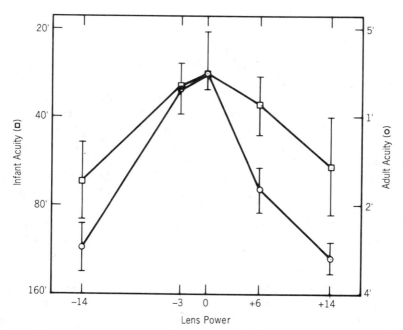

**FIGURE 3.** Relative change in acuity with lens power. □, Mean acuity of 6-week-old infants; ○, mean acuity of adults; Left ordinate, infant acuity; right ordinate, adult acuity. Curves have been shifted vertically to coincide at plano lens power. Abscissa: lens power in diopters. Bars are ±2 standard errors of the mean. Ten adults and ten infants were tested. Reprinted with permission from Powers and Dobson (1982).

relative reduction in acuity produced by defocusing lenses was less in infants than in adults, presumably because of their poor acuity and resulting inability to detect defocus.

## E. Temporal Properties of the Stimulus

It has been suggested that use of temporally modulated stimuli might improve the acuity scores obtained from infants tested behaviorally, because temporal modulation might increase the salience of the stimuli, which, in turn, should increase the probability that infants would look at them. In two early studies in which young infants were tested, it appeared that temporal modulation of acuity targets had no effect on the acuity results obtained. Atkinson, Braddick, and Moar (1977b) found no difference in the acuity of 1- to 3-month-old infants for gratings drifting at 3 Hz as compared with that for stationary gratings. Similarly, we (Dobson, Teller & Belgum, 1978) found no difference in the acuity of 2-month-olds for stationary gratings versus phase-alternated checkerborads. Recently, however, Sokol, Moskowitz, and Augliere (1987) reported for older infants, in the 3- to 8-month-old age range, that acuity scores obtained with gratings phase-alternated at 14 reversals/sec were at least one octave better than acuity scores obtained with stationary gratings.

## F. Grating Orientation

In general, both adults (Appelle, 1972) and children (Gwiazda, Scheiman, & Held, 1984; M. J. Mayer, 1983) show better acuity for vertical and horizontal gratings than for obliquely oriented gratings. This differential acuity for grating orientations, known as the oblique effect, does not appear to be present in infants less than 6 months of age (Gwiazda et al., 1978, 1980; Teller et al., 1974); however, it has been demonstrated by Gwiazda et al. (1978, 1980) to be present in older infants.

Astigmatism is a defect of the optics of the eye, in which the refractive power is different in different meridia. The meridia of maximum and minimum power, referred to as the axes of the astigmatism, are at right angles to one another. This means that orthogonally oriented gratings will never be in good focus on the retina at the same time. As a result, astigmatic individuals who do not receive glasses at an early age may grow up without experiencing well-focused stimuli in many orientations. When tested as adults with grating stimuli of different orientations, astigmats show an acuity versus orientation function that is quite different from the oblique effect of nonastigmatic adults. Specifically, astigmats show a difference in acuity for orthogonal gratings, for example, vertical versus horizontal, even when they are tested with optical correction. This difference in acuity for orthogonal orientations has been termed "meridional amblyopia" and is predictable from the axis and power of the astigmatism (Mitchell, Freeman, Millidot, & Haegerstrom, 1973).

Although a large percentage of infants have astigmatism (Fulton, Dobson, Salem, Mar, Petersen & Hansen, 1980; Howland, Atkinson, Braddick, & French, 1978; Ingram & Barr, 1979; Mohindra, Held, Gwiazda, & Brill, 1978), they do not show meridional amblyopia when they are tested with optical correction (Gwiazda, Mohindra, Brill, & Held, 1985; Teller, Allen, Regal, & Mayer, 1978). These results suggest that the meridional amblyopia of older astigmats may take months or even years to develop. This suggestion is supported by the fact that the earliest age at which meridional amblyopia has been detected is 2 years 10 months (Mohindra, Jacobson, & Held, 1983). Additional support is provided by the longitudinal results of Gwiazda, Bauer, Thorn, and Held (1986) and Gwiazda et al. (1984), which showed that the pattern of meridional amblyopia seen in nonastigmatic children at 5 to 11 years of age could be predicted from knowledge of their astigmatism when they were between 6 and 24 months of age, even though at these early ages they showed no meridional amblyopia.

## G. Monocular versus Binocular Testing

Adults show slightly better (0.1 to 0.2 octave) acuity under binocular, as compared with monocular, test conditions, a result that is likely to be due to a combination of probability summation and neural summation (see Blake, Sloane, & Fox, 1981, for review). A similar difference between monocular and binocular acuity results has been reported for infants (Atkinson et al., 1982). It is unclear, however, how much of the decrease in infants' acuity during monocular testing is due to sensory differences in monocular versus binocular acuity and how much is due to nonsensory factors, for example, the decreased cooperation that occurs when infants have one eye patched, or the increased difficulty of making FPL judgments when only one eye of the infant is visible to the observer.

## H. Gestational Age

The role of the environment in visual development has been of considerable interest in recent years (see Boothe, Dobson, & Teller, 1985, for summary). Because infants born prior to term are afforded visual experience at a time when they would normally have none, it is of interest to know if their added visual experience affects their visual development. Although there is some evidence from visual evoked potential studies that preterm infants' early experience may enhance their visual development (Norcia, Tyler, Piecuch, Clyman, & Grobstein, 1987; Sokol & Jones, 1979), acuity results from behavioral studies fail to find such enhancement of development. FPL results indicate, instead, that visual acuity development appears to be primarily maturationally determined, so that regardless of how much visual experience an infant has had, his or her visual acuity is similar to that of other infants of the same age from conception (Brown &

Yamamoto, 1986; Dobson et al., 1980; Fantz, Fagan, & Miranda, 1975; Shepherd, Fagan, & Kleiner, 1985; van Hof-van Duin & Mohn, 1986; van Hof-van Duin, Mohn, Fetter, Mettau, & Baerts, 1983).

## I. Summary of Parametric Studies of Acuity

From the preceding description of studies that have examined parameters that influence the results of behavioral assessment of acuity in infants, it is clear that work in this field is only just beginning. The data collected to date do not even approach the large body of results available for adults (for reviews, see Lit, 1968; Riggs, 1965; Westheimer, 1965). Furthermore, with infants and young children, acuity varies with age, and therefore, variations in stimulus parameters and observer characteristics may have different effects on acuity at different ages. This means that not only are there many parameters to be studied, but also that each parameter must be studied at a variety of ages. The difficulty of conducting parametric studies of infants and young children is further complicated by the fact that the subjects are getting older while they are being tested. Thus, it is not possible to conduct a detailed study of, for example, the effect of stimulus size on FPL acuity in a single 1-month-old infant, because by the time enough trials were completed to reveal the effect of a wide variety of stimulus sizes on acuity, the infant would no longer be 1 month old.

In conclusion, it is clear that there is much that we do not know about the factors that influence acuity results in infants and young children. It is likely that much of the research in the field of behavioral acuity testing over the next few years will be concerned with investigating the effects of various factors, as well as with developing methodology to improve the efficiency with which parametric data can be obtained in young human subjects.

## V. IMPLICATIONS

In addition to providing descriptive data concerning visual development, the results of visual acuity studies of infants and young children have both practical and theoretical implications. The practical implications center around the assessment of visual acuity in clinical settings. Laboratory studies of visual acuity have provided (a) methodology with which acuity can be assessed in infants and young children who have or are at risk for ocular pathology, (b) normative data against which to compare the results of patients tested clinically, and (c) data concerning the degree to which test parameters, for example, stimulus luminance or test distance, must be controlled during testing. As a result, behavioral procedures have been used to assess visual acuity in infants and young children with a variety of ocular disorders, including strabismus, high refractive errors, and cataracts. Reviews of clinical applications of visual acuity testing can be found in a number of publications, including Jacobson, Mohindra, and Held (1983), Mohindra et al. (1983), Boothe et al. (1985), and Maurer et al. (in press).

More relevant to this chapter are the theoretical implications of studies of visual acuity in infants. Behavioral studies of visual acuity play an important role in the interpretation of the results of anatomical and physiological studies of the developing visual system. In addition, behavioral studies of visual acuity allow constraints to be placed on models of the developing visual system, thereby suggesting future research directions for behavioral, anatomical, and physiological studies.

One of the most fruitful applications of infant visual acuity results has been in the exploration of mechanisms that limit visual acuity in infants. As illustrated in Fig. 2, the visual acuity of newborn infants is much poorer than that of adults, and it is not until early childhood that acuity reaches adult levels. Numerous hypotheses have been advanced to account for the poor acuity shown by infants. These hypotheses include optical defocus, anatomical immaturity of the foveal photoreceptors, physiological immaturity of the foveal photoreceptors, use of nonfoveal photoreceptors, immaturity of the postreceptoral visual or motor pathways, inattentiveness, and the lack of motivation. The following pages describe the manner in which behaviorally obtained visual acuity results have been used to reduce the number of plausible hypotheses concerning the mechanisms that limit acuity in infants.

## A. Optical Defocus

Anyone who wears glasses knows that optical defocus can produce a reduction in acuity. Studies of refractive error and studies of accommodation in young infants have indicated that optical defocus is more prevalent among young infants than among children and young adults, and therefore it has been suggested that optical defocus may play a significant role in producing infants' poor acuity scores.

Mohindra and Held (1981) refracted large numbers of infants and young children and examined the change in the distribution of refractive errors that occurred with age. Their data were reported as spherical equivalents, which indicate whether an individual is hyperopic (far-sighted), myopic (near-sighted), or emmetropic (neither hyperopic nor myopic). Mohindra and Held found that refractive errors were normally distributed in young infants, but that the distribution became peaked around emmetropia as children got older. This meant that infants were more likely than adults to be hyperopic or myopic. Another aspect of refractive errors is the astigmatic component, which is a measure of the sphericity of the optics. Several studies have shown that the prevalence of astigmatism is significantly higher in infants than in adults (Fulton et al., 1980; Howland et al., 1978; Ingram & Barr, 1979; Mohindra et al., 1978). Thus, both the spherical and the astigmatic components of the refractive error are more likely to deviate from emmetropia in infants than in adults. This means that the retinal image is less likely to be in good focus in infants than in adults.

In addition to any refractive errors that they may have, infants less than 3 months of age may also have retinal defocus due to an immature accommodative mechanism. Haynes et al. (1965), Banks (1980), and Brookman (1983) have shown that young infants do not change their focus appropriately to changes in stimulus distance. This means that changes in stimulus distance produce changes in the optical focus of images on the retina. Thus, both the increased prevalence of high refractive errors in infants and the poor responsiveness of the infant's accommodative mechanism suggest that the quality of the retinal image is often poorer in infants than in adults. This has led to the hypothesis that retinal defocus may contribute to the poor acuity scores obtained in young infants.

As described above in Section IVD, the results of behavioral studies of visual acuity in infants suggest that, contrary to this hypothesis, optical defocus does not play a significant role in causing the poor acuity shown by young infants, probably because the acuity of young infants is too poor to detect all but very large amounts of optical defocus (see Green et al., 1980; Salapatek et al., 1976).

## B. Immaturity of the Photoreceptors

**1. Anatomical Immaturity of Foveal Cones.** In adults, visual acuity varies as a function of retinal location, with the best acuity obtained with targets that are viewed foveally. Anatomical studies (Abramov et al., 1982; Hendrickson & Yuodelis, 1984; Mann, 1964; Yuodelis & Hendrickson, 1986) have shown that the fovea of the young infant is immature, in that the diameter of the foveal photoreceptors, the cones, is three to four times larger than the diameter of adult foveal cones, and the photopigment-bearing portion of the cone, the outer segment, is much shorter than that of the adult. Furthermore, the inner cell layers of the retina (the inner nuclear layer and the ganglion cell layer) have not yet migrated to produce the foveal pit characteristic of the retina of older infants and adults.

Each of these anatomical immaturities of the infant fovea would be expected to reduce the visual acuity of foveally presented targets. First, the presence of cell bodies overlying the photoreceptors would introduce scatter in the light reaching the photoreceptors. Second, the increased diameter of the foveal cones would reduce the upper limit on resolution. Banks, Bennett, and Schefrin (1987) and Brown et al. (1987), however, have calculated that the upper limit on resolution, termed the Nyquist limit, would be about 15.5 cycles/deg in the 2-month-old infant, which is far above the visual acuity obtained either behaviorally (approximately 2 cycles/deg) (summarized in Dobson & Teller, 1978) or with the visual evoked potential (5.5 cycles/deg) (Norcia & Tyler, 1985) in 2-month-olds.

Third, the finding that the outer segments of infants' foveal cones are shorter than those of adults suggests that the density of optical pigment is probably low, so that the quantum catch is less than that of adults. Fourth, the combination of fat, squarish inner segments and short, narrow outer segments suggests that it is unlikely that infants' foveal cones act as very effective waveguides. Based on these last two characteristics of infants' cones, Brown et al. (1987) calculated that from 14 to 55 times fewer quanta and Banks et al. (1987) calculated that approximately 33 times fewer quanta are caught by the foveal cones of infants than are caught by the foveal cones of adults. This decrease in quantum catch, however, is not sufficient to account for the poor acuity shown by young infants (Banks et al., 1987).

Thus, even though there are several anatomical features of foveal cones that would be predicted to have significant effects on visual acuity, the predicted deficits are not severe enough to account for the 3- to 4-octave difference between the acuity of young infants and that of adults.

**2. Use of Peripheral Cones Instead of Physiologically Immature Foveal Cones.** Another hypothesis is that the foveal cones are so physiologically immature that the infant, instead, uses peripheral cones, which are anatomically similar to those of adults (Abramov et al., 1982). This hypothesis is supported by qualitative similarities between infant visual behavior and the results of psychophysical studies of the adult peripheral retina (e.g., Bronson, 1974; Packer et al., 1984).

With respect to visual acuity results, this hypothesis suggests that there should be some location in the adult extrafoveal retina for which both the absolute acuity scores obtained and the variation in acuity scores that occurs as a function of changes in stimulus parameters are similar to the acuity scores shown by infants under the same conditions. Recently, Brown et al. (1987) evaluated the hypothesis by measuring the effect of variations in luminance on visual acuity in 2-month-old infants and adults. Infants were tested under free-viewing conditions, with the FPL procedure. Adults were tested under three conditions: (1) free viewing, (2) with the stimuli at 21° eccentricity, and (3) with the stimuli at 51° eccentricity. The results, shown in Fig. 4, indicated that under free-viewing conditions, both the shape of the acuity versus luminance function and the absolute value of the acuity scores obtained under light-adapted luminances were different for adults and infants. When adults were tested with stimuli at 21°, the shape of the acuity versus luminance function was similar to that of infants, but the absolute acuity scores differed by about 2.5 octaves. When adults were tested with stimuli at 51° under light-adapted conditions, acuity scores were similar to those of infants. However, the adult acuity versus luminance function at 51° was quite different from that of free-viewing infants. Although this study was not exhaustive in the areas of the peripheral retina tested, the results suggest that there is no one area of the adult peripheral retina that can serve as an adequate model for the visual

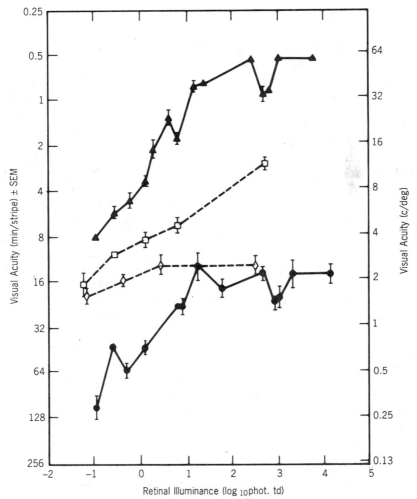

**FIGURE 4.** Average visual acuity as a function of retinal illuminance for free-fixating adults (▲), adults with the stimuli falling 21° to the right or left of the fixation point (□), adults with the stimuli falling 51° in the horizontal meridian of the temporal visual field (◇), and free-fixating 2-month-old infants (●). Error bars are ±1 SEM. Reprinted with permission from Brown, Dobson, & Maier (1987).

acuity response of the young human infant. Banks (1976) came to a similar conclusion in a comparison of infant versus adult contrast sensitivity functions.

**3. Use of Rods instead of Physiologically Immature Foveal Cones.** A third hypothesis related to foveal immaturity is that grating stimuli are detected by rods rather than by cones, even at middle and high luminances. This hypothesis is suggested by the finding that the acuity of the young infant under light-adapted conditions is similar to that of adults at a luminance of $-2.6 \log \text{cd}/\text{m}^2$, at which luminance adult acuity is mediated solely by rods (Hecht, 1934). At first glance, this hypothesis seems unlikely since other studies have shown that cones in young infants are able to catch quanta and to mediate some aspects of color vision (e.g., Dobson, 1976; Moskowitz-Cook, 1979; Peeples & Teller, 1978). However, studies that have shown evidence of functional cones in young infants have used stimuli containing relatively large spatial patterns, in contrast to the relatively finer gratings used in visual acuity studies. Therefore, the question of whether rods or cones are responsible for detection of spatial patterns has not been resolved.

Two features of the results of Brown et al. (1987) rule out rod detection of gratings as a realistic hypothesis to explain the poor visual acuity of young infants. First, the data show a substantial decrease in acuity at low luminances (see Fig. 4). If infants' acuity was being mediated by an adult-like rod system, one would expect acuity to remain constant at low to middle luminances, and to have an absolute value similar to that of adults at low luminances. Second, if an adult-like rod system were mediating infants' acuity, one would expect a decrease in acuity at the highest luminances used by Brown et al., since those luminances were above the luminance at which adult rods saturate (Aguilar & Stiles, 1954). However, as shown in Fig. 4, infants' acuity remains constant at luminances above about $1.0 \log \text{cd}/\text{m}^2$.

**4. The "Dark Glasses" Hypothesis.** The "dark glasses" hypothesis (e.g., MacLeod, 1978) is a rather general hypothesis that suggests that the changes in sensitivity produced by light adaptation are due to an overall reduction in the efficiency of all stimuli, as if the individual were wearing sunglasses. In the comparison of infant versus adult acuity results, a modified version of the "dark glasses" hypothesis would suggest that the reason infants show poor acuity is because light is a less effective stimulus for the visual system of infants than it is for the visual system of adults. The anatomical or physiological mechanisms that might underly such a reduction in sensitivity include cloudy optic media (a condition seen in young kittens (Bonds & Freeman, 1978) but not human infants), reduced optical density of the photoreceptors, or reduced photosensitivity of the photoreceptors.

Although considerable speculation would be possible concerning which parts of the developing visual system might actually produce reduced sensitivity in infants, empirical acuity data from infants make this sort of speculation unnecessary, because these data allow us to reject the "dark glasses" hypothesis in its entirety. The logic is as follows: If the limitation on infants' acuity is a lack of sensitivity of the visual system, then the way to overcome this limitation would be to provide more light to the system. Thus, the "dark glasses" hypothesis would predict that if stimuli were made bright enough to compensate for the reduced sensitivity of the visual system, infants would show acuity scores as good as those of adults. As can be seen in Fig. 4, the data of Brown et al. (1987) allow us to reject this hypothesis, since increases in luminance above about 10 cd/m² did not result in improvements in infants' acuity results.

### C. Immaturity of Postreceptoral Factors

The results summarized in the preceding pages suggest that neither the optics of the infant eye nor the immaturity of infant foveal photoreceptors can be the primary reason that infants do not see as well as adults. Thus, it is likely that some factor or factors central to the photoreceptors play a major role in limiting the visual acuity of young infants. The possibilities can be grouped under three broad headings: (1) Limitations in the postreceptoral sensory visual pathways, (2) immaturity of visual–motor pathways, and (3) motivational/attentional factors.

**1. Limitations in the Postreceptoral Sensory Visual Pathways.** There are a number of characteristics of the postreceptoral visual system that might be responsible for infants' limited acuity. For example, the pathways responsible for processing high spatial frequencies may not be functional or, if functional, they may be tuned to lower spatial frequencies than are the same pathways in adults. Alternatively, the pathways responsible for processing high spatial frequencies may be fully functional, but may be much less sensitive or have lower gain than those of adults.

The behavioral acuity data described in this chapter do not allow us to decide among these various possibilities and, therefore, no speculation concerning the functioning of these higher pathways will be included in this chapter. Further information on limitations imposed on infant vision by postreceptoral visual pathways can be found in behavioral studies of the characteristics of channels that process spatial information in infants (e.g., Banks, Stephens, and Hartmann, 1985; Braddick, Wattam-Bell, & Atkinson, 1986) and in single unit studies of the spatial characteristics of individual cells in the visual pathways of infant monkeys (e.g., Blakemore & Vital-Durand, 1979).

**2. Immaturity of Visual–Motor Pathways.** A second hypothesis is that it is not immaturity of the sensory visual pathways that limit infant acuity, but rather the immaturity of the visual–motor pathways required for the behavioral response required of the infant in FPL and OKN studies of acuity. This hypothesis suggests that the visual pathways are mature enough to allow the infant good spatial vision, but that the visual–motor connections are too immature to allow the infant to make a behavioral response to indicate that he or she can resolve fine gratings.

For the 2-month-old infant, two types of data argue against this hypothesis. First, Regal (1981) has shown that 2-month-olds tested with the FPL procedure can reveal nearly adult levels of sensitivity to flicker, suggesting that it is not the inability of the infant to make an FPL response that produces poor FPL acuity scores in young infants. Second, visual evoked potential studies of 2-month-olds show that their acuity is poorer than that of adults (for reviews, see Dobson & Teller, 1978; Norcia & Tyler, 1985), suggesting that at least part of what limits the acuity of 2-month-olds occurs early in the sensory visual pathway, prior to the locus of the visual evoked potential.

In contrast, visual evoked potential acuity results from 6- to 12-month-olds show nearly adult levels of resolution (see Dobson & Teller, 1978; Norcia & Tyler, 1985). This result suggests that, for older infants, the factors that limit behavioral acuity occur at higher levels of the visual or visual–motor system.

**3. Immaturity of Motivational and/or Attentional Mechanisms.** The final hypothesis concerning the mechanism underlying infants' poor visual acuity scores is the motivational/attentional hypothesis. This hypothesis suggests that infants may see as well as adults, but the immature motivational and/or attentional mechanisms prevent them from producing acuity scores as good as those of adults in behaviorally measured acuity tasks. This hypothesis is given credence by the results of visual evoked potential studies of acuity, which show that infants' brains have much better spatial resolution than the behavioral results indicate.

Two studies have examined the issue of infants' motivation in behavioral acuity testing. In both, it was hypothesized that if lack of motivation were a major factor limiting infants' acuity, then rewarding the infant for looking at the stimulus should improve performance. D. L. Mayer and Dobson (1982) compared acuity results obtained with the operant preferential looking procedure in 5- and 6-month-old infants with acuity results obtained with unrewarded FPL testing in infants of the same age tested by Allen (1979). No differences between the results of rewarded versus unrewarded FPL acuity testing were found. In the second study, Stephens and Banks (1988) compared the acuity results of 3-, 5-, and 7-month-old infants during rewarded FPL acuity testing with the results of infants of the same age tested in the same setting without reward. No significant differences in acuity between the rewarded versus the unrewarded groups of the same age were found. These results suggest that lack of motivation is probably not the primary factor underlying the poor acuity shown by young infants; however, one could always argue that a different type of reward (other than the animated toy used by both D. L. Mayer & Dobson, 1982, and Stephens & Banks, 1988), might result in an improvement in performance.

It could also be argued that even if infants are motivated to perform well in an FPL task, they may have difficulty doing so because the stimuli used in acuity testing are not stimuli to which infants readily attend. For example, it has been suggested that use of a face rather than a grating stimulus might increase the infant's attention to the stimuli and thus produce improved acuity scores. To test this hypothesis, two groups of researchers have used face stimuli to measure acuity in young infants. In one study, the results showed no difference in acuity for faces versus gratings in 1-, 3-, and 5-month-olds (Harris, Hansen, & Fulton, 1984), while in the other study, 1-, 2-, and 3-month-olds showed better acuity for gratings than for faces (Atkinson, Braddick, & Moar, 1977c). Thus, although face-like stimuli might be expected to have more experiential significance for infants than would a grating pattern, use of face-like acuity stimuli does not result in improved acuity scores.

It has also been suggested that use of temporally modulated rather than stationary gratings might increase the likelihood that infants will look at a grating target. As described above, neither Dobson, Teller & Belgum (1978) nor Atkinson et al. (1977b) found an improvement in acuity when temporally modulated stimuli were used to measure acuity in infants less than 3 months of age. However, Sokol et al. (1987) reported that an improvement in acuity of approximately 1 octave could be obtained when infants in the 3, to 8-month range were tested with gratings that were phase-alternated at 14 reversals per second. Thus, increasing the salience of the stimulus may result in improvements in acuity of as much as 1 octave. However, this increase is not sufficient to account for the 3- to 4-octave differences found between the acuities of young infants and those of adults.

### D. Summary of Potential Limiting Factors That Influence Infants' Visual Acuity

From an examination of the anatomical and physiological characteristics of the developing visual system, one can formulate many hypotheses to account for the finding that behaviorally measured visual acuity in infants is considerably poorer than behaviorally measured visual acuity in adults. The strength of empirical acuity measurements is that they allow us to test and refine these hypotheses.

One of the best examples of the interaction between theory and empirical results can be found in the section above on optical defocus. Based on an initial empirical report that young infants did not accommodate to targets at varying distances (Haynes et al., 1965), it was predicted that infants' acuity should vary as a function of target distance because of optical defocus. Subsequently, empirical results (e.g., Salapatek et al., 1976) showed that acuity did *not* vary as a function of distance in young infants. This contradiction between theory and empirical results led to a rethinking of theoretical predictions concerning the effects of defocus on accommodation and acuity in infants (e.g., Green et al., 1980).

Similarly, the anatomical immaturity of the infant fovea, the adult-like appearance of infant extrafoveal photoreceptors, and the finding that infants' acuity at high luminance levels is more similar to adults' acuity in their peripheral rather than in their foveal retina, led to the hypothesis that infants' vision can be modeled by the adult peripheral retina. This hypothesis was tenable when the only data available were from studies in which the outcome measure was a single acuity value obtained under fixed stimulus conditions. However, as described above, more comprehensive empirical testing involving measurement of contrast sensitivity at a variety of spatial frequencies (Banks, 1976) and measurement of acuity under varying luminance conditions (Brown et al., 1987) revealed the inadequacy and inaccuracy of using adult peripheral vision as a model of infants' vision.

In summary, empirical studies of visual acuity have allowed revision and refinement of theories concerning the mechanisms that limit visual acuity in infants. As a result of behavioral acuity studies, researchers have been able to rule out a number of mechanisms as playing a significant role in limiting infants' acuity. These include optical defocus, anatomical immaturity of the foveal photoreceptors, use of adult-like peripheral cones rather than foveal cones for detection of spatial patterns, use of rods rather than cones for spatial detection, or a reduction in the overall sensitivity of the infant's visual system. Because these mechanisms have been ruled out, researchers can now concentrate on other possible mechanisms, including limitations in the postreceptoral visual pathways and immaturity of infants' motivational and attentional systems.

## VI. SUMMARY AND CONCLUSIONS

The goal of this chapter has been to describe three phases in the study of visual acuity in infants. The first phase, which, of necessity, was also the initial phase chronologically, was the methodological phase, in which procedures for acuity assessment were developed. Difficulties inherent in this phase centered around two problems: (1) Infants, like animals, are nonverbal, and (2) infants grow older rapidly, so that it can be difficult or impossible to obtain a complete set of psychophysical data on an individual infant, especially since, unlike laboratory animals, infants are available for testing only when parents are willing to bring them to the laboratory. Once methods for acuity assessment were developed, it was possible to begin the second phase of the study of visual acuity in infants. In this phase, normative data were obtained and studies were conducted to examine the effect of parametric variations on acuity results. The third phase, the incorporation of empirical results into theories of visual development, followed upon the generation of normative and parametric data in the second phase. As data accumulated, it was possible to begin to place constraints on hypotheses about visual development that were based on anatomical and physiological results, and to use these refined theories to direct further empirical studies.

Although these phases evolved chronologically, it is clear from this chapter that all three are continuing to progress simultaneously. Methodologically, the emphasis continues to be on increasing the quantity of data collected and decreasing the time requried to collect it. Parametric studies to date have included only a small number of variations, tested over limited age ranges. Many of these studies have, in fact, raised more research questions than they have answered. Finally, it is obvious that as the quantity of empirical data increases, the usefulness of the data in theoretical formulations will increase significantly. Thus, even though the study of visual acuity is one of the most advanced areas of research on infant sensory development, it is clear that the field has only begun to explore the many issues related to this basic aspect of visual development.

## ACKNOWLEDGMENTS

The author thanks Martin Banks, Angela Brown, and Maureen Powers and the editors of this book for helpful comments on an earlier version of this manuscript. During writing of the manuscript, the author was supported in part by NIH Grant EY 05804.

## REFERENCES

Abramov, I., Gordon, J., Hendrickson, A., Hainline, L., Dobson, V., & LaBossiere, E. (1982). The retina of the newborn human infant. *Science, 217,* 265–267.

Aguilar, M., & Stiles, W. S. (1954). Saturation of the rod mechanism of the retina. *Optica Acta,* **1,** 59–65.

Allen, J. L. (1979). *The development of visual acuity in human infants during the early postnatal weeks.* Unpublished doctoral dissertation, University of Washington, Seattle.

Appelle, S. (1972). Perception and discrimination as a function of stimulus orientation: The "oblique effect" in man and animals. *Psychological Bulletin,* **78,** 266–278.

Atkinson, J. (1984). Human visual development over the first 6 months of life. A review and a hypothesis. *Human Neurobiology,* **3,** 61–74.

Atkinson, J., & Braddick, O. (1981). Acuity, contrast sensitivity, and accommodation in infancy. In R. N. Aslin, J. R. Alberts, & M. R. Petersen (Eds.), *Development of perception: Vol. 2. The visual system* (pp. 245–277). New York: Academic Press.

Atkinson, J., Braddick, O., & Braddick, F. (1974). Acuity and contrast sensitivity of infant vision. *Nature (London),* **247,** 403–404.

Atkinson, J., Braddick, O., & Moar, K. (1977a). Development of contrast sensitivity over the first 3 months of life in the human infant. *Vision Research, 17,* 1037–1044.

Atkinson, J., Braddick, O., & Moar, K. (1977b). Contrast sensitivity of the human infant for moving and static patterns. *Vision Research, 17,* 1045–1047.

Atkinson, J., Braddick, O., & Moar, K. (1977c). Infants' detection of image defocus. *Vision Research, 17,* 1125–1126.

Atkinson, J., Braddick, O., & Pimm-Smith, E. (1982). 'Preferential looking' for monocular and binocular acuity testing of infants. *British Journal of Ophthalmology, 66,* 264–268.

Atkinson, J., French, J., & Braddick, O. (1981). Contrast sensitivity function of preschool children. *British Journal of Ophthalmology, 65,* 525–529.

Atkinson, J., Pimm-Smith, E., Evans, C., & Braddick, O. J. (1983). The effects of screen size and eccentricity on acuity estimates in infants using preferential looking. *Vision Research, 23,* 1479–1483.

Banks, M. S. (1976). *Infant form vision: The modulation transfer function.* Unpublished doctoral dissertation, University of Minnesota, Minneapolis.

Banks, M. S. (1980). The development of visual accommodation during early infancy. *Child Development, 51,* 646–666.

Banks, M. S., Bennett, P. J., & Schefrin, B. (1987). Foveal cones and spatial vision in human neonates. *Investigative Ophthalmology & Visual Science, 28,*(Suppl.), 4.

Banks, M. S., & Salapatek, P. (1978). Acuity and contrast sensitivity in 1-, 2-, and 3-month-old human infants. *Investigative Ophthalmology & Visual Science, 17,* 361–365.

Banks, M. S., & Salapatek, P. (1981). Infant pattern vision: A new approach based on the contrast sensitivity function. *Journal of Experimental Child Psychology, 31,* 1–45.

Banks, M. S., & Salapatek, P. (1983). Infant visual perception. In P. H. Mussen, M. M. Haith, & J. J. Campos (Eds.), *Handbook of child psychology* (4th ed., Vol. 2, pp. 435–571). New York: Wiley.

Banks, M. S., Stephens, B. R., & Hartmann, E. E. (1985). The development of basic mechanisms of pattern vision: Spatial frequency channels. *Journal of Experimental Child Psychology, 40,* 501–527.

Berlyne, D. E. (1958). The influence of the albedo and complexity of stimuli on visual fixation in the human infant. *British Journal of Psychology, 49,* 315–318.

Birch, E. E., Gwiazda, J., Bauer, J. A., Jr., Naegele, J., & Held, R. (1983). Visual acuity and its meridional variations in children aged 7 to 60 months. *Vision Research, 23,* 1019–1024.

Blake, R., Sloane, M., & Fox, R. (1981). Further developments in binocular summation. *Perception & Psychophysics, 30,* 266–276.

Blakemore, C., & Vital-Durand, F. (1979). Development of the neural basis of visual acuity in monkeys. *Transactions of the Ophthalmological Society of the United Kingdom, 99,* 363–368.

Bloom, M., & Berkley, M. A. (1977). Visual acuity and the near point of accommodation in cats. *Vision Research, 17,* 723–730.

Bonds, A. B., & Freeman, R. D. (1978). Development of optical quality in the kitten eye. *Vision Research*, **18**, 391–398.

Boothe, R. G., Dobson, V., & Teller, D. Y. (1985). Postnatal development of vision in human and nonhuman primates. *Annual Review of Neuroscience*, **8**, 495–545.

Braddick, O. J., Wattam-Bell, J., & Atkinson, J. (1986). Orientation-specific cortical responses develop in early infancy. *Nature (London)*, **320**, 617–619.

Brookman, K. E. (1983). Ocular accommodation in human infants. *American Journal of Optometry and Physiological Optics*, **60**, 91–99.

Bronson, G. (1974). The postnatal growth of visual capacity. *Child Development*, **45**, 873–890.

Brown, A. M. (1986). Scotopic sensitivity of the two-month-old human infant. *Vision Research*, **26**, 707–710.

Brown, A. M., Dobson, V., & Maier, J. (1987). Visual acuity of human infants at scotopic, mesopic, and photopic luminances. *Vision Research*, **27**, 1845–1858.

Brown, A. M., & Yamamoto, M. (1986). Visual acuity in newborn and preterm infants measured with grating acuity cards. *American Journal of Ophthalmology*, **102**, 245–253.

Cornell, E. H., & McDonnell, P. M. (1986). Infants' acuity at twenty feet. *Investigative Ophthalmology & Visual Science*, **27**, 1417–1420.

Dayton, G. O., Jones, M. H., Aiu, P., Rawson, R. A., Steele, B., & Rose, M. (1964). Developmental study of coordinated eye movements in the human infant. *Archives of Ophthalmology (Chicago)*, **71**, 865–870.

Dobson, V. (1976). Spectral sensitivity of the 2-month infant as measured by the visually evoked cortical potential. *Vision Research*, **16**, 367–374.

Dobson, V. (1983). Clinical application of preferential looking measures of visual acuity. *Behavioral Brain Research*, **10**, 25–38.

Dobson, V., Mayer, D. L., & Lee, C. P. (1980). Visual acuity screening of preterm infants. *Investigative Ophthalmology & Visual Science*, **19**, 1498–1505.

Dobson, V., Salem, D., & Carson, J. B. (1983). Visual acuity in infants—The effect of variations in stimulus luminance within the photopic range. *Investigative Ophthalmology & Visual Science*, **24**, 519–522.

Dobson, V., Salem, D., Mayer, D. L., Moss, C, & Sebris, S. L. (1985). Visual acuity screening of children 6 months to 3 years of age. *Investigative Ophthalmology & Visual Science*, **26**, 1057–1063.

Dobson, V., Schwartz, T. L., Sandstrom, D. J., & Michel, L. (1987). Binocular visual acuity in neonates: The acuity card procedure. *Developmental Medicine and Child Neurology*, **29**, 199–206.

Dobson, V., & Teller, D. Y. (1978). Visual acuity in human infants: A review and comparison of behavioral and electrophysiological studies. *Vision Research*, **18**, 1469–1483.

Dobson, V., Teller, D. Y., & Belgum, J. (1978). Visual acuity in human infants assessed with stationary stripes and phase-alternated checkerborads. *Vision Research*, **18**, 1233–1238.

Dobson, V., Teller, D. Y., Lee, C. P., & Wade, B. (1978). A behavioral method for efficient screening of visual acuity in young infants. I. Preliminary laboratory development. *Investigative Ophthalmology & Visual Science*, **17**, 1142–1150.

Droste, P. J., Archer, S. M., & Helveston, E. M. (1987). Quantification of low vision in children. *Investigative Ophthalomology & Visual Science*, **28,**(Suppl.), 154.

Fantz, R. L. (1958). Pattern vision in young infants. *Psychological Record*, **8,** 43–47.

Fantz, R. L., Fagan, J. F., III, & Miranda, S. B. (1975). Early visual selectivity. In L. B. Cohen & P. Salapatek (Eds.), *Infant perception: From sensation to cognition: Vol. 1. Basic visual processes* (pp. 249–345). New York: Academic Press.

Fantz, R. L., Ordy, J. M., & Udelf, M. S. (1962). Maturation of pattern vision in infants during the first six months. *Journal of Comparative and Physiological Psychology*, **55,** 907–917.

Fiorentini, A., Pirchio, M. & Sandini, G. (1984). Development of retinal acuity in infants evaluated with pattern electroretinogram. *Human neurobiology*, **3,** 93–95.

Fiorentini, A., Pirchio, M., & Spinelli, D. (1980). Scotopic contrast sensitivity in infants evaluated by evoked potentials. *Investigative Ophthalmology & Visual Science*, **19,** 950–955.

Fulton, A., Dobson, V., Salem, D., Mar, C., Petersen, R. A., & Hansen, R. M. (1980). Cycloplegic refractions in infants and young children. *American Journal of Ophthalmology*, **90,** 239–247.

Gorman, J. J., Cogan, D. G., & Gellis, S. S. (1957). An apparatus for grading the visual acuity of infants on the basis of opticokinetic nystagmus. *Pediatrics*, **19,** 1088–1092.

Green, D. G., Powers, M. K., & Banks, M. S. (1980). Depth of focus, eye size and visual acuity. *Vision Research*, **20,** 827–835.

Gwiazda, J., Bauer, J., Thorn, F., & Held, R. (1986). Meridional amblyopia *does* result from astigmatism in early childhood. *Clinical Vision Sciences*, **1,** 145–152.

Gwiazda, J., Brill, S., Mohindra, I., & Held, R. (1978). Infant visual acuity and its meridional variation, *Vision Research*, **18,** 1557–1564.

Gwiazda, J., Brill, S., Mohindra, I., & Held, R. (1980). Preferential looking acuity in infants from two to fifty-eight weeks of age. *American Journal of Optometry and Physiological Optics*, **57,** 428–432.

Gwiazda, J., Mohindra, I., Brill, S., & Held, R. (1985). Infant astigmatism and meridional amblyopia. *Vision Research*, **25,** 1269–1276.

Gwiazda, J., Scheiman, M., & Held, R. (1984). Anisotropic resolution in children's vision. *Vision Research*, **24,** 527–531.

Hansen, R. M., & Fulton, A. B. (1981). Behavioral measurement of background adaptation in infants. *Investigative Ophthalmology & Visual Science*, **21,** 625–629.

Harris, S. J., Hansen, R. M., & Fulton, A. B. (1984). Assessment of acuity in human infants using face and grating stimuli. *Investigative Ophthalmology & Visual Science*, **25,** 782–786.

Hartmann, E. E., & Banks, M. S. (1984). Development of temporal contrast sensitivity. [Special ICIS Issue]. *Infant Behavior and Development*, **7,** 163.

Haynes, H., White, B. L., & Held, R. (1965). Visual accommodation in human infants. *Science*, **148,** 528–530.

Hecht, S. (1934). Vision. II. The nature of the photoreceptor process. In C. Murchison (Ed.), *A handbook of general experimental psychology* (pp. 704–828). Worcester, MA: Clark University Press.

Held, R. (1979). Development of visual resolution. *Canadian Journal of Psychology*, **33**, 213–221.

Hendrickson, A. E., & Yuodelis, C. (1984). The morphological development of the human fovea. *Ophthalmology*, **91**, 603–612.

Hertz, B. G. (1987). Acuity card testing of retarded children. *Behavioral Brain Research*, **24**, 85–92.

Howell, E. R., & Hess, R. F. (1978). The functional area for summation to threshold for sinusiodal gratings. *Vision Research*, **18**, 369–374.

Howland, H. C., Atkinson, J., Braddick, O., & French, J. (1978). Infant astigmatism measured by photorefraction. *Science*, **202**, 331–332.

Ingram, R. M., & Barr, A. (1979). Changes in refraction between the ages of 1 and 3½ years. *British Journal of Ophthalmology*, **63**, 339–342.

Jacobson, S. G., Mohindra, I., & Held, R. (1983). Monocular visual form deprivation in human infants. *Documenta Ophthalmologica*, **55**, 199–211.

Kohl, P., Rolen, R. D., Bedford, A. K., Samek, M., & Stern, N. (1986). Refractive error and preferential looking acuity in human infants: A pilot study. *Journal of the American Optometric Association*, **57**, 290–296.

Lashley, K. S. (1930). The mechanism of vision. I. A method for the rapid analysis of pattern-vision in the rat. *Journal of Genetic Psychology*, **37**, 453–460.

Lipsitt, L. P. (1967). The concepts of development and learning in child behavior. *UCLA Forum in Medical Sciences* 4(6), 211–248.

Lit, A. (1968). Visual acuity. *Annual Review of Psychology*, **19**, 27–54.

MacLeod, D. I. A. (1978). Visual sensitivity. *Annual Review of Psychology*, **29**, 613–645.

Mann, I. C. (1964). *The development of the human eye*. London: British Medical Association.

Manny, R. E., & Klein, S. A. (1984). The development of vernier acuity in infants. *Current Eye Research*, **3**, 453–462.

Maurer, D., Jobson, S., & Lewis, T. L (1986). The development of peripheral acuity [Special ICIS Issue]. *Infant Behavior and Development*, **9**, 245.

Maurer, D., Lewis, T. L., & Brent, H. P. (in press). The effects of deprivation on human visual development: Studies of children treated for cataracts. In F. J. Morrison, C. E. Lord, & D. P. Keating (Eds.), *Applied Developmental Psychology* (Vol. 3). New York: Academic Press.

Mayer, D. L. (1986). Acuity of amblyopic children for small field gratings and recognition stimuli. *Investigative Ophthalmology & Visual Science*, **27**, 1148–1153.

Mayer, D. L., & Dobson, V. (1980). Assessment of vision in young children: A new operant approach yields estimates of acuity. *Investigative Ophthalmology & Visual Science*, **19**, 566–570.

Mayer, D. L., & Dobson, V. (1982). Visual acuity development in infants and young children, as assessed by operant preferential looking. *Vision Research*, **22**, 1141–1151.

Mayer, D. L., Fulton, A. B., & Hansen, R. M. (1982). Preferential looking acuity obtained with a staircase procedure in pediatric patients. *Investigative Ophthalmology & Visual Science*, **23**, 538–543.

Mayer, D. L., Fulton, A. B., & Hansen, R. M. (1985). Visual acuity of infants and children with retinal degenerations. *Ophthalmic Pediatrics and Genetics, 5,* 51–56.

Mayer, D. L., Morrell, A. S., & Rodier, D. W. (1986). Breakthrough acuity test for the pediatric eye clinic: The acuity card procedure. *Investigative Ophthalmology & Visual Science,* **27,**(Suppl.), 147.

Mayer, M. J. (1983). Non-astigmatic children's contrast sensitivities differ from anisotropic patterns of adults. *Vision Research, 23,* 551–559.

McDonald, M. A., Ankrum, C., Preston, K., Sebris, S. L., & Dobson, V. (1986). Monocular and binocular acuity in 18-to-36-month-olds: Acuity card results. *American Journal of Optometry and Physiological Optics, 63,* 181–186.

McDonald, M. A., Dobson, V., Sebris, S. L., Baitch, L., Varner, D., & Teller, D. Y. (1985). The acuity card procedure: A rapid test of infant acuity. *Investigative Ophthalmology & Visual Science,* **26,** 1158–1162.

McDonald, M. A., Sebris, S. L., Mohn, G., Teller, D. Y., & Dobson, V. (1986). Monocular acuity in normal infants: The acuity card procedure. *American Journal of Optometry and Physiological Optics, 63,* 127–134.

McKee, S. P., Klein, S. A., & Teller, D. Y. (1985). Statistical properties of forced-choice psychometric functions: Implications of probit analysis. *Perception & Psychophysics, 37,* 286–298.

Milewski, A. (1976). Infants' discrimination of internal and external pattern elements. *Journal of Experimental Child Psychology, 22,* 229–246.

Mitchell, D. E., Freeman, R. D., Millidot, M., & Haegerstrom, G. (1973). Meridional amblyopia: Evidence for modification of the human visual system by early visual experience. *Vision Research, 13,* 535–558.

Mitchell, D. E., Griffin, R., Wilkinson, F., Anderson, P., & Smith, M. L. (1976). Visual resolution in young kittens. *Vision Research, 16,* 363–366.

Mohindra, I., & Held, R. (1981). Refraction in humans from birth to five years. *Documenta Ophthalmologica Proceedings Series, 28,* 19–27.

Mohindra, I., Held, R., Gwiazda, J., & Brill, S. (1978). Astigmatism in infants. *Science,* **202,** 329–331.

Mohindra, I., Jacobson, S. G., & Held, R. (1983). Binocular visual form deprivation in human infants. *Documenta Ophthalmologica, 55,* 237–249.

Mohn, G., & van Hof-van Duin, J. (1986). Rapid assessment of visual acuity in infants and children in a clinical setting, using acuity cards. *Documenta Ophthalmologica Proceedings Series, 45,* 363–372.

Moore, J. M., & Wilson, W. R. (1978). Visual reinforcement audiometry (VRA) with infants. In S. E. Gerber & G. T. Mencher (Eds.), *Early diagnosis of hearing loss* (pp. 177–213). New York: Grune & Stratton.

Moskowitz-Cook, A. (1979). The development of photopic spectral sensitivity in human infants. *Vision Research, 19,* 1133–1142.

National Research Council–National Academy of Sciences Committee on Vision (1980). Recommended standard procedures for the clinical measurement and specification of visual acuity. *Advances in Ophthalmology, 41,* 103–148.

Norcia, A. M., & Tyler, C. W. (1985). Spatial frequency sweep VEP: Visual acuity during the first year of life. *Vision Research, 25,* 1399–1408.

Norcia, A. M., Tyler, C. W., Piecuch, R., Clyman, R., & Grobstein, J. (1987). Visual acuity development in normal and abnormal preterm human infants. *Journal of Pediatric Ophthalmology and Strabismus*, **24**, 70–74.

Odom, J. V., Maida, T. M., Dawson, W. W., & Romano, P. E. (1983). Retinal and cortical pattern responses: A comparison of infants & adults. *American Journal of Optometry and Physiological Optics*, **60**, 369–375.

Owens, D. A. (1980). A comparison of accommodative responsiveness and contrast sensitivity for sinusiodal gratings. *Vision Research*, **20**, 159–167.

Packer, O., Hartmann, E. E., & Teller, D. Y. (1984). Infant color vision: The effect of test field size on Rayleigh discriminations. *Vision Research*, **24**, 1247–1260.

Peeples, D. R, & Teller, D. Y. (1978). White-adapted photopic spectral sensitivity in human infants. *Vision Research*, **18**, 49–53.

Powers, M. K., & Dobson, V. (1982). Effect of focus on visual acuity of human infants. *Vision Research*, **22**, 521–528.

Powers, M. K., Schneck, M., & Teller, D. Y. (1981). Spectral sensitivity of human infants at absolute visual threshold. *Vision Research*, **21**, 1005–1016.

Preston, K. L., McDonald, M. A., Sebris, S. L., Dobson, V., & Teller, D. Y. (1987). Validation of the acuity card procedure for assessment of infants with ocular disorders. *Ophthalmology (Rochester, Minnesota)*, **94**, 644–653.

Regal, D. M. (1981). Development of critical flicker frequency in human infants. *Vision Research*, **21**, 549–555.

Riggs, L. A. (1965). Visual acuity. In C. H. Graham (Ed.), *Vision and visual perception*, (pp. 321–349). New York: Wiley.

Robson, J. G., & Graham, N. (1981). Probability summation and regional variation in contrast sensitivity across the visual field. *Vision Research*, **21**, 409–418.

Salapatek, P., Bechtold, A. G., & Bushnell, E. W. (1976). Infant visual acuity as a function of viewing distance. *Child Development*, **47**, 860–863.

Schneck, M. E., Hamer, R. D., Packer, O. S., & Teller, D. Y. (1984). Area-threshold relations at controlled retinal locations in 1-month-old infants. *Vision Research*, **24**, 1753–1763.

Sebris, S. L., Dobson, V., McDonald, M. A., & Teller, D. Y. (1987). Acuity cards for visual acuity assessment of infants and children in clinical settings. *Clinical Vision Sciences*, **2**, 45–58.

Shepherd, P. A., Fagan, J. F., III, & Kleiner, K. A. (1985). Visual pattern detection in preterm neonates. *Infant Behavior and Development*, **8**, 47–63.

Shimojo, S., Birch, E. E., Gwiazda, J., & Held, R. (1984). Development of vernier acuity in infants. *Vision Research*, **24**, 721–728.

Sireteanu, R., Kellerer, R., & Boergen, K.-P. (1984). The development of peripheral visual acuity in human infants. A preliminary study. *Human Neurobiology*, **3**, 81–85.

Sokol, S., & Jones, K. (1979). Implicit time of pattern evoked potentials in infants: An index of maturation of spatial vision. *Vision Research*, **19**, 747–755.

Sokol, S., Moskowitz, A., & Augliere, R. (1987). Preferential looking acuity is temporally tuned. *Investigative Ophthalmology & Visual Science*, **28**,(Suppl.), 5.

Stephens, B. R., & Banks, M. S. (1988). The effect of reinforcement of infants' performance in a preferential looking acuity task. *American Journal of Optometry and Physiological Optics,* **65,** 637–643.

Teller, D. Y. (1979). The forced-choice preferential looking procedure: A psychophysical technique for use with human infants. *Infant Behavior and Development,* **2,** 135–153.

Teller, D. Y. (1983). Measurement of visual acuity in human and monkey infants: The interface between laboratory and clinic. *Behavioral Brain Research,* **10,** 15–23.

Teller, D. Y. (1985). Psychophysics and infant vision: Definitions and limitations. In G. Gottlieb & N. A. Krasnegor (Eds.), *Measurement of audition and vision in the first year of postnatal life* (pp. 127–143). Norwood, NJ: Ablex.

Teller, D. Y., Allen, J. L., Regal, D. M., & Mayer, D. L. (1978). Astigmatism and acuity in two primate infants. *Investigative Ophthalmology & Visual Science,* **17,** 344–349.

Teller, D. Y., & Bornstein, M. H. (1987). Infant color vision and color perception. In P. Salapatek & L. Cohen (Eds.), *Handbook of infant perception: Vol. 1. From sensation to perception* (pp. 185–236). Orlando, FL: Academic Press.

Teller, D. Y., McDonald, M. A., Preston, K., Sebris, S. L., & Dobson, V. (1986). Assessment of visual acuity in infants and children: The acuity card procedure. *Developmental Medicine and Child Neurology,* **28,** 779–789.

Teller, D. Y., Morse, R., Borton, R., & Regal, D. (1974). Visual acuity for vertical and diagonal gratings in human infants. *Vision Research,* **14,** 1433–1439.

van Hof-van Duin, J., & Mohn, G. (1986). The development of visual acuity in normal fullterm and preterm infants. *Vision Research,* **26,** 909–916.

van Hof-van Duin, J., Mohn, G., Fetter, W. P. F., Mettau, J. W., & Baerts, W. (1983). Preferential looking acuity in preterm infants. *Behavioral Brain Research,* **10,** 47–50.

Volkmann, F. C., & Dobson, M. V. (1976). Infant responses of ocular fixation to moving visual stimuli. *Journal of Experimental Child Psychology,* **22,** 86–99.

Volkmann, F. C., & Prizer, D. C. (1973, May). *Thresholds for visual tilt in two-and-three-year old children.* Paper presented at the meeting of the Eastern Psychological Association, Washington, DC.

von Noorden, G. K., Dowling, J. E., & Ferguson, D. C. (1970). Experimental amblyopia in monkeys. I. Behavioral studies of stimulus deprivation amblyopia, *Archives of Ophthalmology (Chicago),* **84,** 206–214.

Westheimer, G. (1965). Visual acuity. *Annual Review of Psychology,* **16,** 359–380.

Westheimer, G. (1979). The spatial sense of the eye. *Investigative Ophthalmology & Visual Science,* **18,** 893–912.

Yuodelis, C., & Hendrickson, A. (1986). A qualitative and quantitative analysis of the human fovea during development. *Vision Research,* **26,** 847–855.

# INDEX